Nasal Toxicity and Dosimetry of Inhaled Xenobiotics

Nasal Toxicity and Dosimetry of Inhaled Xenobiotics: Implications for Human Health

Cosponsors

Chemical Industry Institute of Toxicology
Agency for Toxic Substances and Disease Registry
Basic Acrylic Monomer Manufacturers Association
DuPont Haskell Laboratory for Toxicology and Industrial Medicine
Environmental Criteria and Assessment Office/RTP, U.S. Environmental Protection Agency
European Centre for Ecotoxicology and Toxicology of Chemicals
Methacrylate Producers Association
National Institute of Environmental Health Sciences
National Institute for Occupational Safety and Health
Rohm and Haas Company
The Society of the Plastics Industry, Inc., Ad Hoc Vinyl Acetate Task Force

Organizing Committee

Chairman: F. Miller, CIIT
G. Boorman, NIEHS
C. Frederick, Rohm and Haas Co.
J. Haartz, NIOSH
J. Harkema, ITRI
J. Kimbell, CIIT
N. Krivanek, E. I. du Pont de Nemours and Company
R. McClellan, CIIT
K. Morgan, CIIT
J. Morris, University of Connecticut
P. Scherer, University of Pennsylvania
C. Shoaf, U.S. EPA

Published simultaneously as a special issue of *Inhalation Toxicology*.

Taylor & Francis
Publishers since 1798

USA	Publishing Office:	Taylor & Francis 1101 Vermont Avenue, N.W., Suite 200 Washington, DC 20005-3521 Tel: (202) 289-2174 Fax: (202) 289-3665
	Distribution Center:	Taylor & Francis 1900 Frost Road, Suite 101 Bristol, PA 19007-1598 Tel: (215) 785-5800 Fax: (215) 785-5515
UK		Taylor & Francis Ltd. 4 John St. London WCIN 2ET Tel: 071 405 2237 Fax: 071 831 2035

NASAL TOXICITY AND DOSIMETRY OF INHALED XENOBIOTICS: Implications for Human Health

The contents of this book also appear as a special issue of the Taylor & Francis journal *Inhalation Toxicology*.

1 2 3 4 5 6 7 8 9 0 B R B R 9 8 7 6 5 4

This book was set in Omega by Publishers Prep House. The editors were Elizabeth L. Dugger and Caroline Schweiter; the production supervisor was Peggy M. Rote. Cover design by Michelle Fleitz. Printing and binding by Braun Brumfield Inc.

A CIP catalog record for this book is available from the British Library.
⊗ The paper in this publication meets the requirements of the ANSI Standard Z39.48-1984 (Permanence of Paper)

Library of Congress Cataloging-in-Publication Data

Nasal toxicity and dosimetry of inhaled xenobiotics: implications for human health / cosponsors, Chemical Industry Institute of Toxicology . . . [et al.]; organizing committee, chairman, F. Miller; G. Boorman . . . [et al.].
p. cm.
"Inhalation toxicology, volume 6, supplement"—Table of contents p.
Includes index.

1. Gases, Asphyxiating and poisonous—Toxicology—Congresses. 2. Gases, Asphyxiating and poisonous—Toxicology—Dose-response relationships—Congresses. 3. Nose—Effect of chemicals on—Congresses. 4. Environmental Exposure—adverse effects—congresses. 5. Xenobiotics—congresses. I. Miller, Fred J. II. Boorman, Gary A. III. Chemical Industry Institute of Toxicology. IV. Inhalation toxicology. V. 6 (Supplement).
[DNLM 1. Nasal Cavity—physiopathology—congresses. W1 IN4495F v.6 1994 Suppl. / WV 300 N241 1994]
RA1245.N37 1994
615.9'02—dc20
DNLM/DLC
for Library of Congress 94-29943
CIP

ISBN 1-56032-366-3
ISSN 0895-8378

CONTENTS

Section 3. Chemically Induced Nasal Response in Humans and Rodents

Nasal Effects Seen in Animals:

Spectrum of Responses Seen in Humans:

Section 4. Implications for Human Health and Exposure Limits

I PREFACE

The articles and extended abstracts in this publication address a number of topics concerning nasal dosimetry and toxicity of inhaled xenobiotics. There has been an ever-increasing interest in the location and severity of toxic responses in the noses of laboratory animals following inhalation of various compounds. This interest has arisen not only due to the types of responses observed, but also because of the increased use of animal toxicological data for assessing human health risks from exposure to these compounds.

A symposium was therefore organized by the Chemical Industry Institute of Toxicology (CIIT) at which scientists and regulatory officials came together to discuss nasal dosimetry and toxicity issues. The symposium, entitled Nasal Toxicity and Dosimetry of Inhaled Xenobiotics: Implications for Human Health, was held September 20–22, 1993, in Durham, N.C. This meeting attracted attendees with a wide range of scientific and regulatory backgrounds. The broad goals were (1) to assess the current state-of-knowledge concerning nasal dosimetry and toxicity in humans and other animals, (2) to examine issues encountered in using animal data to assess human risks, and (3) to identify research areas that would improve future interspecies comparisons of nasal dosimetry and toxicity of inhaled chemicals.

To aid in achieving the above goals, the symposium's Organizing Committee, which was comprised of individuals from academic, governmental, and industrial institutions, developed the following specific goals to be addressed by invited speakers and other attendees:

- Dosimetry—What level of knowledge about dose is needed (regional to site-specific) to understand and extrapolate nasal toxicity responses?
- Toxic Responses—What lesions are significant? How do severity, persistence, and homology interrelate? What is adverse?
- Which are the important interspecies extrapolation issues?
- Case studies for human exposure limits—What are the major issues that have surfaced in establishing exposure limits thus far?

The organization of this publication basically follows that of the symposium. Overview articles cover such topics as end points and responses of concern in humans, the nasal structure of laboratory animals, end points seen in nasal toxicity studies, approaches to use in examining dosimetry of inhaled xenobiotics, and ways to integrate available information to assess human risk from nasal toxicants. These articles provide the reader with a perspective for integrating the information contained in later articles that treat these subjects in greater depth.

Next, mathematical and physical models are examined that can be used to study the deposition of inhaled compounds in the upper respiratory tracts of humans and animals. Approaches for experimental dosimetry studies are discussed. A number of case studies are provided that illustrate various aspects of dosimetry, ranging from the effects physiological and physicochemical factors can play to studies on the fate of specific compounds. These articles are followed by a series of articles that examine in depth chemically induced nasal responses in animals and humans. Treatment of lesion distribution and toxicity is organized according to the major types of nasal epithelium—squamous, transitional, respiratory, and olfactory. With these and an article on effects seen in accessory nasal structures of animals, the reader is provided with the basic information needed to compare the effects seen in animals with the spectrum of responses seen in humans. Two articles discuss the nasal responses observed in humans and the techniques available for such studies.

An integrating theme of the symposium concerned the use of nasal dosimetry and toxicity data for assessing human health implications and establishing appropriate exposure limits. Exposure limits are needed for workplace and environmental settings. The populations at risk and the nature (averaging time and concentration) of the exposure limit can vary, depending upon the legislative or philosophical mandate of the organization. A series of articles deals with approaches taken by various organizations (governmental, professional) to set exposure limits. In addition, use of the U.S. Environmental Protection Agency's inhalation reference concentration methodology for establishing exposure limits is presented for two compounds. These case studies illustrate the kind of data needed and the types of issues and questions that can arise in establishing exposure limits protective of human health.

Articles from the plenary sessions are followed by extended abstracts from the poster session. Inclusion of the abstracts provides a sense of completeness of the symposium—acknowledging not only the valuable information presented at the poster session, but also the need for an interdisciplinary exchange of information if significant progress is to be made in the future in assessing human risks from nasal toxicants. Topics addressed in the extended abstracts include structure and function of the upper respiratory tract, methods and techniques for exposure and biological response assessment, toxicity studies in animals and humans, dosimetry of the upper respiratory tract, and aspects of risk assessment.

On behalf of the Organizing Committee and Cosponsors of the symposium, I hope that you find the information in this publication enlightening and useful.

Frederick J. Miller, PhD
Chairman, Organizing Committee

Section 1. Overview

OUR UPPER AIRWAYS AND OUR AMBIENT AIR: A HISTORICAL PERSPECTIVE

Donald F. Proctor

Johns Hopkins Medical Institutions, Baltimore, Mary and, USA

Over two centuries ago, John Arbuthnot (1773) published an essay on the subject to which this symposium is addressed, "The Effects of Air on Human Bodies." He described the "Air of cities . . . which is replete with sulphurous steams of fuel," and lamented the lack of inquiries into the effects of the air, "this substance that we take inwardly every moment." As we now know, little attention was paid to his warnings.

In 1898, nearly one century ago, another challenge to scientific investigators went unheeded. John MacKenzie's (1898) exposition on the relations between the nose and the sexual apparatus remains in need of our attention. That article by Mackenzie, first professor of otolaryngology at Hopkins, published in the *Johns Hopkins Hospital Bulletin*, is both illuminating and entertaining. There are other relatively unexplored areas related to our upper respiratory defenses and other events in the more recent past that form a background, if not a foundation, to today's research on the health effects of air pollution.

Let me remind you of some aspects of the state of the art 33 years ago, when I first turned my own attention to the nose as a defense organ. At that time, in 1950, I had been practicing otolaryngology for 23 years, but had been so absorbed by the problems of diagnosis and treatment of nasal and sinus disease that I gave little thought to the role of the nose as a defense against airborne hazards. My research had been directed at the lungs, with special attention to pediatric pulmonary problems.

But after the London episode of 1952 (with thousands of excess respiratory deaths) and the subsequent awakening of interest in air pollution, I rather suddenly found myself drawn away from the lung to its portal of entry, the target of airborne noxious agents and the first line of defense against them, the nose. Thus, I ask those of you who are old enough to remember that far back to recall that, when this transition came in my own work in 1960, the little work being done in this field then was largely riveted

This article is based upon a speech given by Dr. Proctor at the symposium banquet at which he was recognized for his "pioneering and enthusiastically sustained contributions to our understanding of the structure, function, and toxicity of the upper airways nasal toxicity and dosimetry of inhaled xenobiotics."

Address correspondence to Dr. Donald F. Proctor, Johns Hopkins School of Hygiene, Baltimore, MD 21205.

on the measurement of sulfur dioxide and, occasionally, dust at a few monitoring stations, and the incidence of pulmonary disease.

So little thought was given to the nose. Physiologists and clinicians alike used the term "upper airways" to refer to the trachea and major bronchi, and simply ignored the fact that there were airways intervening in the path of inspired air. They also ignored the importance of "indoor" air as compared to the "outdoor" air.

My decision to attack the problems of the physiology of the nose was a somewhat unwelcome and certainly unheralded approach to a terra incognita. Of course, rhinologists were surgically assaulting the region rather ruthlessly. Except for a few individuals, such as Arthur Proetz and Andy Hilding, there was almost no interest in nasal physiology. I must modify that statement to mention the work of Catlin, Sharpey, Purkinje, Goodale, Holmes, and others. But, in general, at the time, the little investigation being done on the possible deleterious effects of our ambient air on nasal surfaces, or nasal defenses against those effects, was largely confined to such mistaken conceptions as the evil effects of dry air. Parenthetically, I should tell you that the studies that I shall describe have been done with my right arm, David Swift, in collaboration with some 18 young associates from nine countries other than the United States.

My first efforts were bent on learning the basic rules of that esoteric subject, fluid flow, and then applying them to the problem of airflow through the complex passages of the nose. Instrumentation for measurement of airflow within the living human nose was simply not available, and I resorted to the use of models.

I had considerable experience in carving models of the human torso out of wood. But for the nose, I chose to cast nasal passages in fresh cadavers at the morgue, and then to use those casts to carve in lucite (Figure 1) with an old Army surplus dental drill.

Using casts to study airflow patterns (Figure 2), it became readily apparent that both the direction of inspired air currents and the velocity of the stream indicated the middle meatus as the region most susceptible to injury from inhaled air, as well as the major target of its unfriendly character or constituents. Later on, the question of oro-nasal breathing came to our attention. Here the work of Philip Cole and Niinimaa, and later still of Rodenstein and Stanescu, did much to clarify that difficult problem. Scans with radiolabeled aerosols revealed how deposition patterns vary with nasal breathing (Figure 3A) compared to oral breathing (Figure 3B).

But even now many questions remain, especially in relation to occupational exposures. It is my belief that toxicological studies of the upper airways cannot be relied upon in experimental animals, but this is certainly the fact in relation to oro-nasal breathing. Animals do not ordinarily resort to mouth breathing, and even in panting, there is little similarity to mouth breathing, as in the exercising human. For example, note the position of the epiglottis and palate in the barking deer (Figure 4). Extrapolation from ani-

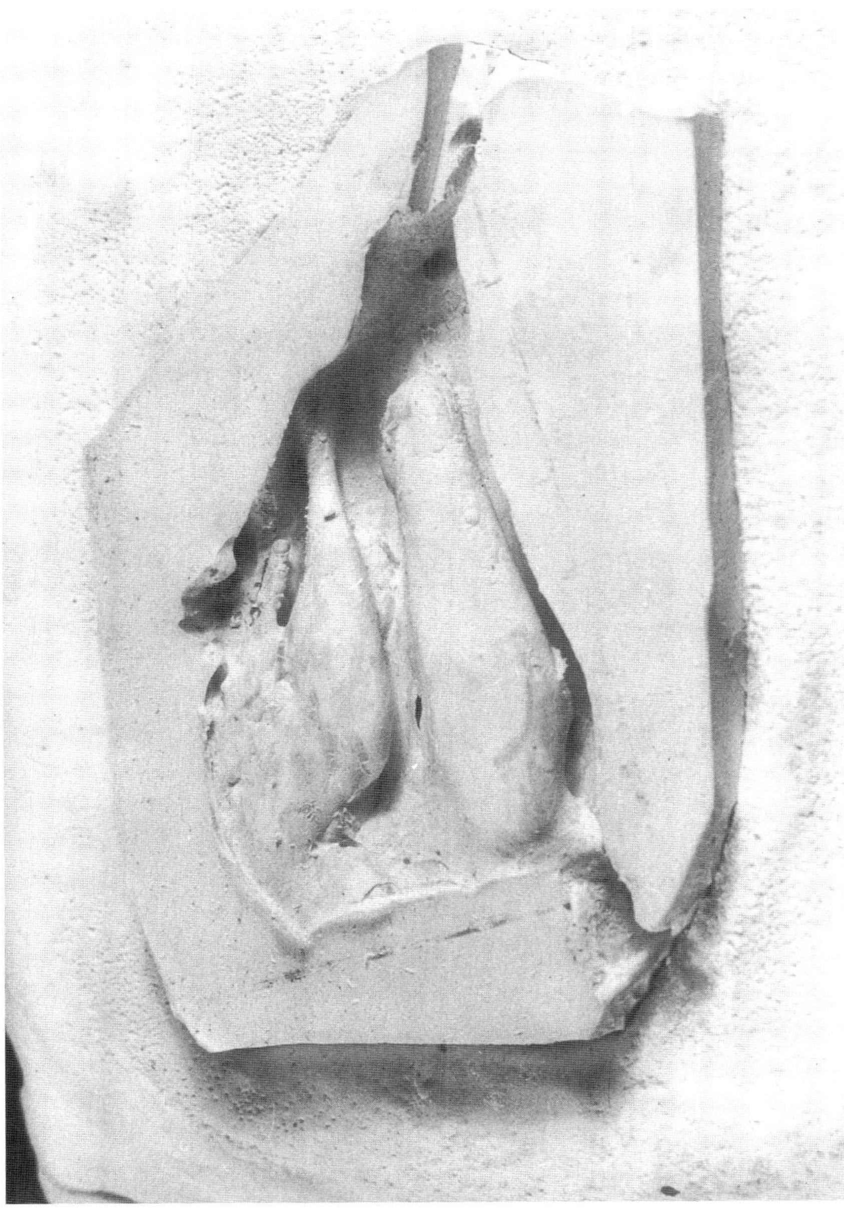

(A)

FIGURE 1. (A) Cast made from casting of human nose taken at the morgue.

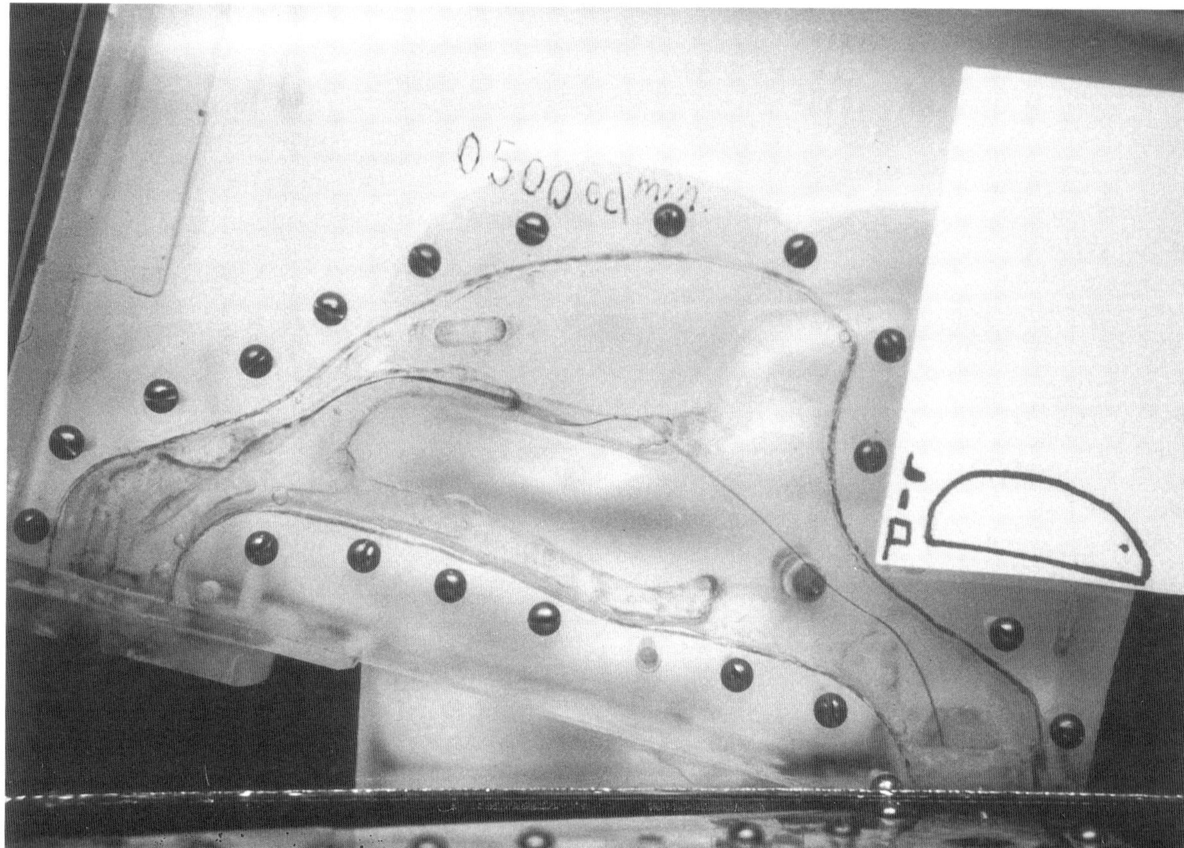

(B)

FIGURE 1. *(Continued)* (B) Lucite model taken after cast. The fine line shows a passage of dye. Inspiratory path begins on right.

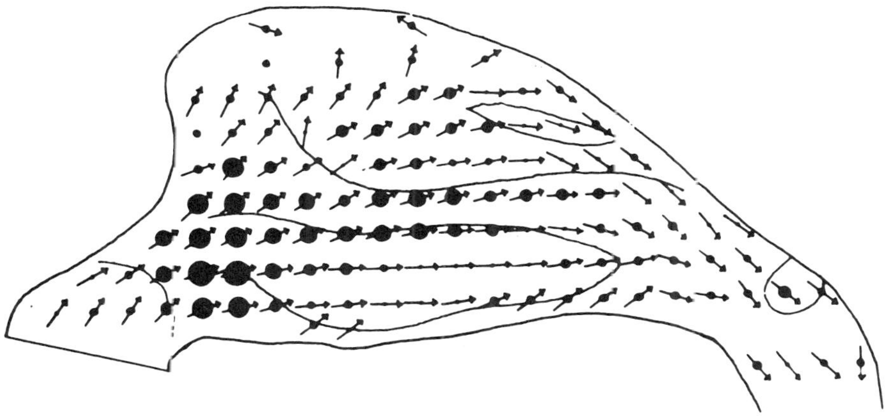

FIGURE 2. Schematic showing linear velocity and direction of air upon inspiration. Velocity indicated by the size of the dot. Reprinted from Swift and Proctor (1977) with permission.

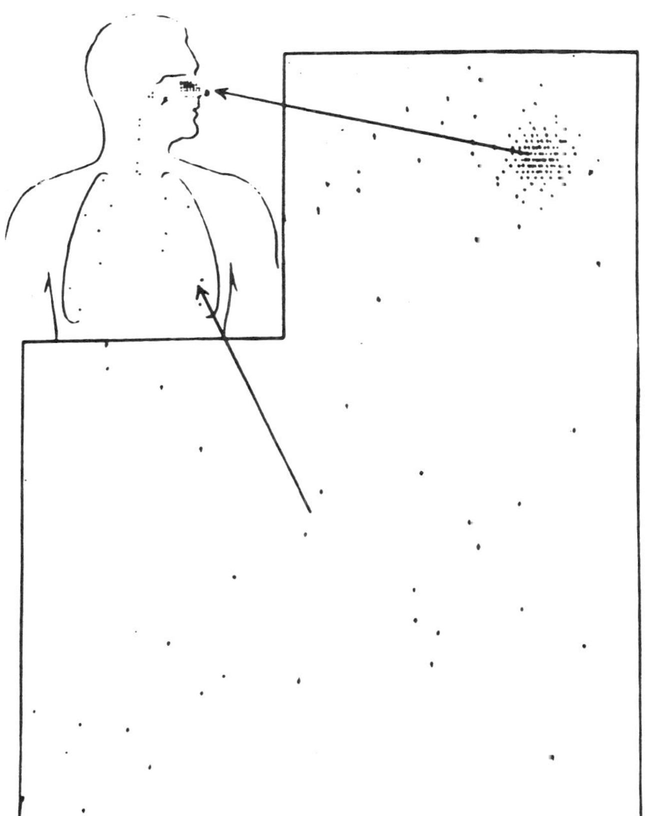

FIGURE 3. (A) Scan of inhaled isotope showing that with nasal breathing almost all of the inhaled material is stopped in the nasal passage, shown on the upper right.

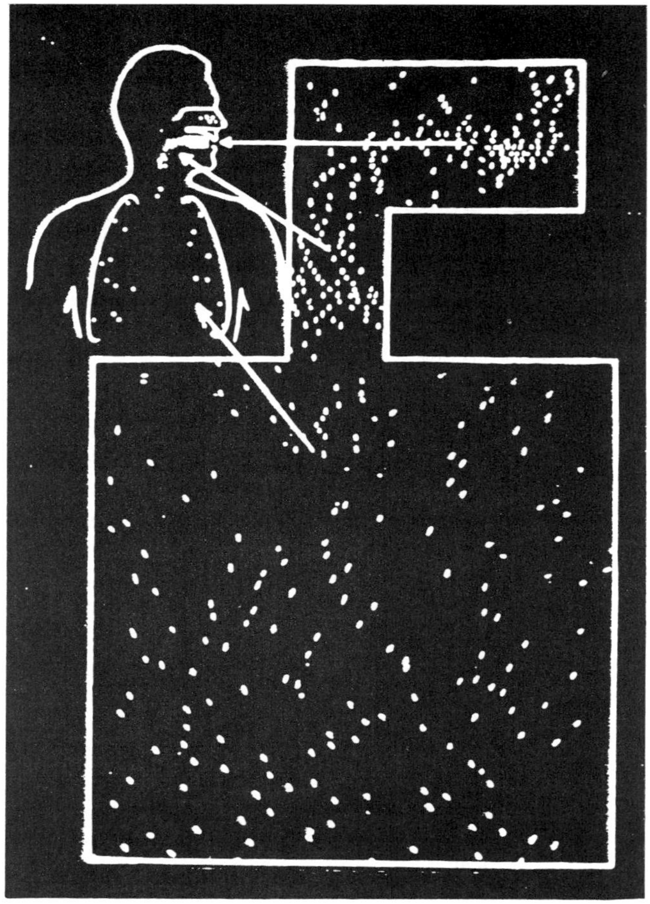

FIGURE 3. (*Continued*) (B) In contrast, when the material is taken through the mouth, although there is some deposition in the mouth and pharynx (shown upper right), a great deal passes through into the lungs.

mals to humans in the nose is open to question. Meanwhile, we began working on the means by which nasal surfaces are cleaned, mucociliary function. Sharpey, Hilding, and Messerklinger had all done useful work in that field. Sharpey's (1830) article has always fascinated me. His first sentence goes, "In the course of some investigations on the development of the tadpole in which I was lately engaged, I was accidentally led to observe, that the surface of that animal possessed the power of exciting currents in the water contiguous of it, in a constant and determinate direction." Would that more of us could be so "accidentally led."

But no method for measuring nasal clearance exactly in humans had been developed by 1967. An Australian working in our lab, Michael Quinlan (1969), came up with the ingenious idea of tagging a single minute particle and following its motion in the mucociliary stream with external detectors (Figure 5). I will return to our application of that method a bit later.

FIGURE 4. Head of the barking deer. Note the interdigitation of the epiglottis and soft palate, allowing nasal airway continuation with the larynx and separation from the mouth. Reprinted from Negus (1958) with permission.

At that time, 1968, two young Danes visited our laboratory. They were just completing a "climate chamber" in Aarhus, Denmark, planned for the investigation of human health effects of indoor air. On seeing the progress we were making on methods for measuring nasal function in man, they invited me to join them in their studies of indoor air. Shortly afterward they obtained financial support from NATO, and in 1970 we began our collaboration in Aarhus.

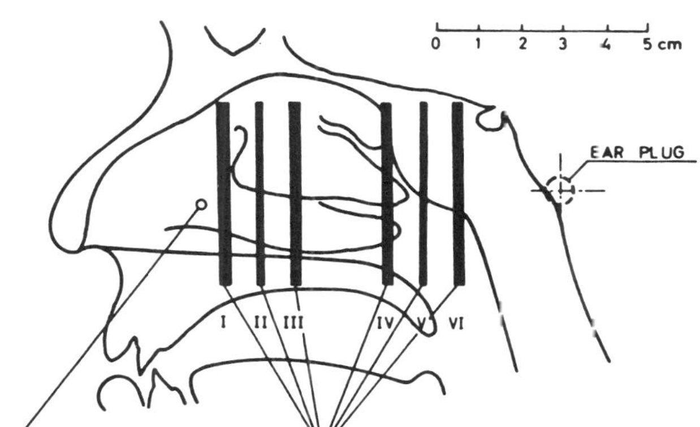

FIGURE 5. Diagram of the collimation used to detect particle transport. Depicted are the collimator slits; particle placement is often more distal (i.e., between slits 1 and 2) than shown in the diagram. Reprinted from Andersen et al. (1974) with permission.

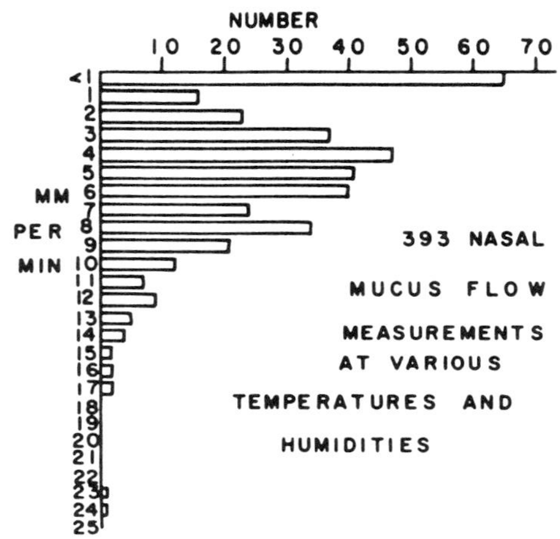

FIGURE 6. Distribution of flow rates of mucus. Note that of the 393 separate measurements, about 65 were found to be <1 mm/min. Reprinted from Proctor (1977) with permission.

Between 1970 and 1982 we studied nasal function in relation to controlled ambient temperature and relative humidity, and then went on to examine other aspects of indoor air with investigations in identical twins in Stockholm, the common cold in Baltimore, and in Salisbury England, sulfur dioxide, dust, and toluene. Our studies of the effects of an inert dust showed surprisingly little effect, even at levels of exposure that left our subjects looking as though they had just emerged from a coal mine. The results of these collaborative studies were reported in 16 articles.

During the course of these investigations it occurred to us that a simpler method for measurement of nasal clearance would be useful in field studies and for studies in children. I knew from past personal experience that a substance that has a distinctive taste, when placed in the nose, would after a short time be detected by the taste buds in the throat. We decided to try a small particle of saccharine and found it to be ideal, yielding a sharp end point. When the saccharine was colored with a dye, a simple look at the pharynx served as a check on the otherwise purely subjective response of the subject.

I'm sure that many major advances have been made in the physiology and toxicology of nasal exposure to our environment during the past decade, but I believe that a few problems remain unsolved, or perhaps not even yet thought of. I will mention only a few of those that seem to be in need of examination. As you see from Figure 6, about 17% (100 × 65/393) of all of our "normal" subjects were found to be what we came to call "slow movers." That is to say that their nasal clearance rate was less than 1 mm/min. We should try to learn more about the significance of this peculiarity.

I was perfectly serious when I referred earlier to MacKenzie's article. Although some useful studies have since been done by Wolf, Hardy, and

Goodell, we still need to learn much more about the effects of a wide variety of simple or complex sources of life stress. Finally, especially in relation to various occupational exposures, there should be investigations of the occurrence, nature, and effects of oro-nasal breathing at work.

But now is a good time for me to remember my promise that my remarks would be brief. So I'll take George Catlin's (1869) advice and "shut my mouth."

REFERENCES

Andersen, I. B., Lundqvist, G. R., Jensen, P. L., and Proctor, D. F. 1974. Human response to controlled levels of sulfur dioxide. *Arch. Environ. Health* 28:31–39.

Arbuthnot, J. 1773. An Essay Concerning the Effects of Air on Human Bodies. London: J. Tonson.

Catlin, George. 1869. *The breath of life or mal-respiration. Shut your mouth.* New York: John Wiley & Son.

MacKenzie, J. N. 1898. On the physiological and pathological relations between the nose and the sexual apparatus in man. *Bull. Johns Hopkins Hosp.* 9:10–17.

Negus, V. E. 1958. *The comparative anatomy and physiology of the nose and paranasal sinuses.* Edinburgh: Livingstone.

Proctor, D. F. 1977. Nasal physiology and the defense of the lungs. *Am. Rev. Respir. Dis.* 115:97–129.

Quinlan, M., Salman, S., Swift, D. L., Wagner, H. N., Jr., and Proctor, D. F. 1969. Measurement of mucociliary function in man. *Am. Rev. Respir. Dis.* 99:13–23.

Sharpey, W. 1830. On a peculiar motion excited in fluids by the surfaces of certain animals. *Ednab. Med. J.* 34:113–122.

Swift, D. L., and Proctor, D. F. 1977. Access of air to the respiratory tract. In *Respiratory defence mechanisms,* Chap. 3. New York: Marcel Dekker.

RESEARCH STRATEGY FOR ASSESSING HUMAN RISK FROM INHALED NASAL TOXICANTS

Roger O. McClellan

Chemical Industry Institute of Toxicology, Research Triangle Park, North Carolina, USA

Information exists on the nasal toxicity in humans of only a few xenobiotic compounds. Thus, as for many other materials, it is necessary to conduct studies in laboratory animal species and extrapolate the findings to people. This article advocates a strategy for conducting such research that emphasizes the need for obtaining a mechanistic and quantitative understanding of the linkages between exposure and dose to critical biological targets and adverse responses. Central to the strategy is the conduct and integration of data obtained from physical and chemical characterization of potential toxicants, in vitro studies with cells and tissues from humans and laboratory animals, in vivo studies in laboratory animals, and human data when they exist.

The estimation of human risk from inhaled xenobiotics is a complex process that includes direct observation on exposed people and, more typically, extrapolations from studies on other species. This includes in vitro investigations that utilize cells and tissues from humans and laboratory animal species and in vivo studies with a range of laboratory animal species, although the most extensive observations are on rats. This research is directed toward elucidation of the mechanistic linkages among exposure, dose to critical tissues, cells, and biomolecules, and the resultant biological responses, including those characterized as disease. The successful conduct of such research and, ultimately, the characterization of human risk require the talents of individuals trained in a wide range of disciplines from the engineering and physical sciences to the biomedical sciences. This article reviews the issues involved and the approaches being taken to provide a scientific basis for estimating the human health risks of inhaled nasal toxicants. It provides an integrated structure for considering the other articles included in this volume. Additional details on the mechanism-based approach to risk assessment are contained in two complementary papers (McClellan, 1994; Conolly, 1994).

Paper presented at International Workshop on Nasal Toxicity and Dosimetry of Inhaled Xenobiotics: Implications for Human Health, September 20–22, 1993, Durham, NC.

I gratefully acknowledge the many valuable discussions with my colleagues at the CIIT and with members of various advisory committees on which I have served that have helped mold the views expressed here.

Address correspondence to Dr. Roger O. McClellan, President, Chemical Industry Institute of Toxicology, P.O. Box 12137, Research Triangle Park, NC 27709, USA.

ISSUES

The foundation for our research and related risk characterization efforts is the shared societal goal of protecting the health of workers, consumers, and the public, while still making available new technologies and products to society. It is recognized that a wide range of xenobiotics is found in the air breathed by people in the workplace, the home, and the general environment.

Past experience has shown that some of these exposures result in responses, some of which are characterized as disease. Examples drawn on human experience are nasal irritation with sulfur dioxide (Andersen et al., 1974), nasal septal necrosis with chromic acid fumes (Maran & Lund, 1990), and nasal cancers in nickel workers (Boysen et al., 1984). In other cases concern for human risk is based on laboratory animal experience, such as nasal irritation with aldehydes (Kane et al., 1979) and nasal cancer resulting from long-term, high-level exposure of rats to formaldehyde (Kerns et al., 1983). Thus, there is recognition that some materials arising either as by-products or intended products of societal activities may pose a risk to people.

There is a collective need to take a proactive approach to identifying and characterizing potential risks at an early stage in the development of new technologies and new products before significant human exposure has occurred. The research arising from such proactive approaches can then be used to establish appropriate standards for the workplace, consumer environment, and the general environment to minimize potential health risks to workers, consumers, and the public.

There are many challenges to implementation of such an approach for potential air pollutants. Some of the most vexing challenges relate to the extent to which the air environment typically involves a multitude of chemical compounds, of which the individual concentrations (other than oxygen, ozone, nitrogen and carbon dioxide) are quite low; the exposure gradient between the most exposed and least exposed populations is frequently quite low; the effects of confounding factors, such as cigarette smoking, are substantial; the diseases of concern are not uniquely related to specific environmental exposures; and there is a paucity of well-developed and validated animal models.

Because our concerns extend to exposures beyond those for which direct human observations are available, the need arises to extrapolate. Extrapolation is the science, and some would add, the art, of estimating what is not directly observed from what can be observed. In the easiest of circumstances, this involves extrapolation using human data from high levels of exposure to lower levels of exposure and attendant much lower levels of risk. However, this circumstance is rarely the case. More typically, human data do not exist or are too fragmentary to use with confidence. In the absence of adequate human data, a range of extrapolation issues arise: from rats and other laboratory animal species to people, from high levels of exposure to low, from short periods of time to lifetimes, from one compound to another, from in vitro observations to the in vivo situation, and from disaggregated systems, such as cells in culture, to complex integrated systems, such as the intact mammal with its wide array of homeostatic mechanisms.

The airborne materials of concern for human risk assessment range from solid particles or liquid droplets to gases. The physical and chemical characteristics of the material in the air can have a profound effect on whether the material will have the potential for deposition in the nose or portions of the respiratory tract distal to the nose. Deposition of the material and subsequent retention or translocation will depend on multiple factors, including for droplets and physical particles their aerodynamic and diffusion diameters, and for gases their solubility and reactivity. Once deposited, other parameters, such as their potential for being metabolized, being complexed with other molecules, or being solubilized, become of concern in influencing retention or translocation.

If the material is in the form of a solid particle or liquid droplet, the aerodynamic diameter and diffusion diameter will have a major influence on deposition. The largest particles capable of being inspired will have the greatest probability of being deposited in the nose and lowest probability of being drawn through the nose and deposited in more distal locations in the respiratory tract. As the aerodynamic diameter of the particle or liquid droplet decreases from on the order of 10 mm to 1 mm, the probability of deposition both in the nose and at deeper sites in the respiratory tract is reduced. From about 1 mm to 0.5 mm aerodynamic diameter (where aerodynamic size ceases to be the major determinant of deposition), there is little nasal deposition but increasing deposition in the more distal portions of the respiratory tract. At about 0.5 mm diameter real size and below, the diffusion properties of particles and droplets become dominant in influencing deposition. The smaller the particle or droplet, the greater the influence of diffusion. At the smallest particle sizes, nasal deposition via diffusion will remove a substantial portion of the particles and droplets from the inspired air, leaving very little to be deposited at more distal locations.

For materials that are in the form of gases, solubility and reactivity greatly influence deposition. The higher the reactivity and solubility, the greater the potential for deposition in the nose. Once deposited, the reactivity, solubility, and ease of metabolism influence whether the chemical remains at the site of deposition, is metabolized to another form, or is translocated in the parent form or as a metabolite to other locations.

It is next appropriate to consider what the functions are of the nasal passages and tissue that should be of concern from the viewpoint of the well-being of the individual. The nasal passages and tissue have multiple functions, the most important being to serve as a passageway for movement of gases to and from the lungs. Other functions include filtration and removal of particles, fibers, and droplets so they do not reach the lungs, and conditioning of the temperature and humidity of incoming air before it reaches the more delicate structures of the lung, a site for immunological reactions and a key point of interaction for purposes of olfaction. Metabolic alteration of inhaled materials can be a part of the process of olfaction.

It is desirable to protect the ability of the nasal passages and tissues to carry out these functions. In addition to direct alterations in function, there is concern for development of neoplasms that may indirectly influence the ability of the nasal tissues to carry out normal functions, as well as the potential for neoplasms to be life-threatening. The foregoing leads to our initial risk question: Do air pollutants, individually or in mixtures, cause respiratory disease? It follows from this that our concern extends to a broader question: What are the exposure-response relationships on a population basis for air pollutants, individually or in mixtures? Are there thresholds of exposure below which effects are not seen? What is the estimated prevalence in the population? What is the severity of the condition in affected individuals? Are there sensitive subpopulations that are affected? Beyond concern for overt disease, there is a need to ask the question: When is a change in function/structure an adverse health effect? This question will be increasingly posed as more sensitive indicators of subtle changes in structure and function are developed.

AN APPROACH TO CHARACTERIZATION OF HUMAN RISK

The approach advocated here for characterization of human risk involves the joining of a mechanistic understanding of dosimetry and pathogenesis for inhaled xenobiotics with a risk assessment orientation that quantitatively links exposure to response. The approaches must be integrated if scientifically defensible risk characterizations are to be developed.

One leg of this approach, that of risk assessment, was codified by a National Academy of Sciences/National Research Council Committee in 1983 as involving research, risk assessment and risk management (NAS/NRC, 1983). The risk assessment component was further subdivided into (1) hazard identification, (2) exposure-dose-response assessment, (3) exposure assessment, and (4) risk characterization. The exposure–dose-response assessment element was originally shown as dose-response assessment. However, it is clear for airborne materials that exposure and dose are definitionally distinct, so the expanded statement that makes exposure an explicit element is more appropriate. More recently, another committee of the NAS/NRC (1994) and a report of the Office of Technology Assessment (U.S. Congress, 1993) have emphasized the importance of a feedback loop from risk assessment to help guide research. This results in a broadened paradigm, as shown in Figure 1.

The mechanism-based leg of the approach is depicted graphically in Figure 2. Emphasis is given to establishing mechanistic linkages between exposure and the dose to critical biological targets and from these doses ultimately to the biological responses, whether they be reversible or irreversible changes and involve cancer or other diseases. Each mechanism's arrow in Figure 2 is in reality a complex set of interrelated interactions that is shown graphically in Figure 3. Much progress has been made in recent years in the development of structured mathematical models to describe the linkage between exposure and dose as reviewed by Ultman (1994) in this volume. Less, but very important, progress has been made in the develop-

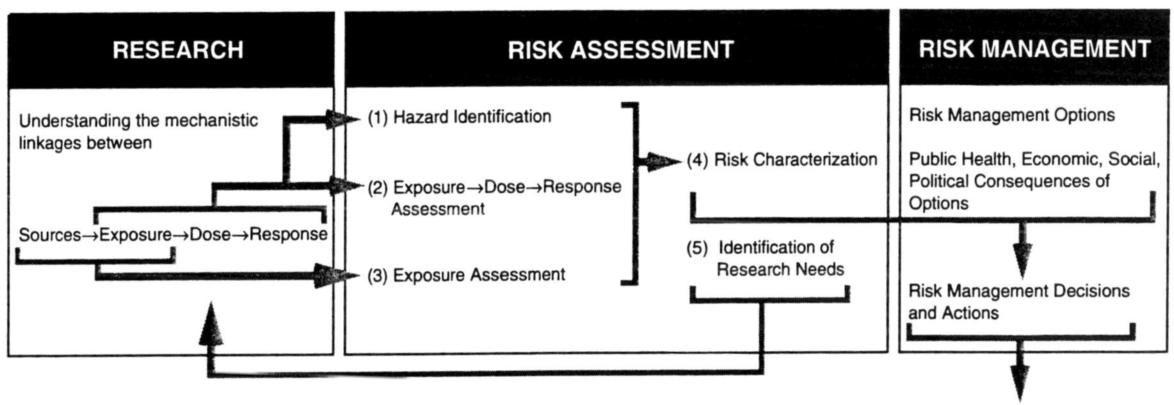

FIGURE 1. Research, risk assessment, and risk management paradigm. Adapted from NRC/NAS (1983, 1994).

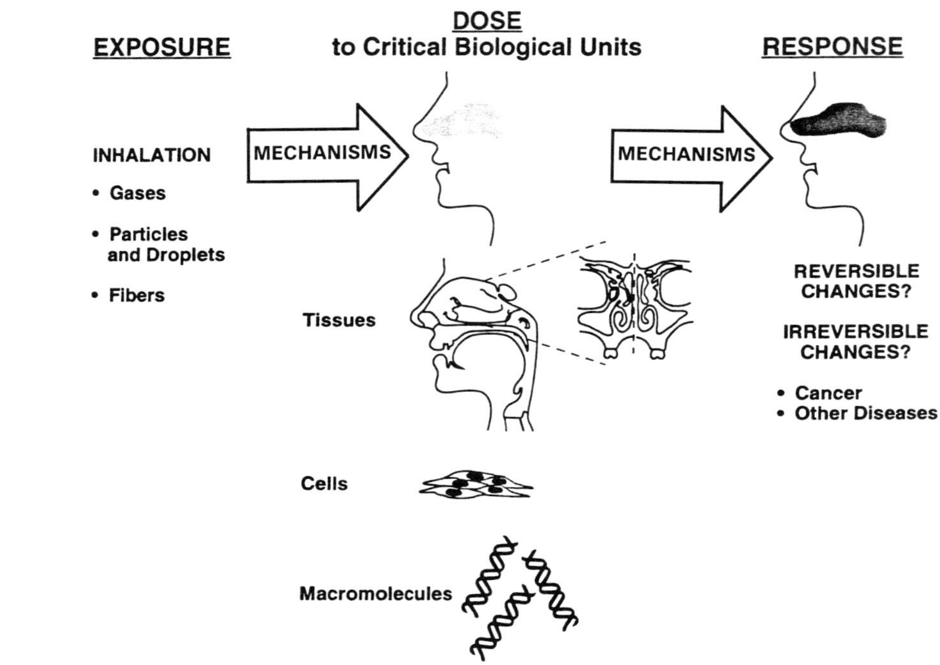

FIGURE 2. Mechanism-based approach to understanding and assessing nasal toxicity.

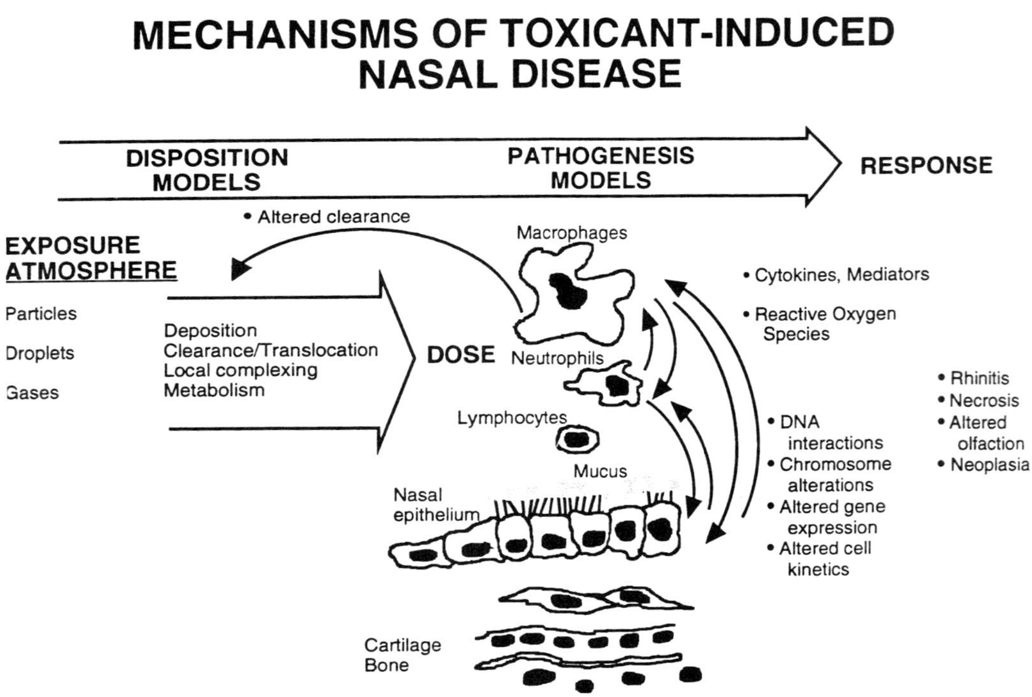

FIGURE 3. Mechanisms of toxicant-induced nasal disease.

ment of models describing the relationship between dose and neoplastic responses (Conolly & Andersen, 1991). An example, for the nasal toxicant formaldehyde, has been advanced by Conolly et al. (1992) and Conolly and Andersen (1993) and is discussed in-depth elsewhere (Conolly, 1994). The Conolly et al. model is an extension of the model of Moolgavkar and Knudson (1981) that explicitly considers not only DNA reactivity and resultant mutational events, but also the role of cell death and division as occurring naturally, as well as induced by mitogenic and cytotoxic effects of toxicants. Such models are valuable for integrating and synthesizing our mechanistic understanding of toxicologic processes and for providing a means for assessing whether the information "fits together," rather than allowing individual pieces of information to be viewed in isolation.

THE SYMPOSIUM CONTENTS

The various papers in the symposium were grouped into sessions that can be viewed from an integrative perspective (Figure 4). One set of papers emphasized the role of models, a second set presented the results of in vitro studies with cells and tissues, a third set described the effects of nasal toxicants in laboratory animal species elucidated from in vivo investigations, and a fourth set of papers described the spectrum of human responses observed in vivo. A fifth set of papers integrated this information and discussed implications for human health and, especially, the setting of human exposure limits.

It is clear that the aggregate value of the information presented in the individual papers at the symposium was more than a simple summing of

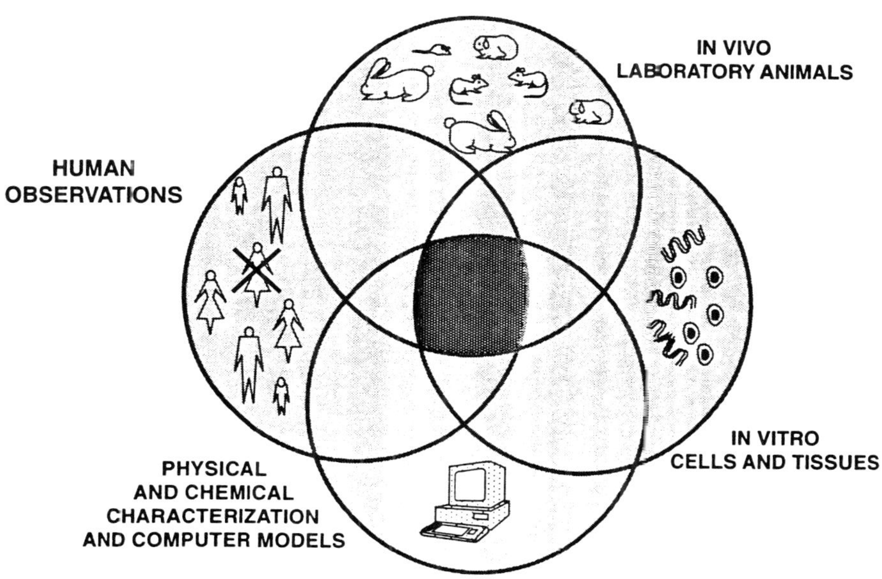

FIGURE 4. Multifaceted approach to evaluating health risks of chemicals (McClellan, 1993).

the value of the individual contributions. Some of the value derived from synergy was achieved through the many informal discussions that took place at the workshop and will be manifest in future collaborative endeavors, and, ultimately, in publications. Much of this value will come from having brought together scientists of varied disciplinary backgrounds and users of different research approaches. Hopefully, the interactions among the attendees will result in more attention being given in the future to comparisons between results obtained in cells and tissues from humans versus laboratory animals, as well as for results from in vitro studies compared to those obtained in vivo in the intact organism. And finally, there will be value gained from placing results of acute exposure-response studies into perspective as a part of the pathogenesis of effects arising from chronic exposures.

A PERSPECTIVE FOR CONSIDERING THE SYMPOSIUM PAPERS

There are a couple of important points to keep in mind when reading the various symposium papers. First and foremost, I think it is appropriate to consider how the information presented helps in characterizing human risk within a mechanistic exposure-dose-response framework. In considering the mechanistic information, two questions should be kept in mind. The first is whether the exposure-dose-response relationships are likely to be observed uniformly across the human population. Or are they responses that might be expected to be nonuniform across the population, with a sen-

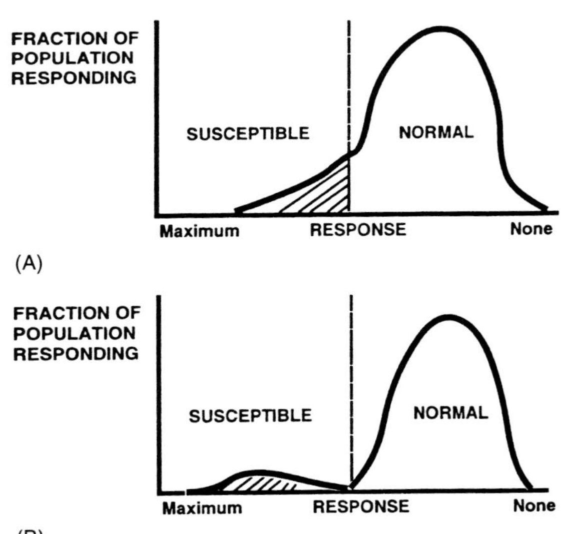

FIGURE 5. Schematic representation of distribution of responses in populations with (A) discrete susceptible subpopulation, and (B) susceptible subpopulation tail on heterogeneous distribution of response to chemical exposure (McClellan, 1989).

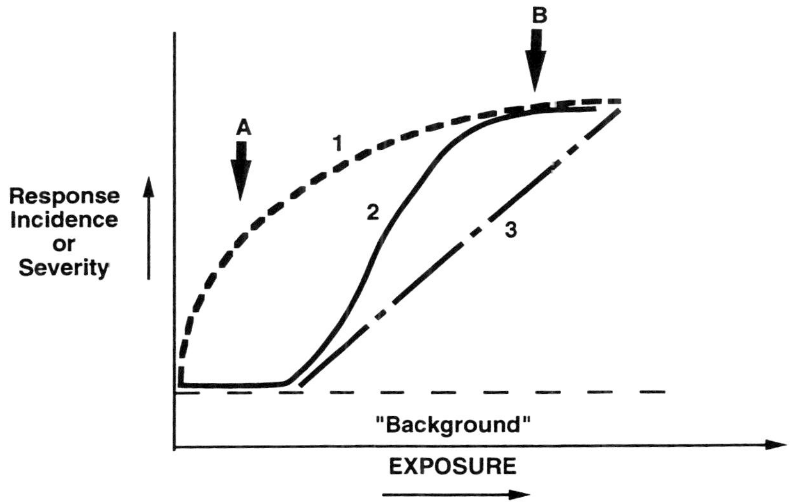

FIGURE 6. Hypothetical exposure-response relationships (curves 1, 2, and 3) and two exposure levels (A and B) used for mechanistic studies.

sitive subpopulation being apparent, as depicted in Figure 5? In Figure 5A, the susceptible subpopulation is shown as a discrete subpopulation, while in Figure 5B, the susceptible individuals are a "tail" on the distribution. The existence of either susceptible subpopulation has important implications for the setting of standards for air pollutants. The importance of variations in interindividual susceptibility is apparent when considering susceptibility to aeroallergen-induced reactions, a specialized topic that, because of time limitations, could not be addressed in detail in this symposium.

The second question to be asked is the relevance of the mechanistic data for understanding responses that might arise in exposure circumstances encountered by people. It is easy to envision responses observed at high levels of exposure that may play a dominant role in the overall response at that level but are absent at lower levels of exposure. This situation is shown schematically in Figure 6, where three hypothetical exposure-response curves (1, 2, and 3) and two exposure levels (A and B) for mechanistic studies are shown. If mechanistic studies were to be conducted only at the highest exposure level (B), they might be used inappropriately to infer mechanisms of action at lower exposures (level A). This would be especially misleading if the true exposure-response relationship were as shown for 2 or 3 in Figure 6. This emphasizes the need to study multiple exposure levels in mechanistic studies. It can also be argued that multiple time points should be examined, especially when the mechanistic results are to be used to evaluate the pathogenesis of diseases, such as cancers that occur late in life and, frequently, are elevated in incidence by chemicals only when there is chronic exposure.

The studies of the pathogenesis of nasal cancer in rats produced by chronic exposure to high levels of formaldehyde took into account these key points (Monticello, 1990; Monticello et al., 1991; Morgan & Monticello, 1990). As a result, it was shown that with chronic exposure to 15 ppm formaldehyde, persistent tissue damage, increased cell proliferation, and a high prevalence of nasal cancer were associated. At the lowest exposure levels studied, there was an absence of persistent damage and compensatory cell proliferation. At this time, it is not known if there is any elevated level of mutations in nasal cells at any of the exposure levels. This is a matter that will be addressed at CIIT in the near future. The availability of such data will provide insight into the relative importance of cell proliferation versus mutational changes in the pathogenesis of formaldehyde-induced nasal cancers and the most appropriate form of the exposure-response curve for extrapolation to levels of formaldehyde below those for which direct observations can be made. Use of mechanistic models of the type described by Conolly (1994) and McClellan (1994) will be helpful in assessing the validity of these extrapolations.

SUMMARY

A strategy for characterizing the health risks of exposure to airborne materials was briefly reviewed. The strategy builds on a targeted approach to developing mechanistic data that will help elucidate quantitative linkages between exposure, dose, and response. It emphasizes the need to integrate information obtained from studies at all levels of biological organization, from macromolecules and cells to intact mammals to populations of individuals.

REFERENCES

Andersen, I., Lundqvist, G. R., Jensen, P. L., and Proctor, D. F. 1974. Human response to controlled levels of sulfur dioxide. *Arch. Environ. Health* 28:31–39.

Boysen, M., Solberg, L. A., Torjussen, W., Poppe, S., and Hogetveit, A. C. 1984. Histological changes, rhinoscopical findings and nickel concentration in plasma and urine in retired nickel workers. *Acta Otolaryngol. (Stockh.)* 97:105–115.

Conolly, R. B. 1994. Cancer and non-cancer risk assessment: Not so different if you consider mechanisms. Proceedings of 1993 U.S. Environmental Protection Agency/Health Effects Research Laboratory Symposium on Biological Mechanisms and Quantitative Risk Assessment. *Toxicology*, in press.

Conolly, R. B., and Andersen, M. E. 1991. Biologically-based pharmacodynamic models: Tools for toxicological research and risk assessment. *Annu. Rev. Pharmacol. Toxicol.* 31:503–523.

Conolly, R. B., and Andersen, M. E. 1993. An approach to mechanism-based cancer risk assessment: Formaldehyde. *Environ. Health Perspect. Suppl.* 101(Suppl. 6):169–176.

Conolly, R. B., Monticello, T. M., Morgan, K. T., Clewell, H. J., III, and Andersen, M. E. 1992. A biologically-based risk assessment strategy for inhaled formaldehyde. *Comments Toxicol.* 4:269–293.

Kane, L. E., Barrow, C. S., and Alarie, Y. 1979. A short-term test to predict acceptable levels of exposure to airborne sensory irritants. *Am. Ind. Hyg. Assoc. J.* 40:207–229.

Kerns, W. D., Pavkov, K. L., Donofrio, D. J., Gralla, E. J., and Swenberg, J. A. 1983. Carcinogenicity of formaldehyde in rats and mice after long-term inhalation exposure. *Cancer Res.* 43:4382–4398.

Maran, A. G. D., and Lund, V. J. 1990. *Clinical rhinology*, 121 pp. New York: Thieme.

McClellan, R. O. 1989. Reflections on the symposium: Susceptibility to inhaled pollutants. In *Susceptibility to inhaled pollutants*, ASTM STP 1024, eds. M.J. Utell and R. Frank, pp. 224–231. Philadelphia: American Society for Testing and Materials.

McClellan, R. O. 1993. Occupational and environmental health: The importance of animal research. *CIIT Activities* 13(1):1.

McClellan, R. O. 1994. Risk assessment and biological mechanisms: Lessons learned, future opportunities. Proceedings of U.S. Environmental Protection Agency/Health Effects Research Laboratory Symposium on Biological Mechanisms and Quantitative Risk Assessment. *Toxicology*, in press.

Monticello, T. M. 1990. Formaldehyde-induced pathology and cell proliferation. PhD dissertation, Duke University, Durham, NC.

Monticello, T. M., Miller, F. J., and Morgan, K. T. 1991. Regional increases in rat nasal respiratory epithelial cell proliferation following acute and subacute inhalation of formaldehyde. *Toxicol. Appl. Pharmacol.* 111:409–421.

Moolgavkar, S. H., and Knudson, A. G. 1981. Mutation and cancer: A model for human carcinogenesis. *JNCI* 66:1037–1052.

Morgan, K. T., and Monticello, T. M. 1990. Formaldehyde toxicity: Respiratory epithelial cell injury and repair. In *Biology, toxicology, and carcinogenesis of respiratory epithelium*, eds. D. G. Thomassen and P. Nettesheim, pp. 155–171. New York: Hemisphere.

National Academy of Sciences/National Research Council. 1983. *Risk assessment in the federal government; Managing the process*. Washington, DC: National Academy Press.

National Academy of Sciences/National Research Council. 1994. *Science and judgment in risk assessment*. Washington, DC: National Academy Press.

Ultman, J. S. 1994. Dosimetry modeling: Approaches and issues. *Inhal. Toxicol.* 6(suppl):59–71.

U.S. Congress, Office of Technology Assessment. 1993. Researching Health Risks, OTABBS570. Washington, DC: U.S. Government Printing Office, November.

NASAL TOXICITY: END POINTS OF CONCERN IN HUMANS

Donald A. Leopold

Johns Hopkins Medical Institutions, Baltimore, Maryland, USA

The nose is perhaps the most used portal of the human body, with gases, particles, and aerosols flowing in and out of it many times each minute of life It should be no surprise that some of these inhaled materials are chemically and biologically active and that their effects may be either beneficial or harmful. Gaseous oxygen is certainly needed to sustain life, while cyanide can quickly end it. Similarly, while pure water aerosols can moisten and help heal the nasal mucosa, viral and bacterial particles in the aerosol can infect that same mucosa. The anatomy of the nasal cavities and the path of airflow through them determine many of the characteristics of the nasal response to inhaled xenobiotics. For example, carcinomas caused by the inhalation of wood dust often form on the anterior head of the middle turbinate, where one of the main air streams in the nasal cavity hits it. It is at the level of the nasal mucosa, however, where inhaled xenobiotics interact with the cells of the nasal lining, whether they be squamous, respiratory, transitional, or olfactory. This interaction can have local effects in the nasal cavity, such as mucosal inflammation or crusting caused by cilia dysfunction. There can also be distant or systemic effects, such as decreased athletic performance, when inhaled ozone is not detoxified by the metabolically active nasal mucosa. There are many ways to evaluate the effects of inhaled materials on the nose. These can be used in response to individual complaints after a particular exposure, or as part of a testing protocol. Direct visual examination of the nasal cavities, especially with endoscopy, can reveal changes in the nasal epithelium. Other studies can assess airflow resistance, nasal cavity shape, and nasal mucosal thickness while nasal lavage assesses the mucosal secretions. Neural function can be evaluated with tests of olfactory and trigeminal ability. If the nose has been damaged by inhaled xenobiotics, the traditional medical techniques of humidification, antibiotics, steroids, and surgery can sometimes restore health. More importantly, in specific areas of high pollution, prevention and avoidance programs have been successful in decreasing the levels of morbidity.

The inhalation of xenobiotic materials into and through the nasal cavity can have significant effects on the human organism, from obvious anatomical changes and death to annoying symptoms that debilitate a person's daily life. In this review, the human nasal anatomy and physiology are summarized so that the biological effects of pollutants can be understood. These effects, both on the nose and on distant parts of the body, are listed and discussed. When evaluating these end points of human–toxin interaction, many standard methods of clinical investigation are applicable, from interview techniques to examination methods to tests of physiological function. The end result of all these endeavors will hopefully be a better understanding of basic nasal function, how that function is damaged by inhaled xenobiotics, and methods to avoid the damage.

The author would like to thank Keyvan Keyhani of the University of Pennsylvania for providing the drawings in Figure 2.

Address correspondence to Donald A. Leopold, MD, Johns Hopkins Otolaryngol-HNS, Suite 595A, 4940 Eastern Ave., Baltimore, MD 21224, USA.

HUMAN NASAL ANATOMY AND AIRFLOW

The nasal cavity through which inhaled gases, particles, and aerosols flow starts as a 1-cm oval opening, but rapidly becomes a 7-cm narrow crevasse that is only 1 or 2 mm wide. The framework of the nose anterior to the face is cartilaginous, while the remainder is a bony cavity. While most of the cartilage in the anterior nose is relatively rigid, the lateral walls are flexible and move with respiration. This changing anatomy creates a complicated airflow situation, making accurate mathematical and physical modeling of the nasal cavity difficult (Keyhani et al., 1993).

Inside the nasal cavity, the nasal septum has multiple ridges, is usually deviated to one side, and has an engorgable "swell body" just in front of the middle turbinate (Figure 1) (Wvstrow, 1951). The turbinates, which jut out from the lateral wall, are also lined with this engorgable mucosa. Blood flow to this specialized mucosa actively controls the size of the nasal aperture and thus the nasal airflow pattern (Cole et al., 1989). Without blood flow, such as in a cadaver, the anterior nasal cavity may be more than double its width during life! Measurements made from cadaveric models, therefore, are likely to overestimate the width of the anterior nasal cavity, and airflow assessments based on these data may be inaccurate (Keyhani et al., 1993).

Air that enters the nostrils is initially traveling up through them (Figure 2). The pattern of flow quickly turns horizontally (posteriorly) within 1 cm of the nostrils, as it enters the narrowest section of the whole nasal cavity, the nasal valve (Swift, 1982). In the nasal valve area, the flow velocity is rapid, and the flow characteristic is not steady (Keyhani et al., 1993). Just posterior to the valve area, the cross-sectional area of the nasal cavity increases markedly, resulting in a slowing of airflow. This slowing of flow will result in deposition of suspended particles and aerosols in this area, which is near the anterior end of the middle turbinate (Figure 3) (Cole, 1994). A second major bend of the extra thoracic human airflow occurs in the nasopharynx, where the flow pattern turns down toward the larynx and trachea. This is also a site for particle and aerosol deposition and may explain the location of the immunologically active adenoid and tonsil tissue in this area.

NASAL MUCOSA

The nasal cavity is lined with a blanket of mucosa from the anterior nares to the pharynx. As is described in greater detail in other sections of this volume, the mucosa anteriorly is squamous and without cilia. Within 1 cm of the nostrils, a transitional epithelium is found that quickly leads to respiratory epithelium. This ciliated respiratory epithelium is the major lining of the human nasal cavity and has great importance in cleaning the mucosa through the mucociliary system. Olfactory epithelium is located in the very superior aspect of the nasal cavity and occupies approximately 2

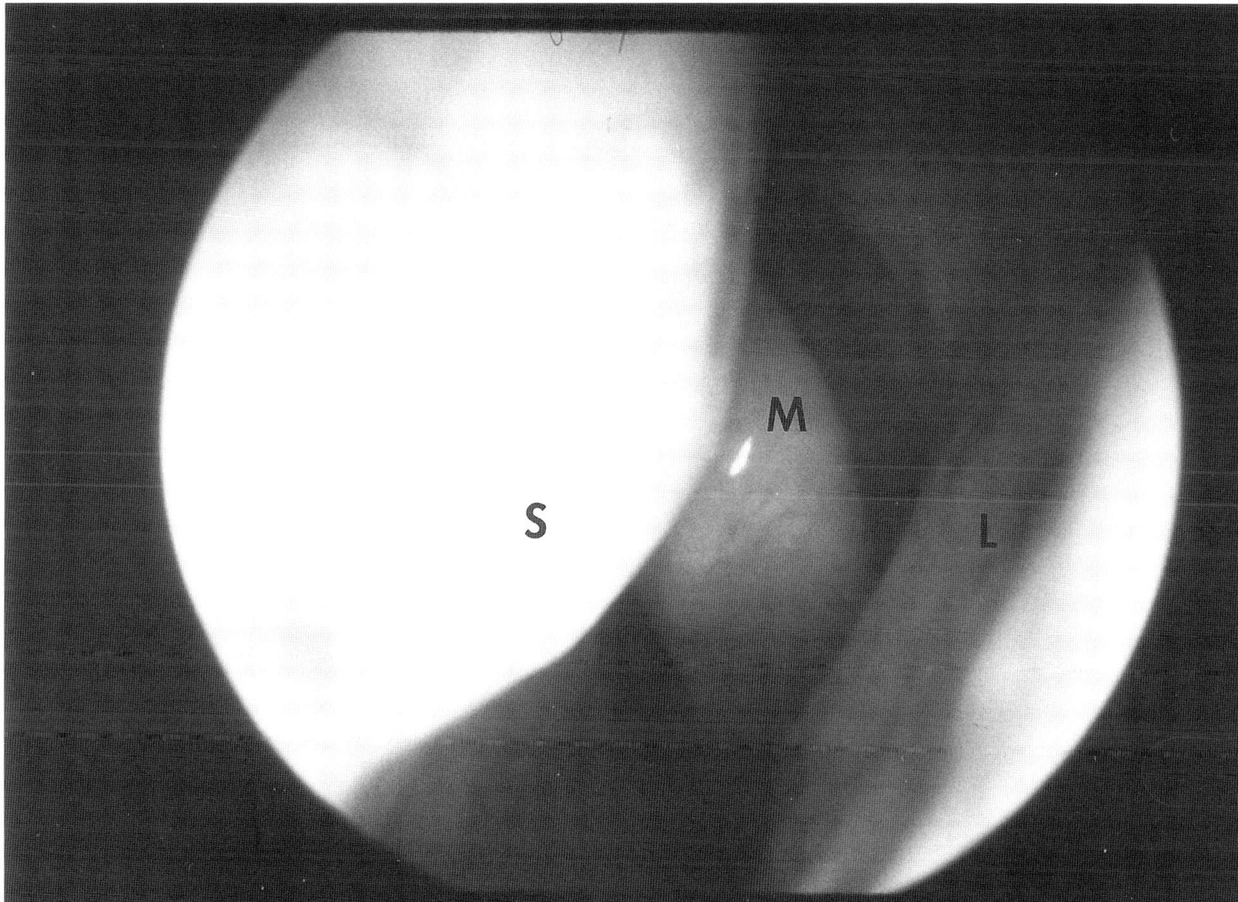

FIGURE 1. Middle of the left nasal cavity demonstrating swell body of septum and curved airway (S, septum; M, middle turbinate; L, lateral nasal wall).

A

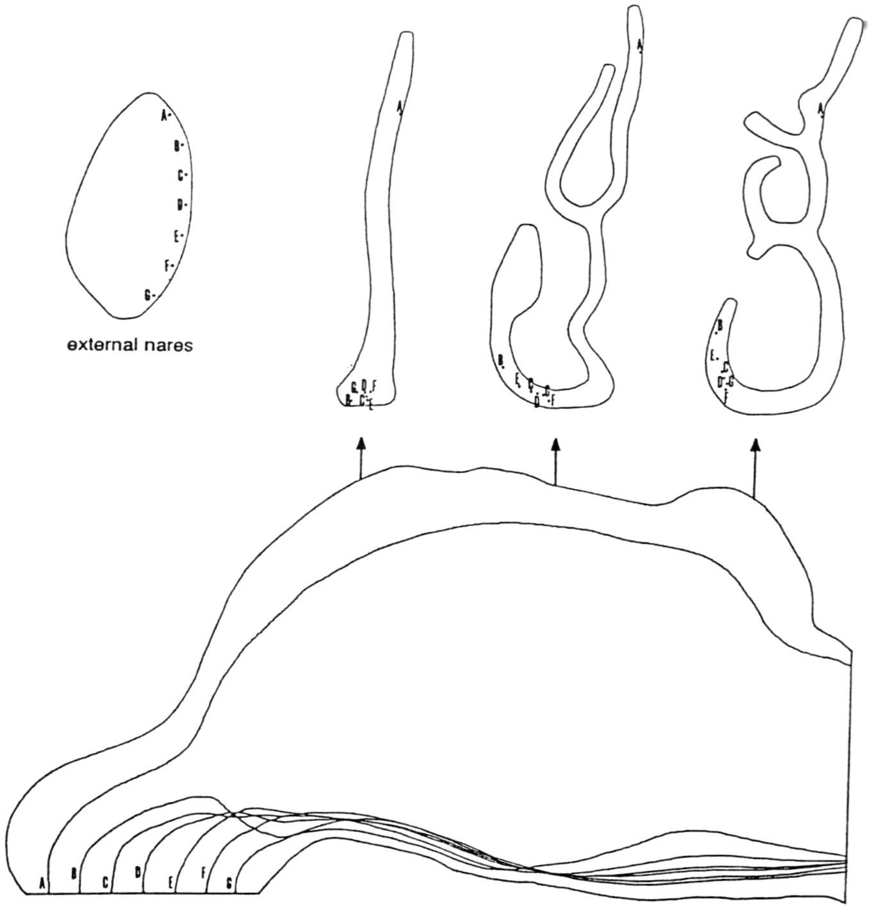

FIGURE 2. Numerical streamline plots obtained on a 20× normal-size right human nasal cavity model. The path from the streamlines from the septal side view and the points where they pass on three coronal sections (indicated by arrows) are shown. The locations of point particle release sites are shown on the external nares: in (A) particles are released 0.5 mm from the septal side.

cm^2 of each nostril. In addition, in all but neonates, there is a patchy infiltration of respiratory epithelium throughout the olfactory cleft (Nakashima et al., 1984; Paik et al., 1992). All of these olfactory and respiratory tissues are covered by a mucus layer that separates and protects them from the air blowing through the nasal cavities. For interactions to occur between inhaled materials and the mucosal tissues, the materials must penetrate the mucus. This is true for odorants that become dissolved in the mucus prior to interaction with the olfactory neuron and for many xenobiotics that chemically react with the mucus layer.

TRIGEMINAL INNERVATION

Trigeminal nerve fibers innervate the entire nasal epithelium. Their response to a variety of volatile chemical stimuli is called the common chemical sense. At least some of these trigeminal fibers contain neuropeptides and may be of the nociceptor type (polymodal or mechano-heat-sensitive) (Silver, 1992). These chemoreceptive fibers are thought to act as *affectors* that relay information about pungent and nonpungent odorants to the trigeminal nucleus of the brainstem. From the brainstem, not only are the impulses transferred to the odor appreciation areas for integration with

B

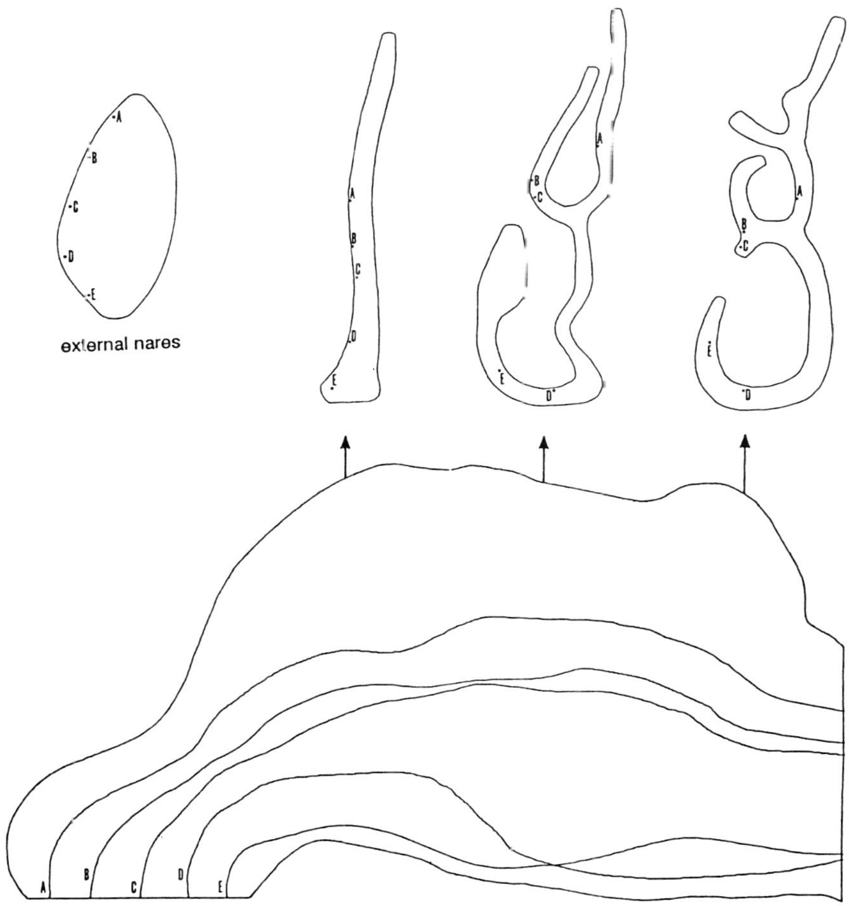

FIGURE 2. Numerical streamline plots obtained on a 20× normal-size right human nasal cavity model. The path from the streamlines from the septal side view and the points where they pass on three coronal sections (indicated by arrows) are shown. The locations of point particle release sites are shown on the external nares (*Continued*): in (B) particles are released 0.5 mm from the lateral side.

C

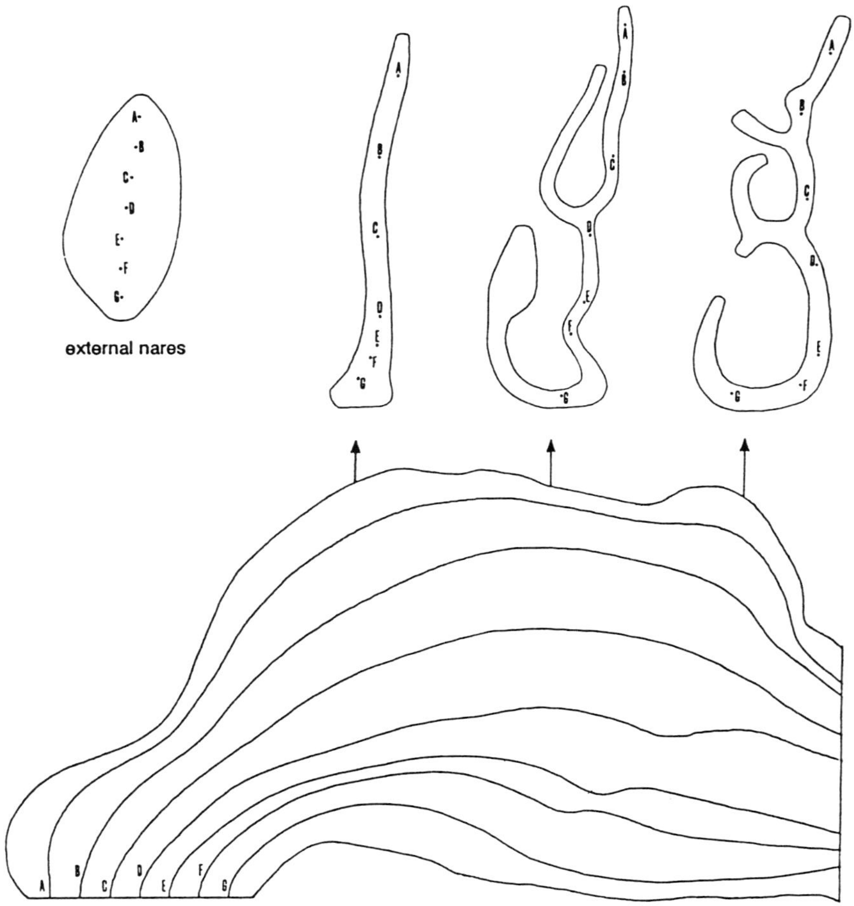

FIGURE 2. Numerical streamline plots obtained on a 20× normal-size right human nasal cavity model. The path from the streamlines from the septal side view and the points where they pass on three coronal sections (indicated by arrows) are shown. The locations of point particle release sites are shown on the external nares (*Continued*): in (C) particles are released from the middle of the nerves.

the olfactory input, but protective reflexes are initiated to limit exposure to a possibly dangerous chemical. These reflexes include cessation of breathing, slowing of the heart rate, and closure of the glottis. In addition, Silver notes that trigeminal chemoreceptive fibers also act as *effectors*, giving rise to a variety of peripheral axon reflexes. The result of these reflexes could be an alteration in the reception, transmission, or processing of information by the olfactory system (Silver, 1992).

Psychophysical studies have explored the differences between olfactory and pungent (common chemical sense) stimulation. In general, pungent

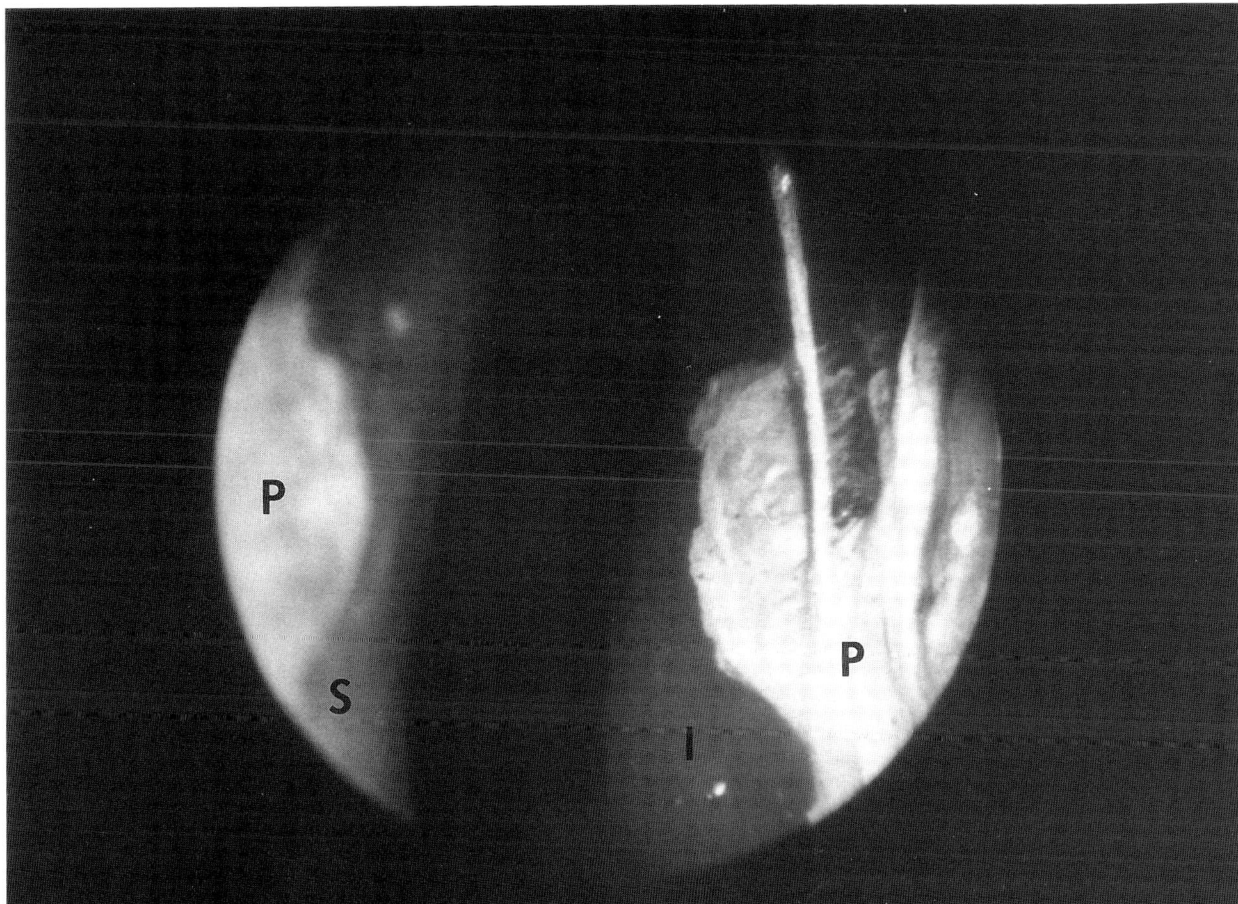

FIGURE 3. Left anterior nasal airway on a 28-yr-old man who had recently been spray painting. Note paint dried on anterior septum and lateral nasal wall. Also note darkened, inflamed mucosa just posterior to paint on inferior turbinate. (S, septum; P, paint; I, anterior end of inferior turbinate).

sensations are achieved at higher concentrations than odor sensations, they increase in intensity at a steeper rate, they show a higher degree of addition in mixtures, and they are more resistant to adaptation (Cometto-Muniz & Cain, 1992). In spite of these differences, however, individuals exposed to pollution can have difficulty describing the sensations they perceive. Because of the differences, common chemical sense stimuli can be more bothersome. For these reasons, Berglund and Lindvall (1992) suggested that when asking about a patient's response to a particular environment, such as a building, care must be exercised to "discern the response characteristics of the occupant before we use the occupant in diagnosing the building."

EFFECTS ON THE NOSE AND SINUSES—OLFACTION

It is known that inhaled chemicals can be damaging to olfactory ability. In their study of 750 patients from the University of Pennsylvania Smell and Taste Center, Deems and his colleagues (1991) found that 18 patients, or 2%, had olfactory losses due to toxins. In 1986, Amoore described nearly 120 chemicals thought to cause olfactory loss (Amoore, 1986). More recently, Schiffman and Nagle (1992) have listed over 50 compounds, dusts, and processes that are associated with permanent decreases in olfactory ability. Their article also details many of the suspected pathophysiological mechanisms of olfactory damage, most of which occur in the olfactory neuroepithelium. Most of these toxin-caused olfactory losses occur over a period of months to years. By the time the loss is noticed, a biopsy of the olfactory cleft usually reveals only the respiratory epithelium, which has taken the place of the destroyed olfactory neurons. Animals exposed to olfactory toxins have been shown to have severe, but sometimes reversible, changes in olfactory histology (Uriah et al., 1990; Hurtt et al., 1988). The animal behavioral result of these changes, however, is poorly known.

Humans who experience a loss of olfactory ability as a result of a toxin exposure can have either a partial (hyposmia) or total (anosmia) loss. Sometimes, especially in those who have only hyposmia, there can be a distortion of the remaining olfactory ability. This is usually manifest as a disagreeable or foul distortion to familiar stimuli (e.g., bread having the odor of rotted meat). This distortion may be related to the decreased numbers of olfactory receptors, which are unable to form a complete olfactory "picture" of the stimulus.

There are three theoretical ways that toxins could affect olfactory ability. In the first method, the access of odorants to the olfactory cleft could be blocked. This could happen with a toxin-caused inflammation of the respiratory epithelium inferior to the olfactory cleft, which would swell the narrow airway shut. If the reaction to the toxic material included the growth of nasal polyps, a fungal ball, or a tumor, access could also be blocked. While this is not an effect directly on the olfactory neurons, it would clinically fall into the category of olfactory loss after a substance exposure.

Without any other damage to the olfactory epithelium, this type of loss should be totally reversible. Leopold (1992) has reported on over 25 materials that can cause nasal inflammation, and these would be suspect in causing this type of loss.

The second method by which toxins could cause olfactory loss is by interfering with the function of the olfactory neurons. This could be a temporary or permanent dysfunction, and may involve neurons in the nasal cavity or those in the central nervous system. An example of this may be the increased dopamine concentration found in the olfactory epithelium after fungicide exposure (Schiffman & Nagle, 1992).

The third method, which is unfortunately the most common, involves a destruction of the primary olfactory neurons. These cells in the olfactory cleft are accessible to inhaled xenobiotic materials. The toxic chemical may rapidly destroy all of the neurons, or the loss may occur a few cells at a time for many months.

EFFECTS ON THE NOSE AND SINUSES—MUCOSAL INFLAMMATION

Nasal and sinus mucosal inflammation can be initiated in many ways (Fairbanks & Raphael, 1993). One of the main effects of the inflammation is thickening of the mucosa. If the thickened mucosa contacts other mucosally lined structures, and especially if this contact area closes off another area from ventilation, further inflammatory changes can ensue (Draf, 1983). Thus, for many diverse stimulating conditions, such as allergens, viruses, cold air, emotional states, or hormonal changes, the nose responds in a similar way. Inhaled materials can also cause this same reaction. One apparent mechanism by which this occurs is a decrease in upper airway immunoglobulin A (IgA), allowing pathologic bacteria to colonize the nasal mucosa (Marth et al., 1989). A listing of some reported materials that can cause nasal disease is provided in Table 1.

Most of the published reports of nasal or sinus irritation from inhaled materials are case reports or surveys from specific industries where one might expect injury to occur (e.g., borax or grain dust, Table 1) (Leopold, 1992). There are some studies, however, that suggest that children and adults may be at risk in their usual living conditions from inhaled pollutants. In Singapore, 3216 primary school children, who lived in industrial, urban, and rural areas, were surveyed about respiratory symptoms (Goh et al., 1986). Measurements of sulfur dioxide, nitrogen oxides, and suspended particulates were made in each of these three living areas. The children in the more polluted areas had increased symptoms of cough (which especially in children can be a symptom of sinus disease), eye irritation, rhinitis, and sinusitis, and these symptoms were associated with increased levels of sulfur dioxide. In a similar survey study of 488 adults living in severely polluted, moderately polluted, and nonpolluted regions of southeastern Finland, statistically significant associations were found between nasal, eye, and respiratory symptoms and

TABLE 1. Pollutant Effects on the Nose and Sinuses

Exposure of occupation	Nasal/sinus symptoms	Limitation	Probable mechanism of effect			BHC	MHC
			SIR	DMF	XM		
Grain dust	Stuffy nose Catarrhal rhinitis	X	X	X	?		
Cigarette smoke	Congestion Rhinorrhea Sneezing	X	○	X	?	?	
Acrylic-fingernail sculptors	Irritation	X		?	?	?	
Spray painters	"Nonspecific"	?		?	?	X	
Spice grinders	Sneezing Rhinorrhea	X	?	?	?		
Tobacco factory	Congestion	X	X	?	?	?	
Newspaper dust	Sneezing Rhinorrhea Obstruction	x	?	?	?		
Wool textile workers	Rhinitis Nosebleeds	X	?	?	?		
Green coffee bean processing	Sneezing Rhinorrhea	X		?			
Chili grinders	Irritation	X	X	?			
Indoor air pollution	Irritation	X	?	?	?	X	
Diesel exhaust particles	Irritation	?		?	?	X	?
Boric oxide	Nosebleeds	X		X	X		
Boric acid dust	Dryness						
Borax dust	Dryness Nosebleeds	X	?	?	?	?	
Cotton dust	Irritation Rhinorrhea	X	X	?	?	X	
Hair spray	Irritation	?		X	?		
Sulfur dioxide	Rhinorrhea Irritation	X	X	?	X	?	
Fiberglass dust	Itching	X	○			?	○
Ozone	Irritation	?		○	?	X	?
Cadmium	Rhinorrhea	?		X	?	?	?
Nickel workers	Rhinitis Polyps Carcinoma Congestion	X	?	X	?	X	X
Freon	None	?		○	?	?	
Ionizing radiation	None	?		?		X	X
Wood dust	Rhinorrhea Rhinitis Obstruction	X	X	X	X	X	X
Leather dust	Irritation Rhinorrhea	X		?	?	X	X
Formaldehyde	Irritation Rhinorrhea	X		?	?	X	?

Note. Listing of multiple types of pollutant effects on the nose and sinuses from different materials: SIR, specific immune reaction; DMF, decreased mucociliary function; XM, xenobiotic metabolism; BHC, benign histopathologic changes; MHC, malignant histopathologic changes; X, probable association; ○, proved non-association; ?, possible association (Leopold, 1992).

pollution levels of hydrogen sulfide, methyl mercaptan, and methyl sulfites previously thought to be acceptable (Jaakkola et al., 1990).

In addition to irritation, acute nasal and sinus infections may be related to air pollution. In southwestern Finland, the rate of acute frontal sinus infection at one hospital more than tripled from the 1977–1981 period to the 1982–1986 period (Suonpaa & Antila, 1990). The authors suspect that increased levels of air pollution in that region of Finland contributed to this increase. A prospective study that adds further evidence that air pollution and upper respiratory infection are related was conducted on 625 randomly chosen children ages 0–5 yr in two Swiss cities (Braun-Fahrlander et al., 1992). Indoor and outdoor air pollution was monitored, and diaries were completed by the parents on respiratory symptoms, including cough, rhinorrhea, and upper respiratory episodes. Twenty percent of the diaries were validated by the attending pediatrician's case notes and found to be in good agreement. The authors concluded that the incidence and duration of the upper respiratory episodes were influenced by the total suspended particles measured in the ambient air, and that the duration may be associated with the outdoor (not indoor) levels of nitrogen dioxide.

EFFECTS ON THE NOSE AND SINUSES—STRUCTURAL CHANGES

When the mucosal lining of the nose is broken down, by whatever mechanism, the cartilage and bone that delineate and support the nasal and sinus cavities can be altered or destroyed. The cartilage of the anterior septum is especially vulnerable to this, since it derives all of its nourishment from the perichondrial membranes deep to the mucosa. Without this nourishment, the septal cartilage will necrose and dissolve, leaving a septal perforation. Bone can also be lost when its protective periosteal covering is not intact. Examples of this damage include the septal perforations seen in noses after chronic cocaine application and in chromium workers (Baruthio, 1992; Dingle, 1992).

EFFECTS ON THE NOSE AND SINUSES—NEOPLASTIC CHANGES

An increase in mucosal cellular proliferation has been noted in response to some inhaled xenobiotics (Bolla et al., 1990). It may be that carcinoma develops when this process gets out of control, or that other factors are involved. Human nose and sinus carcinomas have been found in individuals exposed to nickel dust, ionizing radiation, wood dust, formaldehyde, and leather dust (Leopold, 1992; Luce et al., 1993).

SYSTEMIC EFFECTS OF NASALLY INHALED MATERIALS

Many of the inhaled xenobiotics that have systemic effects are absorbed mostly through the trachea and lungs and have little nasal metabolism or absorption. Only those materials that have some nasal interaction are mentioned here.

The human nose contains enzymes and chemical reactivity that make it a good "scrubber" for some chemicals. As an example, the cyanide-metabolizing enzyme rhodanese has been identified in human nasal mucosa (Lewis et al., 1991). These investigators have determined that smokers have less of this enzyme in their nasal tissues, suggesting a greater metabolism of it by smokers. Certainly nasal breathing would be appropriate if one suddenly smelled the "roasted almond" smell of cyanide. In another example, human controlled and observational studies have led to the conclusion that athletic activity is adversely affected by air pollution. Pierson has shown that sulfur dioxide and ozone are particularly likely to decrease pulmonary function (Pierson, 1989). During the heavy breathing associated with athletic performance, these gases can bypass the nose and pharynx, which normally solubilize them, enter the lower respiratory tract, and decrease athletic performance.

Central nervous system (CNS) effects from inhaled pollutants can be very nonspecific, such as that reported after industrial exposure to xylene (Uchida et al., 1993). Alternatively, it is possible that particular conditions, like Alzheimer's and Parkinson's disease, are associated with inhaled xenobiotics (Roberts, 1986; Young, 1992; Calne et al., 1986). Olfactory dysfunction is noted early in both of the conditions, and similar histopathologic changes are noted in both olfactory and CNS tissue (Serby, 1985; Talamo et al., 1989; Doty et al., 1988, 1992).

It may be that many of the symptoms associated with inhaled materials, especially volatile organic chemicals, can be attributed to neurogenic inflammation (Meggs, 1993). This nonimmunogenic pathway is located in sensory nerve C-fibers, where neuropeptides are released when irritants such as volatile organic chemicals stimulate sensory nerves. The trigeminal nerve, which innervates the nasal airway, is a likely candidate for this stimulation because of its ability to transduce pungent stimuli and the nasal irritation patients perceive with them. These neuropeptides can then produce a wide range of physical and mental symptoms. Meggs (1993) has suggested that multiple chemical sensitivity syndrome, sick building syndrome, reactive airways dysfunction syndrome, and reactive upper airways dysfunction syndrome are all mediated through this mechanism.

METHODS TO EVALUATE CLINICAL RESPONSE

Since the clinical response to inhaled xenobiotics can vary considerably, the first task in most evaluations should be a thorough medical history. In this history, not only can details of a particular exposure be recorded, but other aspects of the person's health, diet, medications, environment, family, and occupation that relate to this exposure can be documented. Particular questions should be asked about the nasal airway and sensation, olfactory and taste ability, eye or face irritation, and general neurologic function.

A general physical examination, with special attention to the head, neck, lungs, and skin, will give further information on a person's response to a xenobiotic agent. Examination of the upper respiratory pathways using nasal and hypopharyngeal endoscopy is essential, since this is often the area of toxin-mucosal interaction. In nasal endoscopy, the rigid endoscopes tend to give better images that can be useful with photodocumentation, but the flexible endoscopes can sometimes be more easily maneuvered. For suspected inhalation of smaller particles or less reactive aerosols, tracheobronchoscopy would be indicated. When examining these mucosal surfaces, the color, texture, and thickness of the mucosa, along with the character of the mucus, should be documented.

There are many tests, both subjective and objective, that can be used to assess the nasal status of an individual. These can be divided into nasal cavity, mucosal, and neural categories. Computerized tomography of the nasal and sinus cavities, especially in the coronal plane, gives an excellent image of the nasal bones, cartilage, and mucosa. It can, however, be time-consuming and expensive, and involves a degree of ionizing radiation exposure (Zinreich & Dolan, 1993). Acoustic rhinometry, which measures sound waves bounced off the internal nasal walls, can graph the cross-sectional area of the anterior nasal cavity. It requires little subject cooperation, but cannot detail the exact anatomic shape, nor is it particularly useful in severely deviated nasal cavities (Grymer et al., 1989). Rhinostereometry has been used in Sweden to assess the mucosal thickness in the nose (Ohm et al., 1993). Finally, measurements of total nasal airway pressure and flow can be made with rhinomanometry. From these measurements, resistance to airflow can be calculated for either an individual nostril or the total nose (Lai & Corey, 1993). Unfortunately, there is no clinical test available that will document the pattern or location of individual airflow streams in the nose (Leopold, 1994).

Useful information on nasal mucosal function has been obtained with nasal cavity lavage techniques (Koren et al., 1990). By assaying the nasal mucus, important information on mediators of inflammation and general mucosal function has been obtained (Naclerio et al., 1983; Baroody et al., 1993). The function of the nasal cilia can be assessed with measurements of the time required for saccharine or technesium-99 particles to move through the nose (Liote et al., 1989). The generally harmful influence of pollutants on this transit and the cilia that propel it has been recently reviewed by Pederson (1990).

Olfactory and trigeminal ability is part of general nasal function and should be tested as part of a thorough examination. Although thresholds for the most dilute odorant detectable or identifiable can be tested, generally useful clinical information can be obtained with suprathreshold identification tests (Doty, 1992). A commercially available scratch-and-sniff olfactory identification test is available that uses microencapsulated chemicals representing 40 odorants (Sensonics, Inc., Hadden Heights, NJ; Doty et al., 1984). This test is easily used, portable, and has good test–retest reliability. It can also be used to compare the person's test result with age- and sex-adjusted norms. Single-nostril testing, by holding the opposite nostril closed, is useful in finding unilateral losses.

Tests for trigeminal chemosensory function generally evaluate response to irritants. It is difficult to separate the olfactory from the trigeminal response. Several tests have been proposed or utilized.Some of these tests puff carbon monoxide, carbon dioxide, or cigarette smoke onto the faces of test subjects to measure eye irritation, thereby avoiding olfactory stimulation (Kjaergaard et al., 1992; Winneke, 1992). Other tests in the nose are either done on subjects who are anosmic, or an apnea response can be measured with irritant stimuli in the nose (Cometto-Muniz & Cain, 1992).

TREATMENT/PREVENTION

There have been very few specific efforts to treat or protect the noses of individuals from inhaled toxins. Many examples exist, however, where general respiratory precautions using masks and respirators have preserved nasal health. In one such study, Polish dock workers, who worked in an area polluted with inorganic dusts, used frequent nasal and oral rinses with saline, coated their noses with a protective gel, and used mucolytic drugs. After 2 yr, there was a reduction in the number of respiratory tract diseases of 60% in the sulfur workers and 34.2% in the raw phosphate workers (Wolska-Goszka & Dubaniewicz, 1992). In another study trying to predict toxic exposures, a German foundry company measured serum and nasal discharge levels of IgA (Hummel et al., 1990). This was apparently done because of the reported decrease in IgA of subjects exposed to pollutants. Their results showed no significant differences between the exposed and the control groups. Apparently, at least in this group, the IgA-depressing effect for this level of exposure was not robust enough to show a difference.

CONCLUSION

The xenobiotic materials that accompany respiratory gases into the nasal cavity can have significant and lasting effects on the human nose and the entire body. It is important that more be learned of the pathophysiology of this disease production and that ways to treat and prevent it be advanced (Bonnefoi et al., 1991).

REFERENCES

Amoore J. E. 1986. Effects of chemical exposure on olfaction in humans. In *Toxicology of the nasal passages*, ed. C. S. Barrow, pp. 155–190. Washington, DC: Hemisphere.

Baroody F. M., Wagenmann, M., and Naclerio, R. M. 1993. A comparison of the secretory response of the nasal mucosa to methacholine and histamine. *J. Appl. Physiol.* 74(6):2661–2671.

Baruthio, F. 1992. Toxic effects of chromium and its compounds. *Biological Trace Element Research* 32:145–153.

Berglund, B., and Lindvall, T. 1992. Theory and method of sensory evaluation of complex gas mixtures. *Ann. NY Acad. Sciences* 641:277–293.

Bolla, I., Gariboldi, L. M., Gabrielli, M., Baldo, D., Romanelli, A., Tuberti, E., and Magnani, F. 1990. Nose disease caused by occupational exposure to chromium in the electroplating industry: cytomorphological aspects. *Med. Lav.* 81(5):390–398.

Bonnefoi, M., Monticello, T. M., and Morgan, K. T. 1991. Toxic and neoplastic responses in the nasal passages: Future research needs. *Exp. Lung Res.* 17:853–868.

Braun-Fahrlander, C., Ackermann-Liebrich, U., Schwartz, J., Gnehm, H. P., Rutishauser, M., and Wanner, H. U. 1992. Air pollution and respiratory symptoms in preschool children. *Am. Rev. Respir. Dis.* 145:42–47.

Calne, D. B., McGreer, E., Eisen, A., and Spencer, P. 1986. Alzheimer's disease, Parkinson's disease, and motoneurone disease: Abiotropic interaction between aging and envioronment? *Lancet* 2:1067–1070.

Cole, P. 1994. Nose and sinus airflow. *Current opinion. Otolaryngol-Head Neck Surg.* 2(1):16–21.

Cole, P., Haight, J. S. J., Naito, K., and Kucharczyk, W. 1989. Magnetic resonance imaging of the nasal airways. *Am. J. Rhinol.* 3(2):63–67.

Cometto-Muniz, J. E., and Cain, W. S. 1992. Sensory irritation. Relation to indoor air pollution. *Ann. NY Acad. Sci.* 541:137–151.

Deems, D. A., Doty, R. L., Settle, R. G., Moore-Gillon, V., Shaman, P., Mester, A. F., Kimmelman, C. P., Brightman, V. J., and Snow, J. B. 1991. Smell and taste disorders, a study of 750 patients from the University of Pennsylvania Smell and Taste Center. *Arch. Otolaryngol. Head Neck Surg.* 117:519–528.

Dingle, A. F. 1992. Nasal disease in chrome workers. *Clin. Otolaryngol.* 17:287–288.

Doty, R. L. 1992. Diagnostic tests and assessment. *J. Head Trauma Rehab.* 7(1):47–65.

Doty, R. L., Shaman, P., and Dann, M. 1984. Development of the University of Pennsylvania Smell Identification test: A standardized microencapsulated test of olfactory function. *Physiol. Behav.* 32:489–502.

Doty, R. L., Deems, D. A., and Stellar, S. 1988. Olfactory dysfunction in parkinsonism: A general deficit unrelated to neurologic signs, disease stage, or disease duration. *Neurology* 38(8):1237–1244.

Doty, R. L., Stern, M. B., Pfeiffer, C., Gollomp, S. M., and Hurtig, H. I. 1992. Bilateral olfactory dysfunction in early stage treated and untreated idiopathic Parkinson's disease. *J. Neurol. Neurosurg. Psychiatry* 55:138–142.

Draf, W. 1983. Endoscopy of the paranasal sinsuses. New York: Springer-Verlag.

Fairbanks, D. N. F., and Raphael, G. D. 1993. Nonallergic rhinitis and infection. In *Otolaryngology—Head and neck surgery*, ed. C. W. Cummings, pp. 775–785 Mosby Year Book, Vol. 1.

Goh, K. T., Lun, K. C., Chong, Y. M., Ong, T. C., Tan, J. L., and Chay, S. O. 1986. Prevalence of respiratory illnesses of school children in the industrial, urban and rural areas of Singapore. *Trop. Geogr. Med.* 38:344–350.

Grymer, L. F., Hilberg, O., Elbrond, O., and Pedersen, O. F. 1989. Acoustic rhinometry: Evaluation of the nasal cavity with septal deviations, before and after septoplasty. *Laryngoscope* 99:1180–1187.

Hummel, K., Winkler, U., and Hennebach, U. 1990. pH and IgA determination of nasal secretions—Possible use in occupational medicine-rhinologic diagnosis in relation to function. *Z. Gesamte Hyg. Grezgeb.* 36(6):320–322.

Hurtt, M. E., Thomas, D. A., Working, P. K., Monticello, T. M., and Morgan, K. T. 1988. Degeneration and regeneration of the olfactory epithelium following inhalation exposure to methyl bromide: Pathology, cell kinetics, and olfactory function. *Toxicol. Appl. Pharmacol.* 94:311–328.

Jaakkola, J. J. K., Vilkka, V., Marttila, O., Jappinen, P., and Haahtela, T. 1990. The South Karelia air pollution study. The effects of malodorous sulfur compounds from pulp mills on respiratory and other symptoms. *Am. Rev. Respir. Dis.* 142:1344–1350.

Keyhani, K., Scherer, P. W., Hahn, I., and Mozell, M. M. 1993. Modeling of airflow in the human nasal cavity. Abstract for *Advances in Bioengineering, Proc. ASME*, Winter Annual Meeting, New Orleans, November.

Kjaergaard, S., Pederson, O. F., and Molhave, L. 1992. Sensitivity of the eyes to airborne irritant stimuli: influence of individual characteristics. *Arch. Environ. Health* 47(1):45–50.

Koren, H. S, Hatch, G. E., and Graham, D. E. 1990. Nasal lavage as a tool in assessing acute inflammation in response to inhaled pollutants. *Toxicology* 60(1–2):15–25.

Lai, V. W., and Corey, J. P. 1993. The objective assessment of nasal patency. *Ear, Nose Throat J.* 72(6):395–396, 399–400.

Leopold, D. A. 1992. Pollution: The nose and sinuses. *Otolaryngol. Head Neck Surg.* 106(6):713–719.

Leopold, D. A. 1994. On the importance of the nasal airflow pattern (Commentary). Current opinion. *Otolaryngol. Head Neck Surg.* 2(1):1–3.

Lewis, J. L., Rhoades, C. E., Gervasi, P.-G., Griffith, W. C., and Dahl, A. R. 1991. The cyanide-metabolizing enzyme rhodanese in human nasal respiratory mucosa. *Toxicol. Appl. Pharmacol.* 108:114–120.

Liote, H., Zahm, J. M., Pierrot, D., and Puchelle, E. 1989. Role of mucus and cilia in nasal mucociliary clearance in healthy subjects. *Am. Rev. Respir. Dis.* 140(1):132–136.

Luce, D., Gerin, M., Leclerc, A., Morcet, J. F., and Godberg, M. 1993. Sinonasal cancer and occupational exposure to formaldehyde and other substances. *Int. J. Cancer* 53:224–231.

Marth, E., Mose, J. R., Bencko, V., Tichacek, B., Sixl, W., Kock, M., and Sixl-Voight, B. 1989. *Zentralbl. Hyg.* 188:439–448.

Meggs, W. J. 1993. Neurogenic inflammation and sensitivity to environmental chemicals. *Environ. Health Perspect.* 101(3):234–238.

Naclerio, R.M., Meier, H. L., Kagey-Sobotka, A., Atkinson, N. F., Jr., Meyers, D. A., Norman, P. S., and Sichtenstein, L. M. 1983. Mediator release after airway challenge with antigen. *Am. Rev. Respir. Dis.* 128:597–602.

Nakashima, T., Kimmelman, C. P., and Snow, J. B. 1984. Structure of human fetal and adult neuroepithelium. *Arch. Otol. Head Neck Surg.* 110:641–646.

Ohm, M., Juto, J.-E., and Andersson, K. 1993. Nasal hyperreactivity and sick-building syndrome. *Amer. J. Rhinol.* 7(4):178.

Paik, S. I., Lehman, M. N., Seiden, A. M., Duncan, H. J., and Smith, D. V. 1992 Human olfactory biopsy, the influence of age and receptor distribution. *Arch. Otolaryngol. Head Neck Surg.* 118:731–738.

Pedersen, 1990. M. Ciliary activity and pollution. *Lung* Suppl.:368–376.

Pierson, W. E. 1989. Impact of air pollutants on athletic performance. *Allergy Proc.* 10(3):209–214.

Roberts, E. 1986. Alzheimer's disease may begin in the nose and be caused by aluminosilicates. *Neurobiol. Aging* 7:561–657.

Serby, M., Corwin, J., Conrad, P., and Rotrosen, J. 1985. Olfactory dysfunction in Alzheimer's disease and Parkinsons' disease. *Am. J. Psychiatry* 142:781–782.

Schiffman, S. S., and Nagle, H. T. 1992. Effect of environmental pollutants on taste and smell. *Otolaryngol. Head Neck Surg.* 106:693–700.

Silver, W. L. 1992. Neural and pharmacological basis for nasal irritation. *Ann. NY Acad. Sci.* 641:152–163.

Suonpaa, J., and Antila, J. 1990. Increase of acute frontal sinusitis in Southwestern Finland. *Scand. J. Infect. Dis.* 22:563–568.

Swift, D. L. 1982. Physical principles of airflow and transport phenomena influencing air modification. In *The nose: Upper airway physiology and the atmospheric environment*, eds. D. F. Proctor and I. Anderson, pp. 337–348. Amsterdam: Elsevier.

Talamo, B. R., Rudel, R. A., Kosik, K. S., Lee, V. M.-Y., Neff, S., Adelman, L., and Kauer, J. S. 1989. Pathological changes in olfactory neurons in patients with Alzheimer's disease. *Nature* 337:736–739.

Uchida, Y., Nakatsuka, H., Ukai, H., Watanabe, T., Liu, Y. T., Huang, M. Y., Wang, Y. L., Zhu, F. X., and Ikeda, M. 1993. Symptoms and signs in workers exposed predominently to xylenes. *Int. Arch. Occup. Environ. Health* 64(8):597–605.

Uriah, L. C, Morgan, K. T., and Maronpot, R. R., eds. 1990. Proceedings of a symposium on toxicologic pathology of the upper respiratory tract. *Environ. Health Perspect.* 85:161–352.

Winneke, G. 1992. Structure and determinants of psychophysiological response to odorant/irritant air pollution. *Ann. NY Acad. Sci.* 641:261–276.

Wolska-Goszka, L., and Dubaniewicz, A. 1992. Evaluation of the effectiveness of medical prophylactic procedures applied to workers in ports. *Wiad. Lek.* 45(17–18):655–658.

Wvstrow, F. 1951. Schwellkorper. Am Septum Nasi. *Z. Anat. Entwicklunggesch.* 16:139–142.

Young, J. K. 1992. Alzheimer's disese and metal containing glia. *Med. Hypotheses* 38(1):1–4.

Zinreich, S. J., and Dolan, K. D. 1993. Radiology of the nasal cavity and paranasal sinuses. In *Otolaryngology—Head and neck surgery,* ed. C. W. Cummings, pp. 907–928. Mosby Year Book, Vol. 1.

NASAL DOSIMETRY, LESION DISTRIBUTION, AND THE TOXICOLOGIC PATHOLOGIST: A BRIEF REVIEW

Kevin T. Morgan

Chemical Industry Institute of Toxicology, Research Triangle Park, North Carolina, USA

The distribution of nasal lesions induced by xenobiotics is attributable to regional dosimetry, local tissue susceptibility, or a combination of these factors. Local tissue susceptibility is frequently related to site-specific xenobiotic metabolism, which also plays a role in the dose of parent compound, or toxic metabolite(s), delivered to individual cell populations in the nose. The toxicologic pathologist can provide valuable clues concerning the role of tissue dosimetry in nasal toxicity through recognition of lesion distribution patterns. Detection of such patterns may be achieved by the use of mapping strategies, combined with knowledge of factors that may account for dose-dependent site specificity of the responses observed. The regional dosimetry of xenobiotics is influenced by their physical and chemical properties. In the nose, interactions of xenobiotics with air, mucus, and blood flow, and the local metabolism also play a critical role. Integration of these factors with knowledge of lesion location provides insight for biologically based approaches to interspecies extrapolation and human risk estimation. Identification of target cell populations and characterization of the nature of disease processes observed are also an essential part of this process. During interpretation of nasal lesions and selection of responses for more detailed analyses directed toward quantitative human risk assessment, a disease-oriented approach is recommended. These issues are briefly reviewed here, with special reference to chemicals that induce nasal injury in rodents.

A wide range of xenobiotics has been shown to damage the nasal mucosa of laboratory animals (Barrow, 1986; Buckley et al., 1984; Jiang et al., 1986; Monticello et al., 1990; Uraih et al., 1990), and these lesions generally exhibit characteristic distribution patterns (Mery et al., 1994a; Morgan & Monticello, 1990a). Increased understanding of factors that influence lesion distribution, when combined with dosimetry models for inhaled gases and vapors (Gerde & Dahl, 1991; Kimbell et al., 1993; Morris et al., 1993; Overton & Miller, 1988), provides a valuable adjunct to the limited information available from humans. In light of the expanding database on chemically induced nasal lesions in laboratory animals, studies of lesion distribution patterns in the human nose would clearly be of value for interspecies extrapolation. For ethical and technical reasons, the distribution of lesions in the nasal passages of humans is difficult to determine precisely. This situation prevents direct comparison of lesion location for materials

I wish to thank the many people who have made my research on the nose possible and enjoyable, with special thanks to Betsy Gross and Donald Joyner for their unstinting support. Donald Joyner is also acknowledged for preparation of the diagram of nasal blood supply.

Address correspondence to Kevin T. Morgan, Chemical Industry Institute of Toxicology, Six Davis Drive, P.O. Box 12137, Research Triangle Park, NC 27709.

such as ozone and formaldehyde, chemicals that induce nasal responses in laboratory animals (Harkema et al., 1987; Johnson et al., 1990; Kerns et al., 1983; Monticello et al., 1989) and are also considered to do so in humans (Calderon-Garcidueñas et al., 1992; Holmström & Lund, 1991; Koren et al., 1990).

For quantitative modeling of disease processes, interspecies extrapolation, and human risk assessment based on laboratory animal studies, determination of factors responsible for the location of specific nasal responses is essential (Conolly & Andersen, 1991; Conolly et al., 1992). Toxicologic pathology plays an important role in this process, through characterization of the nature and distribution of nasal lesions and identification of specific target cell populations (Harkema, 1991; Morgan, 1991). This article reviews selected literature reporting nasal lesion distribution in rodent toxicology studies, considers the role of local tissue dosimetry in relation to patterns of nasal responses, and briefly addresses the issue of a "disease-oriented approach" to toxicologic pathology studies designed for human risk estimation. Another potentially important issue, the nose as a portal of entry to the central nervous system, is addressed elsewhere (Hastings & Evans, 1991).

NASAL LESION DISTRIBUTION IN TOXICOLOGY STUDIES

During a number of toxicology studies in this laboratory, chemically induced lesions were found to occur in diverse regions of the nose, in spite of the fact that the majority of these chemicals were administered by the inhalation route. Adequate recording of these patterns of lesion distribution was found to be an important step, and a logical recording approach has been recommended (Morgan, 1991). The issue of differing lesion distribution patterns is exemplified by responses induced in the nasal passages of rats by inhaled formaldehyde and methyl bromide. Formaldehyde-induced lesions are essentially confined to the anterior nose, in regions lined by transitional or respiratory epithelium (Kerns et al., 1983; Morgan et al., 1986a; Woutersen & Feron, 1989). The damage caused by methyl bromide spares these sites, with lesions being specifically found in the olfactory mucosa lining the posterior regions of the nose (Hurtt et al., 1988). Of the increasing number of chemicals examined for nasal toxicity, each one appears to exhibit its own characteristic pattern of lesion distribution (Jiang et al., 1986; Lee et al., 1992; Mery et al., 1994a; Morgan & Monticello, 1990a).

Site specificity of diverse lesions has been reported for each of the four major epithelial (mucosal) types lining the nasal airways of rodents, including the squamous (Gross et al., 1994), transitional (Johnson et al., 1990), respiratory (Morgan et al., 1986a), and olfactory (Genter et al., 1992) regions. Toxicity has been reported in a range of other nasal structures, including Steno's glands (Belinsky et al., 1987), the nasal septum (Wolf et al., 1994b), the vomeronasal organ (Macia et al., 1993), and the ethmoid turbinate bones (Mery et al., 1994b; Morse et al., 1984). To show the location of a number of lesions of the nasal mucosa, selected major internal structures of the nose of the rat are presented in Figure 1A, along with a simplified representation of this diagram (Figure 1B) showing the location of selected lesions induced by exposure to xenobiotics; for details see Mery et al. (1994a). Within the regions exhibiting nasal lesions, specific patterns of cellular damage are also observed. For example, β, β'-iminodiproprionitrile

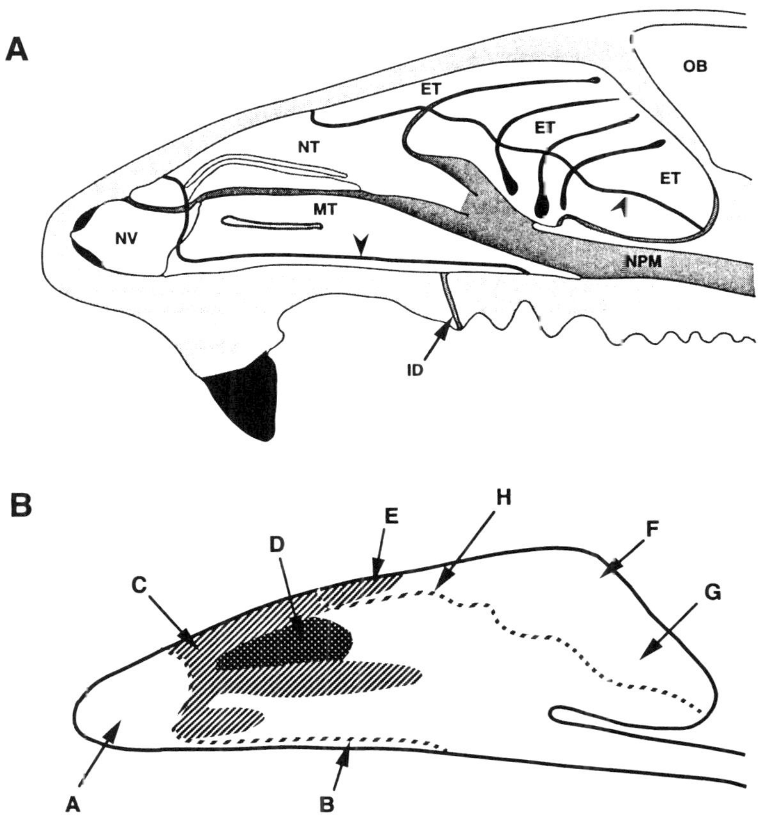

FIGURE 1. (A) Diagrammatic representation of a mid-sagittal section of the nose of a 3- to 6-mo old F344 rat, with the septum removed to show the major internal structures. NV, nasal vestibule; NT, nasoturbinate; MT, maxilloturbinate; ET, ethmoid turbinate; NPM, nasopharyngeal meatus; ID, incisive duct; OB, olfactory bulb. Solid lines indicated by arrowheads show junction of squamous and transitional/respiratory epithelia (anterior) and respiratory and olfactory epithelia (posterior). Modified from Mery et al. (1994a). (B) Outline diagram of the nose of the rat modified from (A) to show location of sites of selected chemically-induced nasal lesions. A, squamous epithelium of nasal vestibule, glutaraldehyde (Gross et al., 1994); B, squamous epithelium of the ventral meatus, hydrogen chloride (Jiang et al., 1986); C, transitional and respiratory epithelia of septum and maxilloturbinate, formaldehyde (Kimbell et al., 1993); D, transitional epithelium of lateral meatus (Kimbell et al., 1993); E, olfactory epithelium of dorsal medial meatus, multiple irritant gases (Buckley et al., 1984); F, dorsal medial olfactory mucosa, IDPN (Genter et al., 1992); G, peripheral olfactory mucosa, chloroform (Mery et al., 1994b); H, subepithelial glands at junction olfactory and respiratory epithelia in the dorsal medial meatus, dimethylamine (Buckley et al., 1985).

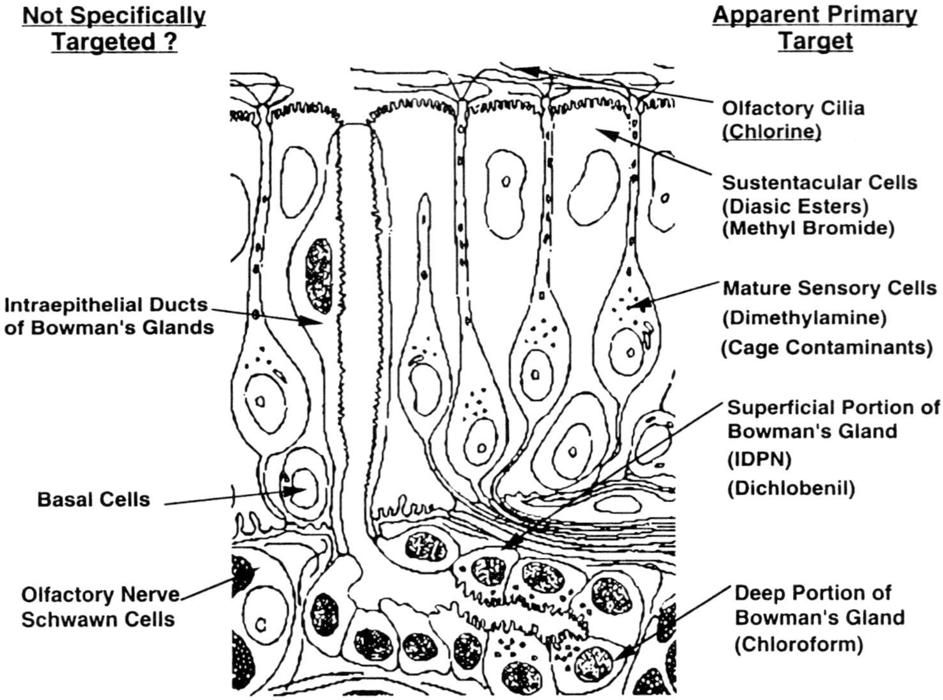

FIGURE 2. Target cell populations in the olfactory mucosa (modified from Reed, 1993, Figure 2). Certain chemicals, as indicated, target specific cell types, see text for details.

(IDPN) targets the olfactory epithelium and superficial Bowman's glands (Genter et al., 1992). In contrast, chloroform-induced olfactory mucosal lesions are found in the deep Bowman's glands and adjacent bone, while the overlying olfactory epithelium is unaffected (Mery et al., 1994b). Other examples of specific target cell populations, using lesions reported in the olfactory mucosa are shown in Figure 2.

Observations concerning the location of nasal lesions and identification of target cell populations are of value when attempting to determine mechanisms responsible for site-specific tissue damage. However, the complex anatomical structure of the nasal airways contributes to difficulties encountered when recording the nature and location of nasal lesions. Maps of the nose have been used to assist this process (Harkema et al., 1987; Jiang et al., 1983; Johnson et al., 1990; Klonne et al., 1987; Lee et al., 1992; Morgan et al., 1986a), but a consistent procedure is needed. Recently, a set of nasal diagrams, designed specifically for recording lesion location, was created for studies in rats and mice (Mery et al., 1994a). It is recommended that these diagrams be used in conjunction with knowledge of the underlying structure and function of the nose with respect to factors that may influence regional dosimetry of the material being assessed for nasal toxicity.

Each of the major factors controlling intranasal xenobiotic dosimetry is next briefly discussed.

REGIONAL DOSIMETRY AND LESION DISTRIBUTION

The distribution of lesions induced in the nasal passages by xenobiotics is attributable to regional deposition (or dose to a critical target site), local tissue susceptibility, or a combination of these factors (Morgan & Monticello, 1990a). Tissue susceptibility is frequently a result of site-specific metabolism (Reed, 1993). Chemicals have been found to induce nasal lesions when delivered by inhalation (Barrow, 1986; Buckley et al., 1984; Harkema, 1990; Uraih et al., 1990) and noninhalation (Barrow, 1986; Brittebo et al., 1991; Genter et al., 1992; Peele et al., 1990; Reznik-Schüller, 1983) routes. Regional dosimetry of inhaled xenobiotics is influenced by many factors, including their physical and chemical properties, airborne concentration, and exposure duration (Bond, 1989). Irrespective of the route of administration, a number of complex flow fields may also influence the ultimate fate of xenobiotics in the nose, including inspired and expired air (Proetz, 1951), nasal mucus (Lucas & Douglas, 1934), and the blood supply (Hebel & Stromberg, 1976). For a review of the subject of fluid flow, its influence on biological systems, and its functional roles in living organisms, see Vogel (1981). In concert with these flow fields, regional metabolism also plays a major role in local dosimetry of toxic materials in the nose through (a) provision of biochemical defenses, which may be saturable, and (b) local generation of toxic metabolites, due to tissue or cell type-specific metabolism (Dahl & Hadley, 1991; Reed, 1993).

Attempts to model regional nasal dosimetry for interspecies extrapolation should ideally consider each of these factors, their relative capacity for removal or translocation of xenobiotics, and how they may interact to determine a potentially toxic outcome (Morris & Cavanagh, 1986). However, there is generally a paucity of detailed quantitative data for such modeling efforts (Morris et al., 1993), especially in humans. Characterization of lesion distribution may therefore provide the first clue that one or more of these factors is playing a critical role in local dosimetry, and thus the toxic outcome. Simplified representations of the air, mucus, and blood flow fields and the locations of selected enzymes responsible for xenobiotic metabolism are presented in Figure 3. There are few accounts of the distribution of chemically induced nasal lesions having been clearly attributed to regional dosimetry as a consequence of the activity of one or more of these flow fields, each of which is now considered separately. The issue of local metabolism in relation to toxicologic pathology is addressed subsequently.

Airflow-Driven Dosimetry

The potential impact of upper respiratory airflow on the deposition of inhaled materials in the nose has been appreciated since much earlier in

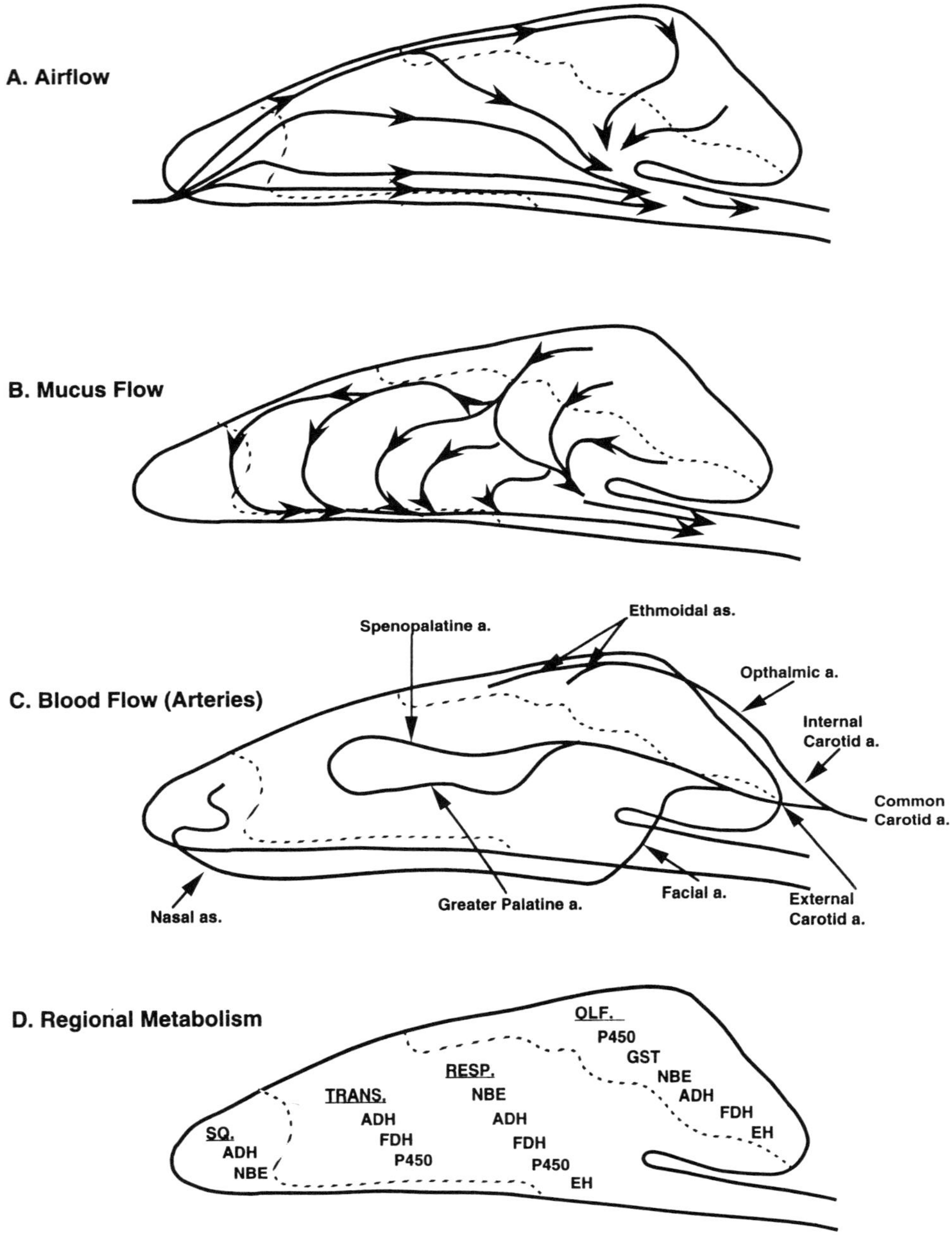

FIGURE 3. Diagrammatic representations of the nose of the rat to show flow fields for (A) inspired air, (B) mucus epiphase, and (C) arterial blood, and (D) the distribution of selected enzymes in the airway lining epithelia. Diagrams modified from published literature: airflow (Morgan et al., 1991), mucociliary function (Morgan et al., 1984b), blood flow (Hebel & Stromberg, 1976), regional metabolism (Bogdanffy et al., 1986b; Keller et al., 1990; Randall et al., 1987; Reed, 1993). SQ, squamous epithelium; TRANS, transitional epithelium; RESP, respiratory epithelium; OLF, olfactory epithelium. a, artery, as, arteries. ADH, aldehyde dehydrogenase; NBE, Naphthyl butyrate esterase; FDH, formaldehyde dehydrogenase; P450, cytochromes P450; EH, epoxide hydrolase; GST, glutathione S-transferase.

this century (Proetz, 1951). A number of extensive reviews of nasal airflow have been published (Cole, 1988; Proetz, 1951). Recent work in this laboratory has focused on nasal airflow in the rat and rhesus monkey (Morgan et al., 1991) and the influence of inspiratory airflow on formaldehyde uptake in the nose of the rat (Kimbell et al., 1993). These studies compared formaldehyde-induced lesion distribution patterns with predicted formaldehyde uptake (local mass flux) patterns in the nose of the rat, demonstrating a high degree of correlation between these end points. This work was considered to indicate a potentially important role for inspiratory airflow patterns in local formaldehyde deposition rate. Recent studies of formaldehyde dosimetry, using DNA-protein cross-links as a dosimeter, indicated high formaldehyde uptake in the lateral meatus (Casanova et al., 1994; Heck et al., 1993) and thus provided further experimental support for this proposal.

On the basis of this work, it would appear that a predictable (formaldehydelike!) pattern of lesions might be expected for materials delivered in high local concentrations as a result of inspiratory airflow. The principal sites of such predicted airflow-driven lesions are the mucosal lining of the lateral, middle medial, and dorsal medial meatuses (Figure 1B). These lesions would also be expected to show a diminishing anterior-to-posterior severity gradient as a result of the scrubbing action of the airway secretions (Kimbell & Morgan, 1991). Inhaled materials exhibiting such a local toxic response pattern would generally be highly soluble in water (or have a high water/air partition coefficient) and/or be highly reactive, resulting in their rapid extraction from air streams passing over the adjacent moist nasal lining. Examples of such gases, which exhibit overall nasal lesion distribution patterns similar to those for inhaled formaldehyde, include dimethylamine (Gross et al., 1987), methyl isocyanate (Uraih et al., 1987), and sulfur dioxide (Giddens & Fairchild, 1972). These studies in rodents, when combined with ongoing simulations of nasal airflow in humans (Hahn et al., 1993), may permit accurate prediction of regional respiratory tract toxicity resulting from high local deposition patterns. Xenobiotics removed from the inspired air during passage through the nose will generally be removed by nasal mucus, which thus provides the next major flow field of interest.

The Mucociliary Apparatus and Regional Dosimetry

With the exception of the squamous epithelial lining of the most anterior regions of the nose, the nasal airway is lined by a watery, sticky material called mucus. This complex viscoelastic fluid exhibits physical and chemical properties that are well suited to its role as an airway defense system. Mucus covering the olfactory mucosa moves very slowly, with a turnover time of probably several days. In contrast, the mucus covering (epiphase) of the transitional and respiratory epithelia is driven along fairly rapidly (1 to 30 mm/min) by ciliary activity, and possibly other forces, with a turnover time of about 10 min in the rat, based on mucus flow rates for this species (Morgan et al., 1984b). Ultimately, the mucus epiphase of both the olfactory and respiratory/transitional regions, with any entrapped materials, is swallowed and passes into the gastrointestinal tract (Proctor, 1982).

The mucociliary system covering the transitional/respiratory regions is known as the mucociliary apparatus. This apparatus exhibits a range of responses to inhaled xenobiotics and can therefore be a sensitive indicator of toxicity (Morgan et al., 1986b). The mucociliary apparatus clearly plays a major role in the removal of deposited particulate matter from the nose (Bang, 1961; Phipps, 1981; Proctor, 1982). However, little is known about the ability of the mucociliary apparatus to trap and remove gases or vapors, to thus protect the underlying epithelia from these materials. On the basis of mucus flow characteristics (Morgan et al., 1984a) and binding to albumin (Bogdanffy et al., 1986a) or other mucus components, it was proposed that the mucociliary apparatus may provide some protection for the underlying epithelium against inhaled formaldehyde (Morgan et al., 1984a). This proposal has not been substantiated. Morris et al. (1986), on the basis of known mucus flow rates and estimated volumes of mucus turnover in the nasal passages of the rat, concluded that the mucociliary apparatus does not provide a significant route for removal for nonreactive gases or vapors, such as acetone and ethanol, from the nose. Cross et al. (1984) proposed that respiratory tract mucus may scavenge reactive oxygen species, and recent theoretical studies indicate that the mucus epiphase may in fact play a significant role in protecting the airway epithelia against inhaled ozone (Pryor, 1992). It is clear that further work is needed to assess the role of the mucus epiphase in nasal toxicity, and the potential for translocation of materials in the epiphase to influence nasal lesion distribution.

Nasal Blood Flow and Disposition of Xenobiotics in the Nose

The nose has a rich and complex vascular system, with each of the airway epithelial regions receiving blood from a separate arterial supply (Hebel & Stromberg, 1976; Warwick & Williams, 1973; Weiss, 1983); see Figure 3. In rats and mice, the nose receives less than 1% of the cardiac output (Stott et al., 1983), but for nonreactive gases nasal uptake is strongly dependent upon nasal blood perfusion rates (Morris et al., 1993). However, little is known about the influence of the nasal blood supply on local deposition or removal of reactive gases in the nose, and the potential impact of the blood supply on regional toxic responses to such xenobiotics. In the case of formaldehyde, it would appear that the parent compound does not reach the systemic circulation to have an impact at remote sites (Heck et al., 1990). The nasal vasculature, however, does represent an important means of removing toxic material from the nose (Morris & Cavanagh, 1986; Morris et al., 1986) and remains a potentially important defense.

For noninhaled materials, and possibly certain inhaled gases or vapors, vascular delivery of toxic chemicals is probably a significant cause of nasal lesions. A wide range of chemicals has been administered to rodents by noninhalation routes with consequent nasal damage, presumably resulting from vascular exposure of this organ to the parent compound or toxic metabolites. These chemicals include numerous nitrosamines (Reznik-Schuller, 1983) and other carcinogens (Brown et al., 1991), certain herbicides (Brittebo et al., 1991), and recently the classic neurotoxin β,β'-iminodiproprionitrile (IDPN) (Genter et al., 1992). The last compound exhibited a characteristic distribution pattern for nasal lesions, which were largely confined to the olfactory mucosa lining the posterior dorsal medial airways. It is possible that local blood perfusion and/or regional metabolism combined to produce this pattern of tissue damage. This proposal remains speculative, as little is known about regional dosimetry of blood-borne chemicals in the nose of the rat. On the basis of nasal anatomy and physiology, Morris et al. (1993) assessed the role of asymmetric blood perfusion patterns in the nose of the rat and concluded that the best fit for their models was obtained when the greatest blood perfusion passed to regions of the nose lined by respiratory epithelium. This work highlighted the need for better understanding of regional blood perfusion patterns in the nose. Such work is critical for integration of histopathology data with blood-flow-driven dosimetry patterns. However, assessment of regional blood perfusion in the nose, using microspheres (Stott et al., 1983) or other approaches, will be challenging due to the structural complexity of nasal vessels within the mucosa, which include multiple anastomoses (Eccles et al., 1987).

Xenobiotic Metabolism and Regional Nasal Toxicity

Whether xenobiotics reach the nasal mucosa via the airways or the blood supply, they are potentially subject to metabolic detoxification or activation in the nose. Extensive reviews are available on the metabolic capacity of nasal enzyme systems, their role in xenobiotic metabolism, and nasal function and toxicity (Dahl, 1988; Dahl & Hadley, 1991; Reed, 1993). Information on tissue or cell-type-specific metabolism may be provided by enzyme histochemistry (Bogdanffy, 1990; Randall et al., 1987), immunocytochemistry (Zupka et al., 1991), or histoautoradiography (Brandt et al., 1990; Brittebo et al., 1990), preferably used in combination with other biochemical approaches. Patterns of lesion distribution in the nose may correlate with patterns of xenobiotic metabolism, and such correlations can provide valuable information for human risk assessment. However, knowledge of regional metabolic patterns is needed before such correlations are possible.

A good example of local metabolism having resulted in site-specific nasal lesions is provided by the nasal cytochromes P-450, a family of enzymes that is responsible for activation of a number of xenobiotics, with consequent nasal toxicity, including chlorthiamid and dichlobenil (Brandt et al., 1990; Brittebo et al., 1991), acetaminophen (Jeffery & Haschek, 1988), and IDPN (Genter et al., 1992). For an excellent review of this topic and other metabolically induced nasal lesions, see Reed (1993). Few xenobiotics have been systematically examined for determination of any relationship between local metabolism and toxicity. An interesting example of such a systematic approach is provided by the literature on a number of dibasic esters. This research has revealed a clear correlation between regional nasal carboxylesterase activity (Bogdanffy et al., 1987) and the primary site of toxicity of these esters, the olfactory epithelial sustentacular cells (Trela et al., 1992). Careful mapping of these lesions (Lee et al., 1992), in addition to studies of nasal deposition of these esters in the nose of the rat (Morris et al., 1991), also indicated a potential role for nasal airflow-driven dosimetry of these esters, for which toxicity was generally confined to the major inspiratory airways. This work provides a valuable example for assessment of risks of olfactory epithelial damage in humans, because data are available for human nasal carboxylesterase activity (Mattes & Mattes, 1992). Precise recording of lesion distribution patterns in laboratory animals and determination of target cell populations was an essential feature of this research.

Another interesting example of the role of regional metabolism in the cellular localization and activation of a nasal toxicant is provided by an excellent series of studies of the herbicide, dichlobenil (Brandt et al., 1990; Brittebo et al., 1991, 1992; Eriksson & Brittebo, 1991). This work also draws attention to the potential importance of the intranasal distribution of cosubstrates or cofactors. These compounds, which may play a critical role in local metabolism of certain xenobiotics, also exhibit characteristic concentration-distribution patterns. For instance, the metabolism of dichlobenil is influenced significantly by the availability of glutathione (Brittebo et al., 1992). This is also true for formaldehyde (Casanova & Heck, 1987). The localization of glutathione, which can be revealed histochemically in tissue sections (Bonnefoi et al., 1991; Keller et al., 1990), may therefore also be compared with lesion distribution in the search for underlying mechanisms (Keller et al., 1990).

DISEASE-ORIENTED APPROACH TO NASAL LESION ANALYSIS

The nose and paranasal sinuses are afflicted by many diseases (Goldman, 1987; Maran & Lund, 1990). The majority of these conditions are a consequence of aging, degenerative disease, or infection by pathogenic bacteria, viruses, and other agents. Exposure of humans to toxic xenobiotics may also be responsible for (or exacerbate) certain clinical conditions of the nasal mucosa, including rhinitis (Calderon-Garcidueñas & Roy-Ocotla, 1993; Koren et al., 1990), dysosmia (Getchell et al., 1991), and cancer (Holmström & Lund, 1991; Reith et al., 1983; Torjussen, 1983). Recent studies have demonstrated severe nasal damage in the human populations living in southwestern Mexico City. The lesions observed in these subjects were considered to be a consequence of exposure to the atmospheric environment, being potentially attributable to recent increases of formaldehyde, ozone, and other air pollutants in this urban environment. A range of nasal lesions, including cancer, has also been related to exposure of human populations in Europe to formaldehyde (Holmström et al., 1989, 1990) and wood dust (Acheson, 1982), indicating the potentially global nature of nasal disease in association with air pollution.

It can be invaluable for the risk assessment process that the pathologist, during interpretation of nasal lesions, provide some insight into the relevance of the lesions observed for human respiratory tract disease. Identification of specific disease concerns may assist interpretation of responses and guide the focus of subsequent studies directed toward human risk assessment. For instance, the olfactory mucosa is an important target site for inhaled methyl bromide in the nose of the rat, for which both histopathology and behavioral studies provided evidence of olfactory mucosal damage in the exposed animals (Hurtt et al., 1988). These findings indicated the potential for human risk of dysosmia as a result of exposure to methyl bromide in the work place. However, the olfactory epithelium is capable of rapid regeneration, even in the face of continued exposure to methyl bromide (Hurtt et al., 1988). This observation highlights the importance of distinguishing acute from chronic exposure scenarios when assessing human olfactory function. This issue is discussed in detail by Foureman and colleagues (1994) in the present volume, with respect to the olfactory toxicity of hexamethylene diisocyanate in rodents.

At present, however, by far the greatest concern during assessment of nasal toxicity lies with cancerous, or potentially cancerous, responses. It is important for the toxicologic pathologist to provide insight, wherever possible, into the relationship between proliferative or metaplastic lesions in the nose and the potential for cancer in this site. For instance, squamous metaplasia is a common nasal lesion found following exposure to nasal carcinogens (Kerns et al., 1983; Woutersen et al., 1986) and to chemicals that are not carcinogenic at this site (Buckley et al., 1985; Gross et al., 1987; Wolf et al., 1994b). The relationship between squamous metaplasia and cancer, and whether this is simply an adaptive response (Burger et al., 1989), remains problematic. However, care should be taken to distinguish "adaptive" squamous metaplasia of the nasal transitional or respiratory epithelia, from squamous metaplasia with prominent keratinization. The presence of such keratinization is potentially a more ominous finding with respect to cancer (Klein-Szanto et al., 1987; Morgan & Monticello, 1990b; Wolf et al., 1994a). The interpretation of such "putative preneoplastic" lesions also remains an important challenge to toxicologic. pathologists, as does the assignment of squamous metaplasia, and other potentially adaptive changes (Burger et al., 1989; Monticello et al., 1990) to a specific disease process.

In rodent toxicology studies, squamous metaplasia is frequently associated with inflammation of the nose following exposure to a wide range of gaseous and other irritants (Buckley et al., 1984; Monticello et al., 1990). Nasal inflammation, or rhinitis, is a major problem in humans, representing a significant cause of acute and chronic clinical distress (Maran & Lund, 1990; Mygind & Weeke, 1985). However, the precise definition of rhinitis as a clinical syndrome is plagued with difficulty, as is true for many disease entities (Mygind et al., 1985). Nasal inflammation may also contribute significantly to chemically induced carcinogenesis (Woutersen et al., 1989). Unraveling the relationship between the wide range of inflammatory and adaptive lesions to be found in the upper respiratory tract of rodents in toxicology studies (Jiang et al., 1986; Monticello et al., 1990) and human diseases, such as rhinitis, remains an important task.

In contrast to squamous metaplasia, secretory metaplasia of the transitional epithelium, characterized by the presence of goblet cells in a region normally devoid of them, is clearly not a neoplastic process. Lesions such as these may be selected for detailed morphologic or morphometric analyses (Hotchkiss et al., 1990), which provide insight into their nature and clarify the concentration-response. Such changes should be considered in relation to disease processes of concern. In the case of nasal secretory metaplasia, such changes indicate concern for excessive nasal secretions in humans. More importantly, however, interspecies differences in regional dosimetry of inhaled gases and vapors need to be considered when interpreting such nasal lesions in rodents. For instance, lesions induced in rats by exposure to 6 ppm formaldehyde are confined to the nose (Monticello et al., 1991), while this same concentration of formaldehyde induced responses as far distally as the carina in rhesus monkeys (Monticello et al., 1989). A similar picture is emerging for chlorine gas (Klonne et al., 1987; Wolf et al., 1994b). These findings indicate that responses in the nose of the rodent may raise concerns about disease risk in the lower respiratory tract of humans. These observations also highlight the importance of caution when attempting extrapolation of rodent data to humans (DeSesso, 1993), and the potential value of mechanistic data for such extrapolations.

CONCLUSIONS

Mapping of nasal lesions and detection of target cell populations by toxicologic pathologists provides an important indicator of dosimetry-driven, site-specific nasal disease. Consideration of underlying nasal structure and physiology may assist this process. Furthermore, it remains the role of the toxicologic pathologist, in spite of frequently limited information, to (a) assess the potential clinical significance of nasal lesions, (b) place them in the context of a defined disease process, and (c) present this information in manner that optimizes opportunities for their subsequent more detailed analysis in order improve the human risk assessment process through the application of dosimetry and other mechanistic data.

REFERENCES

Acheson, E. D. 1982. Seminar to discuss the carcinogenicity and mutagenicity of wood dust. 1, MRC Environmental Epidemiology Unit, South Academic Block, Southampton General Hospital, Southampton, SO9 4XY, UK.

Bang, F. G. 1961. Mucociliary function as protective mechanism in upper respiratory tract, ed. W. McDermott, pp. 228–236. Baltimore: Williams & Wilkins.

Barrow, C. S., ed. 1986. *Toxicology of the nasal passages*. New York: Hemisphere.

Belinsky, S. A., Walker, V. E., Maronpot, R. R., Swenberg, J. A., and Anderson, M. W. 1987. Molecular dosimetry of DNA adduct formation and cell toxicity in rat nasal mucosa following exposure to the tobacco specific nitrosamine 4-(*N*-methyl-*N*-nitrosamino)-1-(3-pyridyl)-1-butanone and their relationship to induction of neoplasia. *Cancer Res*. 47:6058–6065.

Bogdanffy, M. S. 1990. Biotransformation enzymes in the rodent nasal mucosa: The value of a histochemical approach. *Environ. Health Perspect*. 85:177–186.

Bogdanffy, M. S., Morgan, P. H., Morgan, K. T., and Starr, T. B. 1986a. Binding of formaldehyde to human and rat nasal mucus and bovine serum albumin. *Toxicol. Lett*. 38:145–154.

Bogdanffy, M. S., Randall, H. W., and Morgan, K. T. 1986b. Histochemical localization of aldehyde dehydrogenase in the respiratory tract of the Fischer-344 rat. *Toxicol. Appl. Pharmacol*. 82:560–567.

Bogdanffy, M. S., Randall, H. W., and Morgan, K. T. 1987. Biochemical quantitation and histochemical localization of carboxylesterase in the nasal passages of the Fischer-344 rat and B6C3F1 mouse. *Toxicol. Appl. Pharmacol*. 88:183–194.

Bond, J. A. 1989. Factors modifying the disposition of inhaled organic compounds. In *Concepts in inhalation toxicology*, eds. R. O. McClellan and R. F. Henderson, pp. 249–270. New York: Hemisphere.

Bonnefoi, M. S., Monticello, T. M., and Morgan, K. T. 1991. Toxic and neoplastic responses in the nasal passages: Future research needs. *Exp. Lung Res*. 17:853–868.

Brandt, I., Brittebo, E. B., Feil, V. J., and Bakke, J. E. 1990. Irreversible binding and toxicity of the herbicide dichlobenil (2,6-dichlorobenzonitrile) in the olfactory mucosa of mice. *Toxicol. Appl. Pharmacol*. 103:491–501.

Brittebo, E. B., Eriksson, C., and Brandt, I. 1990. Activation and toxicity of bromobenzene in nasal tissues in mice. *Arch. Toxicol*. 64:54–60.

Brittebo, E. B., Eriksson, C., Feil, V., Bakke, J., and Brandt, I. 1991. Toxicity of 2,6-dichlorothiobenzamide (chlorthiamid) and 2,6-dichlorobenzamide in the olfactory nasal mucosa of mice. *Fundam. Appl. Toxicol*. 17:92–102.

Brittebo, E. B., Eriksson, C., and Brandt, I. 1992. Effects of glutathione-modulating agents on the covalent binding and toxicity of dichlobenil in the mouse olfactory mucosa. *Toxicol. Appl. Pharmacol*. 114:31–40.

Brown, H. R., Monticello, T. M., Maronpot, R. R., Randall, H. W., Hotchkiss, J. R., and Morgan, K. T. 1991. Proliferative and neoplastic lesions in the rodent nasal cavity. *Toxicol. Pathol*. 19:358–372.

Buckley, L. A., Jiang, X. Z., James, R. A., Morgan, K. T., and Barrow, C. S. 1984. Respiratory tract lesions induced by sensory irritants at the RD50 concentration. *Toxicol. Appl. Pharmacol*. 74:417–429.

Buckley, L. A., Morgan, K. T., Swenberg, J. A., James, R. A., Hamm, T. E. J., and Barrow, C. S. 1985. The toxicity of dimethylamine in F-344 rats and B6C3F1 mice following a 1 year inhalation exposure. *Fundam. Appl. Toxicol*. 5:341–352.

Burger, G. T., Renne, R. A., Sagartz, J. W., Ayres, P. H., Coggins, C. R. E., Mosberg, A. T., and Hayes, A. W. 1989. Histologic changes in the respiratory tract induced by inhalation of xenobiotics: physiologic adaptation or toxicity? *Toxicol. Appl. Pharmacol*. 101:521–542.

Calderon-Garcidueñas, L., and Roy-Ocotla, G. 1993. Nasal cytology in southwest metropolitan Mexico City inhabitants: A pilot intervention study. *Environ. Health Perspect*. 101:2–8.

Calderon-Garcidueñas, L., Osorno-Velazquez, A., Bravo-Alvarez, H., Delgado-Chavez, R., and Barrios-Marquez, R. 1992. Histopathologic changes of the nasal mucosa in southwest metropolitan Mexico City inhabitants. *Am. J. Pathol*. 140:225–232.

Casanova, M. and Heck, H. 1987. Further studies of the metabolic incorporation and covalent binding of inhaled [^{3}H]- and [^{14}C]formaldehyde in Fischer-344 rats: Effects of glutathione depletion. *Toxicol. Appl. Pharmacol*. 89:105–121.

Casanova, M., Morgan, K. T., Gross, E. A., Moss, O. R., and Heck, H. 1994. DNA-protein cross-links and cell replication at specific sites in the nose of F344 rats exposed subchronically to formaldehyde. *Fundam. Appl. Toxicol*., in press.

Cole, P. 1988. Modification of inspired air. In *Respiratory function of the upper airway*, eds. O. P. Mathew and G. Sant'Ambrogio, pp. 415–445. New York: Marcel Dekker.

Conolly, R. B., and Andersen, M. E. 1991. Biologically based pharmacodynamic models: Tools for toxicological research and risk assessment. *Annu. Rev. Pharmacol. Toxicol.* 31:503–523.

Conolly, R. B., Morgan, K. T., Andersen, M. E., Monticello, T. M., and Clewell, H. J. 1992. A biologically-based risk assessment strategy for inhaled formaldehyde. *Comments Toxicol.* 4:269–293.

Cross, A. E., Halliwell, B., and Allen, A. 1984. Antioxidant protection: A function of tracheobronchial and gastrointestinal mucus. *Lancet* 1:1328–1330.

Dahl, A. R. 1988. The effect of cytochrome P-450-dependent metabolism and other enzyme activities on olfaction. In *Molecular neurobiology of the olfactory system, molecular, membranous, and cytological studies,* eds. F. L. Margolis and T. V. Getchell, pp. 51–70. New York: Plenum Press.

Dahl, A. R., and Hadley, W. H. 1991. Nasal cavity enzymes involved in xenobiotic metabolism: effect on the toxicity of inhalants. *CRC Crit. Rev. Toxicol.* 21:345–372.

DeSesso, J. M. 1993. The relevance to humans of animal models for inhalation studies of cancer in the nose and upper airways. *Qual. Assess. Good Pract. Law* 2:213–231.

Eccles, R., Bende, M., and Widdicombe, J. G. 1987. Nasal blood vessels. In *Allergic and vasomotor rhinitis: Pathophysiological aspects,* eds. N. Mygind and U. Pipkorn, pp. 63–76. Copenhagen Denmark: Munksgaard.

Eriksson, C., and Brittebo, E. B. 1991. Metabolic activation of the herbicide dichlobenil in the olfactory mucosa of mice and rats. *Chem. Biol. Interact.* 7:165–177.

Foureman, G. L., Greenberg, M. M. and Shiotsuka, R. N. 1994. Toxicological issues and questions in derivation of the inhalation reference concentration for hexamethylene diisocyanate (HDI). *Inhal. Toxicol.* 6(suppl):341–355.

Genter, M. B., Llorens, J., O'Callaghan, J. P., Peele, D. B., Morgan, K. T., and Crofton, K. M. 1992. Olfactory toxicity of β,β'-iminodiproprionitrile (IDPN) in the rat. *J. Pharmacol. Exp. Ther.* 263:1432–1439.

Gerde, P., and Dahl, A. R. 1991. A model for the uptake of inhaled vapors in the nose of the dog during cyclic breathing. *Toxicol. Appl. Pharmacol.* 109:276–288.

Getchell, T. V., Doty, R. L., Bartoshuk, L. M., and Snow, J. B. J. 1991. *Smell and taste in health and disease.* New York: Raven Press.

Giddens, W. E. J., and Fairchild, G. A. 1972. Effects of sulfur dioxide on the nasal mucosa of mice. *Arch. Environ. Health* 25:166–173.

Goldman, J. L. 1987. *The principles and practice of rhinology.* Chichester: John Wiley & Sons.

Gross, E. A., Patterson, D. L., and Morgan, K. T. 1987. Effects of acute and chronic dimethylamine exposure on the nasal mucociliary apparatus. of F-344 rats. *Toxicol. Appl. Pharmacol.* 90:359–376.

Gross, E. A., Mellick, P. W., Kari, F. W., Miller, F. J., and Morgan, K. T. 1994. Histopathology and cell replication responses in the respiratory tract of rats and mice exposed by inhalation to glutaraldehyde for up to thirteen weeks. *Fundam. Appl. Toxicol.,* in press.

Hahn, I., Scherer, P. W., and Mozel, M. M. 1993. Velocity profiles measured for airflow through a large-scale model of the human nasal cavity. *J. Appl. Physiol.* 75:2273–2287.

Harkema, J. A. 1990. Comparative pathology of the nasal mucosa in laboratory animals exposed to inhaled irritants. *Environ. Health Perspect.* 85:231–238.

Harkema, J. R. 1991. Comparative aspects of nasal airway anatomy: Relevance to inhalation toxicology. *Toxicol. Pathol.* 19:321–336.

Harkema, J. R., Plopper, C. G., Hyde, D. M., St. George, J. A., Wilson, D. W., and Dungworth, D. L. 1987. Response of the macaque nasal epithelium to ambient levels of ozone. *Am. J. Pathol.* 128:29–44.

Hastings, L., and Evans, J. E. 1991. Olfactory primary neurons as a route of entry for toxic agents in the CNS. *Neurotoxicology* 12:707–714.

Hebel, R., and Stromberg, M. W. 1976. *Anatomy of the laboratory rat.* Baltimore: Williams & Wilkins.

Heck, H., Casanova, M., and Starr, T. B. 1990. Formaldehyde toxicity—New understanding. *CRC Crit. Rev. Toxicol.* 20:397–426.

Heck, H., Casanova, M., Morgan, K. T., Bermudez, E. G., Moss, O. R., Asgharian, B., and James, R. A. 1993. Site-specific formation of DNA-protein cross-links by formaldehyde in pre-exposed rats. *Toxicologist* 13:261.

Holmström, M., and Lund, V. J. 1991. Malignant melanomas of the nasal cavity after occupational exposure to formaldehyde. *Br. J. Ind. Med.* 48:9–11.

Holmström, M., Wilhelmsson, B., Hellquist, H., and Rosen, G. 1989. Histological changes in the nasal mucosa in persons occupationally exposed to formaldehyde alone and in combination with wood dust. *Acta Otolaryngol. (Stockh.)* 107:120–129.

Holmström, M., Wilhelmsson, B., Hellquist, H., and Drettner, B. 1990. Effects of formaldehyde on the nasal mucosa. *Am. J. Rhinol.* 4:105–107.

Hotchkiss, J. A., Harkema, J. R., and Henderson, R. F. 1990. Effect of cumulative ozone exposure on ozone-induced nasal epithelial hyperplasia and secretory metaplasia in rats. *Exp. Lung Res.* 15:589–600.

Hurtt, M. E., Thomas, D. A., Working, P. K., Monticello, T. M., and Morgan, K. T. 1988. Degeneration and regeneration of the olfactory epithellium following inhalation exposure to methyl bromide: pathology, cell kinetics, and olfactory function. *Toxicol. Appl. Pharmacol.* 94:311–328.

Jeffery, E. H., and Haschek, W. M. 1988. Protection by dimethylsulfoxide against acetaminophen-induced hepatic, but not respiratory toxicity in the mouse. *Toxicol. Appl. Pharmacol.* 93:452–461.

Jiang, X. Z., Buckley, L. A., and Morgan, K. T. 1983. Histopathology of toxic responses to chlorine gas in the nasal passages of rats and mice. *Toxicol. Appl. Pharmacol.* 71:225–236.

Jiang, X. Z., Morgan, K. T., and Beauchamp, R. O. 1986. Histopathology of acute and subacute nasal toxicity. In *Toxicology of the nasal passages*, ed. C. S. Barrow, pp. 51–66. New York: Hemisphere.

Johnson, N. F., Hotchkiss, J. A., Harkema, J. R., and Henderson, R. F. 1990. Proliferative responses of rat nasal epithelia of ozone. *Toxicol. Appl. Pharmacol.* 103:143–155.

Keller, D. A., Heck, H., Randall, H. W., and Morgan, K. T. 1990. Histochemical localization of formaldehyde dehydrogenase in the rat. *Toxicol. Appl. Pharmacol* 106:311–326.

Kerns, W. D., Pavkov, K. L., Donofrio, D. J., Gralla, E. J., and Swenberg, J. A. 1983. Carcinogenicity of formaldehyde in rats and mice after long-term inhalation exposure. *Cancer Res.* 43:4382–4398.

Kimbell, J. S., and Morgan, K. T. 1991. Airflow effects on regional disposition of particles and gases in the upper respiratory tract. *Radiat. Dosim. Protect.* 38:213–219.

Kimbell, J. S., Gross, E. A., Joyner, D. R., Godo, M. N., and Morgan, K. T. 1993. Application of computational fluid dynamics to regional dosimetry of inhaled chemicals in the upper respiratory tract of the rat. *Toxicol. Appl. Pharmacol.* 121:253–263.

Klein-Szanto, A. J. P., Boysen, M., and Reith, A. 1987. Keratin and involucrin in preneoplastic and neoplastic lesions, distribution in the nasal mucosa of nickel workers. *Arch. Pathol. Lab. Med.* 111:1057–1061.

Klonne, D. R., Ulrich, C. E., Riley, M. G., Hamm, T. E. J., Morgan, K. T., and Barrow, C. S. 1987. One-year inhalation toxicity study of chlorine in rhesus monkeys (*Macaca mulatta*). *Fundam. Appl. Toxicol.* 9:557–572.

Koren, H. S., Hatch, G. E., and Graham, D. E. 1990. Nasal lavage as a tool in assessing acute inflammation in response to inhaled pollutants. *Toxicology* 60:15–25.

Lee, K. P., Valentine, R., and Bogdanffy, M. S. 1992. Nasal lesion development and reversibility in rats exposed to aerosols of dibasic esters. *Toxicol. Pathol.* 20:376–393.

Lucas, A. M., and Douglas, L. C. 1934. Principles underlying ciliary activity in the respiratory tract. II. A comparison of nasal clearance in man, monkey, and other mammals. *Arch. Otolaryngol.* 70:518–541.

Macia, R. A., Short, B. G., and Kerns, W. D. 1993. Selective damage of the olfactory epithelium (OE) of the vomeronasal organ following intranasal instillation of growth hormone releasing peptide (GH-RP-6) in rats. *Toxicologist* 13:262.

Maran, G. D., and Lund, V. J. 1990. *Clinical rhinology*. New York: Thieme.

Mattes, P. M., and Mattes, W. B. 1992. μ-Naphthyl butyrate carboxylesterase activity in human and rat nasal tissue. *Toxicol. Appl. Pharmacol.* 114:71–76.

Mery, S., Gross, E. A., Joyner, D. R., Godo, M., and Morgan, K. T. 1994a. Nasal diagrams: A tool for recording the distribution of nasal lesions in rats and mice. *Toxicol. Pathol.*, in press.

Mery, S., Larson, J. L., Butterworth, B. E., Wolf, D. C., Harden, R., and Morgan, K. T. 1994b. Nasal toxicity of chlorform in male F344 rats and female B6C3F1 mice following one week inhalation exposure. *Toxicol. Appl. Pharmacol.*125:214–227.

Monticello, T. M., Morgan, K. T., Everitt, J. I., and Popp, J. A. 1989. Effects of formaldehyde gas on the respiratory tract of rhesus monkeys: Pathology and cell proliferation. *Am. J. Pathol.* 134:515–527.

Monticello, T. M., Morgan, K. T., and Uriah, L. 1990. Nonneoplastic nasal lesions in rats and mice. *Environ. Health Perspect.* 85:249–274.

Monticello, T. M., Miller, F. J., and Morgan, K. T. 1991. Regional increases in rat nasal respiratory epithelial cell proliferation following acute and subacute inhalation of formaldehyde. *Toxicol. Appl. Pharmacol.* 111:409–421.

Morgan, K. T., and Monticello, T. M. 1990b. Formaldehyde toxicity: Respiratory epithelial injury and repair. In *Biology, toxicology, and carcinogenesis of respiratory epithelium*, eds. D. G. Thomassen and P. Nettesheim, pp. 155–171. New York: Hemisphere.

Morgan, K. T., Patterson, D. L., and Gross, E. A. 1984a. Frog palate mucociliary apparatus: Structure, function and response to formaldehyde gas. *Fundam. Appl. Toxicol.* 4:58–68.

Morgan, K. T. 1991. Approaches to the identification and recording of nasal lesions in toxicology studies. *Toxicol. Pathol.* 19:337–351.

Morgan, K. T., Jiang, X. Z., Patterson, D. L., and Gross, E. A. 1984b. The nasal mucociliary apparatus: Correlation of structure and function in the rat. *Am. Rev. Respir. Dis.* 130:275–281.

Morgan, K. T., Jiang, X. Z., Starr, T. B., and Kerns, W. D. 1986a. More precise localization of nasal tumors associated with chronic exposure of F-344 rats to formaldehyde gas. *Toxicol. Appl. Pharmacol.* 82:264–271.

Morgan, K. T., Patterson, D. L., and Gross, E. A. 1986b. Responses of the nasal mucociliary apparatus to airborne irritants. In *Toxicology of the nasal passages*, ed. C. S. Barrow, pp. 123–133. New York: Hemisphere.

Morgan, K. T., Kimbell, J. S., Monticello, T. M., Patra, A. L., and Fleishman, A. 1991. Studies of inspiratory airflow patterns in the nasal passages of the F344 rat and rhesus monkey using nasal molds: Relevance to formaldehyde toxicity. *Toxicol. Appl. Pharmacol.* 110:223–240.

Morgan, K. T., and Monticello, T. M. 1990a. Airflow, gas deposition, and lesion distribution in the nasal passages. *Environ. Health Perspect.* 85:209–218.

Morris, J. B., and Cavanagh, D. G. 1986. Deposition of ethanol and acetone vapors in the upper respiratory tract. *Fundam. Appl. Toxicol.* 6:78–88.

Morris, J. B., Clay, R. J., and Cavanagh, D. G. 1986. Species differences in upper respiratory tract deposition of acetone and ethanol vapors. *Fundam. Appl. Toxicol.* 7:671–680.

Morris, J. B., Clay, R. J., Trela, B. A., and Bogdanffy, M. S. 1991. Deposition of dibasic esters in the upper respiratory tract of the male and female Sprague-Dawley rat. *Toxicol. Appl. Pharmacol.* 108:538–546.

Morris, J. B., Hassett, D. N., and Blanchard, K. T. 1993. A physiologically based pharmacokinetic model for nasal uptake and metabolism of nonreactive vapors. *Toxicol. Appl. Pharmacol.* 123:120–129.

Morse, C. C., Boyd, M., and Witschi, H. 1984. The effect of 3-methylfuran inhalation exposure on the rat nasal cavity. *Toxicology* 30:195–204.

Mygind, N., and Weeke, B., eds. 1985. *Allergic and vasomotor rhinitis, clinical aspects*. Copenhagen, Denmark: Munksgaard.

Mygind, N., Änggård, A., and Druce, H. M. 1985. Definition, classification and terminology. In *Allergic and vasomotor rhinitis*, eds. N. Mygind and B. Weeke, pp. 15–20. Copenhagen, Denmark: Munksgaard.

Overton, J. H., and Miller, F. J. 1988. Dosimetry modeling of inhaled toxic reactive gases. In *Air pollution, the automobile, and public health*, eds. A. Y. Watson, R. R. Bates, and D. Kennedy, pp. 367–385. Washington, DC: National Academy Press.

Peele, D. B., Allison, S. D., Bolon, B., Prah, J. D., Jensen, K. F., and Morgan, K. T. 1990. Functional deficits produced by 3-methylindole-induced olfactory mucosal damage revealed by a simple olfactory learning task. *Toxicol. Appl. Pharmacol.* 107:191–202.

Phipps, R. J. 1981. The airway mucociliary system. In *Respiratory physiology III*, ed. J. G. Widdicombe, pp. 213–260. Baltimore: University Park Press.

Proctor, D. F. 1982. The mucociliary system. In *The nose, upper airway physiology and the atmospheric environment*, eds. D. F. Proctor and I. Andersen, pp. 245–278. New York: Elsevier.

Proetz, A. W. 1951. Air currents in the upper respiratory tract and their clinical importance. *Ann. Otol. Rhinol. Laryngol.* 60:439–467.

Pryor, W. A. 1992. How far does ozone penetrate into the pulmonary air/tissue boundary before it reacts? *Free Radical Biol. Med.* 12:83–88.

Randall, H. W., Bogdanffy, M. S., and Morgan, K. T. 1987. Enzyme histochemistry of the rat nasal mucosa embedded in cold glycol methacrylate. *Am. J. Anat.* 179:10–17.

Reed, C. J. 1993. Drug metabolism in the nasal cavity: Relevance to toxicology. *Drug Metab. Rev.* 25:173–205.

Reith, A., Riguat, J.-P., Boysen, M., and Marton, P. 1983. Pseudostratified, metaplastic, dysplastic, and carcinomatous nasal mucosa in nickel workers: a study by scanning electron, transmission electron, and light microscopy. In *Nasal tumors in animals and man, tumor pathology*, eds. G. Reznik and S. F. Stinson, pp. 55–94. Boca Raton, FL: CRC Press.

Reznik-Schüller, H. M. 1983. Nitrosamine-induced nasal cavity carcinogenesis. In *Nasal tumors in animals and man, experimental nasal carcinogenesis*, eds. G. Reznik and S. F. Stinson, pp. 47–77. Boca Raton, FL: CRC Press.

Stott, W. T., Dryzga, M. D., and Ramsey, J. C. 1983. Blood-flow distribution in the mouse. *J. Appl. Toxicol.* 3:310–312.

Torjussen, W. 1983. Nasal cancer in nickel workers. Histopathological findings and nickel concentrations in the nasal mucosa of nickel workers, and a short review of chromium and arsenic. In *Nasal tumors in animals and man, tumor pathology*, eds. G. Reznik and S. F. Stinson, pp. 33–53. Boca Raton, Florida: CRC Press.

Trela, B. A., Frame, S. R., and Bogdanffy, M. S. 1992. A microscopic and ultrastructural evaluation of dibasic esters (DBE) toxicity in rat nasal explants. *Exp. Mol. Pathol.* 56:208–218.

Uraih, L. C., Talley, F. A., Mitsumori, K., Gupta, B. N., Bucher, J. R., and Boorman, G. A. 1987. Ultrastructural changes in the nasal mucosa of Fischer 344 rats and B6C3F1 mice following an acute exposure to methyl isocyanate. *Environ. Health Perspect.* 72:77–88.

Uraih, L. C., Morgan, K. T., and Maronpot, R. R. 1990. Symposium on toxicologic pathology of the upper respiratory system. *Environ. Health Perspect.* 85:161–352.

Vogel, S. 1981. *Life in moving fluids, the physical biology of flow.* Princeton, N.J.: Princeton University Press.

Warwick, R., and Williams, P. L., eds. 1973. *Gray's anatomy*. Philadelphia: W. B. Saunders.

Weiss, L., ed. 1983. *Histology, cell and tissue biology*. New York: Elsevier Biomedical Press.

Wolf, D. C., Gross, E. A., Lyght, O., Bermudez, E., Recio, L., and Morgan, K. T. 1994a. Regional expression of p53, PCNA and TGF-α proteins in formaldehyde-induced rat nasal squamous cell carcinomas: Immunohistochemical studies.

Wolf, D. C., Morgan, K. T., Gross, E. A., Barrow, C., Moss, O. R., James, R. A., and Popp, J. A. 1994b. Two-year inhalation study of chlorine gas toxicity in B6C3F1 mice and F344 rats. *Fundam. Appl. Toxicol.*, in press.

Woutersen, R. A., and Feron, V. J. 1989. Localization of nasal tumors in rats exposed to acetaldehyde or formaldehyde. In *Nasal carcinogenesis in rodents: Relevance to human health risk*, eds. V. J. Feron and M. C. Bosland, pp. 70–75. Wageningen,The Netherlands: Pudoc Press.

Woutersen, R. A., Appelman, L. M., Van Garderen-Hoetmer, A., and Feron, V. J. 1986. Inhalation toxicity of acetaldehyde in rats. III. Carcinogenicity study. *Toxicology* 41:213–231.

Woutersen, R. A., van Garderen-Hoetmer, A., Bruijntjes, J. P., Zwart, A., and Feron, J. 1989. Nasal tumors in rats after severe injury to the nasal mucosa and prolonged exposure to 10 ppm formaldehyde. *J. Appl. Toxicol.* 9:39–46.

Zupka, K., Poria, Y., and Lancet, D. 1991. Immunolocalization of cytochromes *P*-450olf1 and *P*-450olf2 in rat olfactory mucosa. Eur. J. Biochem. 196:51–58.

DOSIMETRY MODELING: APPROACHES AND ISSUES

James S. Ultman

Department of Chemical Engineering, Pennsylvania State University, University Park, Pennsylvania, USA

In this article, mathematical modeling is defined as the use of the physical laws of mass, heat, and momentum conservation to quantify the dynamics of a system of interest. When modeling is used to assess nasal dosimetry, additional information on the anatomy, physiology, and chemistry of the nose and the inhaled xenobiotic substances is required. Each model incorporates a number of simplifying assumptions that determine its individual character. The simpler is the model, the less information is required for its implementation, but also the more restricted is the final formulation. In striking a reasonable balance between simplicity and utility, a modeler must consider both the ultimate application of the model and the availability of experimental measurements required for its validation. This article discusses these issues and presents seven illustrative examples demonstrating progress in this field over the past 35 years.

The purpose of this article is to provide insights into the process of mathematical modeling as it is applied to nasal dosimetry. The article is divided into two parts. First, a general discussion of the motivation, characteristics, and approaches in mathematical modeling is provided. This is followed by a series of seven illustrative examples.

WHY MATHEMATICAL MODELING?

Mathematical models can serve four broad purposes. First, a model proposed before experiments are actually conducted serves as an extended hypothesis that can aid in the experimental design. For example, it is reasonable to hypothesize that inhaled ozone undergoes rapid chemical reaction with the mucosa of the conducting airways of the lung. When this simple hypothesis is fleshed out using a diffusion model, it becomes apparent that the speed of the reaction can be gauged by measuring the relationship between uptake rate and respiratory flow. That is, if reaction between ozone and biochemical substrates is very fast, then uptake will be governed by boundary-layer diffusion in the gas phase, which is directly related to respiratory airflow. On the other hand, if the reaction is slow, then diffusion through the mucosa is important and uptake is relatively insensitive to flow.

Second, a mathematical model can be used to correlate data. In the case of ozone absorption, a steady-state diffusion model suggests that the logarithm of uptake rate should be regressed against the reciprocal of the

This research was supported in part by National Institute of Environmental Health Sciences grant ES06075.

Address correspondence to James S. Ultman, PhD, 106 Fenske Lab, University Park, PA 16802.

airflow (Aharonson et al., 1974). Third, by employing the regression suggested by a mathematical model, outcomes can sometimes be predicted at conditions where measurements were not made. Again referring to ozone absorption, suppose that measurements were made at relatively low respiratory flows characteristic of rest. These uptake data may be applied to exercise conditions by extrapolating the semilogarithmic regression equation to appropriately higher airflows.

Fourth, a mathematical model can be used for simulating observed toxicological phenomena in order to determine their underlying mechanisms. For example, mathematical simulations are being used to determine whether the localization of lesion formation in the nose by formaldehyde inhalation can be explained by a spatial distribution of gas-phase diffusion resistance (Kimbell et al., 1993). Such simulations, in conjunction with computer animation and graphics, provide a powerful tool for visualizing dynamic phenomena in three dimensions. This can lead to a better understanding of dosimetry, even by those who are not well versed in the underlying mathematics.

CHARACTERISTICS OF MODELS

A mathematical model is a set of equations that relate the present state of a system to its present or past inputs. The state of the system is defined by internal variables and parameters that, in an animal or human, are reflective of its anatomy, physiology, and biochemistry. Inputs, on the other hand, are boundary conditions that define those constraints imposed by the surrounding environment. A model of nasal dosimetry might include state variables such as blood flow and regional uptake rates, and parameters such as airway volume, diffusion coefficient, and reaction rate constant. The principal input variable is the exposure concentration of the xenobiotic agent. Although a model can be empirical in nature, drawing only from an intuitive organization of experimental observations, the majority of contemporary models are formulated by using basic scientific principles. In particular, equations that specify conservation of momentum, of mass, and of energy can be solved for the internal velocity, concentration, and temperature distributions, respectively. These equations must be augmented by thermodynamic relationships that define interfacial solubilities and chemical reaction equilibrium, and by rate expressions such as Fick's law of diffusion and Michaelis–Menton enzyme kinetics. Although the conservation equations are universal laws, thermodynamic equilibrium and rate expressions depend on the nature of the inhaled pollutant gas and its interactions with the chemical components of biological tissue.

Although all these equations may, in principle, be utilized in an exact manner, solving the resulting mathematical problem may be virtually impossible. Therefore, all models are simplified by invoking a set of assumptions that constitute approximations to the actual behavior of the system. It is the particular assumptions chosen by a modeler that sets his or her model apart from other models of the same process. When choosing assumptions, the modeler has the responsibility to use the best available information. Nevertheless, the validity of a mathematical model must ultimately be verified by comparing its predictions to experimental data. Ideally, the measurements used to validate the model should be independent of those data used to frame the original assumptions. If the predictions of the mathematical model are not borne out by new observations, this should not be viewed as a failure of the modeling effort. Rather, the exploration for more reasonable alternative assumptions represents an opportunity for refining the model, as well as for better understanding the dynamics of the system.

For example, in a ventilation-perfusion model of upper airway dosimetry proposed by Morris et al. (1986), the diffusion of an inhaled vapor was assumed to be so rapid that its relative concentrations in gas, mucus, and tissue were specified by equilibrium partition coefficients. When applied to acetone uptake during steady inspiratory-directed gas flow through the upper airways of Fischer 344 rats, this model led to an estimate of 0.19 ml/min for the nasal perfusion rate. Using the same perfusion rate to predict ethanol uptake was not successful in the sense that the theoretical values were significantly and consistently less than corresponding experimental measurements. This prompted the investigators to reevaluate their assumptions, and they concluded that, unlike acetone, ethanol is so soluble that its uptake is somewhat limited by diffusion.

CONSTRUCTING A MODEL

Throughout the formulation and validation of a mathematical model, its ultimate application must be clearly defined and kept in mind. Perhaps the most pressing need for modeling is the extrapolation to real-world exposure scenarios from laboratory dose measurements that are constrained by cost, time, and considerations of safety. The most obvious extrapolation issue is the prediction of the dose to people from doses measured in animals. Other related issues include extrapolation from inhaled exposure concentration to target-tissue dose; from high exposure levels to ambient levels; and from acute exposure to subchronic or chronic exposure.

Depending on its designated application, a mathematical model can be made relatively exact by taking into account the detailed anatomy and fluid flow distributions within particular organs. Alternatively, the model can be simplified by representing the body by a parallel-series array of well-mixed compartments. Models that employ well-mixed compartments and are governed by concentration variables that depend on time alone are known as "lumped parameter models." The formulation of these models requires the solution of ordinary differential equations and their accompanying initial conditions. A refinement of lumped parameter modeling that is receiving more and more acceptance in toxicology is the physiologically based pharmacokinetic (PBPK) model. In PBPK models, each compartment corresponds to an actual anatomic entity whose parameters are assigned physiologic values that are often available from independent experiments (e.g., Clewell & Anderson, 1985).

Models that incorporate detailed anatomic information are governed by concentration variables that depend on spatial position as well as on time, and are known as "distributed parameter models" (e.g., Miller et al., 1978). These models are formulated by solving partial differential equations that require the specification of both boundary and initial conditions. The additional spatial information offered by a distributed parameter model relative to a corresponding compartmental model comes at the expense of solving more difficult differential equations, usually by numerical methods that are implemented on digital computers. However, there are continuing improvements in computer products for solving distributed parameter models (e.g., Kimbell et al., 1993), and, at the same time, there is a trend toward formulating more detailed compartmental models (i.e., Morris et al., 1993). Thus, the difference between applying these two types of models is diminishing.

The availability of parameters governing the anatomy, physiology, and chemistry in a biological system of interest is an important consideration in the selection of a model structure and its accompanying simplifying assumptions. The values of some parameters will be known from independent sources, while other parameters must be estimated by regressing data from validation experiments to the model equations. If an excessive number of unknown parameters has been incorporated into a model, then their estimated values may not be unique. Therefore, the modeler must select a formulation that is sufficiently complex and versatile to meet its designated application without demanding the use of too many unknown system parameters.

After scouring the literature for previously measured parameter values, the modeler may elect to carry out in vitro measurements of unknown parameters. For example, vial equilibration measurements on excised tissue can produce solubility and reaction rate constants (Sato & Nakajima, 1979), and evaporation rates in airway casts can be used to evaluate gas-phase diffusion resistances (Hanna & Scherer, 1986). Alternatively, some parameters whose values are not known in biological tissue can be estimated from tabulated values in water or in oils. This is particularly true of solubility and diffusion coefficients.

The nature of chemical reactions between a xenobiotic agent and biological tissues may have a profound influence on the scope of the modeling effort. Some inhaled substances may be confined to the upper respiratory tract because they interact with localized enzymes or because they undergo rapid nonspecific reactions. For these materials, a model whose scope is limited to the upper airways is sufficient. Other inhaled toxics may be relatively inert in the upper respiratory tract but may react in the lower respiratory tract, or they may be eliminated in other organs that they reach via the systemic circulation. For such substances, a model that encompasses the entire respiratory system or, possibly, the entire animal may be necessary. It is also important to consider whether a metabolite of the parent compound can cause a toxic response. For example, an inhaled gas that reacts rapidly with respiratory tissue and is confined to the lungs may produce an active product that is toxic to a systemic target organ, such as the liver.

Another important factor in structuring a model is the definition of a dose metric for quantifying the delivery of the xenobiotic agent. Some alternative choices rank-ordered from the most primitive to the most sophisticated are environmental exposure concentration, inhaled dose (i.e., concentration multiplied by minute ventilation), whole-body uptake, uptake by individual organs, and regional uptake rates within an individual organ. For example, ozone is a highly reactive oxidation agent, and therefore the toxic response of the inhaled gas will be limited to the lung. Studies of ozone-exposed laboratory animals have demonstrated that the proximal alveolar region is a focal site of epithelial cell alterations. A mathematical model capable of explaining this phenomenon on the basis of dosimetry must be capable of simulating the spatial distribution of ozone uptake within the lungs. This explains why Miller and his co-workers developed a one-dimensional distributed parameter model for this purpose (Miller et al., 1978).

EXAMPLES OF DOSIMETRY MODELS

Modeling of the uptake and distribution of gases and vapors in upper airways and its relationship to dosimetry in the lungs, as well as the systemic circulation, is an active area of study in respiratory physiology, toxicology, and anesthesiology. Because it is virtually impossible to review the many different models associated with this research, the discussion below is limited to seven specific examples.[1] These examples are presented in chronological order and illustrate how, with the passing of time, improved insights into underlying transport processes coupled with more extensive asurements of system properties have spawned more comprehensive mathematical models.

Three-Compartment Modified Bohr Model of the Respiratory System by Cander and Forster (1959)

Now more than 100 years old, the Bohr model (1891) subdivides the respiratory system into two compartments. Corresponding to the conducting airways, there is a rigid dead-space compartment in which gas is transported by bulk flow without backmixing; corresponding to the respiratory zone, there is an expandable alveolar compartment that is perfectly mixed. To cir-

[1]To facilitate a comparison between examples, the equations and symbols used in the original publications have been somewhat reorganized.

cumvent the complexity of dealing with a reversing gas flow and cyclic changes in composition, inhalation and exhalation are assumed to occur simultaneously and alveolar gas composition is taken to be constant. According to this model, the volume of inhaled fresh air that ventilates the alveolar region during each breath can be computed as the difference between the tidal volume and the volume of the dead space.

In their pursuit of a noninvasive method of determining pulmonary blood flow by using inhaled inert gases, Cander and Forster modified the Bohr model in order to formulate the decrease in alveolar gas concentration with breathholding time. One major change was the addition of a well-mixed parenchymal tissue compartment adjacent to the well-mixed alveolar gas compartment. Pulmonary diffusion was assumed to be sufficiently rapid that the foreign gas was in equilibrium between these two compartments. The model also included pulmonary blood flow, in which recirculation effects were so small that the gas content of entering venous blood was negligible, and diffusion rates were so rapid that exiting end-capillary blood was in equilibrium with the alveolar and tissue gas contents.

Application of an unsteady-state mass conservation equation to this model indicated that the alveolar concentration of inert gas should be an exponentially decreasing function of breathholding time with an exponential time constant that is inversely related to gas solubility and to pulmonary blood flow, Q_B. This prediction was tested by employing data gathered in healthy people at breathholding times from 0 to 50 s using inert gases and vapors[2] that were poorly to highly soluble. For the lowest solubility gas, sulfur hexafluoride, the exponential time constant was too large to precisely estimate Q_B. For two moderately soluble gases, acetylene and ethyl ether, the same Q_B value of 3.3 L/min was computed, and this compared favorably to independently measured values of blood flow. However, for the most soluble gas, acetone, the Q_B value of 2.3 L/min was unrealistically small.

The authors attributed this discrepancy to the capacity of conducting airway tissue to store dissolved gas, an effect that cannot be simulated without incorporating a dead space tissue compartment. In particular, a soluble gas will absorb into conducting airway tissue during inhalation when high concentrations of the gas flow through the dead space. During exhalation, however, the gas will desorb as low concentration alveolar gas passes through the dead space. Thus, over a complete breath, less gas is actually absorbed into pulmonary blood than is predicted from the model.

Whole-Body Model by Dubois and Rogers (1968)

Dubois and Rogers developed a dosimetry model with the primary goal of assessing toxic gas uptake in the lungs, while still recognizing the need

[2]From this point on, the term "gas" is used to refer to both gases and vapors. The implication is that the dosimetry of the two substances can be modeled in the same fashion provided that the partial pressure of the vapor is substantially below its pure component vapor pressure.

to include systemic elimination. The tracheobronchial tree was subdivided into 16 conducting airway generations, each including gas, tissue, and blood phases; the upper airways were not included in this model. Three alternative models were proposed for diffusion through the tissue and blood phases, reflecting the uncertainty of dealing with bronchial perfusion. In the first model, steady-state diffusion occurred through a continuous slab of tissue with respired gas in contact with one surface of the slab and capillary blood in contact with the opposing surface. In the second model, perfusion was spit between a bloodstream on the outside surface of the slab and a second bloodstream flowing through the center of the slab. In the third model, blood flow was assumed to be uniformly distributed throughout the slab. The third model resulted in the highest uptake and most uniform distribution of toxic gas within the tissue.

The whole-body model that Dubois and Rogers proposed contained many pioneering concepts, such as subdivision of the dead space into bronchial generations, gas diffusion through bronchial tissue, the separate effects of bronchial and pulmonary perfusion, and the role of systemic metabolism. Although a comprehensive mathematical simulation that integrated all of these processes was not presented, this model stands as a useful foundation that other investigators have built upon (e.g., Johanson, 1991).

Distributed Parameter Mass Transfer Coefficient Model of the Upper Airways by Aharonson and Colleagues (1974)

A common experiment in laboratory animals is to surgically isolate the upper airways by tracheostomy and provide a steady inflow of air containing foreign gas at the airway opening (usually the nose) and an outflow of air at the tracheostomy site. Uptake of the foreign gas into tissue can then be determined by measuring the difference between its concentrations at the airway opening and in the trachea.

Aharonson and co-workers modeled this process by assuming that the concentration of the foreign gas varies longitudinally along the upper airways; longitudinal diffusion in the airway is negligible compared to convection; lateral diffusion can be accounted for by using an overall mass transfer coefficient; and somewhere in the tissue, the concentration of the foreign gas is so small that its backpressure is negligible. The last assumption removes the need to consider blood perfusion effects, but requires that the gas be either very reactive or highly soluble. The use of an overall mass transfer coefficient relieves the modeler of the need to describe lateral diffusion phenomenon, but results in an analysis that does not indicate how geometry and physiochemical properties in the gas and tissue phases affect dosimetry.

The uptake rate of foreign gas into airway tissue was found by formulating and solving a steady-state mass conservation equation over a differential length of the airway. The result was:

$$M_i = M_{i0}[1 - \exp(-KaV/\dot{V})] \tag{1}$$

where M_i is the uptake of gas i, M_{i0} is its inhaled dose, V is the airway volume, $\dot{V}$ is the respiratory flow, and a is the surface-to-volume ratio of the airway lumen. This equation was validated by data obtained in dogs with two soluble gases (i.e., ethyl ether and acetone), a rapidly ionizing gas (i.e., sulfur dioxide), and a rapid irreversibly reacting gas (i.e., ozone). In all cases, the model could not fit the data unless Ka was allowed to be an increasing function of $\dot{V}$. The authors offer three possible explanations for this. First, Ka may depend directly on concentration and, with increasing $\dot{V}$, the average concentration of toxic gas in the airway lumen increases. Second, if a substantial portion of the resistance to lateral diffusion is within the gas phase, then increasing V would reduce the thickness of the gas boundary layer and K would increase. Third, at increased $\dot{V}$, there could be increased perfusion in superficial blood vessels of the nose, and this would increase K.

Two-Compartment Equilibrium Model of the Upper Airways by Morris and Colleagues

To interpret measurements of inert gas uptake through isolated upper airways during constant flow, Morris and colleagues (Morris & Cavanaugh, 1986; Morris et al., 1986; Morris, 1990) utilized a model consisting of a well-mixed airway compartment in series with well-mixed tissue and blood compartments. Unlike Aharonson et al. (1974), Morris neglected lateral diffusion resistances so that an equilibrium distribution of foreign gas between the three compartments could be assumed. Moreover, tissue and blood were assumed to have equal Bunsen solubilities so that these two compartments could be combined into a single blood-tissue compartment. By ignoring recirculation effects, the content of foreign gas in the incoming arterial blood could be neglected, while the outgoing venous blood was assumed to be in equilibrium with gas in the tissue compartment.

Applying a steady-state conservation of mass equation to this model resulted in the following formula for the dose of gas i to the tissue:

$$M_i = M_{i0}/[1 + \dot{V}/Q_B\alpha_B P)] \tag{2}$$

where M_{i0} is the inhaled dose, $\dot{V}$ is the volumetric air flow, Q_B is the volumetric blood flow, and α_B is the Bunsen solubility in tissue and blood. It is interesting to compare this equilibrium model to Aharonson's mass transfer coefficient model. In general, Eqs. (1) and (2) exhibit distinctly different relationships with respect to $\dot{V}$. As $\dot{V}$ becomes large, however, Aharonson's model reduces to

$$M_i = M_{i0}(KaV)(1/\dot{V}) \tag{3}$$

and Morris's model reduces to

$$M_i = M_{i0}(Q_B \alpha_B P)(1/\dot{V}) \tag{4}$$

so that M_i is inversely proportional to $\dot{V}$ in both models. In other words, at small uptake rates, it is not possible to distinguish between the two models on the basis of respiratory flow effects alone.

Morris and Cavanagh (1986) validated the equilibrium model by measuring the uptake of two highly soluble gases into the isolated upper airways of Sprague-Dawley rats. As the inspiratory-directed gas flow was decreased from 500 to 70 ml/min, M_i for ethanol increased from 0.50 to 0.89, while M_i for less soluble acetone increased from 0.22 to 0.56. Although the variability of the measurements was large, the investigators indicated on statistical grounds that the experimental data fit the model with a reasonable degree of certainty. Moreover, using values of α_B that were determined in separate in vitro measurements, regressions of the data to Eq. (2) resulted in consistent values of nasal perfusion, 0.31 ml/min for ethanol and 0.33 ml/min for acetone.

In a later publication, Morris (1990) modified his original model to include the possibility of metabolic decomposition of the inhaled gas within the blood-tissue compartment. This was accomplished by including a Michaelis–Menten reaction rate expression as an additional term in the mass conservation equation. This model was tested by measuring the uptake of ethyl acetate in isolated upper airways of F344 rats and Syrian hamsters. It was hypothesized that this foreign gas would be a substrate for the carboxylesterase known to be present in nasal tissue. For the rat, the Michaelis–Menten constants estimated from the model were similar to those obtained in independent in vitro measurements, but this was not true for the hamster. Thus, the appropriateness of Morris's model appears to depend on the animal species.

Distributed Parameter Model of Upper Airways During Tidal Breathing by Gerde and Dahl (1991)

Gerde and Dahl developed a model of foreign gas uptake in the upper airways with two features that set it apart from Aharonson's and Morris's models: There was a tidal rather than a constant gas flow, and the tissue layer forming the gas–blood barrier had a finite resistance to diffusion, as well as a finite capacitance for storing foreign gas. The model idealized the upper airway geometry as a rectangular airway slit bounded by two thin tissue slabs. Respired gas flowed longitudinally within the slit, and blood flowed on the outside surface of the tissue slabs. Blood and gas flows were at right angles to one another, as is expected in the nose. Within the slit, foreign gas concentration was assumed to be laterally well mixed and in equilibrium with tissue at the inside surface of the slab. Entering arterial blood was assumed to be free of foreign gas, while exiting venous blood was equilibrated with tissue at the outside surface of the slab. Bunsen solubilities of the foreign gas in blood and in tissue were assumed to be equal.

Two possible effects in the respired gas that were not accounted for were a lateral diffusion resistance due to boundary-layer formation and longitudinal transport by convective dispersion.

Diffusion within the slabs was formulated with a mass conservation equation that equated the rate of foreign gas storage in a differential volume with the net rate of diffusion into that volume:

$$\partial C_i/\partial t = D_{\text{mol}}\ \partial^2 C_i/\partial y^2 \qquad (5)$$

where C_i is the concentration of the inhaled gas in the tissue slab, which depends on lateral position y and time t, and D_{mol} is the molecular diffusivity of the foreign gas in tissue. Two boundary conditions on y are needed to solve this equation. One condition, at the outer surface of the slab, was a mass balance applied to the flowing blood. A second condition, at the inner surface of the slab, was a differential mass balance on respired gas. It is this condition that imposes a longitudinal variation on C_i. This condition also requires that the input concentration to the slit be known throughout a tidal breath. This poses no problem during inhalation, when the inlet concentration is equal to the exposure concentration. During exhalation, however, the inlet concentration to the slit is the outlet concentration from the lungs, which are not present in the Gerde–Dahl model.

This model was tested by using inert gas concentration data simultaneously measured just external to the nose and in the trachea of beagle dogs. By using a special sampling system, surgical isolation of the upper airways was avoided, the dog breathed through its nose, and inert gas concentrations could be recorded throughout the respiratory cycle. The experiment was repeated for five inert gases of varying solubility. The experimental data indicated that nasal absorption during inspiration increased from 1% to 95% of the inhaled dose for gases that were poorly to highly soluble, while desorption during expiration increased from 1% to 30% for the same set of gases. Thus, over a complete breath, nasal uptake was more and more reduced by storage effects in tissue as the solubility of the inhaled gas increased.

Unlike the case for the simpler mass conservation equations associated with the Aharonson and Morris models, Eq. (5) and its boundary conditions cannot be solved analytically. Rather, these equations must be integrated numerically by using a suitably programmed digital computer. In performing such simulations, Gerde and Dahl selected parameter values for anatomic dimensions, nasal blood flow, diffusion coefficient, and Bunsen solubilities that were available in the literature. Moreover, the inhaled concentration data measured at the nose, as well as the exhaled concentration data measured in the trachea, were used as inputs to the model. Over the entire solubility range of the five test gases, there was good agreement between simulated and experimental values of inspiratory absorption, as well as expiratory desorption.

PKPB Model of Nasal Uptake by Morris and Colleagues (1993)

Recognizing the need to better account for the spatial distribution of ventilation, perfusion, and metabolizing enzymes within the nose, Morris and colleagues recently developed a PBPK model with parallel dosal medial and lateral/ventral airflow paths, each with a separate blood supply and containing different amounts of enzyme. The lateral/ventral path was composed of a series of one mucus, two epithelial, and two submucosal well-mixed compartments representing the respiratory mucosa. The dorsal medial path was composed of respiratory mucosa with the same structure in parallel with olfactory mucosa incorporating three additional epithelial compartments.

In the mass conservation equations, transport of foreign gas between tissue compartments was prescribed by an integrated form of Fick's diffusion law, and accumulation of gas within each compartment was represented by an appropriate time derivative. Therefore, the use of an in-series array of epithelial compartments was equivalent to a distributed parameter model in which spatial derivatives are replaced by finite-difference approximations. This illustrates the blurring distinction between distributed parameter and lumped parameter models that can occur in complex PBPK representations.

The PBPK nasal dosimetry model was validated by using uptake measurements for a collection of five metabolizing and inert gases during steady flow through isolated upper airways of F344 rats. All parameters, including the distribution of enzyme, but excluding the split in respiratory gas flow and the perfusion rates to respiratory and olfactory tissue, were estimated from existing literature values. The data were satisfactorily simulated when it was assumed that 8% of the respiratory gas flow traveled along the dorsal medial path and 6% of the perfusion was directed to olfactory tissue.

Distributed Parameter Model for Gas-Phase Limited Uptake in a Rat Nose by Kimbell and Colleagues (1993)

In trying to explain the site-specific nature of lesion formation in the noses of formaldehyde-exposed F344 rats, Kimbell and co-workers hypothesized that the complex, but reproducible, kinematics of airflow predisposes particular regions in the nasal cavities to tissue damage. In keeping with this hypothesis, these investigators carried out a computer solution to the complete momentum conservation (i.e., Navier–Stokes) equation to determine the velocity field during steady inspiratory flow. This information was then used to solve the complete mass conservation equation and to map out the distribution of formaldehyde uptake in the nasal cavity. The principal input data to this simulation were from an accurate three-dimensional reconstruction of the nasal airways of one rat.

This work was expedited by the availability of a commercial software package for solving the equations and for providing three-dimensional images of the results. The predicted velocity streamlines were validated with laboratory measurements made on airway casts from other F344 rats.

Moreover, the predicted distribution of formaldehyde uptake rates was in reasonable agreement with the lesion distribution presented in previously published pathology reports. The major limitation of this model was the use of a zero-concentration condition at the airway wall. This is equivalent to assuming that chemical reaction within the mucosa is very fast relative to the lateral transport of formaldehyde in the gas phase.

CONCLUSIONS

When constructing a new mathematical model, the modeler should incorporate a reasonable degree of uncertainty, if something is to be learned from the modeling process. By validating the predictions of the model with appropriately designed experiments, these uncertainties are clarified so that the model can be further tested and refined. After sufficient refinement, the model may be used with confidence for performing much-needed extrapolation computations.

With the evolution of faster computers, more reliable commercial software packages, and more versatile graphic interfaces, our ability to develop more complex and, hopefully, more realistic simulations is continuously improving. Simultaneously, the emergence of potent measurement technologies, such as three-dimensional microscopic imaging, fluorescence microscopy, and immunocytochemistry, will provide more detailed data for parameter estimation, as well as for model verification.

REFERENCES

Aharonson, E. F., Menkes, H., Gurtner, G., Swift, D. L., and Proctor, D. F. 1974. Effect of respiratory airflow rate on removal of soluble vapors by the nose. *J. Appl. Physiol.* 37:654–657.

Bohr, C. 1891. Uber die lungenatmung. *Skand. Arch. Physiol.* 2:236–268.

Cander, L., and Forster, R. E. 1959. Determination of pulmonary parenchymal tissue volume and pulmonary capillary blood flow in man. *J. Appl. Physiol.* 14:541–551.

Clewell, H., and Anderson, M. E. 1985. Risk assessment extrapolations and physiological modeling. *Toxicol. Ind. Health* 1:111–131.

Dubois, A. B., and Rogers, R. M. 1968. Respiratory factors determining the tissue concentrations of inhaled toxic substances. *Respir. Physiol.* 5:34–52.

Gerde, P., and Dahl, A. R. 1991. A model for the uptake of inhaled vapors in the nose of the dog during cyclic breathing. *Toxicol. Appl. Pharmacol.* 109:276–288.

Hanna, L., and Scherer, P. W. 1986. Measurement of local mass transfer coefficients in a cast model of the upper respiratory tract. *Trans. Am. Soc. Mech. Eng.* 108:12–18.

Johanson, G. 1991. Modelling of respiratory exchange of polar solvents. *Ann. Occup. Hyg.* 35:323–339.

Kimbell, J. S., Gross, E. A., Joyner, D. R., Godo, M. N., and Morgan, K. T. 1993. Application of computational fluid dynamics to regional dosimetry of inhaled chemicals in the upper respiratory tract of the rat. *Toxicol. Appl. Pharmacol.* 121:253–263.

Miller, F. J., Menzel, D. B., and Coffin, D. L. 1978. Similarity between man and laboratory animals in regional pulmonary deposition of ozone. *Environ. Res.* 17:84–101.

Morris, J. B. 1990. First-pass metabolism of inspired ethyl acetate in the upper respiratory tract of the F344 rat and Syrian hamster. *Toxicol. Appl. Pharmacol.* 102:331–345.

Morris, J. B., and Cavanaugh, D. G. 1986. Deposition of ethanol and aetone vapors in the upper respiratory tract of the rat. *Fundam. Appl. Toxicol.* 6:78–88.

Morris, J. B., Clay, R. J., and Cavanagh, K. G. 1986. Species differences in upper respiratory tract depostion of acetone and ethanol vapors. *Fundam. Appl. Toxicol.* 7:671–680.

Morris, J. B., Hassett, D. N., and Blanchard, K. T. 1993. A physiologically based pharmacokinetic model for nasal uptake and metabolism of nonreactive vapors. *Toxicol. Appl. Pharmacol.* 123:120–129.

Sato, A., and Nakajima, T. 1979. Partition coefficients of some aromatic hydrocarbons and ketones in water, blood and oil. *Br. J. Ind. Med.* 36:231–234.

Section 2. Nasal Dosimetry in Animals and Man: Mathematical and Physical Models

ISSUES IN MODELING DOSIMETRY IN RATS AND PRIMATES

J. S. Kimbell

Chemical Industry Institute of Toxicology, Research Triangle Park, North Carolina, USA

Potential health risks to the human respiratory tract from inhalation exposure to chemicals must be determined, even when no human data are available. When respiratory tract responses in laboratory animals are site specific, interspecies differences in anatomy make accurate extrapolation of animal data to people difficult. A way of incorporating regional dose into the extrapolation process is needed so that potential response sites within the human respiratory tract can be identified. In this article, a strategy for understanding regional dosimetry in nasal toxicity is presented in which a mathematical dosimetry model is developed for the rat and extended to the primate and the human. Central to this strategy, the rat dosimetry model is described for the anterior nasal passages of the F344 rat (Kimbell et al., 1993a). Three-dimensional nasal airflow was simulated using a computer reconstruction of the nasal passages from serial step-sections. Regional formaldehyde gas (HCHO) concentration and uptake were calculated using simulated inspiratory airflow velocities, air-phase diffusion, and uniformly-absorbing airway walls. Evidence was given supporting the hypothesis that airflow patterns are major determinants of regional HCHO dose. This example illustrates the dosimetry model structure underlying the proposed rat–primate–human strategy for understanding the role of regional dosimetry in nasal toxicity. Specific research needs for more accurate uptake simulation are also identified. The dosimetry modeling approach described here will be combined with other quantitative models incorporating tissue and airway lining factors to predict regional dose within the human respiratory tract for assessing risks to human health from inhalation exposures.

Dosimetry, the measurement of dose, plays an important role in determining both benefits and risks to human health from exposure to pharmaceuticals and other chemicals. In some instances, direct measurement of the dose that a person receives can be made. For many substances, however, dose information for people is not available, yet potential health benefits and risks must still be determined. In these cases, models, both physical and mathematical, are used to estimate or predict the information required.

To assess risk to the respiratory tract from inhalation exposure, a frequently

The author thanks Kevin Morgan, Matt Godo, Betsy Gross, Paul Schlosser, Henry Heck, Fred Miller, Rory Conolly, Jack Harkema, Jon Hotchkiss, John Morris, Clay Frederick, Tom Monticello, Gary Hatch, and Jim Ultman for research discussions, advice, and encouragement, Donald Joyner and Regina Richardson for technical assistance, and the staff at Fluid Dynamics International and the North Carolina Supercomputing Program for their expertise.

Address correspondence to Dr. J. S. Kimbell, Chemical Industry Institute of Toxicology, 6 Davis Drive, Research Triangle Park, NC 27709, USA.

used physical model is the exposure of laboratory animals to a substance, followed by the use of a mathematical model to extrapolate the animal data and exposure scenario to people. Mathematical models used for animal-to-human extrapolation often consist of scaling particular exposure levels by body weight or surface area, and uncertainty or safety factors, to allow for interspecies differences in response (Dedrick, 1973; Jarabek et al., 1990). Increased interest in understanding underlying mechanisms of inhalation toxicity has led to consideration of animal-to-human extrapolation models in the context of an exposure-dose-response paradigm (McClellan, 1989). When toxic responses in animals are site-specific within the respiratory tract, measured or predicted regional dose can be correlated with lesion location to evaluate the relative role of dose in the occurrence of the responses. Given a way to predict regional dose in humans, regional dose and knowledge of mechanism can then be used to extrapolate animal data to people more accurately for human health risk assessment.

Inhalation exposure to some chemicals, such as formaldehyde gas (HCHO) and ozone, causes site-specific responses in the nasal passages of laboratory animals, such as rats and primates (Morgan et al., 1986; Monticello et al., 1989; Harkema et al., 1989). A mathematical dosimetry model is needed that can test hypotheses about these responses in animals and can be used to predict potential response regions in humans. Ideally, the model should answer the questions: Where does an inhaled chemical go? How much of an inhaled chemical is deposited, in what time frame, and under what conditions? Is such a dose mitigated by local defenses in these sites? If so, by how much?

Questions such as these can be answered by developing a three-dimensional (3-D) representation of the nasal passages and simulating airflow and uptake processes affecting regional dose. A strategy for developing this model is as follows. The model is first constructed for the rat, a species commonly used in inhalation studies, using experimentally derived rat response data to develop the model structure. The model is extended to the primate, as a species anatomically and physiologically closer to humans. Primate dosimetry model predictions are compared to experimentally derived primate response data to determine the extent to which mechanisms affecting regional dose in the rat also affect dose in the primate. Regional doses associated with toxic response in the primate are then noted, and the model is extended to the human. Locations of dose hot-spots are predicted in the human respiratory tract, and predicted hot-spot dose levels are compared to dose levels designated as toxic in the primate. If experimental data are lacking in the primate, then predicted hot-spot dose levels in the human respiratory tract are compared to dose levels observed as toxic in the rat.

An illustration of the development of this type of model for the rat is given in this article, using inhalation exposure to HCHO as an example. In the case of HCHO, extensive information is available for the F344 rat and rhesus monkey, including dose-response and cell proliferation data, lesion location and severity, and regional dosimetric measurements (Swenberg et al., 1980; Albert et al., 1982; Kerns et al., 1983; Casanova et al., 1989,

1991; Heck et al., 1990; Monticello et al., 1989, 1991). The model described here used simulated 3-D airflow patterns in the F344 rat, HCHO physicochemical characteristics (air-phase diffusion coefficient, high water solubility and reactivity), and rat response data to show that nasal airflow patterns play a major role in determining sites of HCHO-induced toxicity (Kimbell et al., 1993a). The model is being extended to the rhesus monkey to test the hypothesis that nasal airflow patterns in the primate affect locations of HCHO-induced toxicity, using primate HCHO response data. The model will be used to predict locations of HCHO uptake hot-spots in the human nasal passages, and will be generalized to predict human response to other inhaled gases and vapors.

MODEL DEVELOPMENT

Nasal Airflow

Initial information about nasal airflow was gathered from a physical model. Steady-state inspiratory flow patterns were visualized using a water-dye system in hollow casts made from the nasal passages of F344 rats and rhesus monkeys (Morgan et al., 1991). Water flow speeds were scaled to account for viscosity and density differences between air and water. Major inspiratory flow characteristics and flow-stream velocities measured using video analysis were reported (Figure 1). For both rat and primate, the loca-

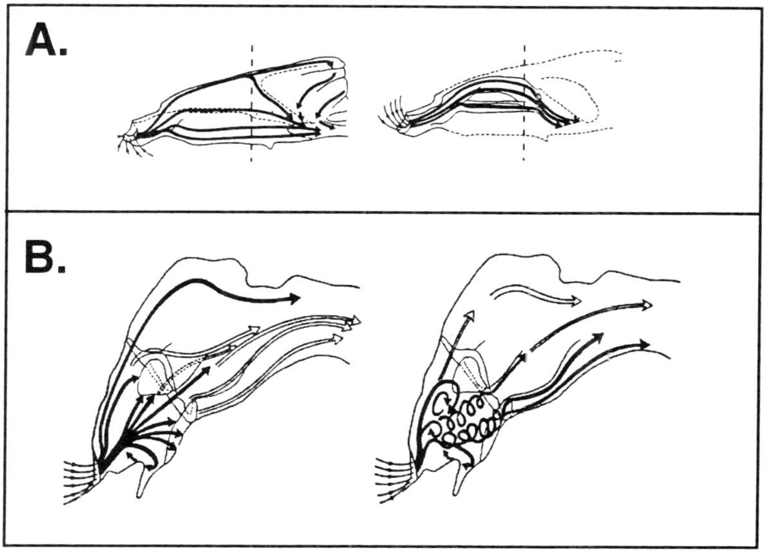

FIGURE 1. Diagram of major inspiratory nasal airflow characteristics in F344 rat and rhesus monkey. In all figures, view is from the side, nostril is to the left. (A) F344 rat; medial streams on left, lateral streams on right. Dotted vertical lines denote extent of nasal passages from nostril used in three-dimensional reconstruction and airflow and uptake simulation. (B) Rhesus monkey; medial streams on left, lateral streams on right. Modified from Morgan et al. (1991).

tions of several flow streams correlated with sites of nasal lesions induced by HCHO exposure, and Morgan et al. (1991) concluded that airflow patterns probably play a major role in determining sites of high HCHO uptake.

The need for quantitative answers to questions about airflow and regional dose led to the development of a mathematical model that could simulate nasal airflow patterns (Kimbell et al., 1993a). In general, airflow patterns can be simulated by solving the Navier–Stokes equations of motion, representing the law of conservation of momentum. In order to solve these equations for nasal airflow, information was needed about the type of flow (laminar or turbulent, compressible or incompressible), the shape or geometry of the nasal passages, how inspired airflow should be specified at the nostril, and appropriate numerical methods. Water-dye experiments showed that over most of the flow rates studied, flow patterns were laminar and consistent with incompressible flow. For simulations, air was assumed to enter through the nostril with the same speed at all points on the nostril surface, and airflow was set to zero at airway walls. The hollow casts showed a high level of complexity in nasal airway geometry. Because of its applicability to complex shapes, the finite-element method, a numerical approach for rendering a region into a finite number of points, was chosen as appropriate for use with the Navier–Stokes equations for nasal airflow simulation.

Since the shape or geometry of the nasal passages is one of the determining factors of nasal airflow patterns, the entire nasal passages of a male F344 rat were embedded in plastic and serially step-sectioned at 50-μm intervals (Kimbell et al., 1993a). Sections were stained and photographed, and selected airway outlines were traced on a digitizing tablet, creating computer files of cross-sectional *x,y*-coordinates. A 3-D reconstruction, or mesh, of the anterior rat nasal passages (Figure 2) was produced from the cross-sectional data using FIDAP (Fluid Dynamics International, Evanston, IL), a finite-element fluid dynamics equation solver and mesh generator, on

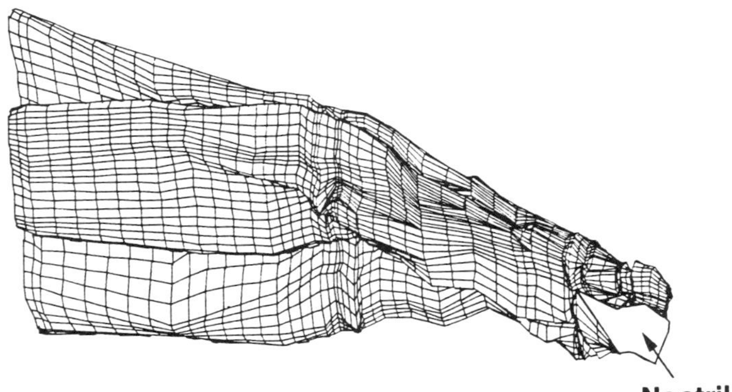

FIGURE 2. Three-dimensional computational mesh of the right side of the anterior F344 rat nasal passages, viewed from lateral side. Modified from Kimbell et al. (1993a).

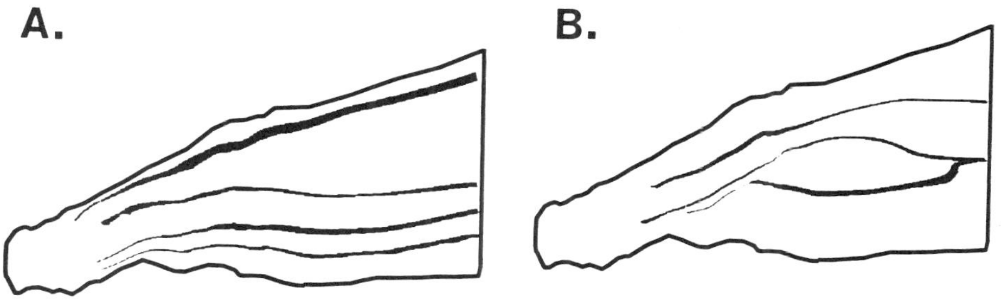

FIGURE 3. Streamlines of simulated inspiratory nasal airflow at 252 ml/min in the anterior nasal passages of the F344 rat. In both figures, nostril is to the left, flow direction is from left to right. Compare with Figure 1A. Simulation performed using FIDAP (FDI, Evanston, IL) on a Cray Y-MP supercomputer (MCNC-North Carolina Supercomputing Program, Research Triangle Park, NC). (A) Medial streams. (B) Lateral streams.

an HP735 workstation (Hewlett-Packard, Palo Alto, CA). Accuracy in reconstructing original nasal airway shape was ascertained by visual comparison with metal molds made of the F344 rat nasal passages (Morgan et al., 1989), as in Godo et al. (1993a).

FIDAP was used to simulate steady-state inspiratory airflow in the reconstructed nasal passages (Figure 3) on a CRAY Y-MP supercomputer (MCNC-North Carolina Supercomputing Program, Research Triangle Park, NC). Simulations were compared to observations from the water-dye visualizations and were found to reproduce experimentally observed flow-stream locations and velocities (Kimbell et al., 1993a; compare Figures 1A and 3).

Regional Uptake

To simulate regional HCHO uptake, mechanisms of transport of inhaled HCHO within the nasal passages, a designated measure of HCHO uptake, and estimates of HCHO concentration at the nostril and nasal passage walls (called boundary conditions) need to be specified. In general, uptake patterns are simulated by solving a mathematical expression for conservation of mass (called a transport equation; Bird et al., 1960). In the model described here, airflow velocities from the simulations described earlier and air-phase diffusion were included in the transport equation as air space transport mechanisms. Chemical reactions occurring within air spaces were neglected. The mass of inhaled HCHO moving across a square millimeter of the airway wall per second (units of regional airway wall mass flux) was used as a measure of HCHO uptake. The imposition of boundary conditions on the HCHO concentration at the nostril simulated the ambient air exposure (in ppm) of the rat to inhaled HCHO. This boundary condition took the form of specifying HCHO concentration at all points in the computational mesh corresponding to the nostril surface. Imposing boundary conditions on HCHO concentration at nasal passage walls simulated the physiological processes of air/airway lining interface interactions with HCHO and disposition of HCHO in the airway lining and tissues. In the simulations described here, this boundary condition took one of two forms: either (1) specific values for HCHO concentration were given, or (2) HCHO wall mass flux across the air/airway lining interface was taken to be proportional to HCHO concentration at the wall.

Uptake of HCHO was simulated using FIDAP on an HP735 workstation (Hewlett-Packard, Palo Alto, CA). All uptake simulations were run with the HCHO concentration at the nostril set to 1 ppm. One set of these simulations was run with HCHO concentration, c, at airway walls set to zero ($c = 0$; first form of boundary conditions described earlier). The $c = 0$ condition was based on the high water solubility and reactivity of HCHO. In other simulations of HCHO uptake, boundary conditions at airway walls were specified using the second form of boundary conditions given earlier, where HCHO wall mass flux was set to be directly proportional to HCHO concentration at airway walls. The constant of proportionality k, called a mass transfer coefficient, was specified to be the same at all airway walls (k = constant; k did not depend on location within the nasal passages), since no information was available to set regional values. The value of k was set so that overall HCHO uptake by the walls of the anterior rat nasal passages model was 93%, as reported by Patterson et al. (1986) for the whole rat nose.

For validation of the mathematical uptake model, a simple system was designed to easily visualize and measure wall mass flux. The comparison of actual wall mass flux to FIDAP simulations of wall mass flux would (1) evaluate the accuracy of the FIDAP software in simulating actual wall mass flux, and (2) determine whether boundary conditions in FIDAP were implemented in an appropriate manner and to a sufficient level of detail to capture essential features identified in wall mass flux visualizations. The physical model used here consisted of a photographic method applied to a laboratory flow tank (Morgan et al., 1992; Dasgupta et al., 1993; Godo et al., 1993b). A simply shaped object (cylinder or upright rectangular block) was placed on the floor of the tank as an obstruction to flow. Dilute photographic developer was recirculated in the tank, and an exposed sheet of photographic film was placed along the flow tank floor in the vicinity of the obstruction. Wall mass flux into the film on the floor of the flow tank in this vicinity was visualized as patterns of development of the film. Morgan et al. (1992) carried out FIDAP simulations of flow and uptake for these geometries. Contour plots of simulated floor mass flux showed overall uptake patterns similar to those seen on photographic films, though numerical comparisons were not made. The conclusion of this study was that FIDAP qualitatively reproduced major features of actual wall uptake patterns. Calibration of the photographic method is needed to make quantitative comparisons.

RESULTS

Regional HCHO concentration values were calculated within nasal air passages, and wall mass flux values were calculated at airway walls. The value of k used in k = constant simulations for 93% overall uptake was 4.6 x 10^{-5} g/mm^2/s. Contour plots of simulated wall mass flux showed similar patterns of high and low uptake along nasal passage walls for both $c = 0$ and k = constant boundary conditions. However, $c = 0$ simulations predicted overall uptake by the anterior nasal passage walls to be 99.9% when simulated volumetric airflow was set to a rate similar to the one used by Patterson et al. (1986). Figure 4 shows simulated HCHO wall mass flux patterns plotted on the airway walls of the nasal passages model, viewed from the lateral side. In this simulation, the volumetric inspiratory airflow rate was 252 ml/min, and $c = 0$ boundary conditions were used. Overall uptake by nasal airway walls in Figure 4 was 98.7%. Locations of hot-spots of uptake predicted by the model, for example the lateral walls of the lateral meatus, as indicated in Figure 4, correlated with reported sites of nasal lesions induced by HCHO inhalation (Kimbell et al., 1993a).

DISCUSSION

In the nasal passages, distributions of site-specific responses are thought to be affected by a combination of factors, including local clearance mechanisms (mucus, blood), cell-specific responses, and regional dose or uptake patterns (Morgan & Monticello, 1990). Regional uptake patterns of an inhaled gas are influenced by nasal airflow patterns, interactions that occur

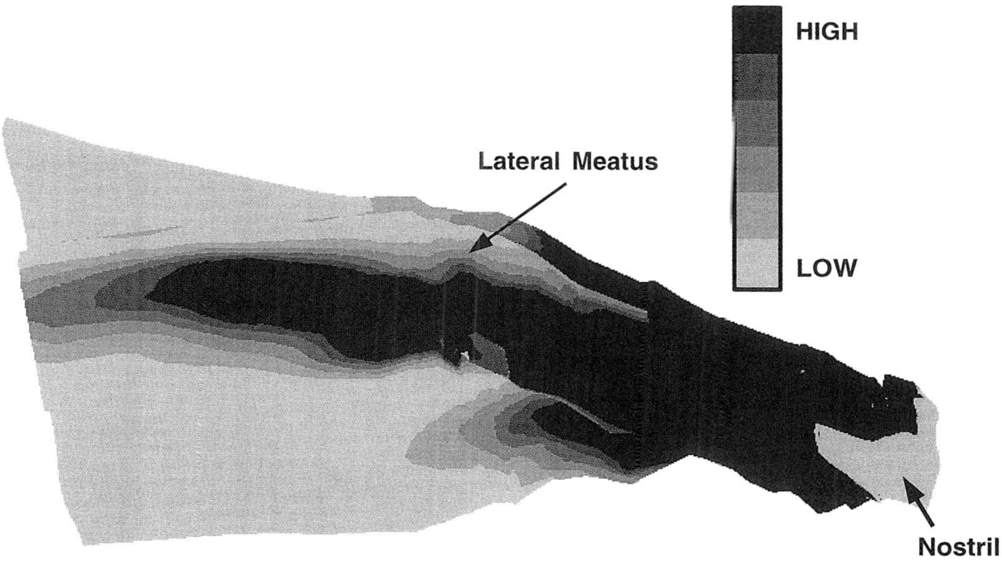

FIGURE 4. Pattern of simulated nasal wall mass flux (uptake) of HCHO in the anterior nasal passages of the F344 rat, viewed from lateral side, using an inspiratory airflow rate of 252 ml/min, and c = 0 boundary conditions.

between the gas and components of the airway lining at the air/airway lining interface, and the disposition of the inhaled substance within airway lining and tissues (Kimbell et al., 1993a). The model described here focused on transport of an inhaled gas through the air phase to the nasal passage walls. The model incorporated actual nasal geometry, air-phase diffusion, 3-D airflow patterns, and evenly absorbing airway walls (represented by either of the $c = 0$ or k = constant boundary conditions) to predict hot-spots of uptake in the rat nasal passages. The assumption of evenly absorbing walls represented the hypothesis that airway geometry, regional airflow patterns, and air-phase diffusion played more of a role in determining lesion distribution than regional air/airway lining interface interactions and tissue factors, such as clearance mechanisms, varying mucus thickness, epithelial cell type, and local metabolism. When this hypothesis was tested for HCHO uptake by comparing evenly absorbing wall simulations with HCHO-induced lesion distributions, correlation between predicted hot-spots and lesion locations indicated that this hypothesis is supported for HCHO (Kimbell et al., 1993a). The same hypothesis was also tested for ozone, and in this case, predicted hot-spots of ozone uptake did not correlate as well with ozone-induced lesion distributions (Kimbell et al., 1993b). The conclusion from this study was that while airflow patterns and air-phase diffusion play some role in determining ozone lesion location, airway lining and wall factors are also important determinants.

In general, a better understanding of how to specify appropriate conditions on inhaled gas concentration at airway walls is needed. For many chemicals, these conditions depend not only on concentration in the air phase, but also on concentration in the tissue phase and blood, and on factors such as the presence and thickness of the mucus lining and the type of underlying epithelium. In the simulation model, conditions such as these could be approximated by using regionally varying values of the mass transfer coefficient k, or by requiring that specific reaction equations hold at airway walls. Because of this, accurate airway wall conditions on inhaled gas concentration for the simulation of air phase delivery may need to be derived by coupling the FIDAP air-phase model with other models focused on airway lining and tissue disposition. In this way, the relevant features of the biological system (air spaces, airway wall lining, tissues, and blood) can each be represented in predictions of, and assessed for their effects on, dosimetry from inhalation exposure to many different types of chemicals.

One way to learn what is needed to specify airway wall conditions on inhaled gas concentration for the FIDAP air-phase model is to study flow tank experiments and simulations. Uptake in the respiratory tract is complicated by anatomy (complex geometry) and not-well-understood air/airway lining interface interactions. These factors make it difficult to measure and compare actual uptake with simulations. Use of a photographic method in a flow tank represents a simple system: simple geometry and flow, and less complex fluid/wall (developer/film) interface processes. Conditions on developer concentration at the film surface in simulations of wall mass flux in a flow tank are analogous to conditions on inhaled gas concentration at airway walls in simulations of respiratory tract uptake. The level of detail that must be used to describe fluid/wall interface interactions for accurate simulations can be studied more easily in a flow tank than in many biological systems. Flow tank simulations can be used to assess the sensitivity of wall mass flux to different transport mechanisms. This approach teaches us how much we must know about transport processes and air/airway lining interface interactions in the respiratory tracts of rats, primates, and people to produce accurate predictions of quantitative chemical uptake.

This article has described a mathematical dosimetry model for nasal toxicity in the rat as a first step in a strategy for developing a dosimetry model for the human respiratory tract. The rat model focused on simulation of regional uptake patterns as a major determinant of dose of an inhaled gas to various regions within the nasal passages. Distributions of site-specific responses are also affected by other factors, such as local clearance mechanisms and cell-specific responses. As extension of the model proceeds to primates and humans, more information will be needed to incorporate these processes, as well as air/airway lining interactions and tissue disposition, into model structure. Physical and mathematical methods are being used to study these issues and others in nasal dosimetry. Overviews and descriptions of many of these methods can be found elsewhere in this volume. The work of these researchers and their colleagues links together information needed to address the complex issues involved in developing accurate models of dosimetry in the respiratory tract. These models make more efficient use of resources by providing a means for posing and testing hypotheses, and by guiding experimental design. Dosimetry models for rats and primates can be used in the risk assessment process by calculating regional dose associated with sites in which lesions develop in laboratory animals. The existence and location of sites in the human respiratory tract at which regional dose is predicted to exceed these levels using similar models will help make more accurate predictions of risk to human health from exposure to inhaled chemicals.

REFERENCES

Albert, R. E., Sellakumar, A. R., Laskin, S., Kirschner, M., Nelson, N., and Snyder, C. A. 1982. Nasal cancer in the rat induced by gaseous formaldehyde and hydrogen chloride. *JNCI* 68:597–603.

Bird, R. B., Stewart, W. E., and Lightfoot, E. N. 1960. *Transport phenomena*. New York: Wiley and Sons.

Casanova, M., Deyo, D. F., and Heck, H. d'A. 1989. Covalent binding of inhaled formaldehyde to DNA in the nasal mucosa of Fischer 344 rats: Analysis of formaldehyde and DNA by high-performance liquid chromatography and provisional pharmacokinetic interpretation. *Fundam. Appl. Toxicol.* 12:397–417.

Casanova, M., Morgan, K. T., Steinhagen, W. H., Everitt, J. I., Popp, J. A., and Heck, H. d'A. 1991. Covalent binding of inhaled formaldehyde to DNA in the respiratory tract of rhesus monkeys:

Pharmacokinetics, rat-to-monkey interspecies scaling, and extrapolation to man. *Fundam. Appl. Toxicol.* 17:409–428.

Dasgupta, A., Guenard, P., Ultman, J. S., Kimbell, J. S., and Morgan, K. T. 1993. A photographic method for the visualization of mass uptake patterns in aqueous systems. *Int. J. Heat Mass Transfer.* 36:453–462.

Dedrick, R. L. 1973. Animal scale-up. *J. Pharmacokinet. Biopharm.* 1:435–461.

Godo, M. N., Gross, E. A., Richardson, R. B., Joyner, D. R., Morgan, K. T., and Kimbell, J. S. 1993a. Comparison of Fischer-344 rat nasal airway computational reconstruction with morphometric data. *Inhal. Toxicol.*, in press.

Godo, M. N., Kimbell, J. S., Ultman, J. S., and Morgan, K. T. 1993b. Comparison between theoretical predictions and experimental observations of unsteady state mass transfer associated with 3-dimensional flow around a submerged parallelepiped. Abstracts, Annual Meeting of the A.I.Ch.E., St Louis, MO, November 7–12, 1993.

Harkema, J. R., Hotchkiss, J. A., and Henderson, R. F. 1989. Effects of 0.12 and 0.80 ppm ozone on rat nasal and nasopharyngeal epithelial mucosubstances: Quantitative histochemistry. *Toxicol. Pathol.* 17:525–535.

Heck, H. d'A., Casanova, M., and Starr, T. B. 1990. Formaldehyde toxicity—New understanding. *Crit. Rev. Toxicol.* 20:397–426.

Jarabek, A. M., Ménache, M. G., Overton, J. H., Dourson, M. L., and Miller, F. J. 1990. The U.S. Environmental Protection Agency's inhalation RfD methodology: Risk assessment for air toxics. *Toxicol. Ind. Health* 6:279–301.

Kerns, W. D., Pavkov, K. L., Donofrio, D. J., Gralla, E. J., and Swenberg, J. A. 1983. Carcinogenicity of formaldehyde in rats and mice after long-term inhalation exposure. *Cancer Res.* 43:4382–4392.

Kimbell, J. S., Gross, E. A., Joyner, D. J., Godo, M. N., and Morgan, K. T. 1993a. Application of computational fluid dynamics to regional dosimetry of inhaled chemicals in the upper respiratory tract of the rat. *Toxicol. Appl. Pharmacol.* 121:253–263.

Kimbell, J. S., Morgan, K. T., Hatch, G. E., Hotchkiss, J. A., and Harkema, J. R. 1993b. Regional nasal uptake and toxicity of ozone: Computational mass flux predictions, ^{18}O-ozone analysis, and cell replication. *Toxicologist* 13:259.

McClellan, R. O. 1989. Reflections on the symposium: Susceptibility to inhaled pollutants. In *Susceptibility to Inhaled Pollutants, ASTM STP 1024,* eds. M. J. Utell and R. Frank, pp. 224–231. Philadelphia: American Society for Testing and Materials.

Monticello, T. M., Morgan, K. T., Everitt, J. I., and Popp, J. A. 1989. Effects of formaldehyde gas on the respiratory tract of rhesus monkeys: Pathology and cell proliferation. *Am. J. Pathol.* 134:515–527.

Monticello, T. M., Miller, F. J., and Morgan, K. T. 1991. Regional increases in rat nasal epithelial cell proliferation following acute and subchronic inhalation of formaldehyde. *Toxicol. Appl. Pharmacol.* 111:409–421.

Morgan, K. T., Jiang, X. Z., Starr, T. B., and Kerns, W. D. 1986. More precise localization of nasal tumors associated with chronic exposure of F-344 rats to formaldehyde gas. *Toxicol. Appl. Pharmacol.* 82:264–271.

Morgan, K. T., Monticello, T. M., Patra, A. L., and Fleishman, A. 1989. Preparation of rat nasal airway casts and their application to studies of nasal airflow. In *Extrapolation of dosimetric relationships for inhaled particles and gases,* eds. J. D. Crapo, E. D. Smolko, F. J. Miller, J. A. Graham, and A. W. Hayes, pp. 45–58. San Diego: Academic Press.

Morgan, K. T., and Monticello, T. M. 1990. Airflow, gas deposition, and lesion distribution in the nasal passages. *Environ. Health Perspect.* 85:209–218.

Morgan, K. T., Kimbell, J. S., Monticello, T. M., Patra, A. L., and Fleishman, A. 1991. Studies of inspiratory airflow patterns in the nasal passages of the F-344 rat and rhesus monkey using nasal molds: Relevance to formaldehyde toxicity. *Toxicol. Appl. Pharmacol.* 110:223–240.

Morgan, K. T., Asgharian, B., Ultman, J. S., Kimbell, J. S., and Dasgupta, A. 1992. Comparison of wall mass flux simulations with flow tank results and application to inhalation toxicology. *Toxicologist* 12:353.

Patterson, D. L., Gross, E. A., Bogdanffy, M. S., and Morgan, K. T. 1986. Retention of formaldehyde gas by the nasal passages of F-344 rats. *Toxicologist* 6:55.

Swenberg, J. A., Kerns, W. D., Mitchell, R. E., Gralla, E. J., and Pavkov, K. L. 1980. Induction of squamous cell carcinomas of the rat nasal cavity by inhalation exposure to formaldehyde vapor. *Cancer Res.* 30:3398–3402.

NASAL DOSIMETRY MODELING FOR HUMANS

Peter W. Scherer, Keyvan Keyhani

Department of Bioengineering, University of Pennsylvania, Philadelphia, Pennsylvania, USA, and Monell Chemical Senses Center, Philadelphia, Pennsylvania, USA

Maxwell M. Mozell

Monell Chemical Senses Center, Philadelphia, Pennsylvania, USA, and Department of Physiology, SUNY Health Science Center at Syracuse, Syracuse, New York, USA

An anatomically accurate, finite element mesh of the right human nasal cavity was constructed from computerized axial tomography (CAT) scans of a healthy adult nose. The equations of motion were solved numerically to determine the steady laminar inspiratory airflow patterns at quiet breathing flow rates. The numerically computed velocity field was compared with the experimentally measured velocity field in a large-scale (20x) model, which was built by scaling up from the same CAT scans. Numerical results showed good agreement with the experimental results throughout the nasal cavity. The numerical velocity field was then used in the solution of the steady convective diffusion equation using the finite element mesh to determine the uptake pattern of inhaled pollutants. The mass transfer boundary condition used at the nasal cavity wall included the effects of solubility, diffusivity, and removal of pollutants by first-order chemical reaction in the mucosal lining. The results showed that about 80% of highly soluble or highly reactive pollutants are absorbed up to the posterior end of the nasal turbinates. For these pollutants, most uptake occured in the anterior and lower half of the nasal cavity. For more insoluble pollutants, uptake was more uniform along the nasal cavity. Finite element and physical models are powerful new tools that, when combined with information on human nasal mucosal histology, blood flow, and biochemistry of the pollutant reactions in the mucosa and throughout the body, can provide valuable information on nasal dosimetry of inhaled pollutants.

Several theoretical models of pollutant uptake in animal nasal cavities (Aharonson et al.,1974; Corn et al., 1976; Morris & Cavanagh, 1986; Stott et al.,1986; Gerde & Dahl, 1991) have been published. In these models, nasal cavity air space geometry is approximated as a straight tube or a slit with tissue and capillary blood flow on the walls. These approximations limit the accuracy of the models in describing local pollutant uptake and the effects on uptake of alterations in nasal cavity geometry. To overcome these limitations and to investigate the human case, we have constructed a three-dimensional finite element model of airflow and pollutant uptake in a

This work was supported by NIH grants CD-00220 and CD-00072.

Address correspondence to Peter W. Scherer, MD, PhD, Department of Bioengineering, University of Pennsylvania, 220 South 33rd Street, Philadelphia, PA 19104-6392, USA.

right human nasal cavity using the software package FIDAP (FDI, Evanston, IL). We have also constructed a large-scale (20× enlarged) model of the nasal cavity, performed velocity profile measurements of airflow in it, and used these measurements to validate the numerically computed velocity field.

METHODS

Experimental Modeling

A large-scale (20×) anatomically correct model of the right human nasal cavity was constructed as follows (Hahn et al., 1993a).

1. Coronal CAT scans of a healthy human nasal cavity were taken 2 mm apart from the tip of the nares to the posterior end of the turbinates (Figure 1).
2. The outlines of the CAT scan cross sections were enlarged 20× and cut out from 4-cm-thick Styrofoam slabs, which were glued together serially.
3. The surface of the airway was smoothed with plaster, and holes were drilled through the sides to insert a hot film anemometer probe (TSI model 1269A). When not in use, the holes were blocked by airtight stoppers.
4. Airflow through the model was produced by a fan placed at the nasopharyngeal end.

The holes in the model are located on six planes (Figure 2). Plane 1 corresponds to the external nares, plane 2 to the posterior section of the nasal valve, plane 3 to the anterior part of the turbinate region, plane 4 to the

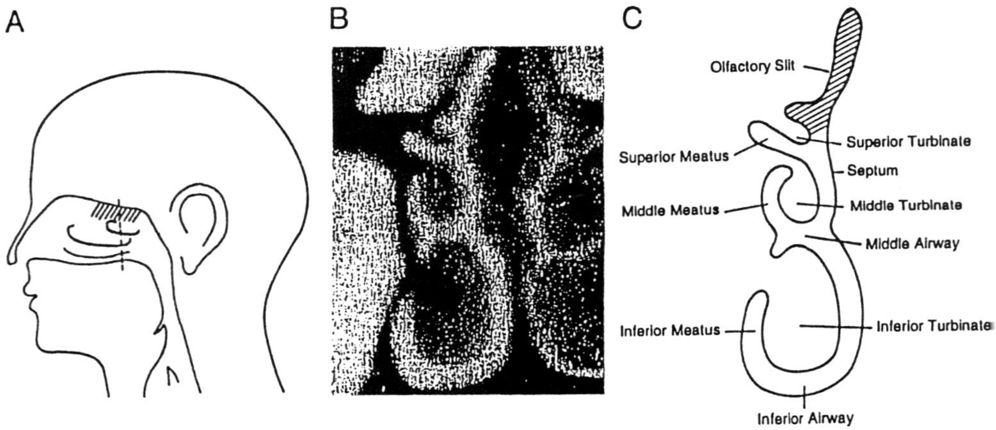

FIGURE 1. (A) Lateral wall of the nasal cavity. The hatched area indicates the olfactory region. (B) A coronal (lateral-to-medial) CAT scan cross section, indicated by dashed line on (A). (C) Various regions of the airway.

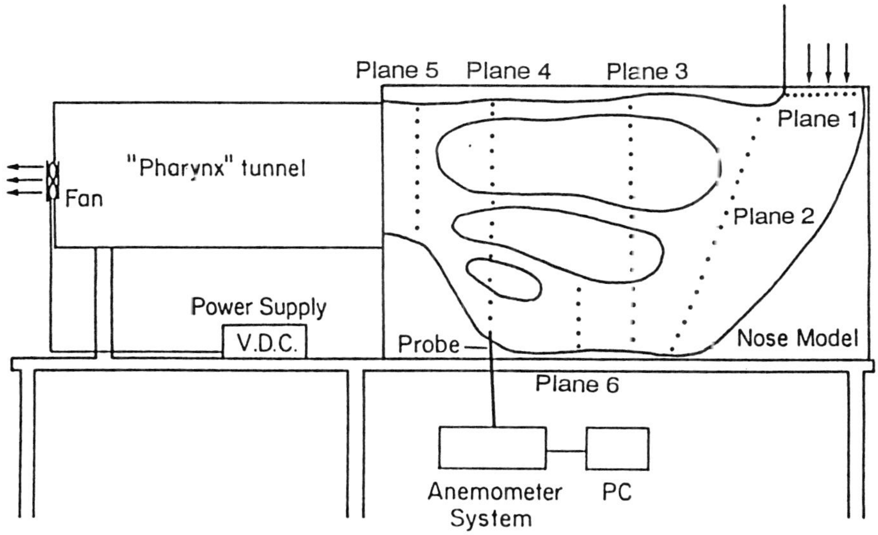

FIGURE 2. Schematic of the experimental apparatus. Velocity measurement cross sections are labeled by the dotted lines (planes 1–6).

superior turbinate region, plane 5 to the posterior end of the turbinate region, and plane 6 to the middle of the olfactory region.

In order for the simulated and real flows to be kinematically similar, airflow through the model was produced such that the Reynolds number, $Re = Ud_h/\nu$, was matched between the model and the real nose, where U is the average speed over a cross section, d_h is the hydraulic diameter of the cross section (4 × area/perimeter), and ν is the kinematic viscosity of air. The importance of unsteadiness of airflow can be determined by the magnitude of the Strouhal number, $S = \omega d_h/U$, where ω is the frequency of breathing. For typical values of d_h (0.3 cm), U (100 cm/s), and ω (2 s^{-1}) during quiet breathing in the turbinate region of the nose, the magnitude of S is about 0.01, which suggests that unsteady effects are not significant in modeling airflow. Therefore, for these studies, the airflow was approximated as being steady state.

The spatial resolution of the probe allowed velocity measurements at increments of 1 mm, which is equivalent to 0.05 mm in the real nose. The probe was positioned in the model such that the component of velocity normal to the planes was measured. The lowest simulated flow rate corresponded to a flow of 125 cm^3/s in the real half nasal cavity (Re = 610 at external nares).

Numerical Modeling

Governing Equations Experimentally measured velocity profiles in the large-scale model indicate that, for quiet resting breathing, flow in the nasal cavity is laminar. The governing equations for the conservation of momentum (Navier–Stokes) and mass (the continuity equation and the convective mass transport equation) for steady incompressible laminar flow are:

$$\rho(\mathbf{u} \cdot \nabla \mathbf{u}) = -\nabla p + \mu \nabla^2 \mathbf{u}$$
$$\nabla \cdot \mathbf{u} = 0$$
$$\mathbf{u} \cdot \nabla c = D_a \nabla^2 c \tag{1}$$

where $\mathbf{u}$ is air velocity, p is pressure, μ is the dynamic viscosity of air, ρ is the density of air, c is the concentration of the inhaled pollutant, and D_a is the diffusion coefficient of the pollutant in air.

For low pollutant concentrations typically inhaled through the nose, it can be assumed that the species concentration has no effect on the velocity field. This allows the equations of momentum and continuity to be uncoupled from the convective diffusion equation. Therefore, the Navier–Stokes and continuity equations were first solved to determine the velocity field, which was then used in the solution of the convective diffusion equation.

Boundary Conditions Since the full set of elliptic equations of motion was solved, it was necessary to prescribe appropriate boundary conditions on each segment of the boundary of the computational domain. For airflow simulations, the walls of the nasal cavity were assumed to be rigid. At the interface between air and the surface of the cavity, the no-slip boundary condition was applied (i.e., $\mathbf{u} = 0$ at walls). At the inlet, an experimentally measured velocity field was imposed for the simulation of inspiratory flow. At the outflow boundary, a stress-free boundary condition that arises naturally from the application of the finite element method was applied.

The boundary conditions for mass transfer are a function of solubility, diffusivity, and removal of pollutants in the mucus lining of the walls by chemical reaction. Several simplifying assumptions are made in order to include these effects in the boundary condition.

First, the sorption–diffusion process across the air–mucus interface was assumed to be quasi-steady and in a state of equilibrium such that the pollutant concentrations on each side of the air–mucus interface are related by:

$$c|_{y=0} = \beta\, c_m|_{y=0} \tag{2}$$

where the subscript m refers to quantities in the mucus phase, β is the equilibrium partition coefficient for a given pollutant molecule, y is the coordinate normal to the boundary, and $y = 0$ at the air–mucus interface (Figure 3).

In addition to Eq. (2), the assumption of quasi-steady equilibrium at the air–mucus interface implies that pollutant flux onto the interface in the air side is exactly balanced by the flux away from the interface into the lower mucus layer. This condition of diffusive flux equality at the interface is given by:

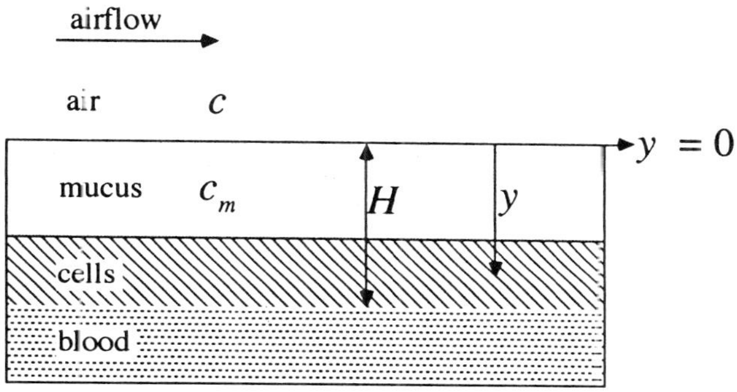

FIGURE 3. Model of mass transfer boundary condition at the air–mucus interface.

$$D_a \frac{\partial c}{\partial y}\bigg|_{y=0} = D_m \frac{\partial c_m}{\partial y}\bigg|_{y=0} \tag{3}$$

The transport of pollutant molecules in the mucosa away from the air–mucus interface is assumed to occur by steady-state diffusion in the y direction with a pseudo-first-order chemical reaction characterized by a single rate constant k_r:

$$D_m \frac{d^2 c_m}{dy^2} = k_r c_m \tag{4}$$

with boundary conditions:

$$c_m|_{y=0} = c_m^0$$
$$c_m|_{y=H} = 0 \tag{5}$$

where we assume the nasal mucosa to be of uniform finite depth H with zero pollutant concentration at the bottom ($y = H$), where the pollutant is assumed to be carried away by blood flow. The solution of Eq. (4) with boundary conditions (5) is:

$$c_m = c_m^0 e^{-\lambda y} + \left(\frac{c_m^0 e^{-\lambda H}}{e^{\lambda H} - e^{-\lambda H}}\right)(e^{-\lambda y} - e^{\lambda y}) \tag{6}$$

where $\lambda = \sqrt{k_r/D_m}$.

Combination of Eqs. (2), (3), and (6) gives the desired mass-transfer mixed boundary condition at the walls ($y = 0$):

$$\frac{\partial c}{\partial y} + \frac{D_m \sqrt{k_r/D_m}}{D_a \beta \tanh(H\sqrt{k_r/D_m})} c = 0 \tag{7}$$

The boundary condition can be made dimensionless by dividing c by the inlet concentration (c_{in}) at the nares and y by the hydraulic diameter (d_h) of the nares to give:

$$\frac{\partial c'}{\partial y'} + Kc' = 0 \tag{8}$$

where $c' = c/c_{in}$, $y' = y/d_h$, and the dimensionless parameter K is given by:

$$K = \frac{d_h D_m \sqrt{k_r/D_m}}{D_a \beta \tanh(H\sqrt{k_r/D_m})} \tag{9}$$

The boundary condition is a linear combination of the pollutant concentration and its normal derivative at the mucus surface (usually referred to as a boundary condition of the third kind or Robin condition). Some limiting cases can arise. First, if the pollutant is highly soluble in mucus ($\beta \rightarrow 0$) or highly reactive ($k_r \rightarrow \infty$), Eq. (8) reduces to $c = 0$ at the mucosal surface. Second, if the pollutant is very insoluble in mucus ($\beta \rightarrow \infty$), Eq. (8) reduces to zero pollutant mass flux ($\partial c/\partial y = 0$) at the surface. Third, if the pollutant is nonreactive ($k_r \rightarrow 0$) the boundary condition reduces to:

$$\frac{\partial c'}{\partial y'} + \frac{D_m d_h}{D_a \beta\, H} c' = 0 \tag{10}$$

The anterior portion of the nasal cavity up to the nasal valve is covered with squamous epithelium. The squamous epithelial lining, which histologically resembles skin, was assumed to be impermeable to inhaled pollutants (Kimbell et al., 1993). Therefore, zero mass flux boundary condition was applied to the anterior 2.5 cm walls of the nasal cavity. A uniform and constant concentration (c_{in} = constant) was imposed at the external nares.

Mesh Generation The finite element mesh was constructed using the same CAT scans that were used to construct the physical model. The coronal CAT scan images were digitized using the software package VIDA (Volumetric Image Display and Analysis). The relative orientation of the coronal sections and the geometric coordinates of points that define the curved edges of the sections were determined. The coordinates were input into the mesh generator module of FIDAP to create a three-dimensional mesh. The mesh contained six-sided brick elements with eight nodes, one at each vertex of the brick.

The mesh was made denser near the walls and in regions where relatively large gradients in velocity exist. The mesh was further refined and optimized after initial simulations, according to geometrical complexity and flow conditions. The final domain contained 75,832 elements and 87,873 nodal points. As examples of the mesh, Figure 4 shows the 3-D mesh for the entire right nasal cavity, while Figure 5 shows a slice through the 3-D mesh at the location of plane 4.

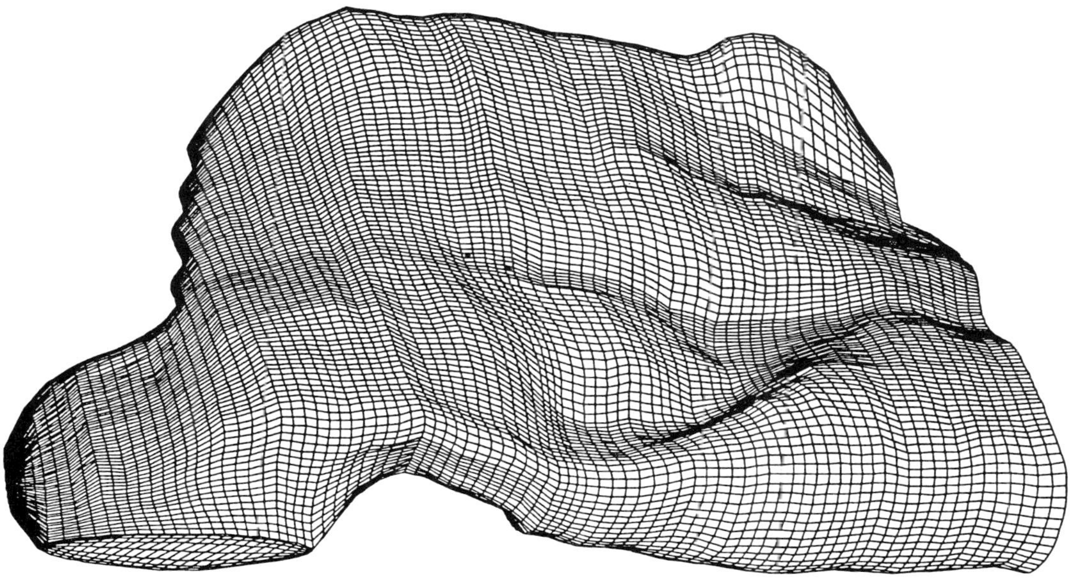

FIGURE 4. Three-dimensional mesh of the right nasal cavity (septum side).

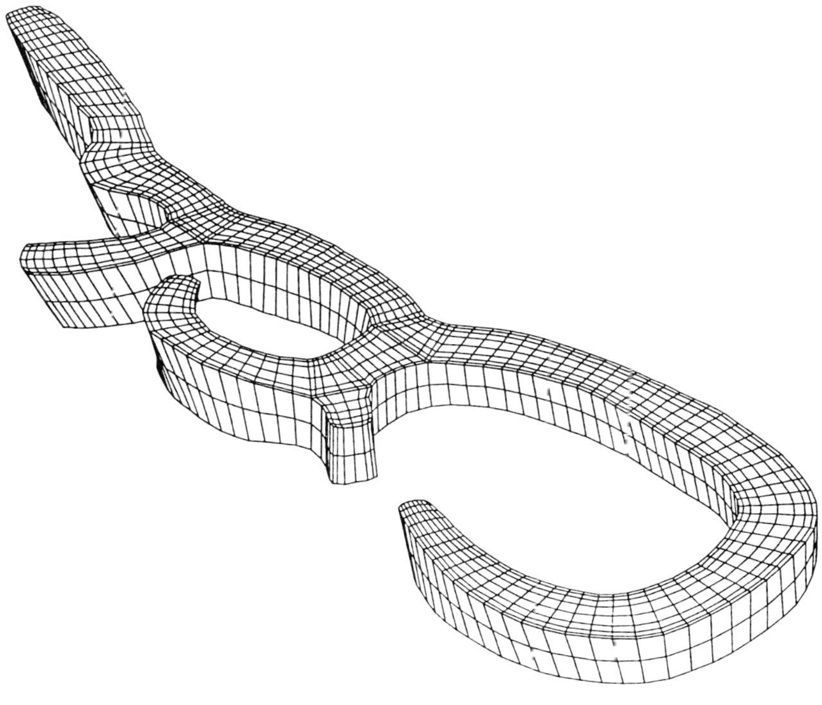

FIGURE 5. A slice through the 3-D mesh at the location of plane 4 (see Figure 2).

Numerical Methods FIDAP utilizes the Galerkin form of the method of weighted residuals (Baker, 1983) to obtain the discrete analog of the transport equations within each element. In the eight-node brick elements, the velocity and concentration components are approximated using trilinear shape functions, and pressure is discretized using a piecewise constant discontinuous approximation.

The discretization of the uncoupled Navier–Stokes and continuity equations results in a nonlinear set of equations. A segregated (pressure projection) algorithm, which creates a set of equations for a single degree of freedom at a time and cycles sequentially through all unknowns at each iteration, was used to solve the nonlinear system (Haroutunian et al., 1993). The linear submatrices at each outer iteration were also solved by iterative methods using conjugate gradient type schemes to further reduce storage requirements. The convergence criterion used for the termination of iterations was:

$$\frac{\|\mathbf{U}_i - \mathbf{U}_{i-1}\|}{\|\mathbf{U}_i\|} \leq \text{tolerance} \tag{11}$$

where **U** is the solution vector of all unknowns at iteration i and i-1, and $\|\cdot\|$ is the root mean square norm. A tighter tolerance was used for the inner iterations (10^{-5}) than the outer iterations (10^{-3}).

Since the problem is highly nonlinear, convergence was not readily obtained at Reynolds numbers of interest using the segregated solver. Therefore, the problem was attacked in stages. As a first estimate of the airflow pattern, a Stokes solution, which neglects the nonlinear convective terms in the momentum equation, was obtained. The Stokes solution was used to specify the values of velocity components for the initial evaluation of the matrix coefficients of the nonlinear problem. The Reynolds number was then successively increased (by increasing the density of the fluid) by increments of about 100, and the previous solution was used as the starting point of the new one, until a converged solution at the minimum Reynolds number used for the experimental simulations was obtained (Re = 610 at external nares). Additionally, it was necessary to use underrelaxation in the segregated method to improve convergence characteristics (Haroutunian et al., 1993).

The velocity field obtained from the solution of the momentum and continuity equations (at Re = 610) was then used in the solution of the convective mass transport equation. The linear system (with one degree of freedom, the pollutant concentration), produced by the discretization of the mass transport equation, was also solved iteratively using conjugate gradient methods. The same convergence criterion as velocity was used for concentration. The mass transport equation was solved for several values of K, with the Schmidt number, Sc = ν/D_a, equal to one.

The computations were performed on a Sun Sparcstation10 computer. Convergence of the velocity field was obtained within 50 iterations at all values of Re. Convergence of the concentration field was obtained within

60 iterations. Conservation of volumetric flow and species mass in the model was checked after the final solutions were obtained. Volumetric flow into the model through the external nares agreed with the flow out of the model through the nasopharynx end to within 0.1%. Species mass entering the nose agreed with the amount absorbed added to the amount that exited to within 1.0%.

RESULTS

Figure 6 shows two examples of comparisons of the experimentally measured velocity profiles in the large-scale model with computations per-

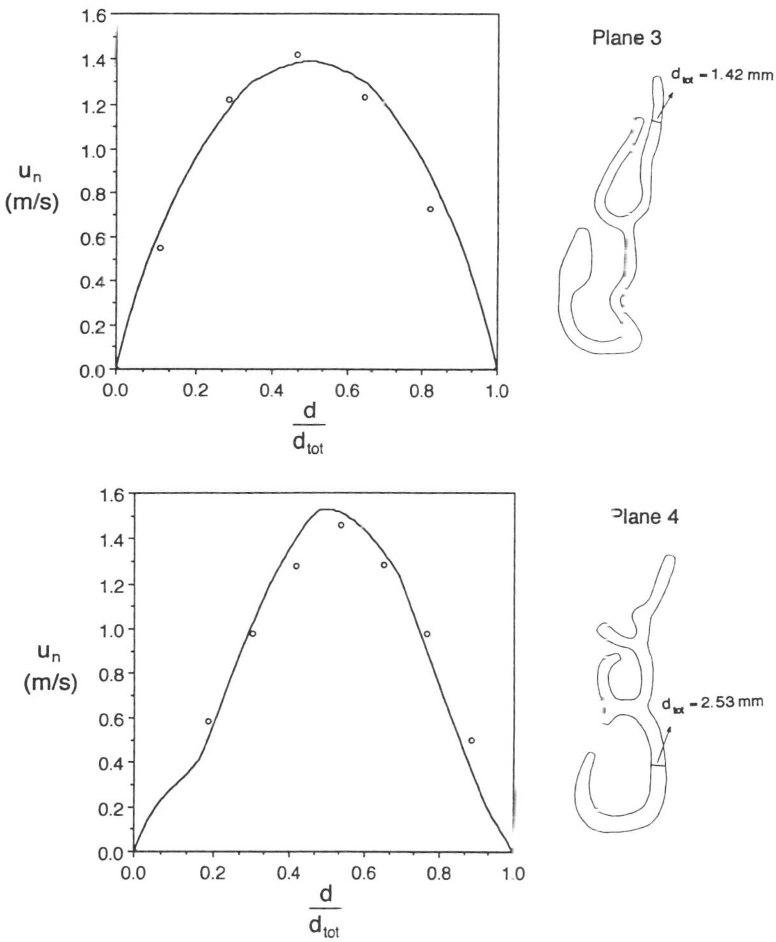

FIGURE 6. Comparison of numerical and experimental results. The solid lines are the numerical predictions, and the open circles are experimental values. The locations of the comparisons are indicated on the planes next to each plot (see Figure 2).

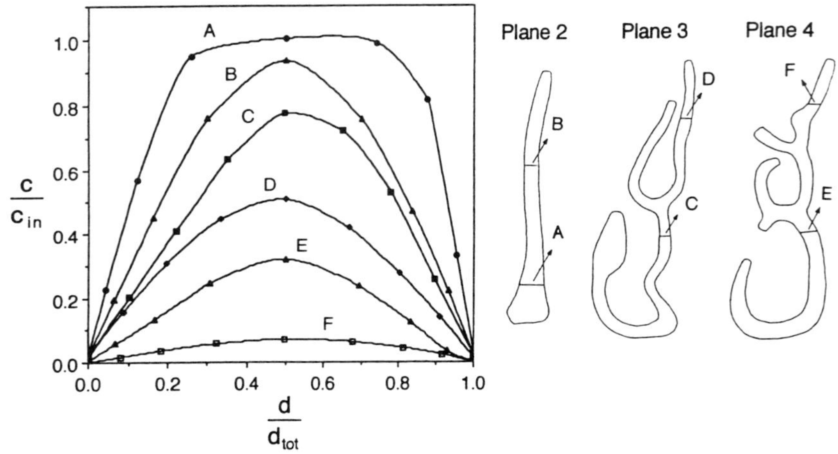

FIGURE 7. Normalized concentration profiles. The locations of the profiles in the airway are indicated on the planes.

formed using the finite element model. The velocity components normal to the planes (u_n) are plotted against cross-sectional distances parallel to the planes (made nondimensional by dividing them by the total length, d_{tot}, between the walls) at the locations shown in the figure. Agreement is quite good for the resting breathing, laminar flow case. With these results in hand, we used the computed velocity field to solve the steady convective diffusion equation and obtain the data shown in Figures 7–10.

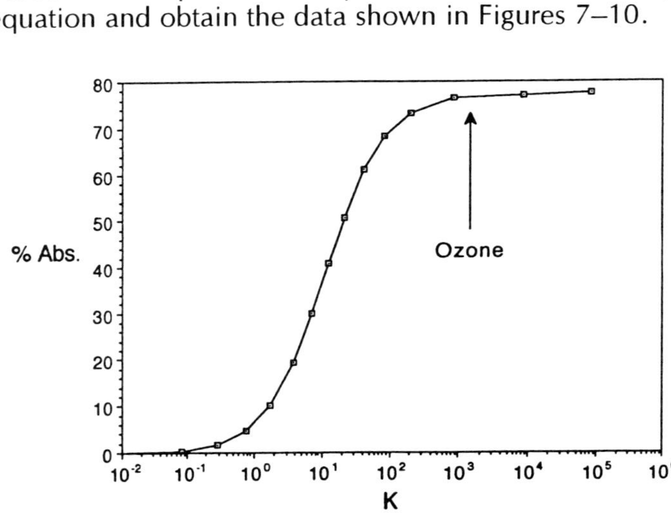

FIGURE 8. Percentage of inhaled pollutant absorption in the entire nasal cavity plotted against the dimensionless *K* parameter for quiet breathing (Re = 610 at the nares) and Sc = 1.0. The total uptake of ozone by the nasal cavity is indicated, using $H = 3.5 \times 10^{-3}$ cm (Hahn et al., 1994), $D_m = 2.66 \times 10^{-5}$ cm^2/s, $\beta = 6.9$, and $k_r = 1198$ s^{-1} (Miller et al., 1985), and assuming a mucus surface reaction with a rate constant equal in magnitude to K_r.

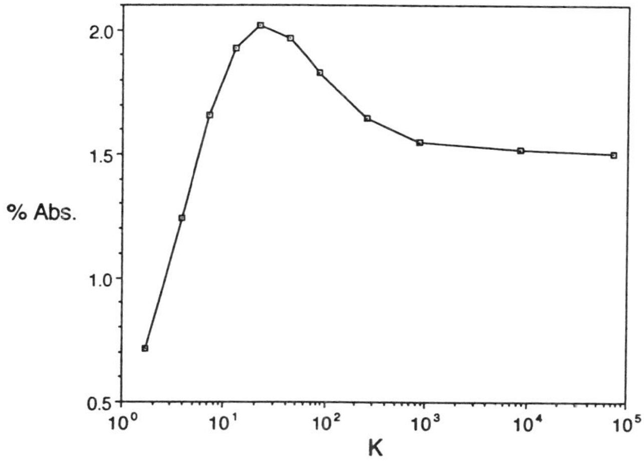

FIGURE 9. Percentage absorption in the olfactory region of the nasal cavity plotted against the dimensionless *K* parameter (Re = 610 at nares and Sc = 1.0).

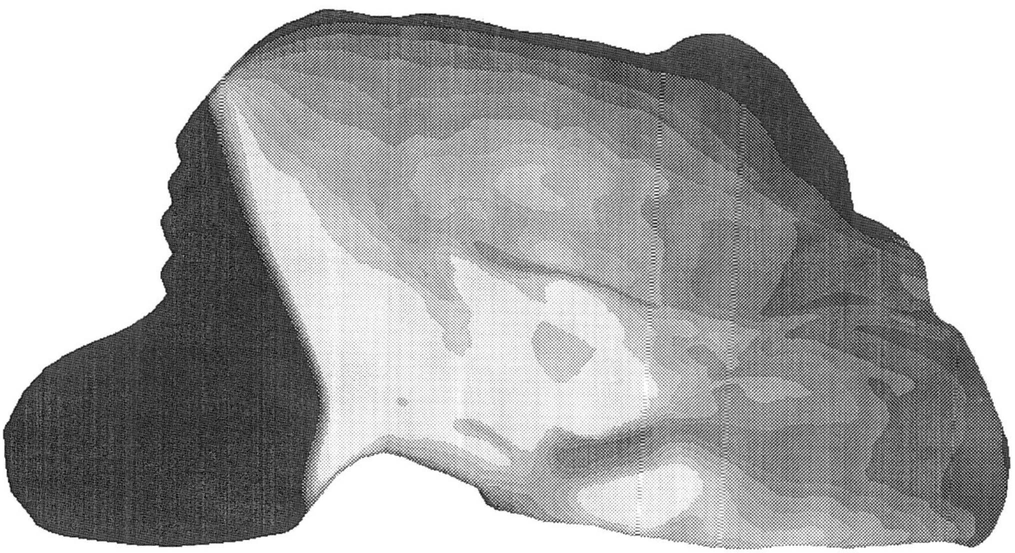

FIGURE 10. Pollutant uptake pattern on the right nasal septum for *K* = 100 (Re = 610 at nares and Sc = 1.0). The normalized wall mass flux increases from black to white (0–0.0315). Nine shades of gray are plotted; each corresponds to an increment of 0.0035 in the normalized flux. To get the actual value of flux, the normalized value must be multiplied by Qx_a ($Q = 0.084$ g/cm^2 · s for quiet breathing, and x_a is the mass fraction of inhaled pollutant).

Figure 7 shows nasal airway concentration profiles for resting breathing of a pollutant, with a value of K equal to 1000, proceeding from anterior to posterior in the nose. As expected, air phase concentration of the pollutant drops in the direction of flow, with the steepest concentration boundary layers occurring in the most anterior part of the nasal cavity.

Figure 8 shows the computed total nasal absorption curve for steady flow resting breathing (Re = 610 at external nares) and a Schmidt number equal to one (Sc = 1.0). For these conditions, the nasal uptake problem is completely characterized by the dimensionless numbers Re, Sc, and K, such that for the boundary conditions assumed at the wall (Figure 3), Figure 8 represents one of a family of universal curves obtained by varying Sc and Re. The total absorption of ozone by the nasal cavity is indicated and is in agreement with recent measurements (Ben-Jebria et al., 1993). This result implies a rapid surface reaction of ozone at the air–mucus interface.

Figure 9 shows an example of regional pollutant absorption in the superior-posterior olfactory region of the human nasal cavity. As K increases, olfactory absorption goes through a maximum due to increasing absorption in the anterior portions of the cavity, leaving less pollutant in the air phase to be absorbed in the olfactory region.

Figure 10 shows the finite element model of regional pollutant uptake on the right nasal septum for a K value of 100. The normalized wall mass flux is plotted at nine increments of equal value. The wall flux was normalized by dividing by the convective flux per unit area at the inlet, Qx_a, where Q is the mass flow rate per unit area and x_a is the mass fraction of the pollutant in air. As can be seen, most uptake occurs in the anterior and lower half of the nasal cavity along the middle and lower turbinates. Uptake patterns for more insoluble pollutants (lower values of K) show a more uniform distribution evenly spread throughout the nasal cavity.

DISCUSSION

The finite element model described in this article is the first to incorporate realistic three-dimensional human nasal cavity geometry. For application to pollutant uptake, however, the model is limited by the accuracy of the nasal wall mucus boundary condition. In particular, much more information is needed concerning human nasal mucosal histology, blood flow, and the solubility and biochemistry of the reactions that specific pollutant molecules undergo in the mucosa. Many reactions may not be first order, making the problem nonlinear, such that overall absorption will not vary linearly with inlet concentration (Hanna et al., 1985).

So far, we have investigated only steady nasal air flow, but it is clear that unsteady effects, including degassing during expiration (Gerde & Dahl 1991), are important. Also, our studies in the large-scale physical model have shown that turbulence is present in the human nasal cavity at all flow rates above resting breathing. This needs to be incorporated into future pollutant uptake modeling. The effects on absorption of anatomic deformities and normal variations, such as mucosal swelling, need to be studied after more accurate boundary conditions are available. The effects on pollutant absorption of heat transfer and water evaporation and condensation from and onto the mucosal surface also need to be studied and tractable ways found to incorporate them into the numerical model.

SUMMARY AND CONCLUSIONS

The physical and finite element numerical models described are valuable new methods of investigating human nasal cavity pollutant uptake that can even be tailored to individual nasal cavities. If these methods can be combined with pharmacokinetic models of overall pollutant distribution and metabolism in the body, a very powerful new tool in pollution research will be created.

REFERENCES

Aharonson, E. F., Menkes, H., Gurtner, G., Swift, D. L., and Proctor, D. F. 1974. Effect of respiratory airflow rate on removal of soluble vapors by the nose. *J. Appl. Physiol.* 37:654–657.

Baker, A. J. 1983. Finite element computational fluid mechanics. Washington, DC: Hemisphere.

Ben-Jebria, A., Kabel, J., and Ultman, J. S. 1993. Regional human lung dosimetry of ozone: the scrubbing efficiency of the nose. *CIIT Conf. on Nasal Toxicity of Inhaled Xenobiotics,* Durham, NC.

Corn, M., Kotsko, N., and Stanton, D. 1976. Mass transfer coefficient for sulphur dioxide and nitrogen dioxide removal in cat upper respiratory tract. *Ann. Occup. Hyg.* 19:1–12.

Gerde, P., and Dahl, A. R. 1991. A model for the uptake of inhaled vapors in the nose of a dog during cyclic breathing. *Toxicol. Appl. Pharmacol.* 109:276–288.

Hahn, I., Scherer, P. W., and Mozell, M. M. 1993a. Velocity profiles measured for airflow through a large scale model of the human nasal cavity. *J. Appl. Physiol.* 75(5): 2273–2287.

Hahn, I., Scherer, P. W., and Mozell, M. M. 1994. A mass transport model of olfaction. *J. Theor. Biol.* 167:115–128

Hanna, L. M., Frank, R., and Scherer, P. W. 1985. Absorption of soluble gases and vapors on the respiratory system. In *Gas mixing and distribution in the lung,* eds. L. A. Engel and M. Paiva, pp. 277–316. New York: Marcel Dekker.

Haroutunian, V., Engelman, M. S., and Hasbani, I. 1993. Segregated finite element algorithms for the numerical solution of large scale incompressible flow problems. *Int. J. Num. Method Fluid.* 17:169–172

Kimbell, J. S., Gross, E. A., Joyner, D. R., Godo, M. N., and Morgan, K. T. 1993. Application of computational fluid dynamics to regional dosimetry of inhaled chemicals in the upper respiratory tract of the rat. *Toxicol. Appl. Pharmacol.* 121:253–263.

Miller, F. J., Overton, J. H., Jr., Jaskot, R. H., and Menzel, D. B. 1985. A model of the regional uptake of gaseous pollutants in the lung. *Toxicol. Appl. Pharmacol.* 109:276–288.

Morris, J. B., and Cavanagh, D. G. 1986. Deposition of ethanol and acetone vapors in the upper respiratory tract of the rat. *Fundam. Appl. Toxicol.* 37:152–157.

Stott, W. T., Ramsey, J. C., and McKenna, M. J. 1986. Absorption of chemical vapors by the upper respiratory tract of rats. In *Toxicology of the nasal passages,* ed. C. S. Barrow, pp. 191–210. New York: Hemisphere.

IN VIVO MEASUREMENTS OF UPTAKE

John B. Morris

Toxicology Program, University of Connecticut, Storrs, Connecticut, USA

To elucidate the key physiologic/biochemical factors that influence the dose of inspired vapors delivered to the nasal cavity during inhalation exposure, upper respiratory tract uptake of a variety of vapors has been investigated in several small rodents, including the mouse, rat, hamster, and guinea pig. For nonreactive vapors, uptake was found to be dependent on the inspiratory flow rate and solubility. Species differences in nasal nonreactive vapor uptake were observed among rodents, with uptake being least efficient in the hamster and guinea pig. Vapors that are substrates for mixed-function oxidase or nasal carboxylesterase were also studied. Pretreatment with metabolic inhibitors significantly decreased uptake of these vapors, indicating that in situ metabolism can exert a strong influence on vapor uptake. These results highlight the importance of nasal metabolism in influencing dosimetry in that site as well as the need to incorporate metabolic terms in mathematic models of nasal vapor uptake.

Dose response is the most fundamental principle of toxicology. The dose of inspired vapors delivered to target sites in the nose during inhalation exposure is dependent on the regional deposition within that site. Thus, knowledge of regional deposition patterns is required for an improved understanding of toxic responses in the nose.

Uptake of vapors in respiratory tract tissues has been described in theoretical terms by many authors and has been reviewed recently by Dahl (1990). Transfer of vapor from the air to tissue phase, termed deposition or uptake, is due to net diffusion of the vapor molecules down a concentration gradient. At the initiation of exposure, the tissue vapor concentration is zero and the uptake rate is maximal. As the exposure progresses, tissue concentration increases, thereby increasing "backpressure" (Aharonson et al., 1974) in the tissue phase, a process that tends to oppose further uptake. Within seconds to minutes after the onset of exposure, nasal uptake attains an apparent steady state that is maintained for as long as 1 h (Aharonson et al., 1974; Stott et al., 1986). The apparent steady state is lost when vapor recirculates in the bloodstream. This occurs when whole-body vapor burdens increase to such an extent that arterial blood entering the nose carries significant amounts of vapor.

Under constant-velocity unidirectional flow conditions, such as those often used to measure upper respiratory tract (URT) vapor uptake experimentally, vapor can be removed from nasal tissue by a limited number of

This research was supported in part by a grant from NIH (R01 ES03676), the Center for Indoor Air Research, E. I. Dupont de Nemours and Company, Rohm and Haas Company, and the Methacrylate Producers Association.

pathways: via the circulation, via metabolism, and/or via direct reaction with water or tissue substrates. Vapor removal via the circulation depends on the nasal blood flow rate, air–blood barrier thickness, and vapor solubility, as measured by the partition coefficient. The higher the partition coefficient, the greater the capacity for blood to remove vapor. Reactive vapors include those that react either directly or enzymatically. Local reaction reduces tissue partial pressures and therefore serves to enhance vapor uptake by increasing the air to tissue concentration gradient. Uptake relationships will differ for nonreactive versus reactive vapors. Nonreactive vapors are removed only via the circulation, whereas reactive vapors are removed by other means as well.

Uptake of vapors during normal breathing is more complex than just described. During inspiration, vapor dissolves in tissues lining the airways, and tissue concentrations increase. During expiration, tissue vapor concentrations decrease due to the loss of vapor into the exhaled air (desorption) (Cander & Forster, 1959; Dahl et al., 1991), removal via the bloodstream, and/or reactivity. Any of these processes will tend to enhance uptake during the subsequent inspiration, as the amount of vapor lost during each expiration is replaced in each subsequent breath. If tissue reactivity is sufficiently rapid, however, tissue concentrations are always low, even during inspiration, and vapor loss during expiration would be expected to have minimal influence on uptake during the inspiratory phase.

The aim of our studies is to provide a systematic investigation of nasal vapor uptake in small rodent species that are typically utilized in inhalation toxicity studies. Measurement of uptake in small rodents requires the use of anesthesia and insertion of an endotracheal tube via surgical manipulation. Thus, it is not possible to conclude with certainty that the measured uptake efficiency is identical to that in animals breathing normally. In recognition of this limitation, our studies have focused on identifying critical physiologic/biochemical mechanisms in the uptake process rather than attempting to measure the "true" uptake efficiency. The underlying hypothesis is that trends seen under defined flow conditions will be reflective of trends occurring in normal animals. Insights provided by these studies will help to identify the key factors influencing the dose of inspired vapor that is delivered to nasal target sites during inhalation exposure. Thus, it is anticipated that these studies will enhance our ability to interpret and extrapolate animal nasal toxicity data.

METHODS

Nasal uptake was measured in the F344 rat, Sprague-Dawley (SD) rat, Syrian hamster, B6C3F1 mouse, and Hartley guinea pig. All animals were obtained from Charles River Laboratories (Wilmington, MA). Complete descriptions of animal lineages, husbandry, and experimental procedures can be found in Morris and Cavanagh (1986), Morris et al. (1986), Morris (1990, 1993) and Morris and Blanchard (1992).

Uptake was measured in the surgically isolated URT of each species. Briefly, animals were anesthetized with urethane, the trachea was split, and an endotracheal tube was inserted in an anterior direction until its tip lay at the larynx. Surgically prepared animals were placed in a supine position in a nose-only inhalation chamber. Exposure chamber air was drawn through the URT under defined flow conditions for periods ranging from 20 to 60 min. Chamber air was heated and humidified to prevent nasal dehydration. To quantitate uptake, vapor concentrations were measured in chamber air and in air exiting the URT. Repetitive air samples were analyzed during the exposure to provide the apparent steady-state uptake values reported herein. Our early studies (Morris & Cavanagh, 1986, 1987; Morris et al., 1986) relied on collection of vapor in midget impingers followed by wet chemical analysis for determination of airborne concentrations; subsequent studies relied on gas chromatographic analysis.

Air was drawn through the URT under constant velocity unidirectional inspiratory, or cyclic flow. The differing species are of different size and have differing minute ventilation rates. To facilitate cross-species extrapolations, the inspiratory flow rates used in the unidirectional flow studies were constant fractions of the minute ventilation rate of each species, as predicted by the equation of Guyton (Phalen, 1984). Estimated minute ventilation rates were 24 ml/min for the mouse (25 g body weight), 74 ml/min for the hamster (110 g), 107 ml/min for the F344 rat (180 g), 147 ml/min for the SD rat (275 g), and 206 ml/min for the guinea pig (430 g).

For cyclic flow studies (F344 rat only), the endotracheal tube was connected to a rodent respirator pump via a T-tube. Air was bled from the T at a rate of 7 ml/min and into the gas chromatograph for analysis. The pump was operated at a frequency of 100/min and a stroke volume of 1.0 ml. The mean inspiratory flow rate, including the 7-ml/min bleed into the gas chromatograph, was 207 ml/min. Since air mixed in the pump, this methodology provided a measure of the average deposition efficiency throughout the inspiratory cycle.

Vapors were selected for the current study on the basis of their physical and biochemical properties. Acetone and ethanol vapors were selected as prototypic nonreactive vapors, as they are not directly reactive and because in vitro studies on nasal homogenates failed to reveal biologically significant metabolism rates. Uptake of metabolized vapors in metabolically inhibited animals has also been used as a surrogate for nonreactive vapors, since limited metabolism would be expected in these conditions (Morris, 1993).

Ester vapor uptake has been studied because esters are substrates for nasal carboxylesterase, and metabolism via this pathway is linked to their olfactory toxicity (Trela & Bodganffy, 1991). The specific esters examined to date include ethyl acetate, ethyl acrylate, and methylmethacrylate. Mixed-function oxidase activity is known to be present in nasal tissues and is implicated in the nasal response to several vapors (Dahl & Hadley, 1991).

Nasal uptake of *o*-xylene, bromobenzene, and ferrocene vapor has been examined to assess potential mixed-function oxidase phenomena. Acetaldehyde uptake has also been studied. In addition to being directly reactive, acetaldehyde is a substrate for nasal aldehyde dehydrogenase (Casanova-Schmitz et al., 1984).

The potential role of nasal metabolism has been investigated by measurement of URT uptake in normal and metabolic inhibitor-pretreated animals. The inhibitors bis-*para*-nitrophenylphosphate (BNPP, 100 mg/kg sc) and metyrapone (150 mg/kg, sc) were used for carboxylesterase and mixed-function oxidase substrate vapors, respectively. To confirm that the inhibitors did not exert nonspecific effects on vapor uptake, their effect on URT acetone deposition was also examined. Neither inhibitor altered acetone uptake (Morris, 1990, 1993).

All data are reported as mean ± SEM unless otherwise stated. Data were analyzed by analysis of variance followed by Newman-Keuls test via statistical software (StatSoft, Tulsa OK). A p level less than .05 was required for statistical significance.

RESULTS AND DISCUSSION

Nonreactive Vapors

Acetone (blood–air partition coefficient of 260) is the most extensively studied nonreactive vapor in our laboratory. In all species examined to date, URT acetone uptake quickly achieved an apparent steady state that was maintained for at least 40 min. This phenomenon has been observed under both unidirectional and cyclic flow conditions. For unidirectional flow, uptake efficiency was highly dependent on the inspiratory flow rate. For example, in the F344 rat, increasing inspiratory flow from 50 to 300 ml/min decreased deposition efficiency from 45 to 12% (Morris et al., 1986). Similar behavior was observed in the hamster (Morris & Cavanagh, 1987), SD rat (Morris & Cavanagh, 1986), mouse (Morris, 1991), guinea pig (Morris et al., 1986), and dog (Aharonson et al., 1974).

As a nonreactive vapor, acetone can only be removed from nasal tissues via the bloodstream during constant-velocity inspiratory flow. Stott and McKenna (1986) estimated, using a microsphere-based methodology, that 1% of the cardiac output enters the nose. This corresponds to approximately 0.4 ml/min in F344 rats (Morris et al., 1986). Dividing the steady-state acetone deposition rate (μg/min) by the nasal blood flow rate (ml/min) provides an estimate of the average nasal blood acetone concentration that is maintained during steady state. Such estimates reveal that, relative to airborne vapor concentrations exiting the nose, nasal blood concentrations equivalent to 30% or more of Henry's law saturation are obtained. Similar values are estimated for other vapors, including *o*-xylene, bromobenzene, and isoamyl alcohol (using data for metabolically inhibited animals; Morris, 1993). These results indicate that physiologically relevant mathematical models for nasal nonreactive vapor uptake must allow for significant backpressure to develop in the nasal vasculature. The nose is specialized for heat and humidification of the airstream. The blood supply is the ultimate source of heat and water vapor; thus it is perhaps not surprising that the air–blood barrier is relatively narrow in this site.

Species differences have been observed in URT acetone uptake. At unidirectional flows approximating the minute ventilation in each species, acetone uptake was most efficient in the F344 rat (27 ± 2%) and mouse (25 ± 2%), and significantly less efficient in the hamster (16 ± 2%) and guinea pig (12 ± 3%) (Morris et al., 1986; Morris & Cavanagh, 1987; Morris, 1991). Uptake in the SD rat URT was of intermediate efficiency (22 ± 3%) and was not statistically different from that in any other species. Similar relationships were observed at other inspiratory flow rates and for ethanol, another nonreactive vapor (Morris et al., 1986). Care should be taken to avoid overinterpreting these data. These results do not directly reflect uptake in normally breathing animals; rather, they suggest that these animal species have different inherent capacities for scrubbing nonreactive vapors from the airstream. Presumably the same rank order would be seen in animals breathing normally.

Acetone uptake in the SD rat URT was less than reported in our previous studies (Morris & Cavanagh, 1986). These early studies used pentobarbital anesthesia, whereas the data reported herein for the SD rat were obtained under urethane anesthesia to facilitate direct comparison to the other species. Perhaps anesthetics that cause respiratory depression, such as pentobarbital, increase nonreactive vapor uptake by increasing nasal perfusion rates.

Data on nonreactive vapor uptake in the human nose are sparse. Landahl and Herrmann (1950) measured human nasal acetone uptake during a 30-s exposure and reported deposition efficiencies of approximately 20% at a unidirectional inspiratory flow rate of 18 L/min. Although precise comparisons are not possible due to the differing methodologies employed, these results suggest that acetone uptake in the human nose may not be very different from that in the rodent. Nasal ethanol uptake was also measured by Landahl and Herrmann (1950) and was similar to that observed in our studies in the rat and guinea pig (Morris et al., 1986), suggesting that species similarity applies to other nonreactive vapors as well.

The mechanism for the rodent species differences is not known. It does not appear to relate to differences in nasal surface area, as morphometric studies reveal the surface area (normalized to the minute ventilation) is fairly similar among these rodents (Gross & Morgan, 1992). Moreover, the surface area to flow ratio in the human nose is about five-fold lower than the rodent, yet nasal acetone and ethanol uptake efficiencies in humans and rodents appear to be similar. Perhaps the species differences relate to species-specific variability among rodents in nasal blood flow rates.

As for alveolar uptake, URT uptake of nonreactive vapors would be expected to depend on the blood–air or tissue–air partition coefficient, with greater uptake being predicted for vapors with greater partition coefficients This result has been observed (Figure 1). A consistent increase in uptake under unidirectional flow was observed with increasing partition coefficient. This effect has been successfully modeled via a physiologically based pharmacokinetic (PBPK) approach (Morris et al., 1993). Uptake in the F344 was consistently higher than in the hamster, thus confirming the species-difference observations made earlier.

Our most recent studies on nonreactive vapor uptake are designed to compare uptake under unidirectional and cyclic flow conditions. Initial studies have focused on acetone and ethanol uptake (Figure 2). A similar dependence of uptake on partition coefficient was seen under both unidirectional and cyclic flow, with deposition for the highly soluble ethanol (blood–air partition coefficient 1800) exceeding that of acetone (blood–air partition coefficient 260) by a factor of 3 under both flow regimes. Uptake of both acetone and ethanol vapors was 1.7-fold higher under cyclic than unidirectional flow ($p < .01$), despite the fact that the mean inspiratory flows were similar under both conditions. The enhanced uptake under cyclic flow suggests that tissue levels of this nonreactive vapor decrease by a physiologically significant amount during expiration and that this process serves to enhance uptake during the inspiratory phase.

It should be recognized that the cyclic flow regime is not truly physiologic. In the spontaneously breathing rat, air passing through the nose dur-

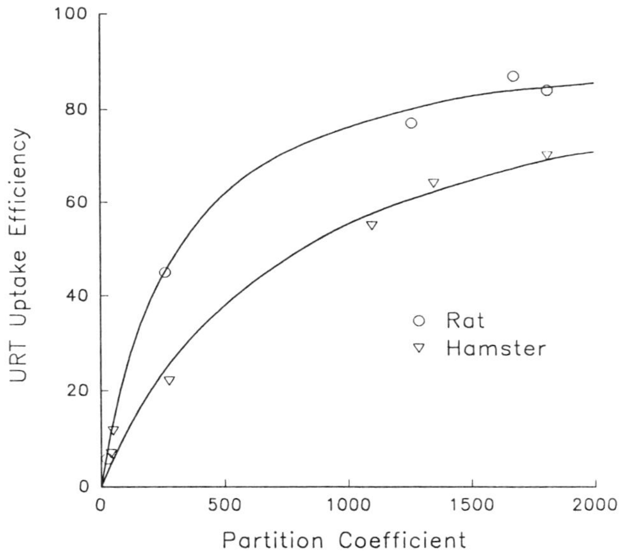

FIGURE 1. URT uptake of nonreactive vapors under constant-velocity unidirectional flow in the rat (50 ml/min) and the hamster (35 ml/min) is plotted against the blood–air partition coefficient. Symbols represent average values (three to five animals per group). The uptake predicted by a PBPK model (Morris et al., 1993) is also shown (solid line).

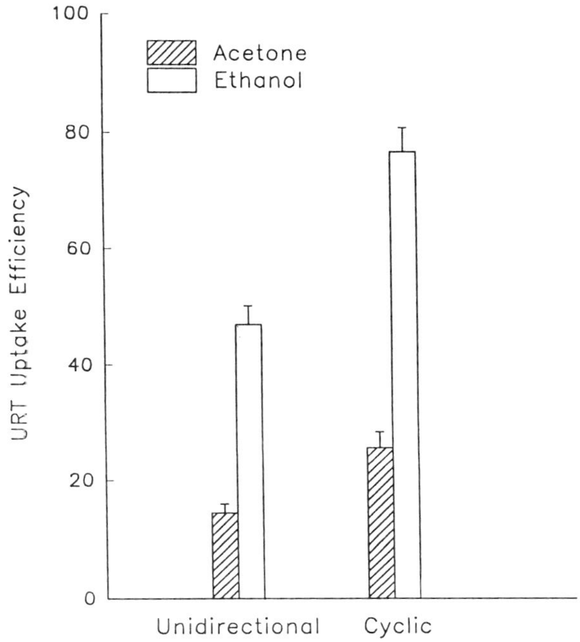

FIGURE 2. Uptake of acetone and ethanol in the URT of the F344 rat under constant-velocity unidirectional (200 ml/min) or cyclic flow (207 ml/min mean inspiratory flow rate). Data are expressed as percent uptake and shown as mean ± SEM (four animals per group). Uptake was significantly higher for ethanol than acetone ($p < .001$) and higher under cyclic than unidirectional flow ($p < .001$).

ing exhalation has been scrubbed of vapor in the lower respiratory tract. In the cyclic flow studies, vapor exiting the URT is directly fed back into the URT from the pump during expiration—no pulmonary absorption occurs. These studies do indicate, however, that uptake of the nonreactive vapor, acetone, is greater under cyclic than unidirectional flow conditions.

Reactive Vapors: Metabolized, But Not Directly Reactive

Nasal tissue is known to contain high levels of xenobiotic-metabolizing enzymes (Dahl & Hadley, 1991). A major effort in our laboratory has focused on examining the potential role of metabolizing enzymes on vapor uptake. A biologically based approach was taken in which uptake was measured in normal animals and in animals pretreated with specific metabolic inhibitors. Every inhibitor used in our studies was shown to be without effect on nonmetabolized vapor uptake, thus minimizing the possibility of nonspecific influences on nasal physiology. It was therefore reasoned that pretreatment with these inhibitors would reduce uptake only if (1) the vapors were metabolized in nasal tissues, and (2) this process served to enhance uptake.

Our initial studies focused on ester vapors (Morris, 1990). The carboxylesterase inhibitor BNPP reduced ethyl acetate uptake in both the F344 rat and hamster (Table 1), with the effect being much greater in the latter species, where the inhibitor decreased uptake on an absolute basis by nearly 30%. This suggests as much as 30% of the inspired burden was metabolized in nasal tissues of this species. Mathematic modeling is needed to estimate the extent of metabolism. The PBPK model developed in our laboratory for nasal vapor uptake and metabolism in the F344 rat (Morris et al., 1993) predicts that at least two-thirds of the ethyl acetate depositing in the nose is metabolized in that site, with only one-third reaching the bloodstream.

It is interesting to note that ethyl acetate uptake in the hamster is much greater than in the F344 rat. This is opposite to the relationship that was observed for nonreactive vapors (Figure 1). This relationship is most likely attributable to species differences in nasal ethyl acetate metabolism rates. While hamster nasal mucosa demonstrates a greater ability to metabolize ethyl acetate in vitro, species differences are nearly abolished by pretreatment with BNPP (Table 1). These results emphasize the need to consider potential species differences in nasal metabolism in the risk assessment process, particularly when a metabolite is implicated in the toxicity. These results also highlight the need for metabolic terms to be incorporated into models of nasal ester vapor uptake.

Ethyl acetate uptake has been measured under unidirectional (200 ml/min) and cyclic (207 ml/min mean inspiratory flow rate) flow in the F344 rat and averaged 0.14 ± 0.03 and 0.14 ± 0.02, respectively. In contrast to the nonreactive vapor acetone, the flow regime is apparently of much less influence on ethyl acetate uptake.

Our studies have also examined the potential influence of metabolism on vapors that are substrates for mixed-function oxidases (Morris, 1993). Pretreatment with the inhibitor metyrapone significantly decreased nasal uptake of *o*-xylene, bromobenzene, and ferrocene, confirming that these vapors were metabolized in situ during the exposure. Uptake of these vapors was 1.5- to 2-fold more efficient in normal than in metabolically inhibited animals, indicating the potential importance of metabolism in influencing URT uptake of mixed-function oxidase substrate vapors.

TABLE 1. Ethyl Acetate Uptake in the URT of Naive and BNPP-Pretreated Hamsters and F344 Rats at Unidirectional Inspiratory Flow Rates Approximating the Minute Ventilation

Species	Flow rate (ml/min)	Percent URT uptake[a]		Enzyme activity[b] V_{max}/K_m (ml/min)
		Normal	BNPP	
F344	100	21 ± 1	16 ± 1	0.7 ± 0.2
Hamster	70	48 ± 2	19 ± 1	1.3 ± 0.2

[a]Data are shown as mean ± SEM; each group contains 8–12 animals (Morris, 1990). Two-factor analysis of variance revealed that BNPP pretreatment significantly reduced URT ethyl acetate uptake in both the F344 rat and the hamster (Morris, 1990).

[b]Ethyl acetate metabolism kinetics were obtained in vitro on homogenates from whole (both olfactory and respiratory mucosa) nasal tissue (Morris, 1990).

TABLE 2. Ethyl Acrylate and Methylmethacrylate Uptake in URT of the F344 Rat Under Unidirectional or Cyclic Flow

	Unidirectional flow		Cyclic flow	
	Naive	BNPP	Naive	BNPP
Ethyl acrylate	22 ± 3	15 ± 3	25 ± 2	16 ± 2
Methylmethacrylate	15 ± 3	10 ± 2	20 ± 2	12 ± 1

Note. URT uptake was measured under unidirectional inspiratory (200 ml/min) or cyclic flow (mean inspiratory flow rate 207 ml/min)and is expressed as percent uptake. See text for details of flow regimes. Ethyl acrylate and methylmethacrylate uptake efficiencies were measured at inspired concentrations of 24 and 23 ppm, respectively. Data are presented as mean ± SEM, with each group containing three to four animals. Multifactorial analysis of variance revealed that BNPP significantly reduced URT uptake of both ester vapors ($p < .002$), and that ethyl acrylate uptake was significantly greater than methylmethacrylate uptake ($p < .01$). Uptake under cyclic and unidirectional flow was not statistically different ($p > .05$).

The mixed-function oxidase studies also revealed the interesting observation that the degree of metabolism in nasal tissues is dependent on the inspiratory flow rate. Low metabolism rates were detected in vivo at low flow (50% of the minute ventilation), compared to high metabolism rates at high flow (300% of the minute ventilation). At the high flows, the estimated in vivo metabolism rate correlated well with that measured in vitro (Morris, 1993).

The mechanism for the flow-rate dependence is not known with certainty. Modeling efforts, however, suggest it is due to regional distribution of enzymatic activity as it relates to the regional air flow patterns within the nose (Morris et al., 1993). Cytochrome P-450 is present in much higher concentrations in olfactory than in respiratory mucosa (Dahl & Hadley, 1991). The recent work of Kimbell et al. (1993) has indicated that the rat nose is not uniformly ventilated. Less than 15% of the total bulk airflow penetrates over the olfactory mucosa, the site of greatest metabolizing capacity. By incorporating regional enzymatic activity and airflow patterns in our PBPK nasal vapor uptake model, we were able to reproduce the inspiratory flow rate dependence on metabolism. At low flows, nasal metabolism is limited by the low delivery of substrate to the olfactory mucosa. At high flows, substrate delivery rates were large enough for the full nasal metabolic potential to be realized (Morris et al., 1993). These results indicate another level of complexity of the nasal cavity. A complete understanding of nasal dosimetry requires consideration not only of the total nasal enzymatic potential, but also of the enzyme distribution between respiratory and olfactory mucosa vis-à-vis regional air flow patterns.

Reactive Vapors: Metabolized and Directly Reactive

Our studies on directly reactive, metabolized vapors included the acrylate esters ethyl acrylate and methylmethacrylate, and the aldehydes acetaldehyde and acrolein. As with ethyl acetate, pretreatment with BNPP significantly reduced URT uptake of acrylate esters by about one-third (Table 2). Similar effects were observed under both unidirectional and cyclic flow. Thus, even directly reactive vapors have the potential to be significantly metabolized in nasal tissue. Deposition under cyclic flow averaged 1.2-fold greater than under unidirectional flow ($p > .05$). Potential uptake differences between unidirectional and cyclic flow appear to be much less marked for the reactive acrylate vapors than for acetone vapor.

Uptake of acetaldehyde vapor has been measured over a span of inspired concentrations to assess potential concentration dependent effects (Morris & Blanchard, 1992). In the F344 rat, uptake of this aldehyde vapor was significantly less efficient at high compared to low inspired concentrations ($p < .001$, Figure 3). Similar concentration-dependent effects have been observed for URT acetaldehyde uptake in the SD rat and hamster (unpublished data). Quantitatively similar effects were seen under both unidirectional and cyclic flow conditions. On the average, uptake was 1.14-fold more efficient under cyclic than unidirectional flow, providing another

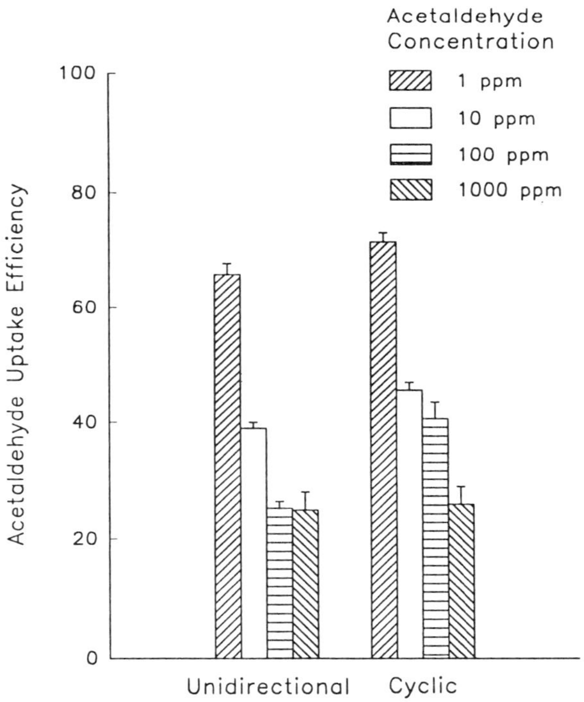

FIGURE 3. URT uptake of acetaldehyde at inspired concentrations of 1, 10, 100, or 1000 ppm is shown for unidirectional (200 ml/min) and cyclic flow (207 ml/min mean inspiratory flow rate). Data were obtained from Morris and Blanchard (1993). Data are expressed as percent uptake and are shown as mean ± SEM (five animals per group). Uptake was significantly influenced by inspired concentration ($p < .001$) and was significantly higher under cyclic than unidirectional flow ($p < .01$)

example of the reduced unidirectional–cyclic flow uptake difference for reactive compared to nonreactive vapors.

The mechanism of the concentration dependence for acetaldehyde uptake is not known. It does not appear to be due to acetaldehyde-induced reduction in nasal perfusion rates, as acetone uptake in the F344 rat URT is not reduced by concurrent exposure to 1000 ppm acetaldehyde (unpublished data). Perhaps acetaldehyde uptake is dependent on reaction with tissue substrates that become depleted at high exposure concentrations. Perhaps deposition is enhanced at low concentrations where effective metabolism of acetaldehyde is possible, but reduced at high concentrations where the metabolic capacity is overwhelmed. In this regard, it is important to note that the absolute deposition rate of acetaldehyde at 1000 ppm (50–100 μg/min) exceeded the V_{max} for nasal acetaldehyde metabolism (roughly 40 μg/min; Casanova-Schmitz et al., 1984). Irrespective of the exact mechanisms, these studies strongly suggest that extreme caution is required in extrapolating effects observed at high exposure concentrations to predict potential outcomes of exposure to much lower concentrations. To be of use for quantitative risk assessment, mathematic models for acetaldehyde uptake must be able to predict this phenomenon.

Humans are rarely exposed to acetaldehyde alone, but rather to a mixture of aldehydes. Often present with acetaldehyde is the potent irritant acrolein. The effect of acrolein on URT acetaldehyde uptake in the F344 rat is shown in Figure 4. Acrolein inhibited acetaldehyde uptake in a concentration-dependent manner. Similar effects were observed under both unidirectional and cyclic flow conditions. Most interestingly, acrolein also influenced the dynamics of acetaldehyde uptake. In animals exposed to acetaldehyde alone, uptake quickly achieved and maintained steady state (Morris & Blanchard, 1992). In contrast, no steady state was achieved in animals simultaneously exposed to acrolein; rather, uptake gradually decreased throughout the exposure. To our knowledge, this type of phenomenon has not been observed previously. The mechanism(s) of this response are not known.

SUMMARY

The results of our studies have elucidated several factors that influence nasal vapor uptake. Nonreactive vapor uptake is influenced by solubility and inspiratory flow rate. Species differences in nonreactive vapor uptake were observed, indicating that comparison of toxic responses among species must be made with knowledge that species-specific differences in nasal dosimetry may exist. In situ metabolism can exert strong effects on the uptake of vapors in the nasal cavity, as evidenced by the effects of metabolic inhibitors on vapor uptake. This phenomenon has been shown for both mixed-function oxidase and carboxylesterase substrate vapors. Carboxylesterase metabolism enhanced uptake of both directly reactive acrylate

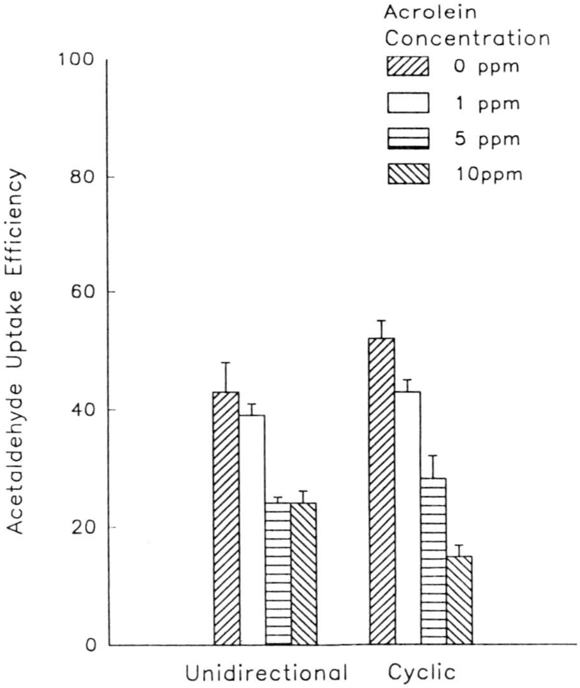

FIGURE 4. The effect of acrolein vapor (1, 5, or 10 ppm inspired concentration) on uptake of acetaldehyde vapor (10 ppm inspired concentration) is shown for unidirectional flow (200 ml/min) and cyclic flow (207 ml/min mean inspiratory flow rate). Data are expressed as percent uptake and are shown as mean ± SEM (four to five animals per group). Acrolein significantly reduced acetaldehyde uptake ($p < .001$). Uptake efficiencies under the two flow regimes were not significantly different ($p > .05$).

esters and the nondirectly reactive ethyl acetate. These observations highlight the need to include consideration of metabolic differences in comparing or extrapolating toxicity among several species following vapor inhalation.

REFERENCES

Aharonson, E. F, Menkes, H., Gurtner, G., Swift, D. L., and Proctor, D. F. 1974. Effect of respiratory airflow rate on removal of soluble vapors by the nose. *J. Appl. Physiol.* 37:654–657.

Cander, L., and Forster, R. E. 1959. Determination of pulmonary parenchymal tissue volume and pulmonary capillary blood flow in man. *J. Appl. Physiol.* 14:541–551.

Casanova-Schmitz, M., David, R. M., and Heck, H. d'A. 1984. Oxidation of formaldehyde and acetaldehyde by NAD^+ dependent dehydrogenases in rat nasal mucosal homogenates. *Biochem. Pharmacol.* 33:1137–1142.

Dahl, A. R. 1990. Dose concepts for inhaled vapors and gases. *Toxicol. Appl. Pharmacol.* 103:185–197.

Dahl, A. R., and Hadley, W. M. 1991. Nasal cavity enzymes involved in xenobiotic metabolism: Effects on the toxicity of inhalants. *Crit. Rev. Toxicol.* 21:345–372.

Dahl, A. R., Snipes, M. B., and Gerde, P. 1991. Sites for uptake of inhaled vapors in beagle dogs. *Toxicol. Appl. Pharmacol.* 109:263–275.

Gross, E. A., and Morgan, K. T. 1992. Architecture of nasal passages and larynx. In *Comparative biology of the normal lung*, ed. R. A. Parent, pp. 7–25. Boca Raton, FL: CRC Press.

Kimbell, J. S., Gross, E. A., Joyner, D. R., Godo, M. N., and Morgan, K. T. 1993. Application of computational fluid dynamics to regional dosimetry of inhaled chemicals in the upper respiratory tract of the rat. *Toxicol. Appl. Pharmacol.* 121:253–263.

Landahl, H. D., and Herrmann, R. G. 1950. Retention of vapors and gases in the human nose and lung. *A.M.A. Arch. Ind. Health* 1:36–45.

Morris, J. B. 1990. First-pass metabolism of inspired ethyl acetate in the upper respiratory tracts of the F344 rat and Syrian hamster. *Toxicol. Appl. Pharmacol.* 102:331–345.

Morris, J. B. 1991. Deposition of acetone vapor in the upper respiratory tract of the B6C3F1 mouse. *Toxicol. Lett.* 56:187–196.

Morris, J. B. 1993. Upper respiratory tract metabolism of inspired alcohol dehydrogenase and mixed function oxidase substrate vapors under defined airflow conditions. *Inhal. Toxicol.* 5:203–221.

Morris, J. B., and Blanchard, K. T. 1992. Upper respiratory tract deposition of inspired acetaldehyde. *Toxicol. Appl. Pharmacol.* 114:140–146.

Morris, J. B., and Cavanagh, D. G. 1986. Deposition of ethanol and acetone vapors in the upper respiratory tract of the rat. *Fundam. Appl. Toxicol.* 6:78–88.

Morris, J. B., and Cavanagh, D. G. 1987. Metabolism and deposition of propanol and acetone vapors in the upper respiratory tract of the hamster. *Fundam. Appl. Toxicol.* 9:34–40.

Morris, J. B., Clay, R. J., and Cavanagh, D. G. 1986. Species differences in upper respiratory tract deposition of acetone and ethanol vapors. *Fundam. Appl. Toxicol.* 7:671–680.

Morris, J. B., Hassett, D. N., and Blanchard, K. T. 1993. A physiologically based pharmacokinetic model for nasal uptake and metabolism of non-reactive vapors. *Toxicol. Appl. Pharmacol.* 123:120–129.

Phalen, R. F. 1984. *Inhalation studies: Foundations and techniques*, pp. 21[illegible]–241. Boca Raton, FL: CRC Press.

Stott, W. T., Ramsey, J. C., and McKenna, M. J. 1986. Absorption of chemical vapors by the upper respiratory tract of rats. In *Toxicology of the nasal passages*, ed. C. S. Barrow, pp. 191–210. Washington, DC: Hemisphere.

Trela, B. A., and Bogdanffy, M. S. 1991. Carboxylesterase-dependent cytotoxicity of dibasic esters (DBE) in rat nasal explants. *Toxicol. Appl. Pharmacol.* 107:285–301.

MODELING THE UPTAKE OF GASES BY THE DOG NASAL-PHARYNGEAL REGION: EFFECTS OF MORPHOMETRIC AND PHYSICOCHEMICAL FACTORS

John H. Overton

Pulmonary Toxicology Branch, Environmental Toxicology Division, Health Effects Research Laboratory, U.S. Environmental Protection Agency, Research Triangle Park, North Carolina, USA

Richard C. Graham

ManTech Environmental Technology, Inc., Research Triangle Park, North Carolina, USA

A dosimetry model is used to investigate the uptake of gases by the nasal-pharyngeal region of the dog. The model accounts for cyclic air flow in the respiratory tract and absorption and desorption by air–blood barriers, diffusion and first-order chemical reactions in the barriers, blood perfusion of and circulation between respiratory tract tissues and tissues of the rest of the body, and loss of inhaled compounds by systemic extraction. The following factors that influence nasal-pharyngeal uptake are studied: length of exposure, blood–air partitioning, respiratory tract tissue first-order reactions, systemic extraction, and lower respiratory tract (LRT) uptake. The simulation results predict the following: (1) The length of time for variables (e.g., regional uptakes) to reach steady state depends on the blood–air partition coefficient and is different for different variables for a given partition coefficient. Even if respiratory tract variables are in steady state, systemic variables may not be. (2) As the first-order tissue rate constant increases, nasal-pharyngeal uptake during both breathing phases increases for a given partition coefficient. During inhalation the nasal-pharyngeal region only absorbs. On exhalation this region may desorb or absorb, depending on the rate constant and partition coefficient; for sufficiently high rate constants, only absorption occurs. (3) Nasal-pharyngeal absorption and desorption processes are not very sensitive to systemic extraction. (4) As the thickness of the nasal-pharyngeal air–blood barrier decreases, nasal-pharyngeal uptake increases during inhalation and desorption decreases during exhalation. (5) Increasing or decreasing LRT uptake increases or decreases, respectively, both nasal-pharyngeal absorption and desorption. These results provide a better understanding of the role of nasal-pharyngeal morphometric and physicochemical properties and the influence of extranasal-pharyngeal factors on the uptake of gases.

Transport within and the uptake of gases by the nasal-pharyngeal region (that part of the respiratory tract from the anterior nares through the larynx) are very complex processes. To describe these processes for modeling purposes requires considerable knowledge and understanding about this region, including the airspace geometry, airflows, and the media responsi-

Address correspondence to John H. Overton, Pulmonary Toxicology Branch, MD-82, Environmental Toxicology Division, Health Effects Research Laboratory, U.S. Environmental Protection Agency, Research Triangle Park, NC 27711, USA.

ble for absorption or desorption, such as fluids and tissues. In addition, the nasal-pharyngeal region is related to the lower respiratory tract (LRT) by airflow and blood perfusion and to the rest of the body by perfusion; both the LRT and the rest of the body can influence nasal-pharyngeal uptake.

The present investigation is concerned mainly with the effects of morphometric and physicochemical factors on the uptake of gases and vapors by the nasal-pharyngeal region. The influence of these factors is explored by the use of a dosimetry model that was developed to better understand the role of the respiratory tract as a portal of entry and as a target organ. The model is applied to the dog because it has been calibrated to the experimental beagle dog data of Dahl et al. (1991) for vapors that are stable in respiratory tract tissue and fluids.

The dosimetry model accounts for the major factors affecting uptake and the delivery of inhaled compounds to systemic blood by the respiratory tract. But some aspects are necessarily simplified in order to obtain a model that is practical. For example, the gas phase (air passages, airway lumens, and airspaces) of the respiratory tract is modeled as a single path in which the airflow rates along surfaces in the same cross section on the path are the same. Thus, gas-phase transport is described by a one-dimensional formulation that cannot account for transverse (perpendicular to main air flow path) variations in dose, and the resulting transverse variations in toxic effects observed in the nasal-pharyngeal region of some laboratory animals (Kimbell et al., 1993). Nevertheless, the model is expected to predict the general features of the uptake and transport of inhaled gases and vapors throughout the respiratory tract.

DESCRIPTION OF MODEL

Figure 1A illustrates the perfusion and ventilation pathways connecting the main regions of the dog used in the dosimetry model. The nasal-pharyngeal, tracheobronchial (TB), and pulmonary regions make up the respiratory tract; all other tissue groups and organs of the body are represented by the systemic compartment. This latter compartment is assumed homogeneous and well mixed, with the capacity to extract inhaled compounds by first-order chemical reactions. Each region of the respiratory tract is represented by a sequence of model divisions that correspond to one or more anatomical divisions. Each model division is composed of three types of transversely arranged compartments (Figure 1B): air or gas-phase network (e.g., air passages), air–blood barrier, and vascular network. Figure 1B illustrates the arrangement of these compartments, as well as indicating the directions of mass transport and the transport processes modeled.

The dog is assumed to breathe only through the nasal-pharyngeal airways. This region is divided into three model divisions, each having approximately the same airway surface area. Volume, surface area, diameter, and length are calculated for each air compartment from the beagle dog data of

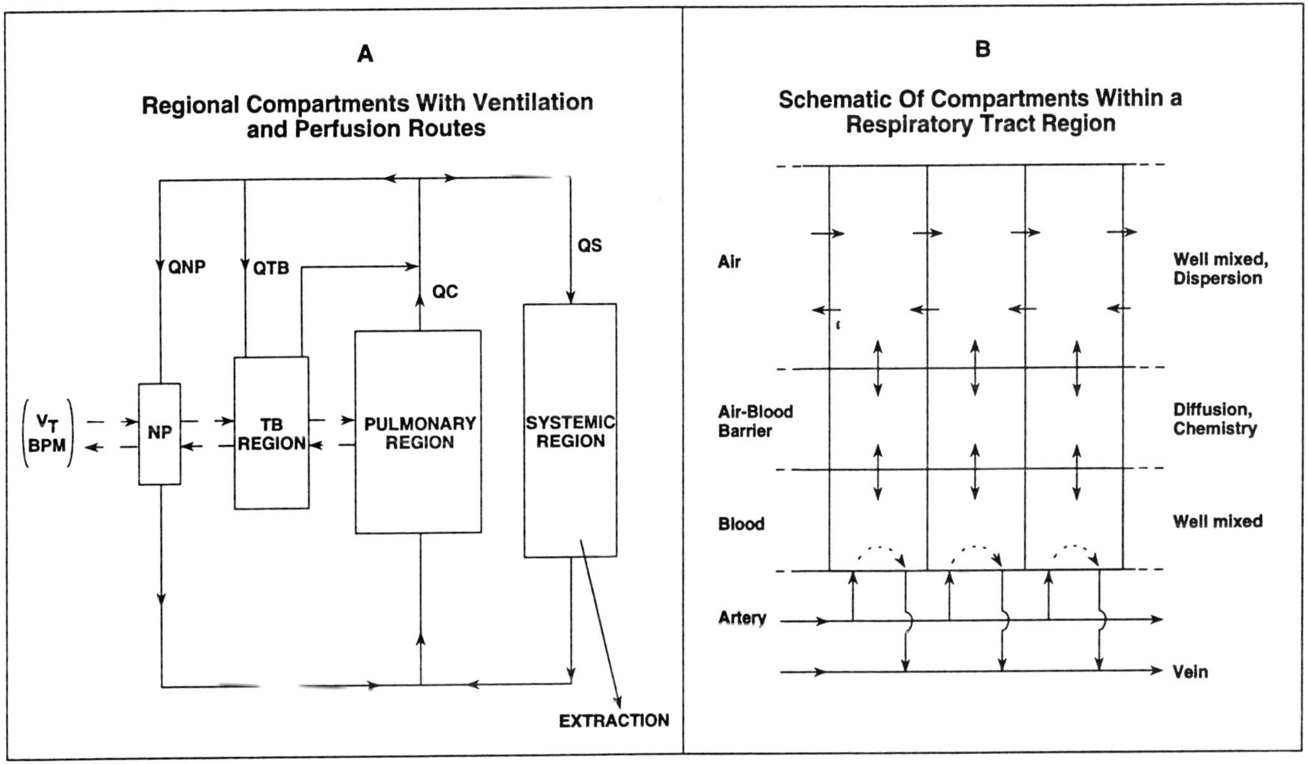

FIGURE 1. (A) Schematic of the relationship between the major model regions and perfusion routes. (B) General configuration of the compartments and their associated mass transfer processes within the respiratory tract regions. QNP, QTB, QC, and QS refer to the perfusion rates of the nasal-pharyngeal (NP), tracheobronchial (TB), pulmonary, and systemic regions or compartments, respectively. The arrows indicate the direction of the flow of blood or the flow of the inhaled compound. V_T and BPM indicate the ventilatory parameters, tidal volume and breathing frequency, respectively.

Schreider and Raabe (1981). The LRT is a modification of the Weibel (1963) model for humans so that the numbers of terminal bronchioles, regional volumes, and pulmonary region surface areas correspond to a 10-kg dog.

The transport model for the air phase of the respiratory tract is based on the compartment-in-series with back-flow model (Wen & Fan, 1975). Each compartment is well mixed; cyclic air flow due to respiration, dispersion, and mass transfer to and from the air–blood barrier are taken into account (Figure 1B). In the air–blood barrier compartments, transport is by time-dependent diffusion with first-order chemical reactions. Transport is assumed perpendicular to the air–liquid lining and the tissue–blood interfaces, which are parallel (Figure 1B). The capillary compartments are well mixed and exchange mass with their associated air–blood barrier compartments. Unlike the gas-phase compartments, there is no transport between adjacent air–blood barrier compartments or between adjacent blood compartments.

Table 1 presents a list of parameters used for the simulations, as well as the source of the parameters, some of which were adjusted to obtain an improved agreement with the experimental data of Dahl et al. (1991). If no source is given, the value depends on other listed parameters. The "assumed" source has no literature or experimental basis; simulation results were not very sensitive to these parameters.

MODEL PREDICTIONS

The following definitions are used:

1. Uptake, absorption, and desorption can refer to processes or to rates. As a rate, uptake is defined as positive for absorption and negative for desorption. Since absorption rate and desorption rate indicate direction, they are both defined as positive.
2. Total uptake or total respiratory tract uptake: Percent of the mass inhaled that is retained per breath by the dog.
3. Regional uptakes: For the specific region, the percent of the mass inhaled that is transferred from the air to the air–blood barrier (absorption) or from the barrier to air (desorption) during the inhalation or the exhalation phase.
4. Total regional uptakes: For the specific region, the percent of the mass inhaled that is transferred from the air to the air–blood barrier or from the barrier to air during a complete breath.
5. Systemic and air–blood barrier uptake: Percent of the inhaled compound that chemically reacts in the respective compartment per breath.
6. Systemic extraction ratio: Fraction of inhaled mass removed from blood by the systemic compartment.
7. Air–blood barrier and respiratory tract "tissue" are used interchangeably.

Figure 2 shows the time course of selected variables for two blood–air partition coefficients, 1 (Figure 2A) and 500 (Figure 2B). The two simula-

TABLE 1. Physiological and Morphometric Parameters for 10-kg Beagle Dog

Parameter	Value	Source
V_T, BPM	300 ml, 12 min^{-1}	Dahl et al. (1991)
Volumes (ml)		
Air (at FRC)		
NP	65.4	Schreider and Raabe (1981)
LRT	590.0	Gerlach et al. (1986)
TB	98.4	Schultz et al. (1992)
PUL	491.6	
Blood		
NP	0.3	Assumed
TB	2.4	Assumed
PUL	20.0	Robinson and Gillespie(1975)
Systemic	9100.0	0.91* (body mass)
Surface area at FRC (cm^2)		
NP	295	Schreider and Raabe (1981)
TB	2396	Present work
PUL	415,516	Siegwart et al. (1971)
Air–blood distance (μm)		
NP	0.5–88.5	Present work
TB	1% of airway diameter	DuBois and Rogers (1968)[a]
PUL	0.4	Siegwart et al. (1971)[a]
Perfusion rates (ml/min)		
NP	15.0	Gerde and Dahl (1991)[a]
TB	46.2	DuBois and Rogers (1968)[a]
PUL	2220.0	Muggenburg and Mauderly (1974)[a]
Systemic	2205.0	
First-order RT tissue rate constants (s^{-1})	0–1000	Present work
Systemic extraction ratio	0–1.0	Present work
Partition coefficients, blood–air = tissue–air	0.1–3160	Present work
Tissue molecular diffusion coefficient (cm^2/s)	5 × 10^{-6}	Present work

Note. V_T and BPM are tidal volume and breathing frequency, respectively. NP, LRT, TB, and PUL indicate, respectively, the nasal-pharyngeal, lower respiratory tract, tracheobronchial, and pulmonary region.
[a]Given value is an adjusted source value giving a better fit to experimental data.

tions illustrate that the length of time for the plotted variables to reach steady state depends on blood–air partition coefficient ("partition coefficient" for short) and is different for different variables for a given partition coefficient. Further, evidence of steady state in the respiratory tract does not imply steady state elsewhere in the dog. In general, the time for a given variable to reach steady state depends not only on partition coefficients but on other parameters as well. Note that for the remaining illustrations, results correspond to steady state in which the averages of all variables, respiratory and systemic, over the time of a breath are independent of time.

Figure 3 illustrates the effect of partition coefficient on regional respiratory tract uptake (Figure 3A) and on nasal-pharyngeal regional (Figure 3B) absorption and desorption processes. Uptakes for inhalation and exhalation are plotted versus partition coefficient. Generally, the respiratory tract regional

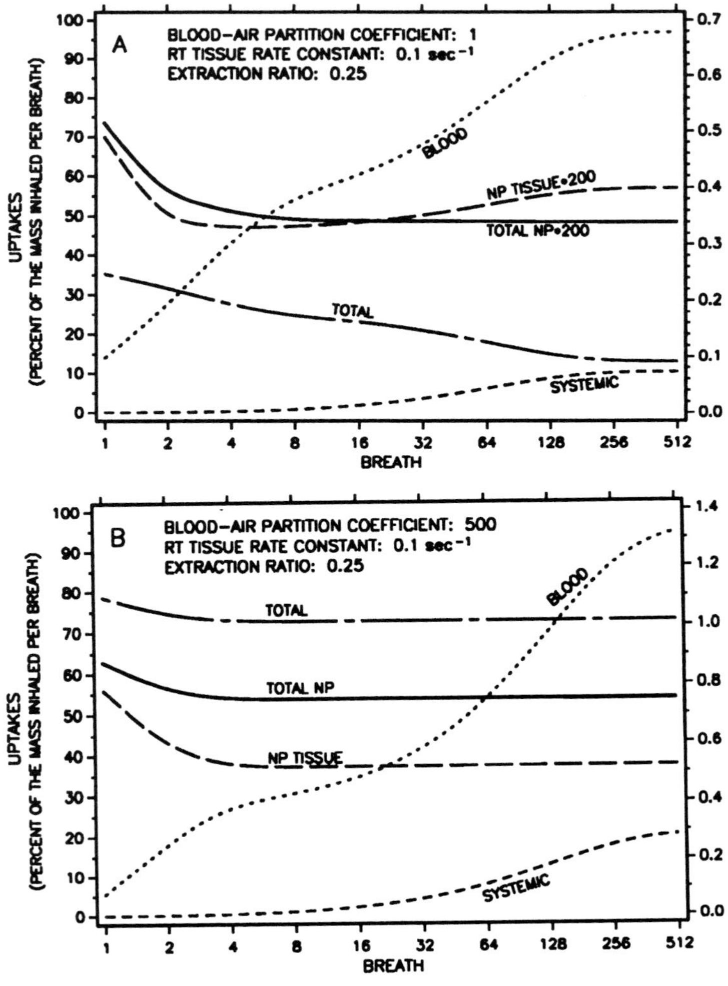

FIGURE 2. Number of breaths to reach equilibrium for two blood–air partition coefficients: (A) 1 and (B) 500. Blood concentration is scaled to the exposure concentration. RT, respiratory tract. NP, nasal-pharyngeal. TOTAL, Percent mass retained by the whole body. TOTAL NP-200, 200 times the percent mass removed from air in the NP region. NP TISSUE-200, 200 times the percent mass retained by the NP region air–blood barrier due to chemical reactions. (In order to graph NP tissue and total NP uptake, the values were multiplied by 200.) SYSTEMIC, percent of mass retained by systemic compartment due to reactions.

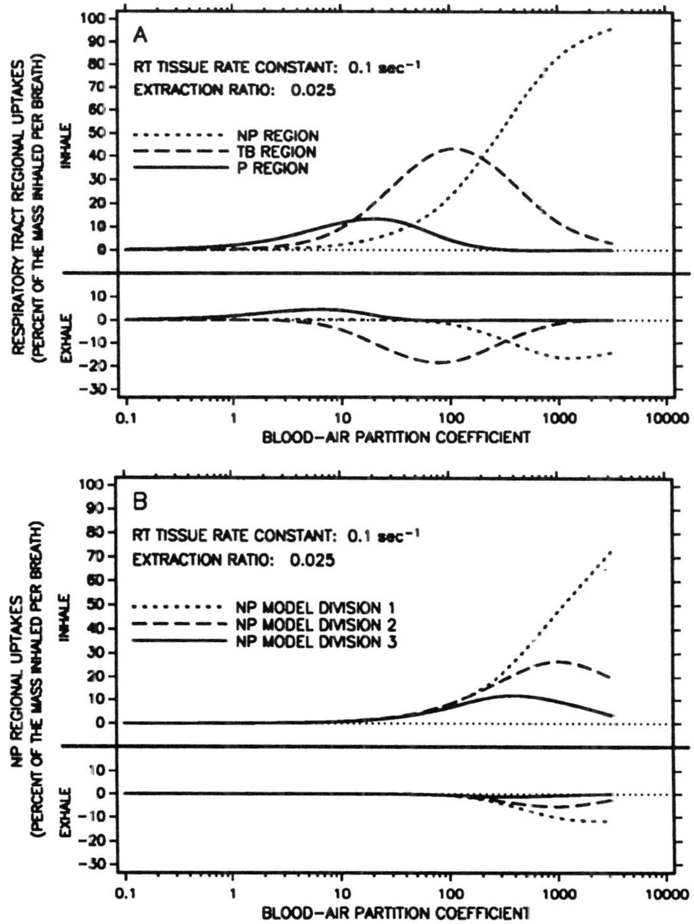

FIGURE 3. Effect of blood–air partition coefficient on (A) regional respiratory tract and (B) nasal-pharyngeal regional absorption and desorption processes. Respiratory Tract (RT) Regional Uptakes are the percent of mass transferred from air to the air–blood barrier or from the air–blood barrier to air in a given region. NP, TB, and P: nasal-pharyngeal, tracheobronchial, and pulmonary, respectively. NP Regional Uptakes are the percent of mass transferred from air to the air–blood barrier or from the air–blood barrier to air in a given NP model division. NP Model Division 1 and NP Model Division 3 are the most proximal and most distal NP model divisions, respectively. Positive and negative uptakes indicate absorption and desorption, respectively.

uptakes behave differently with increasing partition coefficient, and the region of largest absorption and desorption shifts proximally as the partition coefficient increases. The absorption and desorption processes within the nasal-pharyngeal region are illustrated in Figure 3B. This region is divided into three sequential model divisions, with approximately equal surface areas; the uptakes of the three model divisions are plotted versus blood–air partition coefficient. NP model division 1 and NP model division 3 are the most proximal and most distal nasal-pharyngeal model divisions, respectively. Note that the general pattern of the uptakes is similar to that in Figure 3A.

The effect of tissue rate constant and partition coefficient on nasal-pharyngeal absorption and desorption is shown in Figure 4. For several first-order respiratory tract tissue chemical rate constants (k), uptakes for inhalation and exhalation are plotted versus partition coefficient. Generally, as the rate constant increases, uptake during both breathing phases increases for a given blood–air partition coefficient. During inhalation the nasal-pharyngeal region is predicted only to absorb. During exhalation this region may desorb or absorb, depending on the rate constant and partition coefficient; for sufficiently high rate constants ($k > \sim 10$), only absorption occurs during exhalation.

Figure 5 illustrates the effect of systemic extraction ratio (ERS), tissue rate constant, and partition coefficient on nasal-pharyngeal uptake (Figure 3A) and the effect of these parameters on total respiratory tract uptake and total nasal-pharyngeal uptake (Figure 3B). For extraction ratios from 0.025 to 1.0, Figure 5A indicates that the absorption and desorption processes are not very sensitive to extraction ratio for a given tissue rate constant. With k = 0.0 there are very little differences between the uptakes with ERS = 0.025 or 1.0; with k = 1.0, the influence of extraction cannot be discerned from the plots. Although systemic extraction may not have much effect on

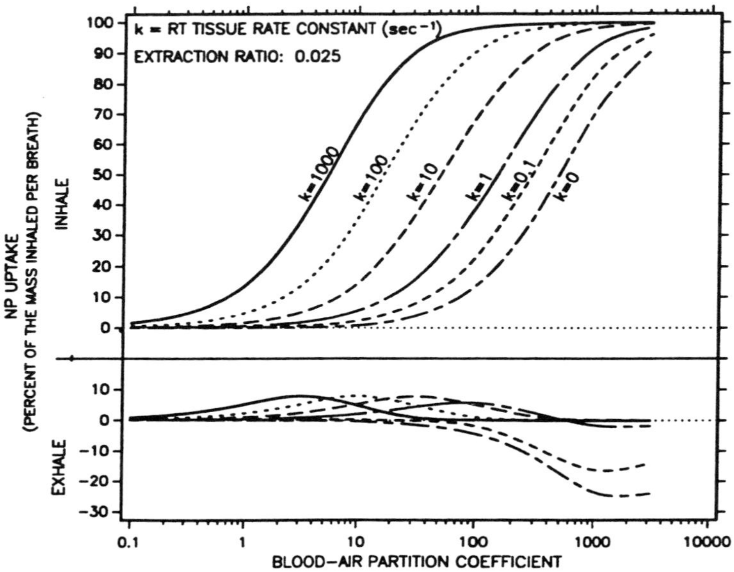

FIGURE 4. Effect of respiratory tract tissue first-order rate constant (k) and blood–air partition coefficient on nasal-pharyngeal (NP) region uptake. NP Uptake, percent of mass transferred from air to the air–blood barrier or from the air–blood barrier to air in the NP region; positive and negative uptakes indicate absorption and desorption, respectively.

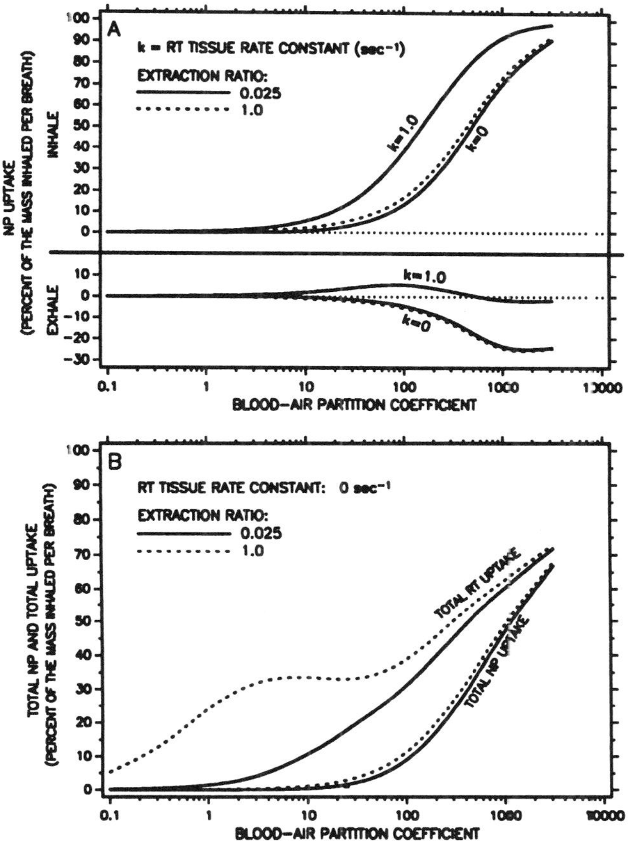

FIGURE 5. (A) Effect of extraction ratio, respiratory tract tissue first-order rate constant, and blood–air partition coefficient on nasal-pharyngeal (NP) region absorption and desorption. RT, respiratory tract. NP Uptake, see Figure 4 for definition. (B) Total respiratory tract and NP total uptake per breath versus blood–air partition coefficient for same two extraction ratios as in (A). Total RT Uptake, percent mass retained by the whole body. Total NP Uptake, percent mass removed from air in the NP region.

processes in the nasal-pharyngeal region, this is not necessarily the case for the respiratory tract as a whole. For $k = 0.0$, Figure 5B shows that the total respiratory tract uptake can be highly influenced by extraction for a partition coefficient less than about 100. This is in contrast to the total nasal-pharyngeal uptake, also shown, for which the extraction ratio has little effect. (In panel B only results for $k = 0.0$ have been illustrated, because the greatest sensitivity to extraction ratio occurs for the smallest rate constant, a property indicated in Figure 5A.)

Nasal-pharyngeal morphometric factors also play a role in nasal-pharyngeal uptake, as illustrated in Figure 6. Plotted are nasal-pharyngeal uptakes versus partition coefficient for air–blood barrier thicknesses of 0.5, 20.0, and 88.5 μm. Uptake on inhalation is predicted to increase as the thickness decreases, but generally this uptake is not very sensitive to thickness; however, desorption is. At the maximum thickness the tissue desorbs regardless of partition coefficient; as the thickness decreases, desorption becomes less and, under some conditions, absorption occurs. For a sufficiently thin barrier, the nasal-pharyngeal region absorbs for partition coefficients less than some value and desorbs for partition coefficients above this value, which depends on the thickness of the air–blood barrier.

Figure 7 illustrates the effect of the LRT on nasal-pharyngeal uptake by contrasting nasal-pharyngeal uptake results for two extreme cases of LRT uptake. For Case 1, the LRT retains approximately 100% of the mass inhaled (by the LRT) during inhalation, so that no mass is exhaled by the LRT to pass through the nasal-pharyngeal region during exhalation; also, none of the gas that is absorbed in the LRT is transferred to blood (the LRT tissue rate constant and gas-phase mass transfer coefficient are very large). For Case 2, the tissues of the LRT absorb no gas, so that inhaled gas is passed back through the nasal-pharyngeal region during exhalation (the gas-phase mass transfer coefficient and the tissue rate constant in the LRT are zero). Three sets of uptake curves are plotted, one set for each of the two extreme conditions and a third set corresponding to a "normal" simulation (the $k = 0.0$ and ERS = 0.025 results depicted in Figure 4). The effect of a large LRT uptake is to increase absorption during inhalation and increase desorption during exhalation, whereas no LRT uptake results in a decrease in absorption during inhalation and an increase in uptake during exhalation.

SUMMARY

A dosimetry model was used to investigate the influence of the following factors on the uptake of gases by the nasal-pharyngeal region of the

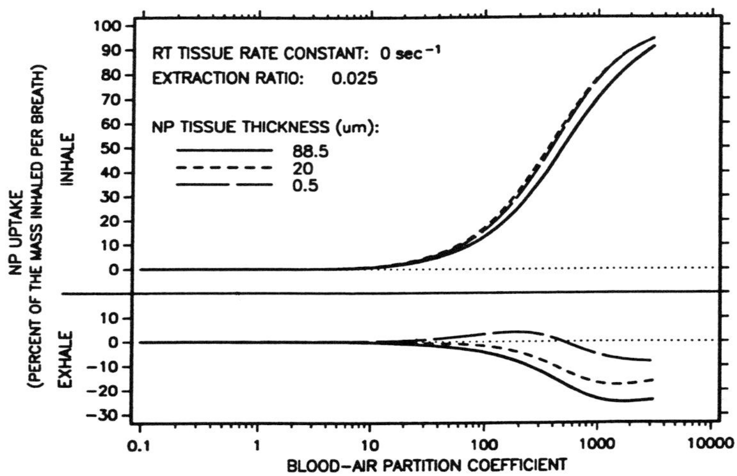

FIGURE 6. Effect of nasal-pharyngeal (NP) tissue thickness and blood–air partition coefficient on NP absorption and desorption. RT, respiratory tract. NP Uptake, see Figure 4 for definition.

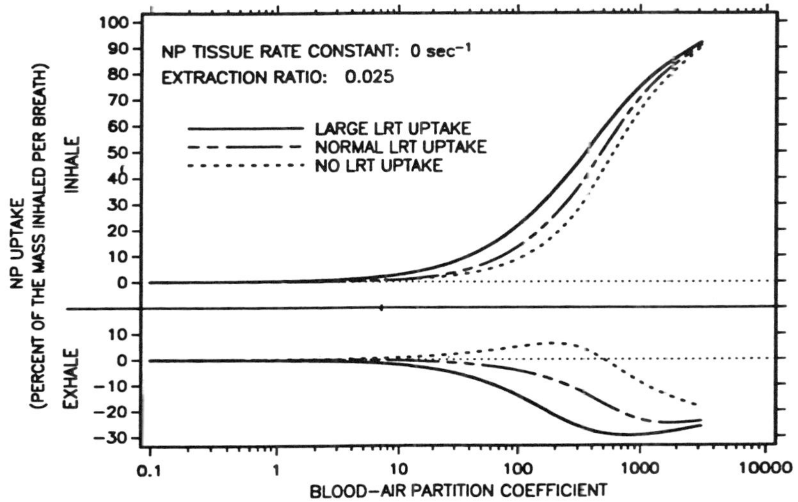

FIGURE 7. Effect of two lower respiratory tract (LRT) absorption scenarios on nasal-pharyngeal (NP) region absorption and desorption. Large LRT Uptake, LRT tissue irreversibly absorbs all of the inhaled mass entering the LRT. No LRT Uptake, LRT absorbs none of the inhaled gas. Normal LRT Uptake, same simulation results as the $k = 0$ plots in Figure 4. NP Uptake, see Figure 4 for definition.

dog: exposure time, blood–air partitioning, respiratory tract tissue first-order reactions, systemic extraction, and lower respiratory tract uptake. Simulation results predicted the following:

1. The length of time for variables (e.g., regional uptakes) to reach steady state depends on the blood–air partition coefficient and is different for different variables for a given partition coefficient. For example, if respiratory tract variables are in steady state, systemic variables may not be.
2. For the uptakes of sequential nasal-pharyngeal regions with the same surface areas and for respiratory tract regional uptakes, the region of largest absorption and desorption generally shifts proximally as the partition coefficient increases.
3. As the first-order tissue rate constant increases, nasal-pharyngeal uptake during both breathing phases increases for a given partition coefficient. During inhalation the nasal-pharyngeal region only absorbs. On exhalation this region may desorb or absorb, depending on the rate constant and partition coefficient; for sufficiently high rate constants, only absorption occurs.
4. Nasal-pharyngeal absorption and desorption processes are not very sensitive to the systemic extraction ratio for a given tissue rate constant, but the total respiratory tract uptake can be highly influenced by extraction ratio.
5. On inhalation, nasal-pharyngeal uptake increases as the thickness of the nasal-pharyngeal air–blood barrier decreases; this uptake is not very sensitive to thickness. However, nasal-pharyngeal desorption and total respiratory tract uptake are.

6. The effect of an increased LRT uptake (due to increasing the LRT tissue rate constant) is to increase nasal-pharyngeal absorption and desorption during inhalation and exhalation, respectively. Reducing LRT uptake by reducing the LRT tissue rate constant results in a decrease in nasal-pharyngeal absorption and desorption during inhalation and exhalation respectively.

The simulation results provide a better understanding of the role of nasal-pharyngeal morphometric and physicochemical properties and the influence of extranasal-pharyngeal aspects on the uptake of inhaled gases by the nasal-pharyngeal region. The model is applicable to humans and other species, if appropriate morphometric and physiological parameters are available. With this information and model validation, interspecies dosimetric extrapolation can be conducted.

REFERENCES

Dahl, A. R., Snipes, M. B., and Gerde, P. 1991. Sites of uptake of inhaled vapors in beagle dogs. *Toxicol. Appl. Pharmacol.* 109:263–275.

DuBois, A. B., and Rogers, R. M. 1968. Respiration factors determining the tissue concentrations of inhaled toxic substances. *Respir. Physiol.* 5:34–52.

Gerde, P., and Dahl, A. R. 1991. A model for the uptake of inhaled vapors in the nose of the dog during cyclic breathing. *Toxicol. Appl. Pharmacol.* 109:276–288.

Gerlach, R. F., Gillett, N. A., Muggenburg, B. A., and Mauderly, J. L. 1986. A cross-sectional study of the effects of age on pulmonary function and lung morphology of beagle dogs. In *Annual Report of the Inhalation Toxicology Research Institute*, eds. B. A. Muggenburg and J. D. Sun. Document LMF-115, December, 382–387.

Kimbell, J. S., Gross, E. A., Joyner, D. R., Godo, M. N., and Morgan, K. T. 1993. Application of computational fluid dynamics to regional dosimetry of inhaled chemicals in the upper respiratory tract of the rat. *Toxicol. Appl. Pharmacol.* 121:253–263.

Muggenburg, B. A., and Mauderly, J. L. 1974. Cardiopulmonary function of awake, sedated and anesthetized beagle dogs. *J. Appl. Physiol.* 37(2):152–157.

Robinson, N. E., and Gillespie, J. R. 1975. Pulmonary diffusing capacity and capillary blood volume in aging dogs. *J. Appl. Physiol.* 38(4):647–650.

Schreider, J. P., and Raabe, O. G. 1981. Anatomy of the nasal-pharyngeal airway of experimental animals. *Anat. Rec.* 200:195–205.

Schulz, H., Heilmann, P., Hillebrecht, A., Gebhart, J., Meyer, M., Piper, J., and Heyder, J. 1992. Convective and diffusive gas transport in canine intrapulmonary airways. *J. Appl. Physiol.* 472(4):1557–1562.

Siegwart, B., Gehr, P., Gil, J., and Weibel, E. R. 1971. Morphometric estimation of pulmonary diffusion capacity IV. The normal dog lung. *Respir. Physiol.* 13:141–159.

Weibel, E. R. 1963. *Morphometry of the human lung*. New York: Academic Press.

Wen, C. Y., and Fan, L. T. 1975. *Models for flow systems and chemical reactors*. New York: Marcel Dekker.

OZONE DOSIMETRY STUDIES IN THE NOSE AND EYE USING OXYGEN-18

Gary E. Hatch

Pulmonary Toxicology Branch, Environmental Toxicology Division, Health Effects Research Laboratory, U.S. Environmental Protection Agency, Research Triangle Park, North Carolina, USA

Jack R. Harkema

Inhalation Toxicology Research Institute, Lovelace Biomedical and Environmental Research Institute, Albuquerque, New Mexico, USA

Charles G. Plopper

Department of Anatomy, School of Veterinary Medicine, University of California, Davis, California, USA

Linda Harris

ManTech Environmental Technology, Inc., Research Triangle Park, North Carolina, USA

Ozone (O_3) causes site-specific morphological changes in the nasal epithelium. Some tissue sites may be more affected than others because of differences in the dose of O_3 to these sites. We have shown previously that tissues exposed to oxygen-18-labeled O_3 ($^{18}O_3$) become labeled with ^{18}O and suggest that this label represents a measure of tissue O_3 dose. The present study seeks to evaluate the dose of O_3 to different sites in the nose of F344 rats and rhesus monkeys using $^{18}O_3$. Rats were exposed to 0.12 or 0.66 ppm $^{18}O_3$ for 6 h, and monkeys were exposed to 1.0 ppm $^{18}O_3$ for 2 h. Nasal epithelial tissues were sampled immediately postexposure by dissection in rats and by a freeze-grinding procedure in the excised nasal tissues of the monkeys. Excess ^{18}O measured mass spectrometrically in the freeze-dried nasal tissues was significantly elevated in both the monkey and rat to 2–10 ng ^{18}O/mg dry weight. Large differences between sites were apparent, and the sites with the highest ^{18}O content appeared to be different in the two species. Since O_3 is thought to penetrate only a short distance into the epithelium (due to its high reactivity), a factor was derived for each sampling method for converting the units of expression from per milligram dry weight to a per mm^2 surface area. Because some sites had undetectable ^{18}O incorporation, a larger sample size will be required to determine statistically significant differences in O_3 dose between different tissue sites and between species.

The authors thank the following for assistance with technical aspects of this project: Brian K. Tarkington, Tim Duval, Viviana Wong, University of California Primate Center, Davis, CA; John McKee, ManTech Environmental Technology, Inc., and Joel Norwood, U.S. EPA, Research Triangle Park, NC.

The research described in this article has been reviewed by the Health Effects Research Laboratory, U.S. Environmental Protection Agency, and approved for publication. Approval does not signify that the contents necessarily reflect the views and policies of the agency, nor does mention of trade names or commercial products constitute endorsement or recommendation for use. The monkeys used were part of an extensive study of ozone dosimetry and toxicity to the lungs and airways.

Address correspondence to Gary E. Hatch, MD-82 U.S. EPA, Research Triangle Park, NC 27711, USA.

Different regions of the nose have been shown to respond differently in terms of severity of the morphological effects of inhaled O_3. In rats, proliferative and cellular remodeling of the epithelium were greatest in the lateral wall of the respiratory epithelium (Johnson et al., 1990). In monkeys, a similar site on the lateral wall composed of "transitional epithelium" has been defined as the site of maximal O_3 effect (Harkema et al., 1987a, 1987b). The nasal epithelium of monkeys shows a larger response than that of rats if the animals are similarly exposed (Plopper et al., 1991). The reason some sites or species seem to be more sensitive to O_3 is not yet known, but may involve the collective effects of O_3 dose due to airflow patterns in the nose (Kimbell et al., 1993), mucus coverage or flow, or tissue antioxidant content. A method that could evaluate the dose of O_3 at different sites along the epithelium may provide an explanation for site-specific effects.

We have shown previously that labeling of the O_3 molecule with heavy oxygen ($^{18}O_3$) is possible and that this label can be detected in tissues by isotope-ratio mass spectrometry. Tissues are dried prior to analysis to eliminate the large diffusable pool of tissue water. Lung tissue from mice has been shown to accumulate ^{18}O in a time-dependent fashion when the mice were exposed by inhalation to $^{18}O_3$ (Santrock et al., 1989). Rats and rabbits exposed to $^{18}O_3$ were shown to have elevated ^{18}O (over background levels) in their nasal tissues (Hatch et al., 1989a, 1989b). However, no effort has been made to quantify and compare the amount of label in different regions of the nose, particularly across species, because satisfactory methods have not been established to accurately sample epithelial tissue nor to express this label as "dose."

The present study examines the feasibility of measuring epithelial tissue levels of excess ^{18}O following $^{18}O_3$ exposure in rats and monkeys. O_3 is believed to react so quickly with the electron-rich matrix of the cell membrane that it penetrates only a very short distance into cells (Pryor, 1992). Reaction products (aldehydes, hydroperoxides, organic and oxygen radicals, etc.) might penetrate a greater distance, but even they would probably not be carried far before reacting. For this reason, particular attention was given to procedures that would allow evaluation of the influence of the tissue surface area/mass ratio on the dose measurement.

METHODS

Animals

Colony-reared rhesus monkeys (*Macacca mulatta*, males, age 3–4 yr, weighing 3–6 kg) were used. On the basis of physical and radiographic examinations and complete blood count analysis, all monkeys were found to be free of clinical respiratory disease. Monkeys were housed outdoors until 2 wk before exposure. Food and water were provided ad libitum. Cesarian-derived Fischer 344 rats (males weighing 164–191 g, 8–10 wk) were purchased from Charles River Breeding Labs (Raleigh, NC). They were housed for 1 wk in plastic cages with heat-treated pine bedding and given food (Ralston Purina Co., St. Louis, MO) and water ad libitum prior to exposures. Animal facilities were AAALAC accredited, and all procedures were designed so as to minimize pain or stress.

$^{18}O_3$ Exposures

$^{18}O_3$ was generated in both studies by substituting isotopically pure $^{18}O_2$ (97%, Isotec, Inc., Miamisburg, OH) for normal oxygen (contains only 0.2% $^{18}O_2$) in a silent arc O_3 generator (Orec Inc., Phoenix, AZ). The $^{18}O_2$ was first diluted to 3% in argon in the monkey study to allow better control of flow rates. Although the conversion of $^{18}O_2$ to $^{18}O_3$ is not complete and results in some excess $^{18}O_2$ in the exposure chamber (about 60 ppm over a background of 426 ppm), this excess is not sufficient to produce a detectable increase in tissue ^{18}O. The concentration of O_3 was monitored continuously using a chemiluminescent O_3 analyzer (Bendix, Lewisburg, WV) calibrated as described previously (Hatch et al., 1986). Chamber concentrations deviated less than 6% from the target levels over the course of the exposure. Rats were exposed to 0.12 or 0.66 ppm (12 per group) $^{18}O_3$ or to filtered air (4 per group) for 6 h. Exposures were in individual stainless steel wire mesh cages in a Rochester-type inhalation chamber. Chamber temperatures ranged from 69 to 75°C and relative humidity ranged from 30 to 50%. Air flow through the chamber was 1 air change (0.3 m^3)/min. Monkeys were exposed individually to 1.0 ppm $^{18}O_3$ or to filtered air for 2 h in Rochester-type inhalation exposure chambers. Chamber air flow was 0.5 change/min.

Rat Nasal Tissue Dissection

Immediately following exposure, rats were anesthetized by with 5% halothane (Aldrich Chemical Co., Milwaukee, WI) using a Frazer-Sweatman vaporizing anesthesia apparatus (Cyprane North America Inc., Tonawanda, NY). Rats were then exsanguinated by severing the abdominal aorta, and nasal lavage was performed on half of the $^{18}O_3$-exposed rats. Warmed (37°C) saline (0.85%, 20 ml/rat) was injected through the oropharynx and collected from the nose.

Following lavage, the nasal cavity was dissected free of the cranium, split longitudinally with a single-edged razor blade, wrapped loosely in aluminum foil, and frozen in liquid N_2. Nasal cavities were warmed as needed under a stream of dry 25°C N_2 and dissected while partially frozen into the regions described previously (Maples et al., 1993). The freezing step was employed to stop mucociliary activity that could sweep away ^{18}O bound to mucus. The dissected tissue was placed into 1.0 ml of ice-cold 3% perchloric acid (PCA) and homogenized (Polytron, Brinkman Instruments Co., Westbury, NY). The PCA step was included to enable assay of antioxicants in the same tissues. The homogenates were centrifuged (20,000 × g, 20 min), the supernatants discarded and replaced with 1 ml deionized water to extract excess PCA, the water was decanted, and the pellets were lyophilized. Dry pellets were stored at 4°C in air-tight plastic containers.

Monkey Nasal Tissue Freeze-Grinding

Monkeys were anesthetized immediately after exposure with an iv injection of ketamine and pentobarbital before euthanizing by exsanguination. The head of each monkey was removed from the carcass and partially skinned. The cranium was removed and the nasal cavity split with a bone saw. Bone fragments were carefully removed and each half of the nose was wrapped in foil and submerged in liquid N_2. The frozen tissues were transported on dry ice and stored at –80°C until freeze-grinding. A 2-mm-diameter burr-tipped drill (Dremmel, Racine, WI) was used for grinding off samples of epithelium while the nasal tissue was half-submerged in liquid N_2. The sites sampled were the same as those described in Harkema et al. (1987a, 1987b). Each site was ground 5 times until the grinder tip was halfway (~1 mm) into the surface. The powdered tissue from the five grinds was collected by evaporating the liquid N_2 until the tissue powder could be transferred with a spatula to a small plastic tube. Corneal tissues of the eye (which was left intact on the section of skull taken) were also sampled in the same manner as the nasal epithelium. This was done in order to determine whether $^{18}O_3$ reacts with the cornea.

Analysis of Dried Tissues for ^{18}O Content

Assay of lyophilized tissues for ^{18}O used a modification of the method of Santrock and Hayes (1987), which employs an elemental analyzer to convert oxygen in dried tissues to carbon monoxide (the oxygen percentage by dry weight was obtained from this CO content), a column filled with I_2O_5 to convert CO to CO_2, and an isotope ratio mass spectrometer for analysis of the resulting CO_2. This procedure was modified by use of an on-line injection system between the elemental analyzer and the mass-spectrometer. The CO_2 mass ratios (46/44 Da) were initially expressed as a delta with respect to a known CO_2 standard. This delta was corrected for time-dependent drift within a sample run. Statistical analyses relating exposed and unexposed tissues were first made on these drift-corrected delta values. The deltas were then converted to $^{18}O/^{16}O$ ratios by use of a standard curve generated with standards included in each sample run. To determine the excess ^{18}O in the exposed tissues, the mean $^{18}O/^{16}O$ ratios of unexposed tissues were subtracted from all samples of a given tissue. The unexposed (background) tissues were included in each mass spectrometer run adjacent to tissues that had been exposed from the same site.

In order to allow similar expression of results across tissues and species and in order to express the ^{18}O incorporation in terms of surface area of epithelium, further unit conversions were made. The molar $^{18}O/^{16}O$ ratio was converted to the units of ng excess ^{18}O/mg total oxygen. Then, using the mean oxygen percentage of the dry weight of each sample, a conversion was made to the units ng excess ^{18}O/mg dry weight. Due to the probability that essentially all of the ^{18}O is concentrated at the air-tissue surface, the mass/surface area ratio for each tissue (units of mg dry/mm^2) was calculated for each tissue. This factor allows easy visualization of the influence of the mass/area ratio on the measurement of ^{18}O.

Statistical Analysis

All analyses were made on drift-corrected delta values of the mass spectrometer (the primary measurement), and unit conversions consisted essentially of multiplications by constants from these base values. Significance was defined as $p < .05$. Analysis of variance tests were first made on $^{18}O_3$-exposed versus air-exposed pooled tissues to determine overall significance of the $^{18}O_3$ exposure effect. Dunnett's test was used in rats to determine significance of changes in individual tissue sites relative to the pooled background. The background measurements made in each tissue in the same analysis sequence as the exposed samples (rather than the pooled background) appeared to provide a reasonable baseline for calculating excess ^{18}O in exposed tissues. Therefore, tables show the latter values rather than the data calculated from the pooled background.

RESULTS

Evaluation of the Tissue Sampling Procedure: Rat Nasal Epithelium

Table 1 shows tissue weights from dissections of nasal tissues from a subset of the rats used in this study. This subset included both $^{18}O_3$-exposed and air-exposed tissues (exposure should not affect these measurements). The wet weights of the different regions ranged from about 14 mg in the anterior septum to 100 mg in the nasal turbinate, and little variation in weights was seen in replicate samples from the same site. The surface areas of the regions sampled by the dissection procedure were determined from histologic sections from the same nasal regions from two rats of the same

TABLE 1. Dissected Rat Nasal Tissue: Weight, Area, and Oxygen Percentage

Tissue	Wet weight (mg)	Epithelial surface area (mm^2)	Oxygen content (% of dry)
Vestibule	20.6 ± 2.0	35.1 ± 3.2	19.62 ± 0.46
Proximal septum (anterior)	13.9 ± 0.8	38.9 ± 3.9	21.34 ± 0.67
Distal septum (posterior)	64.0 ± 3.2	165 ± 7.3	23.70 ± 0.26
Proximal lateral wall and turbinates	101.1 ± 3.0	102 ± 10.3	20.70 ± 0.32
Ethmoid turbinate	29.2 ± 1.7	570 ± 25.9	23.65 ± 0.14

Note. Mean ± SE values are given for 12 samples per tissue in the case of wet weights, 2 samples in the case of epithelial surface areas, and 28 samples in the case of oxygen content.

strain and weight as those used in this study. Values for surface area were in close agreement between the two rats; therefore, further animals were not studied. The elemental oxygen percentage in the dry tissues ranged from 20 to 24%, and little variability was seen between individual site or replicate samples.

Excess ^{18}O in Rat Nasal Epithelium

Table 2 shows the levels of excess ^{18}O in the nasal epithelium of rats exposed to 0.12 and 0.66 ppm $^{18}O_3$ for 6 h. The background values show the variability of the air-exposed samples of each tissue, which was large because of a less than adequate number of background samples for each tissue ($n = 4$). It was thought that background samples from all tissues could be pooled; however, mass spectrometer drift prevented this. Dunnett's test (using pooled background) indicated significantly elevated ^{18}O in the proximal septum and the proximal lateral wall and turbinates at both O_3 concentrations. Using individual tissue backgrounds suggested that the vestibule was also highly labeled. The values calculated from individual tissue backgrounds were thought to be more accurate, since they were assayed in the same sequence as the exposed tissues; therefore, they are shown in Table 2. Negative values of ^{18}O "excess," however, suggested that baseline fluctuations were sometimes of the same magnitude as the small observed changes in ^{18}O content. Statistical analysis of the pooled exposed groups showed a significant elevation in ^{18}O content due to exposure. Comparison of lavaged versus nonlavaged tissues showed no significant difference. In three of the five nasal sites examined, there appeared to be an elevation in ^{18}O content to as high as 10 ng ^{18}O/mg. In the vestibule and proximal lateral wall, there was a higher level of ^{18}O at 0.66 ppm than at 0.12 ppm. It was found that the pooling of background values resulted in different values for excess ^{18}O from those calculated from the individual background values that are shown in Table 2.

TABLE 2. Excess ^{18}O in Different Regions of the Nose of Rats Immediately Following Exposure to $^{18}O_3$ (6 h)

	Excess ^{18}O (ng/mg dry)			
Region of nose	Background	0.12 ppm	0.66 ppm	Factor[a]
Vestibule	0.01 ± 0.77	4.47 ± 0.52	9.17 ± 1.66	0.12
Proximal septum (anterior)	0.00 ± 2.72	9.95 ± 1.78	7.78 ± 2.27	0.07
Distal septum (posterior)	0.00 ± 0.56	−2.35 ± 0.44	0.10 ± 0.44	0.08
Proximal lateral wall and turbinates	0.00 ± 0.18	1.11 ± 0.44	3.97 ± 0.74	0.20
Ethmoid turbinates	0.00 ± 0.91	−0.21 ± 0.35	2.10 ± 0.42	0.01

Note. Mean ± SE values are given for 4 animals/group (background = air–exposed) or 12 animals/group (0.12 or 0.66 ppm $^{18}O_3$).

[a]To obtain the excess ^{18}O in the units of ng $^{18}O/mm^2$, multiply by this factor.

TABLE 3. Excess ^{18}O in Different Regions of the Nose of Rhesus Monkeys Immediately Following Exposure to $^{18}O_3$ (1.0 ppm, 2 h)

Region of nose	Dry weight (mg)	Oxygen content (% of dry)	Excess ^{18}O (ng/mg)		Factor[a]
			Background	1.0 ppm $^{18}O_3$	
Vestibule (squamous)	2.8 ± 0.6	19.6 ± 0.9	0.00	4.71 ± 1.09	0.18
Lateral wall (transitional)	2.3 ± 0.6	19.5 ± 0.6	0.00 ± 1.32	8.15 ± 3.53	0.15
Lateral wall (respiratory)	2.3 ± 0.3	18.1 ± 6.3	0.00 ± 0.82	4.56 ± 1.95	0.15
Maxillary sinus	6.9 ± 1.6	15.6 ± 0.5	0.00 ± 0.26	2.66 ± 0.38	0.44
Ethnoid (olfactory)	5.6 ± 1.8	16.0 ± 0.4	0.00 ± 0.54	9.23 ± 2.84	0.36
Eye cornea	3.8 ± 1.1	19.2 ± 0.9	0.00 ± 0.30	3.97 ± 1.18	0.24

Note. Mean ± SE values are given for two air-exposed (background) and three $^{18}O_3$-exposed animals.
[a]To obtain the excess ^{18}O in the units of ng $^{18}O/mm^2$, multiply by this factor.

A factor was derived, consisting of the mass/surface area ratio (units of ng dry weight/mm^2 surface area) from the mean dry weight (calculated assuming 80% tissue water) and the mean surface area, and is shown in Table 2 for each tissue. This factor ranged over about a 20-fold range, with ethmoid turbinates having the smallest ratio (0.01) and nasal turbinates the highest (0.2). If the mass/area ratio (factor shown in Table 2) was used to convert units to ng excess $^{18}O/mm^2$, values ranged from undetectable (~0.2) to ~0.9 ng/mm^2.

Evaluation of Freeze-Grinding Procedure: Monkey Nasal Epithelium

The freeze-grinding procedure was evaluated by comparing the dry weights of the samples obtained that are shown in Table 3. The theoretical dry weight of each sample (assuming 20% dry/wet and the calculated volumes of 5 half-spherical 2-mm-diameter pits) is 2.1 mg. The theoretical surface area of the 5 pits was 15.7 mm^2, yielding a theoretical mass/surface area ratio of 0.13 mg dry/mm^2. Actual values for tissue dry weights taken for the five nasal sites and one corneal site sampled are shown in Table 3. Overall, these weights ranged from 1.06 to 10.5 mg, with a mean ±SE of 3.5 ± 0.5 (n = 30). There was a three-fold range in the mean values of different tissues (2.3–6.9 mg). Some tissues had higher dry weights than others, apparently due to differences in the softness of the tissues at liquid N_2 temperatures. This made it difficult to control the depth of the grind to 1.0 mm. Assuming that the surface areas remained constant, the individual tissue mass/area ratios would have ranged from 0.15 to 0.44 mg dry/mm^2 (Table 3).

Excess ^{18}O in Monkey Nasal Epithelium

Table 3 shows the excess ^{18}O of epithelium sampled in different regions of the monkey nose following an $^{18}O_3$ exposure that was higher in concentration but of shorter duration than was used in the rat. The tissues sampled in the monkey had not been prelavaged. Five different sites within the nose showed increases in ^{18}O. The excess ^{18}O was highest in the lateral-wall transitional epithelium and ethmoid turbinates (8–9 ng/mg), intermediate in the vestibule and lateral-wall respiratory epithelium (4–5 ng/mg), and lowest in the maxillary sinus (2.7 ng/mg). Epithelium of the eye cornea also had elevated ^{18}O (~4 ng/mg). Pooled data from all nasal sites were statistically elevated in exposed compared to background tissues. Converting to a per square millimeter using the best available mass/area ratios suggests that the ethmoid turbinates have the highest levels of ^{18}O (3.3 ng/mm^2), followed by transitional epithelium and maxillary sinus (1.2 ng/mm^2).

DISCUSSION

A goal of this study was to sample the nasal epithelium in such a way that accurate and meaningful measurements of O_3 dose could be made at different sites and between species. To the extent that incorporation of ^{18}O following $^{18}O_3$ exposure can be considered a measure of O_3 "dose," we partially achieved this goal and gained some methodological insights. Epithelial tissues from both rats and monkeys sampled immediately after exposure to $^{18}O_3$ were found to contain significantly increased ^{18}O. The elevations ranged from 2 to 10 ng ^{18}O/mg dry tissue and appeared to vary according to the site sampled. Some of the nasal sites had barely detectable or undetectable increases in ^{18}O. It is tempting to speculate that the lower ^{18}O in distal regions of the nose of the rat is the result of removal of O_3 at the proximal regions. Instrument drift in the background made it difficult to confidently measure the difference between background and exposed tissues. Negative changes in "excess ^{18}O" were seen in two of the exposed groups of rat tissues. The fact that we encountered only small increases in the epithelial sites challenges the ability to detect differences in ^{18}O accumulation between the different epithelial sites. An increased number of samples could solve part of the problem, but some aspects of the ^{18}O analysis technique have also surfaced as a result of this study because of the small differences in ^{18}O enrichment observed. Better statistical methods are needed for comparing results seen in one mass spectrometer run with another and for controlling the variability of the background.

The influence of the mass/surface area ratio is another issue that could affect results. Differences in the sampling methods used yielded differences in the mass/surface area ratio. When levels of ^{18}O/mg dry weight were converted to per square millimeter, the range of values was about 0.2 to 3 ng/mm^2. The monkey epithelium appeared to reach higher levels of ^{18}O/mm^2 in spite of the fact that the monkey exposures were lower (2 ppm-h) than the rats (4 ppm-h). It is interesting that the monkey nose appears to be more sensitive to O_3 than the rat nose (Plopper et al., 1991). The cornea of the eye of the monkey had increased ^{18}O consistent with $^{18}O_3$ reactivity with the eye. Based on these preliminary data, there did not appear to be a correspondence across species as to which nasal sites accumulated the most ^{18}O. In fact, the ethmoid turbinate in rats had the lowest accumulation, while that of monkeys had the highest.

If the accumulation of ^{18}O follows the highly reactive nature of O_3, most of the deposited ^{18}O label should be localized at the airway surfaces, and any unit for expression of dose should be affected by the mass/surface area ratio of the tissue sample taken. A previous study in rabbits (exposed to 0.4 ppm $^{18}O_3$, 2 h) reported ^{18}O nasal tissue levels within the range of those seen here in rats and monkeys. Bronchoalveolar lavage cells of the rabbits had eight- to ninefold higher levels of ^{18}O than nasal tissues (Hatch et al., 1989a). We suspect that the lavaged cells have a higher surface area/mass ratio than tissues sampled from the nasal lining and that is why they are more highly labeled. Previously, when dissections were performed on lyophilized epithelium, it was difficult to control the depth of the cut into the epithelium and to measure the sample surface area. The method used here for the rat nasal dissection had been reported previously (Maples et al., 1993), although the ratio between surface areas of the regions and the tissue weights sampled had not been determined. One problem we noted in the rat tissue dissection was that homogenization of the more fibrous tissues (nasal vestibule and septum) was often incomplete. This would have caused the small aliquots taken for analysis to not be representative of the whole tissue, which would have influenced our results. We demonstrated that a freeze-grinding method eliminated the homogenization problem, as well as allowing more accurate definition of the areas of epithelium of interest. However, variations between different sites were observed in the dry weights of samples obtained by freeze-grinding (see Table 3). This variation appeared to result from difficultly in controlling the depth of the grind. Some tissues were extremely hard and others quite soft at liquid N_2 temperatures. Perhaps greater care or practice in the grinding step or use of temperatures higher than that of liquid N_2 might reduce these variations. The area of epithelium necessary to make one sample (~25 mm^2) was about the same for freeze-grinding as for dissection. Dissection yielded a larger range of values for the tissue mass/area ratio (Table 1).

In summary, improvements were made in ^{18}O labeling methods for evaluating the nasal tissue dose of O_3 in rats and monkeys. Easily measurable but small increases in ^{18}O content were observable in nasal and ocular tissues, and the amount of increase appeared to depend on the site sampled. Improved data should be possible by increasing the number of samples assayed and through better control of the tissue mass/surface area ratio during sample preparation. A knowledge of incorporation of ^{18}O should enable evaluation of the tissue dose of O_3, which may help explain differences in sensitivity of different epithelial sites and different species to O_3.

REFERENCES

Hatch, G. E., Slade, R., Stead, A. G., and Graham, J. A. 1986. Species comparison of acute inhalation toxicity of O_3 and phosgene. *J. Toxicol Environ. Health* 19:43–53.

Hatch, G. E., Koren, H., and Aissa M. 1989a. A method for comparison of animal and human alveolar dose and toxic effect of inhaled O_3. *Health Phys.* 57:37–40.

Hatch, G. E., Wiester, M. J., Overton, J. H., and Aissa, M. 1989b. Respiratory tract dosimetry of ^{18}O labeled ozone in rats: Implications for a rat-human extrapolation of ozone dose. In *Atmospheric ozone research and its policy implications*, eds. T. Schneider, S. D. Lee, G. T. R. Wolters, and L. D. Grant, pp. 553–560. New York: Elsevier.

Harkema, J. R., Plopper, C. G., Hyde, D. M., St. George, J. A., Wilson, D. W., and Dungworth, D. L. 1987a. Response of the macaque nasal epithelium to ambient levels of ozone: A morphologic and morphometric study of the transitional and respiratory epithelium. *Am. J. Pathol.* 128:29–44.

Harkema, J. R., Plopper, C. G., Hyde, D. M., St. George, J. A., and Dungworth, D. L. 1987b. Effects of an ambient level of ozone on primate nasal epithelial mucosubstances. Quantative histochemistry. *Am. J. Pathol.* 127:90–96.

Kimbell, J. S., Gross, E. A., Joyner, D. R., Godo, M. N., and Morgan, K. T. 1993. Application of computational fluid dynamics to regional dosimetry of inhaled chemicals in the upper respiratory tract of the rat. *Toxicol. Appl. Pharmacol.* 121:253–263.

Johnson, N. F., Hotchkiss, J. A., Harkema, J. R., and Henderson, R. F. 1990. Proliferative responses of rat nasal epithelium to O_3. *Toxicol. Appl. Pharmacol.* 103:143–155.

Maples, K. R., Nikula, K. J., Chen, B. T., Finch, G. L., Griffith, W. C., Jr., and Harkema, J. R. 1993. Effects of cigarette smoke on the glutathione status of the upper and lower respiratory tract of rats. *Inhal. Toxicol.* 5:389–401.

Plopper, C. G., Harkema, J. R., Last, J. A., Pinkerton, K. E., Tyler, W. S., St. George, J. A., Wong, V. J., Nishio, S. J., Weir, A. S., Dungworth, D. L., Barry, B. E., and Hyde, D. M. 1991. The respiratory system of nonhuman primates responds more to ambient concentrations of ozone than does that of rats. In *Tropospheric ozone and the environment*, eds. R. L. Berglund, D. L. Lawson, and D. J. McKee, pp. 137–148. Pittsburgh, PA: Air and Waste Management Association.

Pryor, W. A. 1992. How far does ozone penetrate into the pulmonary air/tissue boundary before it reacts? *Free Radical Biol. Med.* 12:83–88.

Santrock J., and Hayes, J. M. 1987. Adaptation of the Unterzaucher procedure for determination of oxygen–18 in organic substances. *Anal. Chem.* 59:119–127.

Santrock, J., Hatch, G. E., Slade, R., and Hayes, J. M. 1989. Incorporation and disappearance of oxygen-18 in lung tissue from mice allowed to breathe 1.0 ppm $^{18}O_3$. *Toxicol Appl. Pharmacol.* 98:75–80.

COMPARISON OF FOUR BIOLOGICALLY BASED DOSIMETRY MODELS FOR THE DEPOSITION OF RAPIDLY METABOLIZED VAPORS IN THE RODENT NASAL CAVITY

Clay B. Frederick

Toxicology Department, Rohm and Haas Company, Spring House, Pennsylvania, USA

John B. Morris

Toxicology Program, School of Pharmacy, University of Connecticut, Storrs, Connecticut, USA

Julia S. Kimbell, Kevin T. Morgan

Chemical Industry Institute of Toxicology, Research Triangle Park, North Carolina, USA

Peter W. Scherer

Department of Bioengineering, University of Pennsylvania, Philadelphia, Pennsylvania, USA

Various volatile compounds induce toxic effects in the rodent nasal cavity following inhalation exposure. To facilitate interspecies extrapolation for risk assessment, it is necessary to estimate the amount of the compound deposited in the nasal cavity of both the test and target species. Four computer simulation models have been prepared and evaluated for their utility in interspecies extrapolation, description of site-specific toxicity within the nasal cavity, and understanding of systemic toxic effects (or the lack thereof) in rodent bioassays. The simplest model provided an estimate of the fractional penetration of the compound to the lungs and was linked to a one-compartment description of the body. The most complex model used a description of nasal air flow to estimate the amount of compound deposited in various regions of the nasal cavity. This multicompartment nasal cavity was linked to a multicompartment physiologically based pharmacokinetic (PBPK) model describing the systemic tissues. The data for a nasal irritant, ethyl acrylate, were used to evaluate the models. This compound is representative of a variety of organic acids and esters that have induced cytotoxicity in the olfactory epithelium lining the rodent nasal cavity following high-concentration inhalation exposure. No systemic toxicity or carcinogenicity was induced by ethyl acrylate during a chronic inhalation bioassay. The model constructed from a multicompartment nasal cavity linked to a PBPK description of the body was found to have the most utility for adequately describing regional toxicity in the nasal cavity, for evaluating the delivered dose and potential for toxicity at systemic organs, and for interspecies extrapolation.

This study was funded in part by the Basic Acrylic Monomer Manufacturers (BAMM).

Address correspondence to Clay B. Frederick, Toxicology Department, Rohm and Haas Company, 727 Norristown Road, Spring House, PA 19477, USA.

A wide range of volatile organic acids and esters has been demonstrated to be efficiently deposited in the mammalian nasal cavity (Stott & McKenna, 1984; Stott et al., 1986; Morris, 1990; Morris et al., 1991; Morris & Frederick, 1994; Dahl et al., 1991). Following deposition, the esters are rapidly hydrolyzed to the corresponding acids by the abundant esterase activity present in the nasal cavity (Stott & McKenna, 1985; Dahl et al., 1987; Bogdanffy et al., 1987; Olson et al., 1993; Frederick et al., 1994a). Many of these compounds induce toxic effects in the rodent nasal cavity in standard inhalation toxicity evaluations, but no signs of systemic toxicity are observed in organs remote from the respiratory tract. Generally, the acids and esters confine their cytotoxic effects in the nasal cavity to olfactory epithelium and do not induce histopathological effects in respiratory epithelium. Repeated high-concentration acid or ester exposure has often resulted in the replacement of the sensitive olfactory epithelium with resistant respiratory epithelium, a process described as "respiratory metaplasia," but progression to neoplasia in chronic inhalation bioassays has not been observed (Miller et al., 1985; Reininghaus et al., 1991).

The discussion that follows explores the utility of several biologically based dosimetry models to describe the site-specific toxicity of a representative organic ester, ethyl acrylate (EA). This compound is a representative of a large class of monomers used as industrial intermediates to make acrylic polymers. The nasal olfactory toxicity observed following subchronic and chronic inhalation exposure of rats and mice to high concentrations of the methyl, ethyl, and butyl acrylate is typical of the class of organic esters as described earlier (Miller et al., 1985; Reininghaus et al., 1991). A subchronic study with the parent acid for the class, acrylic acid, also induced the same lesion (Miller et al., 1981a). None of the chronic studies induced a dose-related increase in tumors in any tissue or any histopathological indications of toxicity remote from the nasal cavity.

Since human respiratory and olfactory epithelia are histologically similar to the rodent respiratory and olfactory epithelia, it might be reasonable to suggest that human olfactory epithelium might be similarly sensitive to high concentrations of acids and esters. In fact, exposures of humans to high concentrations of approximately 120 different chemicals (including long-term industrial exposures to acrylates and methacrylates) have been reported to cause olfactory dysfunction (Amoore, 1986; Schwartz et al., 1989). To address this range of possible nasal toxicity issues, an understanding of the tissue dosimetry resulting in rodent nasal cavity toxicity has the potential for use in interspecies extrapolation, evaluation of human exposure scenarios, and the prediction and avoidance of potentially toxic effects in humans.

The need to use biologically based dosimetry models for accurate inhalation risk assessment is emphasized by the radically different nasal cavity anatomy (Gross & Morgan, 1992; DeSesso, 1993) and nasal air flow patterns of rodents and humans (Morgan et al., 1991; Kimbell et al., 1993;

Hahn et al., 1993). In particular, most of the posterior region of the rat nasal cavity (approximately 50% of the total surface area) is covered with sensitive olfactory epithelium (Gross et al., 1982; Uraih & Maronpot, 1990). By contrast, only about 10% of the human nasal cavity is covered with olfactory epithelium, and it is confined to the posterior dorsal region of the nasal cavity (Beidler, 1980; Guilmette et al., 1989).

In rodents, an inhaled vapor only has to traverse a few millimeters of resistant respiratory epithelium before reaching sensitive olfactory tissue. In humans, an inhaled vapor has to traverse several centimeters and a much larger surface area of respiratory epithelium to reach the olfactory tissue. In both species, the highly vascularized and thin respiratory epithelium serves to warm and humidify the inhaled air. However, as noted by Beidler (1980), the large anterior region of human respiratory epithelium also provides the possibility of significant absorption of organic odors prior to their reaching olfactory tissue. These conceptual dosimetry differences can, in principle, be modeled, but they are difficult to simulate with standard descriptive toxicological evaluation techniques.

The discussion that follows quantitatively evaluates four dosimetry models of varying complexity against a variety of published experimental observations related to the nasal deposition of vapors (Table 1). These observations suggest a variety of model design criteria related to both the gas-phase mass transport and tissue-phase metabolism and mass transport of rapidly metabolized vapors. A robust model design should be able to adequately describe these previously published experimental observations. In addition, it should be able to predict the dosimetry and biological response of other experimental scenarios. Ultimately, it is hoped that a robust inhalation model derived from this exercise might prove useful for human risk assessment by facilitating interspecies extrapolation, rationalizing site-specific nasal toxicity, and for understanding systemic toxic effects (or the lack thereof) in rodent bioassays.

METHODS AND MATERIALS

Metabolism and Nasal Deposition Data

The details of the in vitro metabolism studies and the in vivo experiments to evaluate the regional depletion of nonprotein sulfhydryl (NPSH) concentration in the nasal cavity will be published separately, although a prior communication has described the regional carboxylesterase activity of the rat nasal epithelium (Frederick et al., 1994a). The deposition of EA in the rat nasal cavity under unidirectional flow conditions (steady, inspiratory flow) at 50 and 105 ml/min has been previously described (Stott & McKenna, 1984; Stott et al., 1986). In addition, a detailed description of the deposition of EA in the rat nasal cavity under unidirectional and cyclic flow at 200 ml/min, the determination of the concentration of desorbed ethanol in the air exiting the nasal cavity, and the effects of carboxyl-

TABLE 1. Criteria for Constructing and Evaluating an Inhalation PBPK Model that Incorporates a Nasal Compartment

Can the model:

1. Respond appropriately to changes in inhalation airflow rate?
   The fraction of inhaled compound that is absorbed in the nasal cavity decreases with increasing flow rate (e.g., Aharonson et al., 1974; Stott & McKenna, 1984; Kimbell et al., 1993; Morris et al., 1993).

2. Incorporate important interspecies differences in nasal anatomy and airflow patterns and adequately describe interspecies differences in fractional deposition?
   The fractional deposition for an inhaled compound varies between species in a manner that is not predictable based on the relative nasal cavity surface area (e.g., Morris et al., 1986; Morgan & Monticello, 1990). This variation appears to be attributable to interspecies differences in anatomy and air flow patterns (e.g., Hahn et al., 1993; Kimbell et al., 1993).

3. Appropriately relate metabolism to nasal deposition?
   Inhibition of nasal epithelial metabolism has been demonstrated to inhibit deposition of inhaled compounds in the nasal cavity (e.g., Stott & McKenna, 1985; Morris et al., 1993).

4. Describe the nasal deposition of the range of compounds demonstrated to deposit with 0.1–99.9% efficiency at physiological flow rates?
   Lipophilic compounds such as *o*-xylene have a very low fractional deposition (<1%), whereas very hydrophilic and ionizable acids (e.g., hydrogen fluoride) are completely deposited (>99%) (Morris & Smith, 1982; Morris et al., 1993). Other compounds (e.g., acetone, ethanol) are deposited with intermediate efficiency (Morris et al., 1986).

5. Describe desorption during cyclic flow?
   Gerde and Dahl (1991) have described the desorption of compounds from the nasal mucosa during exhalation that may limit systemic bioavailability and increase the time required to load systemic tissues with an inhaled compound.

6. Provide tissue dosimetry data to correlate with systemic toxic effects to facilitate route-to-route extrapolation?
   A wide variety of PBPK models have been developed to explore the relationship of tissue dosimetry with toxic effects in tissues remote from the site of exposure, and this approach is useful for extrapolation between different routes of exposure (Gerrity & Henry, 1990).

7. Correlate regional dosimetry in the nasal cavity with regional toxicity?
   Some compounds have been demonstrated to induce a toxic effect in only one of the epithelial tissue types represented in the nasal cavity (e.g., Miller et al., 1981a, 1985; Morgan & Monticello, 1990).

8. Adequately describe the 3 general stages of loading/absorption of an inhaled compound?
   a. Loading of respiratory epithelium and establishment of steady-state uptake in the nasal cavity (< 30 s duration) as observed by Aharonson et al. (1974).
   b. Loading of systemic tissues (hours duration) as modeled by Gerde and Dahl (1991).
   c. Steady-state uptake based on metabolism and systemic excretion (hours to days in duration) as is commonly used for surgical anesthesia.

esterase inhibition on nasal deposition and ethanol desorption will be published elsewhere (Morris & Frederick, 1994).

Computer Modeling

The computer models were prepared and the simulations conducted with either the simulation language SimuSolv (Dow Chemical Co., Midland, MI) on a DECStation 5000/200 or with the related language ACSL/PC for Windows on an IBM/PC-compatible computer. The systemic component of the models was based upon a prior published oral dosing model for EA (Frederick et al., 1992), modified to include a different blood–air partition coefficient and fewer organs. The blood–air partition coefficient for EA was revised based on new laboratory data that indicated that the original determinations had included some residual metabolic activity from an esterase that was resistant to inhibition. The more recent partition coefficient determination with effective metabolic inhibition was 86 ± 21 (±SD; $n = 3$). For comparison, the measured olive oil–air partition coefficient for EA is 454 ± 36 ($n = 3$) and the saline–air partition coefficient is 45 ± 3 ($n = 3$). These values may be used in an equation developed by Gargas et al. (1989) for the prediction of blood–air coefficients to give a similar predicted blood–air partition coefficient for EA of 58. Additional descriptions of all the constants and parameters used in the models will appear in a subsequent publication.

The compartmental model of the nasal cavity was based on a similar model described by Morris et al. (1993), with the substitution of a term to describe the fraction of air flow in the nasal cavity that is assumed to be poorly equilibrated with the underlying mucus layer based on the analysis of Aharonson et al. (1974). The following description of each model will provide additional information on model structure.

RESULTS

Model 1: Nasal Uptake Described by the Model of Aharonson et al. and the Body Described as a Single Compartment with a First-Order Excretion Coefficient

The model of Aharonson et al. (1974) is frequently cited for the flow-dependent deposition of a volatile compound in the nasal cavity. As derived in the original report, this model assumes quasi-steady-state unidirectional airflow and negligible resistance to deposition, "back pressure," from the compound in the blood supplying the nasal cavity; that is, the model assumes that the concentration of the compound in the nasal blood is zero. Originally expressed in terms of the partial pressure of the vapor in the gas phase versus the blood phase (concentration can be substituted for partial pressure via the ideal gas law), the general form of the Aharonson equation as described by Hanna et al. (1989) is

$$\frac{V_f}{RT}\frac{dP_a}{dy} = -K_n p(P_a - P_b)$$

where V_f is airflow rate, p is the perimeter of the airway, K_n is the overall mass transport coefficient, R is the gas constant, T is temperature, y is the distance that air travels through the nasal cavity, P_a is the partial pressure of the compound in the air stream, and P_b is the partial pressure of the compound in the blood. The overall mass transport coefficient (K_n) for the nasal cavity may be determined empirically based on the fractional penetration of the compound in the air stream (F_p) flowing through the nasal cavity:

$$F_p = C_E/C_I = \exp(-K_n A_n/V_f)$$

where C_I and C_E are the concentrations of the compound in the air stream entering and exiting the nasal cavity and A_n is the surface area of the nasal cavity. As noted by Hanna et al. (1989), this overall mass transport coefficient is derived from the air-phase or convective mass transport coefficient (k_c), the Henry's law constant for the compound describing the mucus–air partition coefficient (K_H), and the liquid transport coefficient (k_l) determined by the solubility in the liquid phase, diffusion coefficient, perfusion rate, and the rate of reaction or metabolism in the liquid phase:

$$1/K_n = 1/k_c + 1/K_H k_l$$

The penetration fraction model has been applied to a variety of studies evaluating the deposition of vapors in the nasal cavity, and once empirically calibrated with the fractional deposition at one flow rate, this equation has been found to provide a reasonable estimate of the fractional deposition at other flow rates. Note also that this formulation effectively describes the limit cases of infinite flow rate (near zero fractional deposition) and near-zero flow rate (near total fractional deposition).

The original Aharonson et al. (1974) model can be easily programmed with a simulation language and linked to a one-compartment systemic elimination model (Figure 1) that describes the overall systemic clearance of the compound by all routes. Since rat nasal cavity deposition of EA has been determined at varying flow rates in different laboratories at different times (Stott et al., 1986; Stott & McKenna, 1984; Morris & Frederick, 1994), one data set can be used to calibrate the model and the second data set can be used to validate it. Similarly, the systemic clearance of EA has been described in two laboratories and these data can be used to determine the systemic elimination rate constant (deBethizy et al., 1987; Ghanayem et al., 1987). After calibrating the model with the more recent deposition data, the comparison of the model output to all of the data indicates reasonable correspondence (Figure 2).

As described by the original authors, the strengths of the model are based

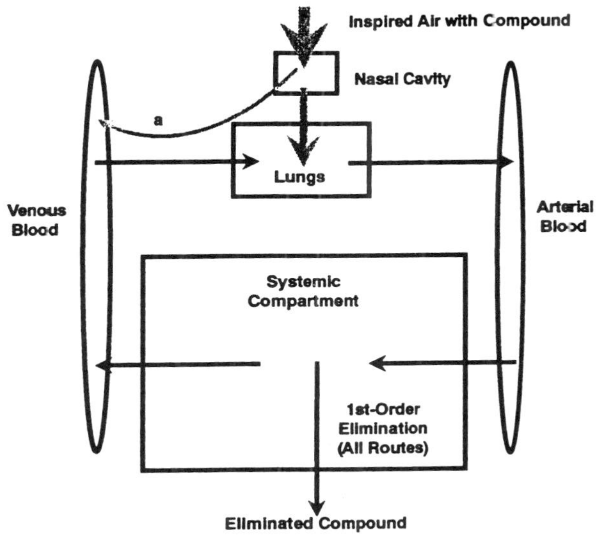

FIGURE 1. Outline of model 1 composed of the unidirectional nasal deposition model of Aharonson et al. (1974) linked to a one-compartment body. This model is designed to estimate the amount of inhaled compound that penetrates the nasal cavity and reaches the lungs. The model is calibrated by determining the fractional deposition of an inhaled compound or vapor in the nasal cavity at a well-defined unidirectional flow rate. The systemic elimination constant is the sum of all elimination rate constants determined in a disposition study. Arrow (a) represents the amount of material deposited in the blood of the nasal cavity.

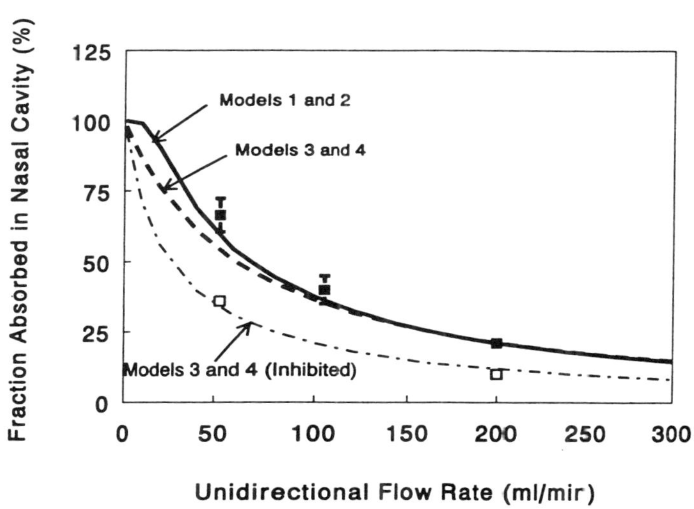

FIGURE 2. Comparison of the output of models 1 and 2 (solid line) and models 3 and 4 (dashed line) to the experimental data (solid squares) collected from two laboratories on the deposition of EA in the rat nasal cavity as a function of flow rate under unidirectional flow conditions (Stott et al., 1986; Morris & Frederick, 1994). The models are calibrated by the fractional deposition at 200 ml/min. The output of models 3 and 4 when simulating 90% inhibition of esterase activity (dashed and dotted line) was compared to data collected from animals pretreated with a carboxylesterase inhibitor (open squares).

on its inherent simplicity for describing the "defense of the lung against polluted air" (Aharonson et al., 1974). The original paper notes several interesting points relative to the model parameters and the experimental data. Since the local uptake coefficients (K) and the perfused area of the nose (A) were not easily determined at various air flow rates, an overall uptake coefficient (R_u) could be calculated based on the experimental data that incorporated the empirical relationship of K and A, even though the exact mathematical relationship was unknown. The data indicated that R_u varied somewhat with changes in flow rate, which prompted the authors to suggest that such factors as variation in the perfused area in the nasal cavity and changes in nasal perfusion with flow rate could affect the overall uptake coefficient.

There are several limitations of the nasal deposition component of this model. Since it is formulated to describe the fractional penetration to the lungs at various flow rates based on an experimental determination of K_n, it does not directly incorporate parameters that facilitate interspecies extrapolation. This is because K_n is dependent on nasal anatomy (well-ventilated vs. poorly ventilated surface area), metabolism, and airflow patterns (including turbulent vs. laminar flow in well-ventilated regions) that vary widely between species; therefore, each species requires a separate experimental determination of K_n. Also, the model does not attempt to describe regional dosimetry within the nasal cavity. This is a problem for interspecies extrapolation of compounds that exhibit marked regional toxicity within the nasal cavity, since the model cannot associate regional anatomic, metabolic, or airflow features with the toxic response. As a consequence, it is not possible to correlate similar anatomical features between species for quantitative dosimetry comparisons. For example, in the case of the acids and esters, which can be toxic to the olfactory epithelium at high concentrations, it is not possible to predict how much compound is absorbed by this sensitive tissue for each species under any given exposure scenario.

The model is based on the simulation of unidirectional flow and cannot describe the cyclic flow of breathing. This can become a critical problem for evaluating the dosimetry of compounds that desorb from the lower respiratory tract during exhalation and are absorbed in the nasal cavity or for those compounds that are desorbed from the nasal cavity during exhalation (Gerde & Dahl, 1991). Similarly, the systemic component of the model has been simplified to the point that interspecies extrapolation of systemic clearance and of the dosimetry between tissues is not possible.

Model 2: Nasal Uptake Described by the Model of Aharonson et al. and with the Body Described by Realistic Physiological, Anatomical, and Metabolic Parameters

This model is the same as model 1, except that the one-compartment body is replaced with multiple compartments designed to simulate one or more critical organs (Figure 3). An effort is made to describe each organ that is modeled with a realistic volume, blood flow, metabolic pathway(s),

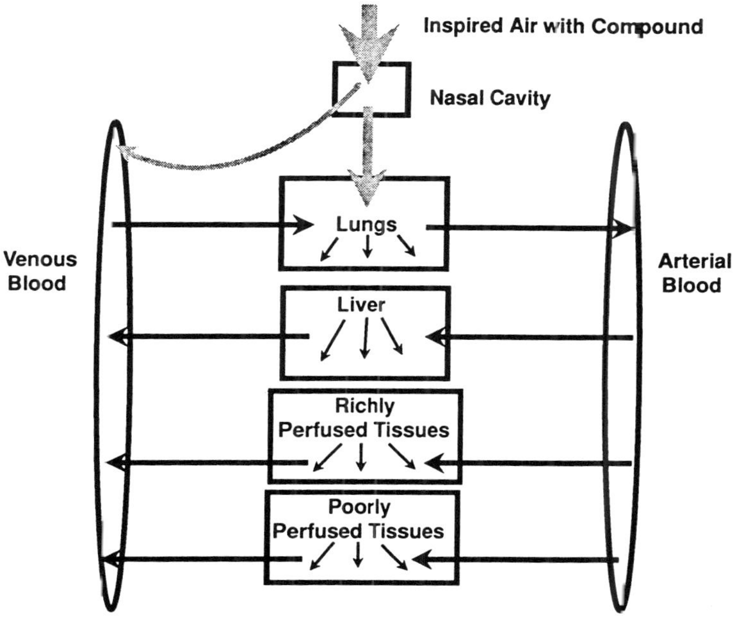

FIGURE 3. Outline of model 2 in which the Aharonson et al. (1974) nasal deposition model is linked to a multicompartment body into which measured metabolic rates are incorporated into various systemic compartments. Although this representative model is built of only four compartments, the modular design of the model allows it to be expanded to include as many systemic organs as necessary to evaluate tissue dosimetry versus a potential toxic effect. This model can also be used for interroute extrapolation relative to systemic toxicity by incorporating similar bioavailability models for the skin and gastrointestinal tract.

and partitioning of the compound. Since the description of the nasal cavity is the same as model 1, the output of the nasal deposition component of the model is also the same. Correspondingly, the strengths and weaknesses of this component of the model are the same as model 1.

However, inasmuch as this model provides more realistic systemic anatomy than model 1, it can provide useful dose-response information on the fraction of the dose delivered to specific tissues. This information can be useful for interroute extrapolation of systemic dose for systemic toxicants. The residual problem of interspecies extrapolation of the fractional penetration of the inhaled compound to the lungs remains.

Model 3: Nasal Uptake Described by a Model Incorporating Species-Specific Anatomy, Air Flow Patterns, and Metabolism with a Body Described as a Single Compartment with a First-Order Excretion Coefficient

Several groups have explored nasal airflow patterns in humans (e.g., Swift & Proctor, 1977; Hanna et al., 1989; Hahn et al., 1993). Similar studies have been conducted of airflow in the rat nasal cavity (Morgan et al., 1991;

Kimbell et al., 1993), since this species is the most common animal model used for inhalation toxicity evaluations. Consistent with the anatomical differences between the human nasal cavity and the rat nasal cavity (reviewed in DeSesso, 1993; Gross & Morgan, 1992), the available data on the airflow patterns in each species are also quite different. These studies assume nasal breathing in both cases, although the rat is an obligate nose breather while humans are not. Given the different anatomy and inhalation rates, it is reasonable to conclude that the calculated regional mass transport coefficients for nasal breathing would be quite different for each species. Comparison of the rodent versus human regional mass transport coefficients may be critical for interspecies dosimetry comparisons.

Although the calculation of local mass transport coefficients for the nasal cavity and their interspecies comparison is not complete, the nasal airflow modeling studies have provided enough information to describe basic airflow patterns in each species. This information may be coupled with biochemical information to construct a biological model of the nasal cavity with realistic airflow patterns and regional metabolism. Depending on the resolution of the data used to validate the model, the complexity of this model may range from fairly simple to very complex.

As an example of this approach, metabolic data were collected for EA

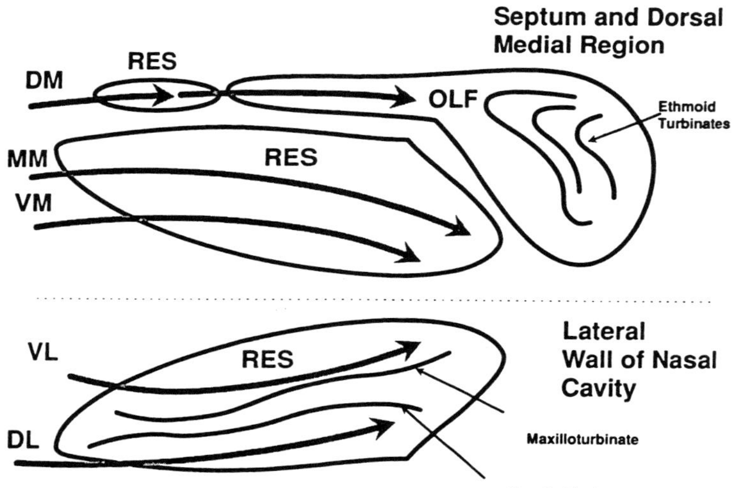

FIGURE 4. Compartmental description of the rat nasal cavity as simulated by models 3 and 4. The rodent nasal cavity is depicted with a sagittal section through one nostril adjacent to the septum and with the lateral wall folded down. The arrows represent the major air streams flowing through the nasal cavity on inhalation as described by Kimbell et al. (1993). The air streams are labeled according to the region over which they flow [DM, dorsal medial meatus; DL, dorsal lateral meatus; MM, middle (medial and lateral) meatuses; VL, ventral lateral meatus; and VM, ventral medial meatuses]. The dominant epithelial type covering each region is designated (RES, respiratory epithelium; OLF, olfactory epithelium), with the exception that the squamous epithelium lining the vestibule region and the floor of the nasal cavity are not shown.

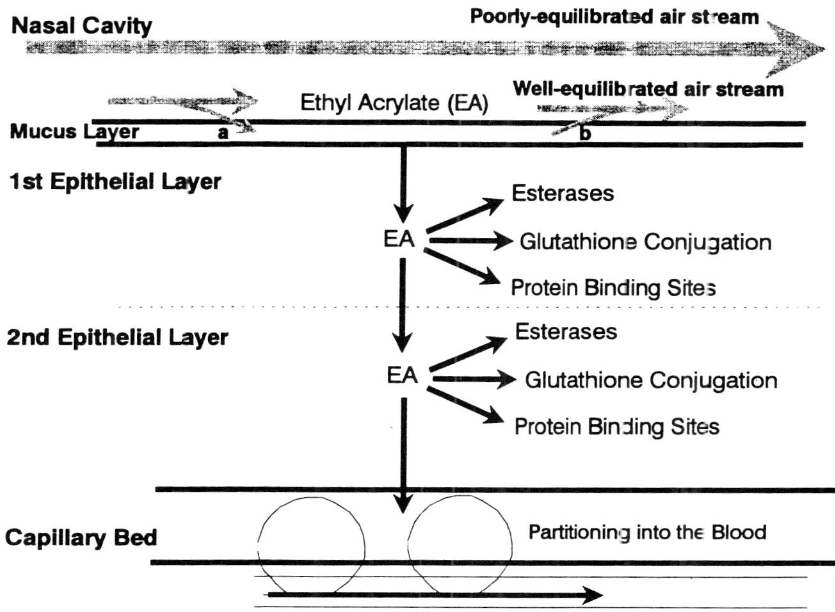

FIGURE 5. Layered compartment description of the epithelium lining the nasal cavity simulated by models 3 and 4. The respiratory epithelium is described by two epithelial layers and the olfactory epithelium by four layers (Morris et al., 1993). The rates of metabolism by each metabolic route in each epithelial layer are assumed to be uniform for models 3 and 4, although a nonuniform distribution of metabolic activity may also be simulated. The rates of metabolism for EA were determined in vitro with homogenates of epithelium removed from various regions of the nasal cavity. These models incorporate a flow-rate-dependent estimate of the fraction of the air stream that is well equilibrated with the underlying epithelium in each region. The rate of deposition of a compound in the mucus layer (arrow a) is assumed to be offset by the partitioning of compound from the mucus into the air stream (arrow b). This allows "back pressure" in the tissue phase of the model that can result in desorption during exhalation.

with olfactory and respiratory epithelium from the rat nasal cavity, as previously described for other tissues (Frederick et al., 1992). The carboxylesterase activity of the nasal tissue toward EA was very high (Frederick et al., 1994a), with an estimated hydrolysis half-life in each region of a fraction of a second. The regional nasal metabolic rates were incorporated into a compartmental model of the rat nasal cavity (Figures 4 and 5) that was based on previously published models for slowly metabolized vapors (Gerde & Dahl, 1991; Morris et al., 1993). The EA model for a rapidly metabolized vapor differed somewhat in that it used the general form of the equation of Aharonson et al. (1974) to estimate the fraction of the air stream that was well equilibrated with the underlying epithelium in each region of the nasal cavity as a function of flow rate (vide infra). Fick's law of diffusion was used to describe mass transport in the epithelium as a process competing with local metabolism. Both the description of air flow equilibration and the description of diffusion in tissue should be viewed as simplifying

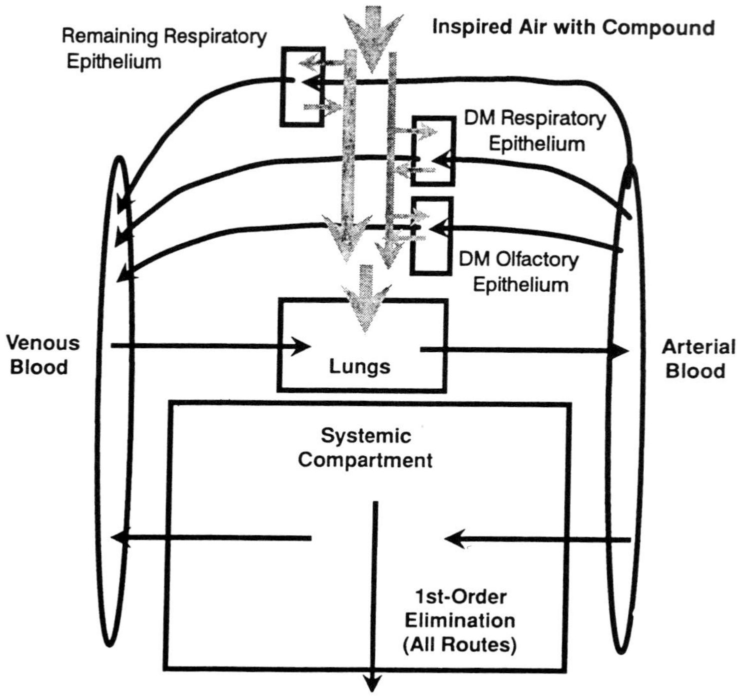

FIGURE 6. Compartmental model for the nasal cavity with a one-compartment model of the body used to construct model 3. This model describes the deposition of a vapor in three regions of the nasal cavity based on the inhalation airflow pattern in the rodent nasal cavity. Two air streams are simulated, with one flowing through the dorsal meatus region (DM) and the other stream (a composite of the remaining air streams) flowing over the remaining respiratory epithelium. Each nasal epithelial compartment is constructed as described in Figure 5. The body is treated as one compartment with a single first-order rate of clearance of the compound once it reaches the systemic circulation.

assumptions, that is, empirical descriptions rather than being firmly grounded in theoretical underpinnings. The nasal cavity was then connected to a one-compartment systemic elimination model (Figure 6).

Although this nasal cavity model was constructed of three epithelial regions, conceptually the model is modular, and could be expanded (or condensed) to accommodate differing anatomical complexities. Here, EA induces cytotoxicity solely in the olfactory epithelium of the nasal cavity. Assuming that the sensitivity of the human olfactory epithelium is equal to that of the rodent olfactory epithelium, quantitation of the amount of compound delivered to the sensitive olfactory epithelium becomes critical for interspecies extrapolation, and a compartmental model was designed to address the problem. Critical features incorporated into the model are the major streams of airflow into sensitive (and nonsensitive) regions of the nasal cavity (Kimbell et al., 1993), as well as an estimate of the scrubbing efficiency of the resistant respiratory epithelium that lies anterior to the sensitive olfactory epithelium.

Since this model incorporates anatomical and biochemical features of the rat nasal cavity, the model outputs can be compared to various indicators of the amount of compound delivered to the nasal epithelium. For example, the predicted overall deposition of EA as a function of unidirectional flow rate is reasonably consistent with the data (Figure 2). In addition to describing overall deposition in the rat nasal cavity, the model can be used to describe the desorption of ethanol released by hydrolysis of EA by the nasal epithelium. The dose-response comparison of model output to the concentration of released ethanol measured in a unidirectional flow experiment with EA (Morris & Frederick, 1994) indicates that the model provides an estimate that is roughly consistent with the data (Figure 7A). Furthermore, the in vivo inhibition of esterase activity with a carboxylesterase inhibitor resulted in a significant decrease in both the deposition of EA (Stott et al., 1984, 1986; Morris & Frederick, 1994) and the release of ethanol by the rat nasal mucosa (Morris & Frederick, 1994). The model provided a fairly reasonable description of the effect of the inhibitor on both endpoints (Figures 2 and 7B).

Future versions of the model are planned to more exactly incorporate local nonuniform mass transport coefficients derived from ongoing airflow dynamics studies with the rat nasal cavity (Kimbell et al., 1993) and to incorporate human metabolism and airflow dynamics (e.g., Hahn et al., 1993). Effectively, the current model uses the general form of the Aharonson equation as an interim description of the published data linking inspiratory airflow rate and deposition in the nasal mucus. These data have been interpreted to indicate that increasing the inspiratory flow rate decreases the fraction of the

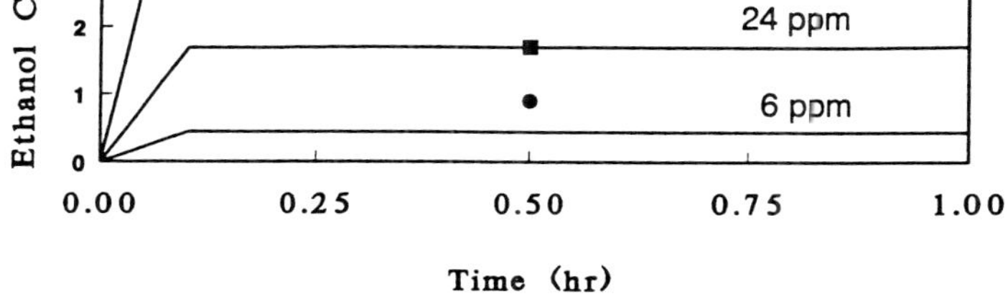

FIGURE 7. Time course and dose response for the release of ethanol from EA by the esterase activity in the rat nasal epithelium during the simulation of a 1-h unidirectional flow deposition experiment. Comparison of the output of models 3 and 4 (solid line) to experimental data at 6 ppm (circle), 24 ppm (square), and 88 ppm (triangle). (A) Normal animals. Although the ethanol concentration was monitored throughout the experiment, only the average concentration is shown for clarity.

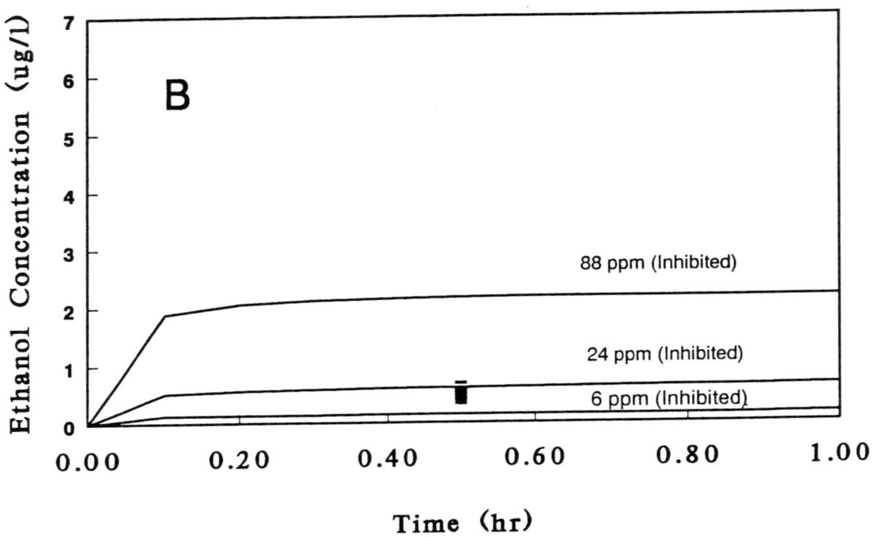

FIGURE 7. Time course and dose response for the release of ethanol from EA by the esterase activity in the rat nasal epithelium during the simulation of a 1-h unidirectional flow deposition experiment. Comparison of the output of models 3 and 4 (solid line) to experimental data at 6 ppm (circle), 24 ppm (square), and 88 ppm (triangle) (*Continued*). (B) Pretreated with a carboxylesterase inhibitor. Although the ethanol concentration was monitored throughout the experiment, only the average concentration is shown for clarity.

air stream that is well equilibrated with the anterior regions of the respiratory tract. This is reasonable if the Aharonson equation is viewed as describing the relationship of two competing, orthogonal gas-phase mass transport vectors in the nasal cavity:

$$F_p = C_E/C_I = \exp(-K_c A_n/U_n A_{cs})$$

where K_c is the convective mass transport coefficient, U_n is the unidirectional flow rate through the nasal cavity, and A_{cs} is the cross-sectional area of the nasal cavity (note that the product of U_n and A_{cs} is V_f). This general mathematical description can be used to describe the regional mass transport of a vapor trying to enter the regional mucus layer (the product of K_c and A_n, assuming the concentration of the compound in the mucus is zero) relative to migrating into the next region of the respiratory tract. In this case, the convective mass transport coefficient is assumed to reflect all factors related to the efficiency of transport from the air phase into the mucus layer (diffusivity, turbulence, shear, etc.). Model 3 assumes that the regional K_c does not change with flow rate and the regional U_n is linearly proportional to total flow rate. The published airflow modeling studies suggest that these simple assumptions on the relationship of the regional K_c and the regional U_n to flow rate are crude, and these relationships are best described by directly encapsulating the results of the airflow modeling studies by curve-fitting their results (e.g., Hahn et al., 1993; Kimbell et al., 1993). The definition of the mass transport coefficient in this equation is different from the original Aharonson equation in that tissue-phase diffusion and reactivity are handled separately in the model. Furthermore, although the equation is used to describe the deposition pressure of the vapor in the gas phase trying to enter the mucus layer, the deposition pressure is offset in the model by Henry's law partitioning of the compound already in the mucus layer back into the gas phase (see Figure 5).

The major weakness of this general approach comes from the body of data needed to parameterize the model and to evaluate it. In addition, some of the airflow dynamics data (including relevant computer simulation output) are not currently in a form that is easily amenable for direct input into a compartmental physiological model. Therefore, some form of "lumping" of the airflow data is necessary to provide an estimate of the amount of compound delivered in the air stream to equilibrate with different regions of the nasal cavity. The computer modeling methods commonly used for airflow dynamics modeling are not easily amenable to modeling of tissues with their associated metabolism, perfusion, and partitioning. In addition, the airflow dynamics models cannot easily simulate the dynamic change in resistance to deposition in the nasal epithelium that results from accumulation of a slowly metabolized compound in the blood during systemic loading of the compound into the tissues. An adequate description of these dynamic processes is necessary to accurately describe the "boundary conditions" of the nasal cavity dosimetry models for simulations of long exposure periods (e.g., 8-h workplace exposure or continuous ambient exposure). Conversely, the methods used for modeling systemic blood flow and the dynamic metabolic processes occurring in tissues are not adequate for modeling air flow dynamics. Given these considerations, it appears that a compartmental nasal model that uses a separate air flow dynamics analysis for its input represents the best technical compromise at this time.

For systemic disposition, model 3 uses a one-compartment model of the body with a single first-order elimination constant. This is sufficient to provide an estimate of systemic loading and elimination, but does not provide organ-specific information to correlate delivered dose with biological effects in systemic tissues.

Model 4: Nasal Uptake Described by a Model Incorporating Species-Specific Anatomy, Airflow Patterns, and Metabolism with a Body Described by Realistic Physiological, Anatomical, and Metabolic Parameters

Model 4 is the same as model 3, except that the one-compartment body is replaced with multiple compartments designed to simulate one or more critical organs (Figure 8) in the same way that a "body" was added to model 1 to make model 2. Since the description of the nasal cavity is the same as model 3, the output of the nasal deposition component of the

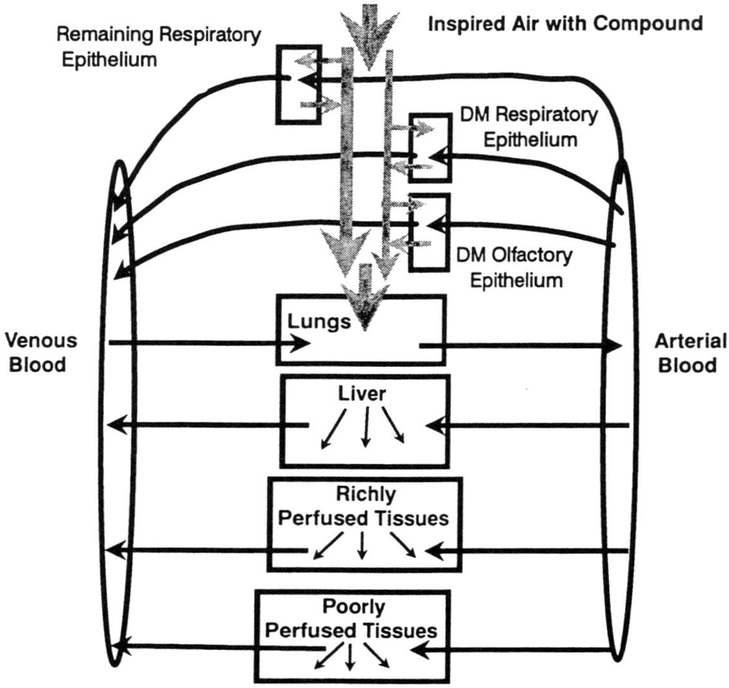

FIGURE 8. Compartmental model of the nasal cavity of model 3 linked to a multicompartment body to give model 4. This model uses measured metabolic rates for the nasal cavity, as well as a variety of systemic tissue compartments. Since its design is modular, the number of compartments can be varied to address specific potential toxicity concerns. Interspecies extrapolation with this model can be conducted by substituting species-specific data on tissue volumes and metabolism, blood flow, epithelial type distribution, and inhalation rates and nasal airflow patterns. Both route-to-route extrapolation for systemic exposure and interspecies extrapolation are possible with this model structure.

model is also the same. Correspondingly, the strengths and weaknesses of this part of the model are the same as model 3.

There are several ways to compare experimental data to the outputs of models 3 or 4 for the purposes of model evaluation or to suggest areas for additional experimentation. For example, an acute nose-only inhalation study was conducted with high doses of EA to correlate local dosimetry in the nasal cavity with NPSH depletion and cytotoxicity observed by histopathology. The histopathological evaluation was conducted by mapping the observed lesions on representative cross sections of the rat nasal cavity to facilitate the correlation of local tissue concentration of the chemical, biochemical effects, and histopathological response based on a mapping strategy developed by Drs. K. T. Morgan, J. S. Kimbell, M. N. Godo, and S. Mery (personal communication, 1993). The histopathological data (Frederick et al., 1994b) confirm previous subchronic and chronic studies that have reported a well-defined localization of the cytotoxic effects of EA to the olfactory epithelium in the dorsal meatus region of the nasal cavity. The lesions appear to be localized along the dorsal medial air stream described by Kimbell et al. (1993) and become more severe and move posteriorly with increasing concentration and exposure time.

The predicted dose response for NPSH depletion in the respiratory epithelium of the rat nasal cavity appears to correlate reasonably well with the data from this study based on a nonenzymatic rate of glutathione conjugation (Figure 9). This tissue demonstrates no histopathological indications of toxicity. The depletion of NPSH in the olfactory epithelium occurs more rapidly than can be explained by simple nonenzymatic conjugation with glutathione. Based on a prior report of glutathione transferase activity for EA in other tissues (Potter & Tran, 1992), the current model incorporates a description of glutathione transferase activity for this region that appears to adequately describe the response. Additional studies are being conducted to explore this effect. At the mid and high doses used in the study, cytotoxicity in this sensitive tissue may have contributed to the loss of NPSH content (Frederick et al., 1994b).

Since models 3 and 4 explicitly describe airflow in the nasal cavity, they can be scaled on a variety of time scales. On a macro time scale (time period of hours or days), a composite of inhalation and exhalation can define the dosimetry associated with biological effects. On a micro time scale (seconds), the model can describe the airflow and deposition of a vapor during inhala-

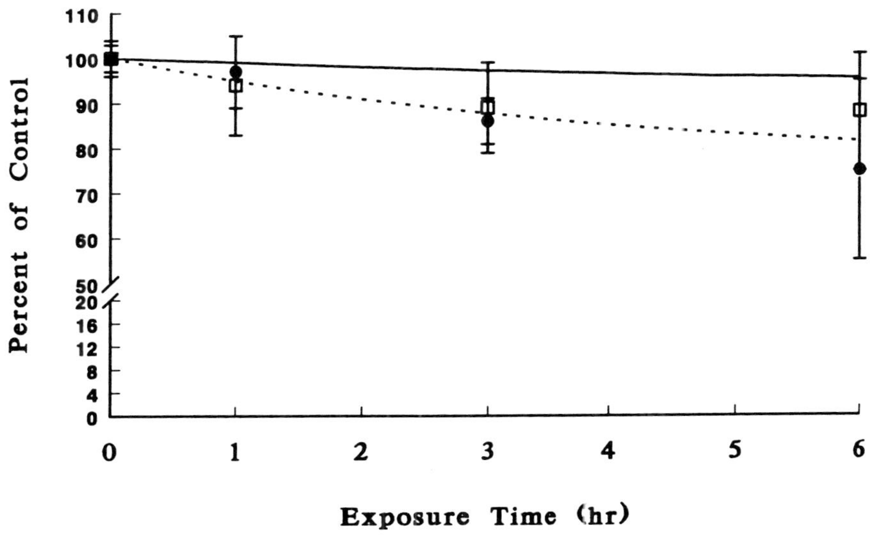

FIGURE 9. Time course for the depletion of NPSH content in the olfactory epithelium relative to the respiratory epithelium during 6 h of nose-only inhalation exposure to EA at 5 ppm with cyclic breathing. The lines are the predicted NPSH concentrations from models 3 and 4 (simulating cyclic breathing) for the respiratory epithelium along the lateral walls (solid line) and olfactory epithelium (dashed line) compared to experimental data for these regions (respiratory, open square; olfactory, solid circle).

tion, as well as the desorption of an absorbed vapor from the lungs or mucus during exhalation. Although we are currently modeling the distribution of airflow during exhalation as the same as inhaled air, this can be changed appropriately based on additional data on exhaled air distribution. An example of modeling the change in concentration of EA in the layers of the olfactory epithelium during a few seconds of breathing demonstrates the rapid clearance of the compound from the tissue due to metabolism (Figure 10). In addition, the model predicts a steep concentration gradient from the mucus layer to the blood exchange region similar to that observed in the skin in dermal absorption studies (Schaefer et al., 1982).

As with model 2, a more complete description of the body can provide useful dose-response information on the fraction of the dose delivered to specific internal tissues. For example, model 4 predicts depletion in rats of lung NPSH concentration to 52% of the control value following 4 h of inhaling 1000 ppm EA versus the experimentally determined value of 60% (Silver & Murphy, 1981). A simulation incorporating esterase inhibition with the same exposure scenario predicted the lung NPSH concentration to be 23% of the control value versus an experimental value of 20%. This information can be useful for interroute extrapolation, since the effects on systemic tissues are generally independent on the portal of entry. By providing the capability to simulate as much normal anatomy and physiology as necessary, this model has the potential for providing a reasonable estimate

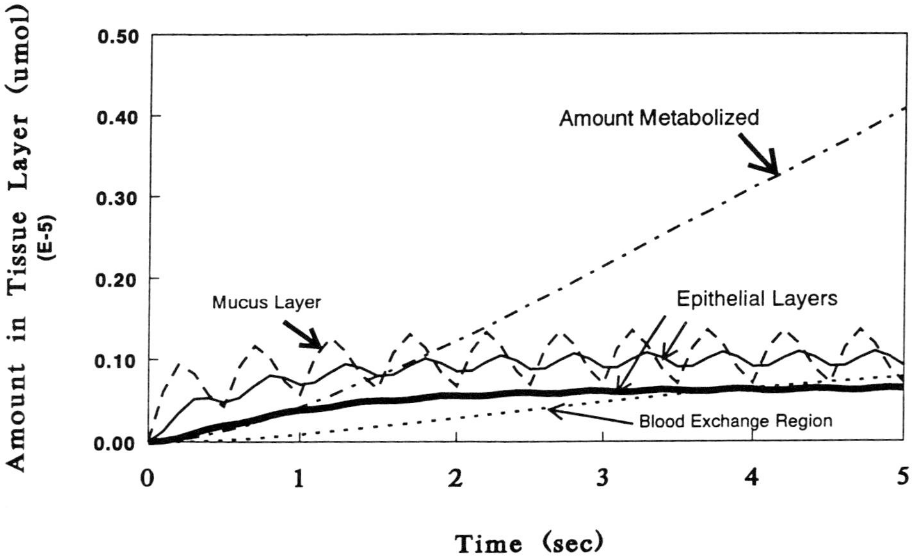

FIGURE 10. Predicted concentration of EA in rat olfactory epithelium during a simulation with models 3 and 4 of normal cyclic breathing in the mucus (dashed line), epithelial layers (solid lines), and blood exchange region (dotted line). The cumulative amount of compound predicted to be metabolized by the epithelium is also shown (dotted and dashed line).

of tissue dosimetry for any organ of interest. The problem of adequately parameterizing this model is increased over that of model 3, since the amount of physiological data on tissue blood flow, partitioning, and metabolism increases linearly with the number of systemic organs that are modeled. Of course, this "investment" in data should be rewarded by increased realism in accurately describing the physiological response (or lack thereof) for various simulated exposures.

DISCUSSION

Biological modeling of compound absorption is not unique to the nasal cavity. Since epithelium with a similar structure covers much of the area that can serve as a portal of entry (e.g., skin, gastrointestinal tract, and respiratory tract with the exception of the lungs), common elements might be expected from the models that have been proposed for each site. In fact, it could be argued that the same basic model could be used to describe absorption across any epithelial tissue, and the only substantive differences between the models should derive from the description of interaction with the phase containing the compound (e.g., air, liquid phase, or gut contents). Obviously, each epithelial tissue has certain specialized elements (e.g., stratum corneum over the skin, mucus over the nasal epithelium, etc.) and differences in thickness and vascularization need to be accounted for, but there is a common basic structure.

This point is worth noting because most of the modeling of epithelial absorption has been with the skin. Various dermal absorption models have been published, and the models may generally be classified as using either a stacked compartmental modeling approach or simple diffusion based on a Fick's law approach. For example, models have been developed using combinations of stacked compartments, diffusion between compartments based on Fick's law, and metabolism or binding competing with diffusion in the viable epithelium (Zatz, 1992; reviewed in Guy & Hadgraft, 1989). Models that have not incorporated epithelial metabolism or binding of the compound in the skin have used a permeation coefficient based on Fick's law to describe the net rate of transport through the skin (e.g., McDougal et al., 1990; Guy & Hadgraft, 1989).

Models 3 and 4 are conceptually based on the stacked compartment dermal models noted earlier linked to an airflow dynamics component to describe the interaction of respiratory tract mucus with the air phase. Various formulations of this general approach can be prepared (e.g., Gerde & Dahl, 1991; Morris et al., 1993). Given that models 3 and 4 have been designed to incorporate a general description of absorption at a contact site tissue, how useful are these models for route-to-route extrapolation of contact site toxicity? The report of a recent workshop on route-to-route extrapolation suggested that compounds that confined their toxic effects to the site of initial contact were poor candidates for route-to-route extrapolation based on dosimetry comparisions between different contact sites (Gerrity & Henry, 1990). Instead, the report suggested that direct experimentation by each exposure route was necessary to provide a reasonably accurate estimate of risk. The data for EA are consistent with this suggestion. In the rat nasal cavity, the results from models 3 and 4 suggest that the respiratory epithelium is exposed to more EA than the olfactory epithelium. However, the respiratory epithelium does not demonstrate any histopathological indications of toxicity at EA concentrations that induce significant cytotoxicity to the olfactory epithelium. Furthermore, subsequent to cytotoxicity in the olfactory epithelium, epithelium that is histologically similar to respiratory epithelium replaces the damaged olfactory epithelium (Miller et al., 1985). Since this replacement respiratory epithelium occupies the same region of the nasal cavity as the original olfactory epithelium, it is exposed to the same amount of inhaled compound as the original tissue, but it does not exhibit any histopathological indications of toxicity nor does it progress to any other cell type on chronic exposure (Miller et al., 1985).

A similar contrast is observed following gavage dosing of EA in the rat stomach. The rat forestomach demonstrates marked cytotoxity and hyperplasia following high gavage doses of EA, but no histopathological indications of toxicity are apparent in the glandular stomach (National Toxicology Program, 1986; Frederick et al., 1990; Ghanayem et al., 1991). Similar to the epithelial tissues of the nasal cavity, this regional difference in gastric toxicity appears to be derived from a difference in tissue sensitivity to EA rather than a substantive difference in regional dosimetry (Frederick et al., 1992).

Although route-to-route extrapolation of contact-site toxicants is problematical, the use of biologically based models such as models 3 and 4 for interspecies extrapolation is feasible as long as the dosimetry in the same tissues is being compared. By incorporating the appropriate anatomy, physiology, metabolism, and airflow description, an estimate of the dosimetry in a human tissue can be compared to an animal that has exhibited a toxic effect and safe exposure limits may be determined. In the case of EA, the dosimetry of the target tissue of concern, olfactory epithelium of the nasal cavity, can be evaluated under a wide range of inhalation exposure scenarios using Monte Carlo sampling methods for model parameters to simulate the distribution of human anatomy, metabolism, and physiology. Our current research is directed toward the construction and validation of a general primate inhalation model for EA to compare to rodent model 4. This comparison should provide a strong scientific basis for the establishment of safe human exposure limits.

REFERENCES

Aharonson, E. F., Menkes, H., Gurtner, G., Swift, D. L., and Proctor, D. F. 1974. Effect of respiratory airflow rate on removal of soluble vapors by the nose. *J. Appl. Physiol.* 37:654–657.

Amoore, J. E. 1986. Effects of chemical exposure on olfaction in humans. In *Toxicology of the nasal passages*, ed. C. S. Barrow, pp. 155–190. Washington, DC: Hemisphere.

Beidler, L. M. 1980. The chemical senses: gustation and olfaction. In *Medical physiology*, ed. V. M. Mountcastle, pp. 586–602. St. Louis, MO: C. V. Mosby.

Bogdanffy, M. S., Randall, H. W., and Morgan, K. T. 1987. Biochemical quantitation and histochemical localization of carboxylesterase in the nasal passages of the Fischer-344 rat and the B6C3F1 mouse. *Toxicol. Appl. Pharmacol.* 88:183–194.

Dahl, A. R., Miller, S. C., and Petridou-Fischer, J. 1987. Carboxylesterases in the respiratory tracts of rabbits, rats and Syrian hamsters. *Toxicol. Lett.* 36:129–136.

Dahl, A. R., Snipes, M. B., and Gerde, P. 1991. Sites of uptake of inhaled vapors in beagle dogs. *Toxicol. Appl. Pharmacol.* 109:263–275.

deBethizy, J. D., Udinsky, J. D., Scribner, H. E., and Frederick, C. B. 1987. The disposition and metabolism of acrylic acid and ethyl acrylate in male Sprague–Dawley rats. *Fundam. Appl. Toxicol.* 8:549–561.

DeSesso, J. M. 1993. The relevance to humans of animal models for inhalation studies of cancer in the nose and upper airways. *Qual. Assur. Good Pract. Regul. Law* 2:213–231.

Frederick, C. B., Hazelton, G. A., and Frantz, J. D. 1990. Histopathologic and biochemical response of the stomach of male F 344/N rats following two weeks of oral dosing with ethyl acrylate. *Toxicol. Pathol.* 18: 247–256.

Frederick, C. B., Potter, D. W., Chang-Mateu, M. I., and Andersen, M. E. 1992. A physiologically based pharmacokinetic and pharmacodynamic model to describe oral dosing of rats with ethyl acrylate and its implications for risk assessment. *Toxicol. Appl. Pharmacol.* 114:246–260.

Frederick, C. B., Udinsky, J. R., and Finch, L. 1994a. The regional hydrolysis of ethyl acrylate to acrylic acid in the rat nasal cavity. *Toxicol. Lett.* 70:49–56.

Frederick, C. B., Brown, D. W., and Lomax, L. G. 1994b. Regional cytotoxicity and nonprotein sulfhydryl depletion in the rat nasal cavity following acute inhalation exposure to ethyl acrylate. *Inhal. Toxicol.* 6(suppl):426–429.

Gargas, M. L., Burgess, R. J., Voisard, D. E., Cason, G. H., and Andersen, M. E. 1989. Partition coefficients of low molecular weight volatile chemicals in various liquids and tissues. *Toxicol. Appl. Pharmacol.* 98:87–99.

Gerde, P. and Dahl, A. R. 1991. A model for the uptake of inhaled vapors in the nose of the dog during cyclic breathing. *Toxicol. Appl. Pharmacol.* 109:276–288.

Gerrity, T. R., and Henry, C. J. 1990. *Principles of route-to-route extrapolation for risk assessment*, pp. 1–12. New York: Elsevier.

Ghanayem, B. I., Burka, L. T., and Matthews, H. B. 1987. Ethyl acrylate distribution, macromolecular binding, excretion, and metabolism in male Fisher 344 rats. *Fundam. Appl. Toxicol.* 9:389–397.

Ghanayem, B. I., Matthews, H. B., and Maronpot, R. R. 1991. Sustainability of forestomach hyperplasia in rats treated with ethyl acrylate for thirteen weeks and regression after cessation of dosing. *Toxicol. Pathol.* 19:273–297.

Gross, E. A., and Morgan, K. T. 1992. Architecture of nasal passages and larynx. In *Comprehensive treatise on pulmonary toxicology*, Vol. I, *Comparative biology of the normal lung*, ed. R. A. Parent, pp. 7–25. Boca Raton, FL: CRC Press.

Gross, E. A., Swenberg, J. A., Fields, S., and Popp, J. A. 1982. Comparative morphometry of the nasal cavity in rats and mice. *J. Anat.* 135:83–88.

Guilmette, R. A., Wicks, J. D., and Wolff, R. K. 1989. Morphometry of human nasal airways in vivo using magnetic resonance imaging. *J. Aerosol Med.* 2:365–377.

Guy, R. H., and Hadgraft, J. 1989. Mathematical models of percutaneous absorption. In *Percutaneous absorption*, eds. R. L. Bronaugh and H. I. Maibach, pp. 13–26. New York: Marcel Dekker.

Hahn, I., Scherer, P. W., and Mozell, M. M. 1993. Velocity profiles measured for air flow through a large-scale model of the human nasal cavity. *J. Appl. Physiol.* 75:2273–2287.

Hanna, L. M., Frank, R., and Scherer, P. W. 1989. Absorption of soluble gases and vapors in the respiratory system. In *Respiratory physiology*, eds. H. K. Chang and M. Paiva, pp. 277–316. New York: Marcel Dekker.

Kimbell, J. S., Gross, E. A., Joyner, D. R., Godo, M. N., and Morgan, K. T. 1993. Application of computational fluid dynamics to regional dosimetry of inhaled chemicals in the upper respiratory tract of the rat. *Toxicol. Appl. Pharmacol.* 121:253–263.

McDougal, J. N., Jepson, G. W., Clewell, H. J., III, Gargas, M. L., and Andersen, M. E. 1990. Dermal absorption of organic chemical vapors in rats and humans. *Fundam. Appl. Toxicol.* 14:299–308.

Miller, R. R., Ayres, J. A., Jersey, G. C., and McKenna, M. J. 1981a. Inhalation toxicity of acrylic acid. *Fundam. Appl. Toxicol.* 1:271–277.

Miller, R. R., Ayres, J. A., Rampy, L. W., and McKenna, M. J. 1981b. Metabolism of acrylate esters in rat tissue homogenates. *Fundam. Appl. Toxicol.* 1:410–414.

Miller, R. R., Young, J. T., Kociba, R. J., Keyes, D. G., Bodner, K. M., Calhoun, L. L., and Ayres, J. A. 1985. Chronic toxicity and oncogenicity bioassay of inhaled ethyl acrylate in Fischer 344 rats and B6C3F1 mice. *Drug Chem. Toxicol.* 8:1–42.

Morgan, K. T., and Monticello, T. M. 1990. Airflow, gas deposition, and lesion distribution in the nasal passages. *Environ. Health Perspect.* 88:209–218.

Morgan, K. T., Kimbell, J. S., Monticello, T. M., Patra, A. L., and Fleishman, A. 1991. Studies of inspiratory airflow patterns in the nasal passages of the F344 rat and rhesus monkey using nasal molds: Relevance to formaldehyde toxicity. *Toxicol. Appl. Pharmacol.* 110:223–240.

Morris, J. B. 1990. First–pass metabolism of inspired ethyl acetate in the upper respiratory tracts of the F344 rat and Syrian hamster. *Toxicol. Appl. Pharmacol.* 102:331–345.

Morris, J. B. and Frederick, C. B. 1994. Upper respiratory tract uptake of acrylate ester and acid vapors. *Inhal. Toxicol.*, in press.

Morris, J. B., and Smith, F. A. 1982. Regional deposition and absorption of inhaled hydrogen fluoride in the rat. *Toxicol. Appl. Pharmacol.* 62: 81–89.

Morris, J. B., Clay, R. J., and Cavanagh, D. G. 1986. Species differences in upper respiratory tract deposition of acetone and ethanol vapors. *Fundam. Appl. Toxicol.* 7:671–680.

Morris, J. B., Clay, R. J., Trela, B. A., and Bogdanffy, M. S. 1991. Deposition of dibasic esters in the upper respiratory tract of the male and female Sprague–Dawley rat. *Toxicol. Appl. Pharmacol.* 108:538–546.

Morris, J. B., Hassett, D. N., and Blanchard, K. T. 1993. A physiologically based pharmacokinetic model for nasal uptake and metabolism of nonreactive vapors. *Toxicol. Appl. Pharmacol.* 123:120–129.

National Toxicology Program. 1986. *Carcinogenesis Bioassay of Ethyl Acrylate*. Technical Report Series 259, Publication (NIH) 82–2515. U.S. Department of Health and Human Services, Public Health Service, National Institutes of Health, Research Triangle Park, NC.

Olson, M. J., Martin, J. L., Larosa, A. C., Brady, A. N., and Pohl, L. R. 1993. Immunochemical localization of carboxylesterase in the nasal mucosa of rats. *J. Histochem. Cytochem.* 41:307–311.

Potter, D. W., and Tran, T–B. 1992. Rates of ethyl acrylate binding to glutathione and protein. *Toxicol. Lett.* 62:275–285.

Reininghaus, W., Koestner, A., and Klimisch, H.–K. 1991. Chronic toxicity and oncogenicity bioassay of inhaled methyl acrylate and *n*-butyl acrylate in Sprague-Dawley rats. *Food Chem. Toxicol.* 29:329–339.

Schaefer, H., Zesch, A., and Stuttgen, G. 1982. *Skin permeability*, pp. 550–639. Berlin: Springer-Verlag.

Schwartz, B. S., Doty, R. L., Monroe, C., Frye, R., and Barker, S. 1989. Olfactory function in chemical workers exposed to acrylate and methacrylate vapors. *Am. J. Public Health* 79:613–618.

Silver, E. H., and Murphy, S. D. 1981. Potentiation of acrylate ester toxicity by prior treatment with the carboxylesterase inhibitor triorthotolyl phosphate (TOTP)[1,2]. *Toxicol. Appl. Pharmacol.* 57:208–219.

Stott, W. T., and McKenna, M. J. 1984. The comparative absorption and excretion of chemical vapors by the upper, lower, and intact respiratory tract of rats. *Fundam. Appl. Toxicol.* 4:594–602.

Stott, W. T., and McKenna, M. J. 1985. Hydrolysis of several glycol ether acetates and acrylate esters by nasal mucosal carboxylesterase in vitro. *Fundam. Appl. Toxicol.* 5:399–404.

Stott, W. T., Ramsey, J. C., and McKenna, M. J. 1986. Absorption of chemical vapors by the upper respiratory tract of rats. In *Toxicology of the nasal passages*, ed. C. S. Barrow, pp. 191–210. Washington, DC: Hemisphere.

Swift, D. L., and Proctor, D. F. 1977. Access of air to the respiratory tract. In *Respiratory defense mechanisms, Part 1,* eds. J. D. Brain, D. F. Proctor, and L. M. Reid, pp. 63–93. New York: Marcel Dekker.

Uraih, L. C., and Maronpot, R. R. 1990. Normal histology of the nasal cavity and application of special techniques. *Environ. Health Perspect.* 85:187–208.

Zatz, J. L. 1992. Simulation studies of skin permeation. *J. Soc. Cosmet. Chem.* 43:37–48.

NASAL DOSIMETRY OF FORMALDEHYDE: MODELING SITE SPECIFICITY AND THE EFFECTS OF PREEXPOSURE

Henry d'A. Heck, Mercedes Casanova

Chemical Industry Institute of Toxicology, Research Triangle Park, North Carolina, USA

The formation of DNA-protein cross-links (DPX) in the nose of F344 rats exposed to formaldehyde (HCHO) is site specific and depends on the prior exposure history of the animal if the exposures were to concentrations ≥6 ppm. The site specificity is interpreted in terms of a pharmacokinetic model in which the rate of DPX formation is proportional to the tissue concentration of HCHO. The model includes both saturable and nonsaturable elimination pathways, and it describes regional differences as being due to differences in nasal airflow rather than to differences in HCHO metabolism. Using a minimal number of adjustable parameters, the model accurately describes DPX formation in two nasal compartments [the lateral meatus (a high-frequency tumor site in a HCHO bioassay) and the medial and posterior meatuses (a low-frequency tumor site in the bioassay)] of F344 rats. The pharmacokinetics of DPX formation in preexposed rats is also describable in terms of the model. The effects of preexposure on DPX yields were predicted by recognizing that concentrations ≥6 ppm induce hyperplasia and squamous metaplasia, which increases the quantity of DNA, whereas concentrations ≤2 ppm do not increase the number of cells. The appropriate pharmacokinetic parameter was scaled in proportion to the quantity of DNA, and the predicted yields of DPX agreed well with experiment. The model suggests a close linkage between the pharmacokinetics of DPX formation and the results of fluid dynamics simulations. The two approaches provide complementary information that is important for an understanding of nasal dosimetry.

Recent studies of the formation of DNA-protein cross-links (DPX) by inhaled formaldehyde (HCHO) in the nose of F344 rats showed that the yield of DPX is site specific, and that the quantity of DPX formed in different regions of the rat nose is correlated with the incidence of nasal squamous-cell carcinomas that were induced in a chronic HCHO bioassay (Casanova et al., 1994). These studies also showed that the concentration of DPX (pmol/mg DNA) formed in the primary target tissue of rats (the lateral meatus, LM) was lower in rats preexposed (PE) to airborne concentrations ≥6 ppm than in naive (N) rats exposed to the same airborne concentration. However, no differences between PE and N rats were detected if the preexposure concentration was ≤2 ppm (Casanova et al., 1994).

The primary purpose of the present investigation was to develop a pharmacokinetic model that can explain (1) the site-specific differences in DPX formation that were observed in the nose of rats and (2) the effects of preexposure that occurred at high concentrations. The model used was

Address correspondence to Dr. Henry d'A. Heck, CIIT, P.O. Box 12137, Research Triangle Park, NC 27709, USA.

essentially identical to one previously developed to account for site specificity in the formation of DPX in the nose of rhesus monkeys (Casanova et al., 1991). In both species, the number of parameters in the model is related to the number of nasal "compartments" (two in the case of rats, three in the case of monkeys). The parameters themselves describe the uptake, elimination, and covalent binding of HCHO in each compartment.

The secondary purpose of this investigation was to explicitly define the relationship between the pharmacokinetics of DPX and nasal airflow. Fluid dynamics simulations offer considerable promise as a means of describing the uptake of toxicants at different sites in the nose of rats (Kimbell et al., 1993) and other species. However, these calculations require validation if they are to be used for quantitative risk assessment. The precise relationship between DPX formation and the results of fluid dynamics simulations is defined later. The pharmacokinetic results provide strong evidence that the site-specific differences in DPX formation are due primarily or exclusively to differences in nasal airflow, not to site-specific differences in HCHO metabolism. Therefore, these results support the use of fluid dynamics simulations to investigate the site specificity of DPX formation in the nasal cavity.

MATERIALS AND METHODS

Materials

The materials used for exposure of male F344 rats to HCHO and to $H^{14}CHO$ and for the determination of DNA-protein cross-links are reported elsewhere (Casanova et al., 1994).

Exposures

The methods used to expose the rats subchronically or acutely to HCHO are described elsewhere (Casanova et al., 1994). Briefly, the nominal exposure concentrations were 0.7, 2, 6, or 15 ppm. Three or four groups of 10 rats each, designated preexposed (PE), were exposed (whole-body, 6 h/day, 5 days/wk, 11 wk + 4 days) at each of these concentrations. Analytical (time-weighted average, TWA) concentrations were 0.70 ± 0.02, 2.05 ± 0.04, 5.88 ± 0.13, or 14.52 ± 0.30 ppm ($\bar{x}$ ± SD). Corresponding groups, designated naive (N), were exposed to room air. On day 5 of wk 12, the PE and N rats were simultaneously exposed (nose-only, 3 h) to $H^{14}CHO$ at the same concentration as that used for preexposure. Analytical TWA concentrations for the nose-only exposures were 0.72 ± 0.02, 2.06 ± 0.03, 6.01 ± 0.12, or 15.8 ± 0.3 ppm ($\bar{x}$ ± SD). Two rats from each PE and N group were reserved for histopathology and were not exposed to $H^{14}CHO$; pathology results are described elsewhere (Casanova et al., 1994).

Tissue Sampling

The rats were killed immediately after exposure by pentobarbital overdose and exsanguination via the abdominal aorta. The head was removed, and the nasal cavity was opened adjacent to the midline. Mucosal tissue was dissected from two regions of the nose denoted as high-frequency tumor and low-frequency tumor areas (Casanova et al., 1994). The high-frequency tumor area consisted of the complete lateral meatus (LM). The low-frequency tumor area consisted of the medial aspects of the naso- and maxilloturbinates, the posterior lateral wall, the posterior dorsal septum excluding the olfactory region, and the nasopharyngeal meatus. This region is denoted hereafter as the medial and posterior meatuses (M:PM).

Weights of tissue collected from the LM and M:PM were similar at all concentrations in N rats and in PE rats that were exposed to 0.7 or 2 ppm (~15 mg each per rat) (Casanova et al., 1994). However, an increase in tissue weight of approximately 1.3-fold (relative to N rats) was observed in the lateral meatus of PE rats exposed to 6 ppm, and an increase of approximately 1.6-fold (relative to N rats) was detected in the lateral meatus and medial and posterior meatuses of PE rats exposed to 15 ppm (Casanova et al., 1994). The increases in tissue weight were due to hyperplasia and squamous metaplasia, which occurred in PE rats exposed to 6 or 15 ppm but not to 0.7 or 2 ppm (Casanova et al., 1994). Proportional increases in the quantity of DNA also occurred, so that the ratio of DNA to tissue weight (3.0 ± 0.6 μg/mg; $\bar{x}$ ± SD) remained approximately constant, even in rats exposed subchronically to 6 or 15 ppm (Casanova et al., 1994).

Analysis of DNA-Protein Cross-Links

Techniques used for the analysis of DPX are described elsewhere (Casanova et al., 1994). These techniques were essentially identical to those previously used to analyze DPX concentrations in the anterior nasal mucosa of rats (Casanova et al., 1989) and in the middle turbinate, anterior wall, and nasopharynx of monkeys (Casanova et al., 1991)

Pharmacokinetics of DPX Formation in the LM and M:PM of Naive Rats

A two-compartment pharmacokinetic model, analogous to a three-compartment model developed previously to describe DPX formation in the nose of naive monkeys exposed to HCHO (Casanova et al., 1991), was used to describe the formation of DPX in the two nasal regions, LM and M:PM, of naive rats. The model is illustrated in Figure 1.

The model describes the absorption of HCHO into tissue and its elimination as metabolites or other products (including DPX) as a function of kinetic parameters that are assumed to be either compartment-invariant or compartment-specific. The compartment-invariant parameters are the *internal* kinetic constants, V_{max} (pmol/mg DNA/h), K_m (pmol/mg DNA), k_{el} (h^{-1}), and

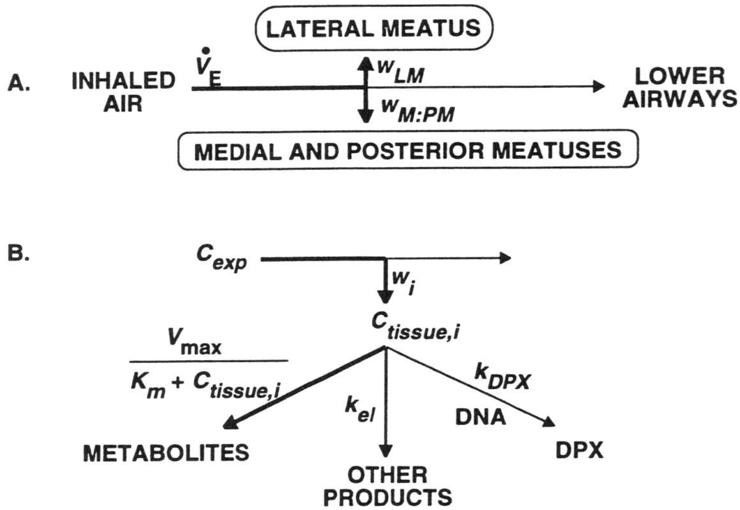

FIGURE 1. (A) Compartmental model of the nasal conducting airways of rats. Symbols used are $\dot{V}_E$, respiratory minute volume; w_i, mass flow rate into compartment i. (B) Diagram showing absorption and elimination of inhaled HCHO in compartment i. Symbols used are C_{exp}, exposure concentration; $C_{tissue,i}$, tissue concentration in compartment i; V_{max}, K_m, k_{el}, and k_{DPX}, kinetic constants for saturable metabolism, nonsaturable elimination, and DNA-protein cross-link formation.

k_{DPX} (h^{-1}), that describe the disposition of HCHO after it has been absorbed into the tissue (Table 1). Justification for the assumption of compartment-invariance of the internal kinetic constants is based on the fact that the tissue lining the two nasal compartments is respiratory epithelium composed of generally similar cell types (Young, 1986; Popp et al., 1986).

The internal kinetic constants cannot be determined from pharmacokinetic measurements alone, but combinations of these constants, P and R (defined in Eqs. 1 and 2, respectively), can be estimated from pharmacokinetics (Casanova et al., 1991):

$$P = V_{max} \frac{k_{DPX}}{k_{el} + k_{DPX}} \tag{1}$$

$$R = \frac{V_{max}/K_m}{k_{el} + k_{DPX}} \tag{2}$$

The units of P are the same as the units of V_{max}, that is, (pmol/mg DNA/h), but the magnitude of P is smaller than that of V_{max} by the fraction $k_{DPX}/(k_{el} + k_{DPX})$. [Calculations suggest (see Discussion) that this fraction is of the order of 10^{-6}, implying that only a minute portion of the inhaled HCHO molecules is bound to DNA, most of it being eliminated by other pathways.] The term R is dimensionless and is equal to the ratio of the limiting, pseudo-first-order rate constant for saturable metabolism that is approached at very low concentrations, V_{max}/K_m, to the sum of the pseudo-first-order rate constants for elimination via all nonsaturable pathways, $k_{el} + k_{DPX}$. (Calculations suggest that this ratio is of the order of 10, i.e., roughly 90% of the inhaled HCHO is probably eliminated via saturable metabolism.) Note that $P/R = k_{DPX}K_m$ is the rate of DPX formation when the tissue concentration of HCHO in compartment i, $C_{tissue,i}$ (i = LM, M:PM), is such that the detoxication pathway is half-saturated. As P and R depend only on internal kinetic constants, they were also assumed to be compartment-invariant (Casanova et al., 1991).

The compartment-specific parameters are (1) the quantity of DNA in compartment i, $M_{DNA,i}$ (mg), and (2) the mass flow rate of HCHO into compartment i, w_i (pmol/h) (Table 1). By definition, $w_i = \Sigma_j n_{ij} a_{ij}$, where n_{ij} is the

TABLE 1. Glossary of Terms in Pharmacokinetic Model

	Compartment-invariant parameters
V_{max}	Zero-order rate constant for maximal rate of elimination of HCHO via a saturable metabolic pathway [pmol/mg DNA/h]
K_m	Michaelis constant for half-saturation of the pathway [pmol (mg DNA)$^{-1}$]
k_{el}	Pseudo-first-order rate constant for elimination of HCHO via nonsaturable pathways other than DPX formation (h^{-1})
k_{DPX}	Pseudo-first-order rate constant for the formation of DPX (h^{-1})
$\dot{V}_E$	Respiratory minute volume (ml/min)
C_{exp}	Exposure concentration (ppm)
t	Exposure time (h)
P	$V_{max}k_{DPX}/(k_{el} + k_{DPX})$ (pmol/mg DNA/h)
R	$(V_{max}/K_m)/(k_{el} + k_{DPX})$ (dimensionless)
	Compartment-specific parameters
w_i	Mass flow rate into compartment i (pmol/h) = mass flux integrated over surface area of compartment i
S_i	Extraction ratio of compartment i (dimensionless) = fractional decrease in airborne concentration due to absorption into compartment i
$M_{DNA,i}$	Amount of DNA in compartment i (mg)
$C_{tissue,i}$	Tissue concentration of HCHO in compartment i [pmol/(mg DNA)]
DPX$_i$	Concentration of DNA-protein cross-links in compartment i [pmol/(mg DNA)]
H_i	$(\dot{V}_E S_i/M_{DNA,i}) \times 2454$ [pmol/mg DNA/h/ppm]
Q_i	$[w_i/M_{DNA,i} \times k_{DPX}/(k_{el} + k_{DPX})]/C_{exp} = H_i k_{DPX}/(k_{el} + k_{DPX})$ (pmol/mg DNA/h/ppm)

Note. The units defined for V_{max} and K_m include "mg DNA" as the extensive variable in the denominator, rather than mg protein, tissue weight, or tissue volume, because mg DNA was used to measure the concentration of DPX. However, mg DNA can be considered to be proportional to mg protein, tissue weight, or volume and therefore can be easily related to the more commonly used variables of enzyme kinetics.

mass flux (pmol/mm^2/h) and a_{ij} is the area (mm^2) of the jth element of compartment i; the summation is taken over the entire surface of the compartment. The mass flux and therefore the mass flow rate can be calculated from fluid dynamics simulations if the partition coefficient of the compound between gas phase and tissue is known or assumed (Kimbell et al. 1993).[1] The mass flow rate depends on the location, shape, and size of compartment i and on other factors, such as the exposure concentration, respiratory minute volume, and physical-chemical properties (solubility, reactivity, vapor pressure) of the compound of interest.

Prior to the advent of fluid dynamics simulations in nasal dosimetry, an alternative expression for the mass flow rate was used (Casanova et al., 1991): $w_i = C_{exp} \times \dot{V}_E \times S_i \times 2454$, where C_{exp} is the exposure concentration (ppm), $\dot{V}_E$ is the respiratory minute volume (ml min^{-1}), S_i is the extraction ratio (Andersen, 1983) of compartment i (dimensionless, representing the decrease in the airborne concentration of inhaled HCHO relative to the exposure concentration due to absorption into compartment i), and the factor, 2454 = 60 (min/h) × 40.9 (pmol/ml/ppm), converts minutes to hours and ppm of airborne HCHO to pmol/ml at ambient temperature and pressure. The extraction ratio, S_i, is a complex parameter dependent on the location, shape, and size of compartment i and on the physical-chemical properties of the compound. The extraction ratio can be determined indirectly from pharmacokinetics (see later) or, alternatively, it can be estimated from fluid dynamics simulations (Kimbell et al., 1993).

The compartment-specific constants, $M_{DNA,i}$ and w_i, cannot be determined from pharmacokinetics alone, but they can be combined with other factors into a single constant, Q_i (pmol/mg DNA/h/ppm), that can be estimated from pharmacokinetics (Casanova et al., 1991). The relationship between Q_i and w_i is given in Eq. 3:

$$Q_i C_{exp} = \frac{w_i}{M_{DNA,i}} \times \frac{k_{DPX}}{k_{el} + k_{DPX}} \tag{3}$$

where C_{exp} is the exposure concentration, and Q_i is the rate constant for DPX formation that is approached at very high HCHO concentrations as the detoxication pathway becomes fully saturated. With appropriate additional information, w_i could provide an estimate of Q_i, or, alternatively, measurements of Q_i could provide a check on the calculated values of w_i. Thus, there is a linkage between the pharmacokinetics of DPX formation and the results of fluid dynamics simulations.

As shown previously (Casanova et al., 1991), the pharmacokinetics of DPX formation in specific areas of the nose is describable in terms of the three parameters, P, R, and Q_i. The relationship between the observed concentra-

[1]In their calculations of HCHO uptake into different regions of the rat nose, Kimbell et al. (1993) assumed that the walls of the nasal passages acted as "sinks," that is, the inhaled gas concentration was equal to zero at the wall surface.

tion of DNA-protein cross-links in compartment i, DPX_i (pmol/mg DNA), the exposure concentration, C_{exp} (ppm), and the time of exposure, t (h) is given by:

$$\frac{DPX_i}{t} = 0.5\left\{\left[\left(Q_iC_{exp} - P\frac{(1+R)}{R}\right)^2 + \frac{4PQ_iC_{exp}}{R}\right]^{0.5} + Q_iC_{exp} - P\frac{(1+R)}{R}\right\} \quad (4)$$

A nonlinear least-squares curve-fitting procedure was used to estimate the parameters P, R, and Q_i in the two nasal compartments, LM and M:PM, of naive rats. The procedure was an iterative three-step process essentially identical to one used previously (Casanova et al., 1991). In the first step, common values of P and R were assumed for both compartments, and the two individual Q_i values were estimated by nonlinear regression of the average rate of DPX formation in each compartment, DPX_i/t, on C_{exp} (Eq. 4). In the second step, the estimated Q_i values were multiplied by the exposure concentrations, C_{exp}, to define compartment-specific exposure variables, Q_iC_{exp} (Eq. 3). In the third step, the values of DPX_i/t were regressed on the products, Q_iC_{exp}, to obtain simultaneous estimates of P and R in both compartments (Eq. 4).[2] The revised estimates of P and R were then used as starting values for the next cycle of calculations. This procedure was carried out a number of times on a VAX/VMS computer, until the changes between successive cycles were judged to be sufficiently small (<0.01% for each estimated parameter). To facilitate the calculations, the regressions were incorporated into a program written for the RS/3 data analysis system (Bolt, Beranek, and Newman, Cambridge, MA). Final estimates for the parameters were then obtained using the program NLIN (SAS Institute, Cary, NC), which provided a check on the results of RS/3 and also yielded the 95% confidence limits on the parameter estimates.

Pharmacokinetics of DPX Formation in the LM of Preexposed Rats

The pharmacokinetic constants P, R, and Q_i estimated for N rats may not be appropriate for PE rats. For example, if enzymes responsible for the detoxication of HCHO were induced by preexposure, the values of P and R in PE rats would differ from those in N rats. However, the induction of HCHO metabolic enzymes in PE rats is not considered likely, inasmuch as earlier studies showed that the specific activities of formaldehyde and aldehyde dehydrogenase in the rat nasal mucosa, two enzymes that could play an important role in HCHO detoxication, were changed only slightly, if at all, by a 10-day exposure to 15 ppm of HCHO (Casanova-Schmitz et al., 1984). No data are available regarding possible effects of preexposure on the other internal kinetic constants, k_{DPX} and k_{el}, but as these are pseudo-first-order rate constants that primarily describe covalent binding reactions

[2]Note in Eq. 4 that Q_i always appears as a product of C_{exp}. Regarding this product as the independent variable in Eq. 2 results in an equation that contains only two adjustable parameters, P and R, that are the same in both compartments. Thus, both compartments can be described by the same mathematical function. This enables the simultaneous estimation of P and R in two compartments.

of HCHO with cellular nucleophiles, they are unlikely to be changed significantly by preexposure. Therefore, as a first approximation, P and R were assumed to be invariant in N and PE rats.

In contrast, Q_i may vary at different concentrations in PE rats if either the mass flow rate from air to tissue, w_i, or the quantity of DNA, $M_{DNA,i}$, is altered by preexposure (Eq. 3). No data are available regarding possible changes in w_i due to preexposure; as a first approximation, this parameter was assumed to be constant at all concentrations. However, as indicated earlier and shown elsewhere (Casanova et al., 1994), $M_{DNA,i}$ was concentration dependent in PE but not in N rats. Specifically, the value of $M_{DNA,i}$ was constant in PE rats exposed to low HCHO concentrations (0.7 or 2 ppm) and was equivalent to that observed in N rats, but the value was increased in PE rats exposed to 6 or 15 ppm (see earlier) (Casanova et al., 1994). As shown in Eq. 3, increases in $M_{DNA,i}$ would be expected to cause a proportional decrease in the magnitude of Q_i if other factors remain unchanged.

Thus, given that Q_i is expected to be concentration dependent in PE rats at high concentrations, the following procedure was used to estimate the pharmacokinetic constants. The constants, P, R, and Q_i (i = LM) appropriate to PE rats exposed to 0.7 or 2 ppm were first estimated by nonlinear regression after combining the DPX data for these rats with those for N rats exposed to 0.7, 2, 6, or 15 ppm (Eq. 4).[3] As noted above, the values of P and R estimated for PE rats exposed to 0.7 or 2 ppm were also assumed to be valid for PE rats exposed to 6 or 15 ppm. The value of Q_i estimated for PE rats exposed to 0.7 or 2 ppm was then multiplied by the ratio of tissue weights observed at low and high concentrations, thus scaling Q_i inversely as the tissue weight. This procedure is justifiable, because the tissue weight was directly proportional to $M_{DNA,i}$ (Casanova et al., 1994). In this manner, predictions of Q_i (corrected for changes in $M_{DNA,i}$) for PE rats exposed to 6 or 15 ppm were derived.

In addition to the pharmacokinetic constants P, R, and Q_i, the curve-fitting procedure for PE rats exposed to 0.7 or 2 ppm yielded the correlation coefficients r_{PQ}, r_{PR}, and r_{QR}, and the 95% confidence limits about the parameter estimates. Using these data, confidence limits about the mean of the fitted curve for DPX formation in PE rats exposed to low concentrations were calculated (Meyer, 1975). Standard methods of error propagation (Meyer, 1975) were then used to estimate 95% confidence limits about the mean values of Q_i predicted for PE rats exposed to high concentrations (6 or 15 ppm). The confidence limits about the parameter estimates and the values

[3]Parameter estimation was performed using the program NLIN (SAS Institute, Cary, NC). Equation 4 was fitted to the DPX data for the lateral meatus of PE rats exposed to 0.7 and 2 ppm together with the DPX data for the lateral meatus of N rats exposed to 0.7, 2, 6, and 15 ppm. Justification for combining the DPX data for PE rats exposed to concentrations $\leq$2 ppm with those for N rats exposed to concentrations $\leq$15 ppm is based on the facts that (1) $M_{DNA,i}$ was invariant in these rats and (2) the yields of DPX were essentially identical in N and PE rats at concentrations $\leq$2 ppm (Casanova et al., 1994). Hence, PE rats exposed to concentrations $\leq$2 ppm and N rats exposed to concentrations $\leq$15 ppm were regarded as kinetically equivalent.

of the correlation coefficients (which were assumed to be invariant with concentration) were used to calculate confidence limits about the means of the predicted curves for DPX formation in PE rats exposed to 6 or 15 ppm.

RESULTS

Naive Rats

The formation of DNA-protein cross-links in the LM and M:PM of naive F344 rats exposed to selected concentrations of HCHO is shown in Figure 2. Yields of DPX were between five- and sixfold higher in the LM (high tumor site) than in the M:PM (low tumor site) at all exposure concentrations. The experimental data in each compartment were described very well by Eq. 4, despite the restrictive assumption that P and R are compartment invariant. Estimated values of the model parameters, P, R, and Q_i (i = LM, M:PM), are summarized in Table 2.

The curves shown in Figure 2 demonstrate that DPX_i/t is a nonlinear function of the airborne concentration, but that it approaches linearity at very low (<1 ppm) and very high (>10 ppm) concentrations. The limiting form of Eq. 4 that is approached asymptotically at high concentrations is

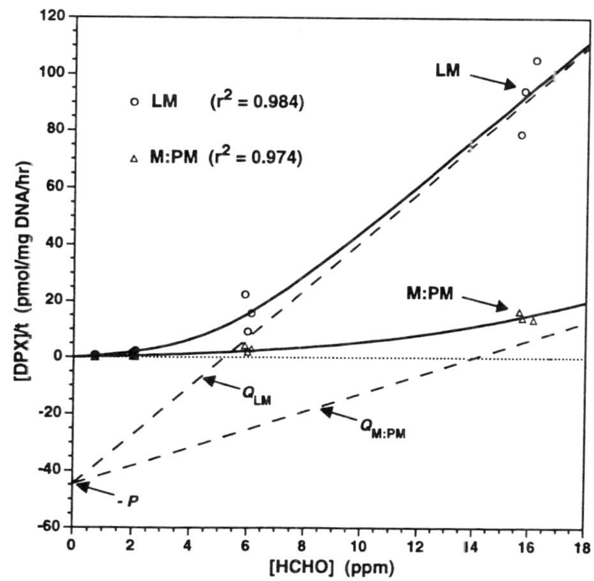

FIGURE 2. Concentrations of DNA-protein cross-links formed per unit time, DPX_i/t, in the lateral meatus (LM) and medial and posterior meatuses (M:PM) of naive F344 rats versus the airborne HCHO concentration (0.72, 2.06, 6.01, or 15.8 ppm). The rats were exposed for 3 h. The computer-generated curves were calculated after iterative estimation of the constants, P, R, Q_{LM}, and $Q_{M:PM}$ (Table 2) by the method described in the text. Dashed lines are the limiting slopes for each tissue approached at high concentrations of HCHO.

TABLE 2. Pharmacokinetic Constants for DPX Formation by HCHO in the Lateral Meatus (LM) and Medial and Posterior Meatuses (M:PM) of Naive F344 Rats

Constant	Estimated value ($\bar{x}$ ± asymptotic SE)	Asymptotic 95% confidence limits	
		Lower	Upper
P (pmol/mg DNA/h)	44.6 ± 2.4	39.7	49.6
R	11.9 ± 2.0	7.7	16.1
Q_{LM} (pmol/mg DNA/h/ppm)	8.58 ± 0.21	8.1	9.0
$Q_{M:PM}$ (pmol/mg DNA/h/ppm)	3.19 ± 0.07	3.0	3.3

Note. Values were estimated using the iterative nonlinear regression procedure described in the text.

$$\frac{DPX_i}{t} = Q_i C_{exp} - P \qquad (C_{exp} \rightarrow \infty) \tag{5}$$

Equation 5 is plotted for each compartment as a dashed line in Figure 2. The slope of the concentration-response curve that is approached at high concentrations (Q_i) is compartment specific, but the intercept of this slope on the ordinate ($-P$) is compartment invariant (Figure 2), in accordance with the model assumptions.

The limiting form of Eq. 4, as the HCHO concentration approaches zero, is given in Eq. 6. The latter equation is not plotted in the figure for reasons of clarity.

$$\frac{DPX_i}{t} = \frac{Q_i}{1 + R} C_{exp} \qquad (C_{exp} \rightarrow 0) \tag{6}$$

However, by comparing Eqs. 5 and 6, it can be seen that the ratio of limiting slopes at high and at low concentrations is equal to ($1 + R$). Thus, the magnitude of R defines the extent of nonlinearity in the concentration-response curve for DPX formation.

Preexposed Rats

Figure 3 illustrates the yields of DPX obtained in the lateral meatus of N rats and of PE rats that were preexposed to 0.7, 2, 6, or 15 ppm of HCHO. At 0.7 or 2 ppm, there were no measurable differences between N and PE rats with respect to DPX yields, and no differences between these rats were detected with respect to tissue weight, quantity of nasal mucosal DNA, or histopathology (Casanova et al., 1994). Therefore, PE rats that were preexposed to HCHO concentrations ≤2 ppm were regarded as kinetically equivalent to N rats. Accordingly, the pharmacokinetic constants appropriate to such rats were estimated by fitting the DPX data for all of them (Figure 3, solid line, with dotted lines showing upper and lower 95% confidence limits). The estimated constants are summarized in Table 3.

At 6 or 15 ppm, the DPX yields observed in PE rats were smaller than those in N rats (Figure 3). As noted under Methods, rats preexposed to these concentrations had an increased number of cells and larger quantities of DNA in the LM, which should decrease the value of Q_i (Eq. 3). However, no major changes were expected to occur in the values of *P* and *R*. The value of Q_i estimated for N rats and for PE rats exposed to concentrations ≤2 ppm was therefore scaled in inverse proportion to the tissue weight to obtain predictions of Q_i at 6 and 15 ppm. The predicted values are summarized in Table 3, and the corresponding concentration-response curves are shown as the dashed lines in Figure 3. The predicted curve at each concentration agrees well with the experimental data, suggesting that the decreased yields of DPX in PE relative to N rats at concentrations ≥6 ppm can largely be explained by an increase in the quantity of nasal mucosal DNA.

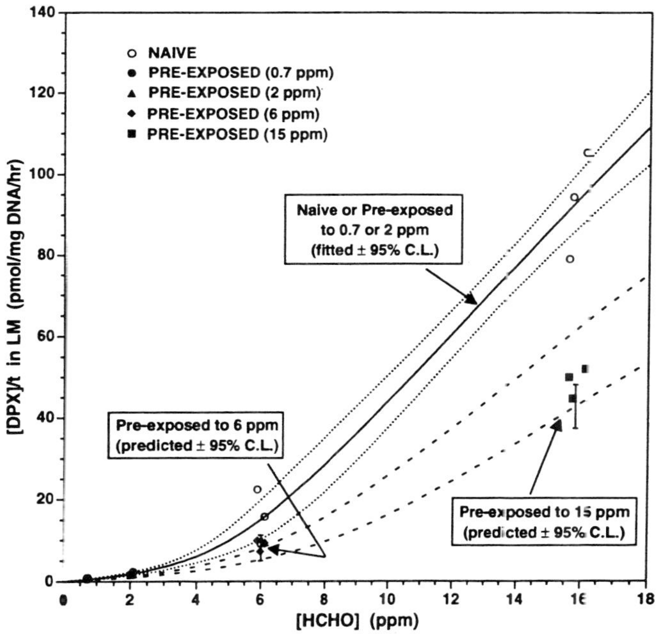

FIGURE 3. Concentrations of DNA-protein cross-links formed per unit time, DPX_i/t, in the lateral meatus (LM) of naive F344 rats (open symbols) or of rats preexposed to 0.70, 2.05, 5.88, or 14.52 ppm (6 h/day, 5 days/wk, 11 wk + 4 days) (filled symbols at each concentration) during a 3-h exposure to HCHO (0.72, 2.06, 6.01, or 15.8 ppm). The solid line is the computer-generated curve calculated after estimating the constants, *P*, *R*, and Q_{LM}, for naive rats or for rats preexposed to concentrations ≤2 ppm (Table 3). Dotted lines are upper and lower 95% confidence limits about the mean. Dashed lines are the computer-generated curves predicted for rats preexposed to either 6 or 15 ppm. The predicted curves were calculated by assuming that *P* and *R* for rats preexposed to 6 or 15 ppm are the same as for naive rats or for rats pre-exposed to concentrations ≤2 ppm, and that Q_{LM} is inversely proportional to the tissue weight (Table 3). Error bars are 95% confidence limits about the predicted mean at concentrations of 6.01 and 15.8 ppm.

TABLE 3. Pharmacokinetic Constants for DPX Formation by HCHO in the Lateral Meatus (LM) of Naive and Preexposed F344 Rats

Constant		Estimated value ($\bar{x}$ ± asymptotic SE)	Asymptotic 95% confidence limits: Lower	Asymptotic 95% confidence limits: Upper
Naive or preexposed rats (preexposure concentration ≤ 2 ppm)[a]				
P	(pmol/mg DNA/h)	46.3 ± 12.2	20.6	72.0
R		12.5 ± 1.7	9.0	16.0
Q_{LM}	(pmol/mg DNA/h/ppm)	8.69 ± 0.79	7.0	10.4
Correlation coefficients				
	r_{PQ}	.9653		
	r_{PR}	−.0768		
	r_{QR}	.0356		
Preexposed rats (preexposure concentration = 5.88 ppm)				
P	(pmol/mg DNA/h)	Assumed same as N or PE to [HCHO] ≤ 2 ppm		
R		Assumed same as N or PE to [HCHO] ≤ 2 ppm		
Q_{LM}	(pmol/mg DNA/h/ppm)[b]	6.63 ± 0.95	3.6	9.6
Preexposed rats (preexposure concentration = 14.5 ppm)				
P	(pmol/mg DNA/h)	Assumed same as N or PE to [HCHO] ≤ 2 ppm		
R		Assumed same as N or PE to [HCHO] ≤ 2 ppm		
Q_{LM}	(pmol/mg DNA/h/ppm)[b]	5.39 ± 0.58	3.9	6.8

[a]Pharmacokinetic constants were estimated by nonlinear regression using combined DPX data of naive rats and of rats preexposed to concentrations ≤2 ppm.

[b]Q_{LM} values for rats preexposed to 5.88 or 14.5 ppm were estimated by scaling the Q_{LM} value for naive rats and for rats preexposed to concentrations ≤2 ppm in inverse proportion to the tissue weights (see text). The ratios of tissue weights (PE/N) for the lateral meatus at 5.88 and 14.5 ppm were 1.31 ± 0.14 and 1.61 ± 0.09, respectively ($\bar{x}$ ± SE) (Casanova et al., 1994).

DISCUSSION

Pharmacokinetics of DPX Formation in Naive and Preexposed Rats

As shown in Figure 2, four adjustable parameters, P, R, and Q_i (i = LM, M:PM), were sufficient to describe DNA-protein cross-link formation in the lateral meatus and medial and posterior meatuses of naive rats. A basic assumption of this model was that the site-specific differences in DPX formation are due to differences in Q_i, rather than to differences in the internal kinetic constants, V_{max}, K_m, k_{el}, and k_{DPX}. Differences in Q_i can be caused by differences in the mass flow rate, w_i, and/or in the quantity of DNA, $M_{DNA,i}$, in each compartment. However, the values of $M_{DNA,i}$ in the two compartments of the naive rat nose were very similar (Casanova et al., 1994), hence in this case the site-specific differences in DPX formation can be ascribed almost entirely to differences in the mass flow rate.

By scaling Q_i in inverse proportion to the tissue weight, the changes in

DPX yields that were observed in preexposed rats at 6 and 15 ppm were accurately predicted. As shown in Figure 3, rats preexposed to either 6 or 15 ppm and then exposed once to HCHO are predicted to have a lower concentration of nasal DPX (due primarily to an increase in $M_{DNA,i}$) than naive rats exposed to the same concentration. However, rats preexposed to a low concentration (≤ 2 ppm) followed by a single exposure to HCHO are expected to have the same concentration of nasal DPX as naive rats; that is, there were no effects of preexposure on the formation of DPX at low preexposure concentrations.

It is useful to compare the pharmacokinetic constants estimated for rats exposed for 3 h to $H^{14}CHO$ (this work) with those previously estimated for DPX formation in the anterior nasal mucosa of rats exposed for 6 h to concentrations ranging from 0.3 to 10 ppm (Table 5 of Casanova et al., 1991). The pharmacokinetic constants estimated from both sets of experiments are essentially identical. This result provides support for the pharmacokinetic estimates, demonstrates that exposures of either 3- or 6-h duration yield essentially the same pharmacokinetic information, and validates the steady-state assumption that was made in deriving Eq. 4 (Casanova et al., 1991). Furthermore, this result implies that the yield of DPX produced in any single exposure is directly proportional to the exposure time.

Evaluation of Internal Kinetic Constants

The pharmacokinetic constants in Tables 2 and 3 can be combined with nonpharmacokinetic information to calculate relative magnitudes of the internal kinetic constants, V_{max}/K_m, k_{el}, and k_{DPX}, and the absolute magnitude of V_{max}. We may define a quantity, H_i, which includes the necessary nonpharmacokinetic data, as follows:

$$H_i = \frac{\dot{V}_E S_i}{M_{DNA,i}} \times 2454 \tag{7}$$

where H_i is the rate constant for the total uptake of HCHO per mg of DNA by compartment i. The value of H_i for the rat lateral meatus (i.e., H_{LM}) can be estimated from published data.[4] We obtain that $H_{LM} \cong (8.5 \pm 2.1) \times 10^6$ pmol/mg DNA/h/ppm ($\bar{x} \pm$ SD).

[4] H_{LM} was evaluated as follows (values are $\bar{x} \pm$ SD unless otherwise indicated):

1. The overall extraction efficiency of HCHO in the rat nose is approximately 93% (Patterson et al., 1986). Assuming that the uptake of HCHO is distributed between the LM (Q_{LM} = 8.58 pmol/mg/DNA/h/ppm) and the M:PM ($Q_{M:PM}$ = 3.19 pmol/mg/DNA/h/ppm) (Table 2), we obtain that the extraction ratio of the LM, $S_{LM} = 0.93 \times 8.58/(8.58 + 3.19) = 0.678 \pm 0.014$ (95% CL).
2. The minute volume can be estimated from the data of Mauderly (1986), who measured $\dot{V}_E$ of male F344 rats at different ages. The rats used in the present study were approximately 22 wk (>5 mo) of age (Casanova et al., 1994). Based on the data of Mauderly (1986), we estimate that $\dot{V}_E \cong 290 \pm 50$ ml/min.
3. The quantity of DNA in the LM of naive rats can be calculated from the average wet weight of tissue (16.7 ± 1.7 mg) and the average tissue concentration of DNA (3.4 ± 0.5 μg/mg) (Casanova et al., 1994). Hence, $M_{DNA,LM} = 56.8 \pm 10.2$ μg.

From the definitions of H_i and Q_i, it can be shown that $H_i/Q_i = (k_{el} + k_{DPX})/k_{DPX}$. Therefore, for the lateral meatus we may write that

$$\frac{k_{el}}{k_{DPX}} = \frac{H_{LM} - Q_{LM}}{Q_{LM}} \tag{8}$$

Comparison of the values of H_{LM} and Q_{LM} (Table 2 or 3) shows that $H_{LM}/Q_{LM} \cong 1 \times 10^6$. Therefore, to a very good approximation, $H_{LM} - Q_{LM} \cong H_{LM}$, and Eq. 8 can be rewritten as $k_{el}/k_{DPX} \cong H_{LM}/Q_{LM}$.

Similarly, from the definitions of R and Q_i, it can be shown that:

$$\frac{V_{max}/K_m}{k_{el}} = R\frac{H_{LM}}{H_{LM} - Q_{LM}} \tag{9}$$

and since $H_{LM} \gg Q_{LM}$, it follows that $(V_{max}/K_m)/k_{el} \cong R$.

The appropriate values can be substituted in Eqs. 8 and 9 to obtain relative values of the constants, V_{max}/K_m, k_{el}, and k_{DPX} (Table 4). The calculations suggest that the rate constant for DPX formation, k_{DPX}, is much smaller than the rate constants for HCHO elimination via saturable metabolism or by nonsaturable pathways other than DPX formation.

The absolute magnitude of V_{max} can be calculated from Eq. 10:

$$V_{max} = \frac{PH_{LM}}{Q_{LM}} \tag{10}$$

Substituting from Table 3 we obtain that V_{max} is 45 ± 14 μmol/mg DNA/h/. To convert V_{max} to the more usual units of nmol/mg protein/min, we make use of the fact that the rat nasal mucosa contains approximately 9% protein by weight (Casanova-Schmitz et al., 1984). The resulting value of V_{max} is shown in Table 4. The in vivo estimate is about 40-fold larger than the in vitro estimate of V_{max} for formaldehyde dehydrogenase at 23°C (Heck & Casanova, 1987) and about 7-fold larger than the in vitro estimate of V_{max} for aldehyde dehydrogenase at 25°C (Casanova-Schmitz et al., 1984) in rat nasal mucosal homogenates. The in vivo–in vitro differences may be due, at least in part, to temperature.

The pharmacokinetic measurements do not provide an estimate of K_m or of the absolute magnitudes of the other rate constants. However, the results can be used to calculate the product, $k_{DPX}K_m = P/R$, which is the rate of DPX formation when the tissue concentration of HCHO is such that the detoxication pathway is half-saturated. The magnitude of $k_{DPX}K_m$ is shown in Table 4. In comparison, the rate of HCHO metabolism at half-saturation is $V_{max}/2 = PH_{LM}/2Q_{LM}$, and the rate of elimination via nonsat-

TABLE 4. Kinetics of HCHO Disposition in the Nasal Mucosa of F344 Rats

Ratio, rate constant, or rate		Estimated value
$V_{max}/K_m : k_{el} : k_{DPX}$	(relative rate constants)	12.5 : 1.0 : 1.0 × 10^{-6}
V_{max}	nmol/mg protein/min	28.5 ± 9.8[a]
Absolute rates at $C_{tissue,i} = K_m$ (half-saturation of detoxication pathway)[a]		
1. Saturable detoxication ($V_{max}/2$)	μmol/mg DNA/h	23 ± 7
2. Nonsaturable pathways other than DPX formation ($k_{el}K_m$)	μmol/mg DNA/h	3.6 ± 1.3
3. DPX formation ($k_{DPX}K_m$)	pmol/mg DNA/h	3.7 ± 1.2

[a]$\bar{x}$ ± SD.

urable pathways other than DPX formation is $k_{el}K_m = (k_{el}/k_{DPX}) \times (k_{DPX}K_m) \cong PH_{LM}/RQ_{LM}$. These rates are also presented in Table 4. The two latter routes of elimination account for approximately 90% and 10%, respectively, of the inhaled HCHO. Only a minute portion is converted to DNA-protein cross-links.

When the detoxication pathway is half-saturated, the airborne concentration is $[(2 + R)/2R]P/Q_i$.[5] Hence, for the lateral meatus, half-saturation should occur at 3 ppm. If the detoxication pathway were completely inhibited at this concentration ($V_{max} = 0$), the tissue concentration of HCHO (and, therefore, the rate of DPX formation) would be increased by the factor, $(R/2 + 1) = 7.3 \pm 0.9$.[6] Data are available on the effect of an inhibition of formaldehyde dehydrogenase (by depletion of glutathione, GSH) on the formation of DPX over a wide range of airborne concentrations (Casanova & Heck, 1987). The data suggest that at 3 ppm, the DPX yield increases by about a factor of 4 when the GSH concentration is depleted to 10% of normal (see Figure 5 of Casanova & Heck, 1987). The observed increase was about 4/7.3 × 100 = 55% of that expected for complete inhibition of the saturable pathway. This result suggests that most of the inhaled HCHO is detoxified by GSH-dependent metabolism, but that a significant portion of the HCHO is apparently detoxified by pathways that do not require GSH.

[5]Equation 4 can be rewritten as $Q_iC_{exp} = \{1 + P/[(P/R) + (DPX_i/t)]\}(DPX_i/t)$. Since the detoxication pathway is half-saturated when $DPX_i/t = P/R$, the airborne concentration at half-saturation is that given in the text.

[6]At half-saturation of the detoxication pathway, the tissue concentration of HCHO is $C_{tissue,i} = K_m = (Q_iC_{exp} - V_{max}/2)/(k_{el} + k_{DPX}) \cong (Q_iC_{exp} - V_{max}/2)/k_{el}$, since $k_{el} \gg k_{DPX}$. The latter expression can be rewritten as $C_{tissue,i} = (Q_iC_{exp}/k_{el})2/(R + 2)$. If the saturable detoxication pathway were completely inhibited at the same exposure concentration, the tissue concentration would be increased to $C'_{tissue,i} = Q_iC_{exp}/k_{el}$. Therefore, the ratio of tissue concentrations (inhibited/noninhibited), $C'_{tissue,i}/C_{tissue,i} = R/2 + 1$.

CONCLUSIONS

The pharmacokinetic model presented here suggests that the site specificity of DPX formation in the nose is mainly a result of the airflow characteristics of the nasal passages and is not a consequence of site-specific differences in HCHO metabolism. Hence, the model suggests a close linkage between the pharmacokinetics of DPX formation and the results of fluid dynamics simulations. The two approaches provide complementary information that is important for an understanding of nasal toxicology. It may be possible to use DPX data to validate the results of fluid dynamics simulations, or by combining fluid dynamics simulations with estimates of the pharmacokinetic constants, to predict DPX concentrations at specific sites in the nose of other species (Kimbell et al., 1993). To extrapolate reliably to other species, however, a better understanding of the magnitudes of the internal kinetic constants in different species is essential.

A previous pharmacokinetic study of DPX formation in the nose of rhesus monkeys exposed to HCHO also led to the conclusion that nasal airflow is the primary determinant of the site specificity of DPX (Casanova et al., 1991). Thus, the pharmacokinetic model appears to be generally applicable to both rodents and primates. Further investigations are needed, however, to relate the pharmacokinetic constants to in vitro parameters, which are required for extrapolation to humans. In addition, it is important to understand the genotoxic implications of DPX to ensure that the formation of this "adduct-dosimeter" is used appropriately in cancer risk assessment.

REFERENCES

Andersen, M. E. 1983. Flow-limited clearance. In *Modeling of inhalation exposure to vapors: Uptake, distribution, and elimination*, ed. V. Fiserova-Bergerova, pp. 67–95. Boca Raton, FL: CRC Press.

Casanova, M., and Heck, H. d'A. 1987. Further studies of the metabolic incorporation and covalent binding of inhaled [^{3}H]- and [^{14}C]formaldehyde in Fischer-344 rats: Effects of glutathione depletion. *Toxicol. Appl. Pharmacol.* 89:105–121.

Casanova, M., Deyo, D. F., and Heck, H. d'A. 1989. Covalent binding of inhaled formaldehyde to DNA in the nasal mucosa of Fischer 344 rats: Analysis of formaldehyde and DNA by high-performance liquid chromatography and provisional pharmacokinetic interpretation. *Fundam. Appl. Toxicol.* 12:397–417.

Casanova, M., Morgan, K. T., Steinhagen, W. H., Everitt, J. I., Popp, J. A., and Heck, H. d'A. 1991. Covalent binding of inhaled formaldehyde to DNA in the respiratory tract of rhesus monkeys: pharmacokinetics, rat-to-monkey interspecies scaling, and extrapolation to man. *Fundam. Appl. Toxicol.* 17:409–428.

Casanova, M., Morgan, K. T., Gross, E. A., Moss, O. R., and Heck, H. d'A. 1994. DNA-protein crosslinks and cell replication at specific sites in the nose of F344 rats exposed subchronically to formaldehyde. *Fundam. Appl. Toxicol.*, in press.

Casanova-Schmitz, M., David, R. M., and Heck, H. d'A. 1984. Oxidation of formaldehyde and acetaldehyde by NAD^+-dependent dehydrogenases in rat nasal mucosal homogenates. *Biochem. Pharmacol.* 33:1137–1142.

Heck, H. d'A., and Casanova, M. 1987. Isotope effects and their implications for the covalent binding of inhaled [^{3}H]- and [^{14}C]formaldehyde in the rat nasal mucosa. *Toxicol. Appl. Pharmacol.* 89:122–134.

Kimbell, J. S., Gross, E. A., Joyner, D. R., Godo, M. N., and Morgan, K. T. 1993. Application of computational fluid dynamics to regional dosimetry of inhaled chemicals in the upper respiratory tract of the rat. *Toxicol. Appl. Pharmacol.* 121:253–263.

Mauderly, J. L. 1986. Respiration of F-344 rats in nose-only inhalation exposure tubes. *J. Appl. Toxicol.* 6:25–30.

Meyer, S. L. 1975. *Data analysis for scientists and engineers*, pp. 39–45. New York: Wiley.

Patterson, D. L., Gross, E. A., Bogdanffy, M. S., and Morgan, K. T. 1986. Retention of formaldehyde gas by the nasal passages of F-344 rats. *Toxicologist* 6:55 (abstr. 217).

Popp, J. A., Monteiro-Riviere, N. A., and Martin, J. T. 1986. Ultrastructure of the rat nasal mucosa. In *Toxicology of the nasal passages*, ed. C. S. Barrow, pp. 37–49. Washington, DC: Hemisphere.

Young, J. T. 1986. Light microscopic examination of the rat nasal passages: Preparation and morphologic features. In *Toxicology of the nasal passages*, ed. C. S. Barrow, pp. 27–36. Washington, DC: Hemisphere.

Section 3. Chemically Induced Nasal Response in Humans and Rodents

NASAL LESION DISTRIBUTION AND TOXICITY IN THE SQUAMOUS EPITHELIAL REGION

Thomas M. Monticello

Bristol-Myers Squibb Pharmaceutical Research Institute, Princeton, New Jersey, USA

Kevin T. Morgan

Chemical Industry Institute of Toxicology, Research Triangle Park, North Carolina, USA

There is an increasing awareness of the importance of the nasal epithelium as a target site for chemically induced toxicity and carcinogenicity. Extrapolating inhalation toxicology results from animal studies to humans remains a difficult task, however, due to varied and marked interspecies differences. Studies of the distribution of nasal lesions may provide insight into the mechanisms involved in toxicant-induced nasal pathology and improve the human risk assessment process. It becomes important to continually expand the database on nasal toxicologic pathology by providing a careful description and mapping of lesions from inhalation toxicology studies. In this brief review, the commonly encountered distribution of nasal lesions in the rodent and the factors that influence this distribution are presented. In addition, histology and toxicant-induced lesions of the nasal squamous epithelium, which lines the entrance to the upper respiratory tract, are addressed.

Rodents are used extensively for inhalation toxicologic studies that are directed toward the prediction of human risks to environmental toxicants. The extrapolation of results from rodent nasal toxicity studies to humans remains a difficult task, however, due to marked differences in anatomy, mucociliary clearance patterns, and airflow and regional uptake patterns. As the portal of entry for the respiratory tract, the nasal passages are a common site for particle deposition and the absorption of many gases and vapors, and often are a prime site for toxicant-induced pathology. Like other regions of the respiratory tract, the cells lining the airway luminal surfaces are the targets for inhaled toxicants.

The nasal cavity and mucosa warm, moisten, and filter incoming air and contain the receptors for olfaction (Sorekin, 1977). The surface of the nasal cavity is composed of squamous, transitional, respiratory, and olfactory epithelium, with generally distinct lines of demarcation between the epithelial types. Consistency in trimming nasal tissues, in addition to an adequate number of levels for examination, is essential for a through evaluation of the nasal air-

Address correspondence to Dr. Thomas M. Monticello, Bristol-Myers Squibb Pharmaceutical Research Institute, PO Box 4000, F1.4102, Princeton, NJ 08543, USA.

ways and for documenting lesions. For detailed descriptions of nasal anatomy, histology, and trimming of the nasal cavity, the reader is referred elsewhere (Young, 1981; Uriah & Maronpot, 1990; Harkema, 1991; Morgan, 1991).

Over the past decade, there has been a large volume of work establishing the rodent nasal cavity as an important target site for chemically-induced toxicity and carcinogenicity. In part, this increasing interest in the nasal passages of laboratory animals is the result of an expanded database obtained by a more complete assessment of the upper respiratory tract of rodents following acute and chronic inhalation exposures. However, an accurate assessment of pathological findings and the recognition of frequently encountered site-specific changes cannot be obtained unless there is a good knowledge of the anatomy and histology of the nasal cavity. This article reviews the commonly encountered distribution of certain nasal lesions and the factors that influence this distribution. Histology and toxicant-induced lesions of the squamous epithelium, which lines the nasal vestibule, are also described. The remaining nasal epithelial types (i.e., transitional, respiratory, and olfactory), in addition to the accessory nasal structures, are addressed by others (this volume).

NASAL LESION DISTRIBUTION

Often there are marked interspecies differences in the distribution and severity of lesions induced in the respiratory tract following exposure to xenobiotics. An increased understanding of the factors responsible for interspecies differences is necessary, since the basis of human health risk assessments is based chiefly on results of animal studies. The toxicologic pathologist can play an important role in this research process by delineating and reporting the nature and location of upper respiratory tract lesions in experimental animals used (Morgan, 1991).

Nasal epithelial lesions develop in specific sites of the nasal passages for many inhaled toxicants (Morgan & Monticello, 1990). For example, in rodents exposed to sufficiently high concentrations of an irritant gas, lesions are frequently present on the margins of the atrioturbinates, the epithelium lining the walls of the lateral meatus, and regions bounding the anterior portion of the middle meatus (Morgan & Monticello, 1990). A common location for olfactory epithelial lesions is the narrow extension of this epithelium into the anterior dorsal meatus (Buckley et al., 1984).

The distribution of chemically induced lesions in the respiratory tract is believed to be influenced by one or more of the following: (1) regional uptake patterns, (2) cellular susceptibility to a given dose of chemical, and (3) local clearance processes by the mucosa and vasculature (Morgan & Monticello, 1990; Kimbell et al., 1993). For certain gases or vapors, regional uptake patterns or cellular susceptibility appears to play the primary role in the determination of lesion distribution.

Formaldehyde-induced nasal lesion mapping studies, performed in both the F344 rat and rhesus monkey, suggest that the distribution of lesions is closely related to regional uptake patterns for this chemical and not tissue susceptibility (Morgan & Monticello, 1990). Using nasal molds, studies of inspiratory airflow patterns in the nasal passages of these species have determined that regional uptake patterns are a direct consequence of nasal airflow characteristics (Morgan et al., 1991). Formaldehyde-induced lesion distribution in the rhesus monkey (Monticello et al., 1989) attributed to nasal airflow patterns is further supported by molecular dosimetry studies in which formaldehyde-induced DNA-protein cross-links correlated with sites of lesions (Casanova et al., 1991). Interspecies differences in airflow patterns, sometimes due to marked differences in nasal anatomy, may account for the species-specific lesion distribution (e.g., rodent vs. nonhuman primate) observed following inhalation exposure to certain highly water-soluble gaseous irritants. Recently, more quantitative studies evaluating airflow patterns with regional uptake of inhaled gases have been performed using computational fluid dynamics (Kimbell et al., 1993; Kimbell, 1994).

The various epithelial types that line the nasal passages each have specific morphological, biochemical, and physiological properties that may influence their susceptibility to toxicant-induced insult (Morgan & Monticello, 1990). For example, the squamous epithelium of the nasal vestibule is fairly resistant to inhaled gaseous irritants (Jiang et al., 1986). In the nonhuman primate, the transitional epithelium, which is located between the anterior squamous epithelium and more posterior respiratory epithelium, is a target site for ozone toxicity (Harkema et al., 1987). Another example of site-specific toxicity attributed to cellular susceptibility is olfactory toxicity induced by methyl bromide exposure (Hurtt et al., 1988). Inhalation studies with methyl bromide gas have demonstrated that this chemical is specifically toxic to olfactory epithelium in the rat, while other nasal epithelia are unaffected. The olfactory site specificity of methyl bromide is believed to be due to a biochemical lesion in the olfactory sustentacular and sensory cells (Hurtt et al., 1988).

STRATIFIED SQUAMOUS EPITHELIAL TOXICITY

Squamous epithelium covers approximately 3.5% of the entire nasal cavity in the F344 rat, and 7.5% of the nasal cavity of the B6C3F$_1$ mouse (Gross et al., 1982). This epithelium lines both the nasal vestibule (Figure 1), the most anterior portion of the nasal cavity, and, in the rodent, the floor of the ventral meatus anterior to the nasopalatine ducts. The nasopalatine ducts, also lined by stratified squamous epithelium, are normal anatomical structures that lie in the boney incisive canals and extend from the floor of the ventral meatus through the hard palate, exiting into the oral cavity at the incisive papilla, providing direct communication between the nasal and oral cavities (Negus, 1958).

Nasal vestibular squamous epithelium consists of two to six-cell layers composed of basal cells, located along the basal lamina, and one or more layers of stratified squamous cells that become progressively flatter toward the luminal surface (Figure 2). Superficial keratinization is sometimes

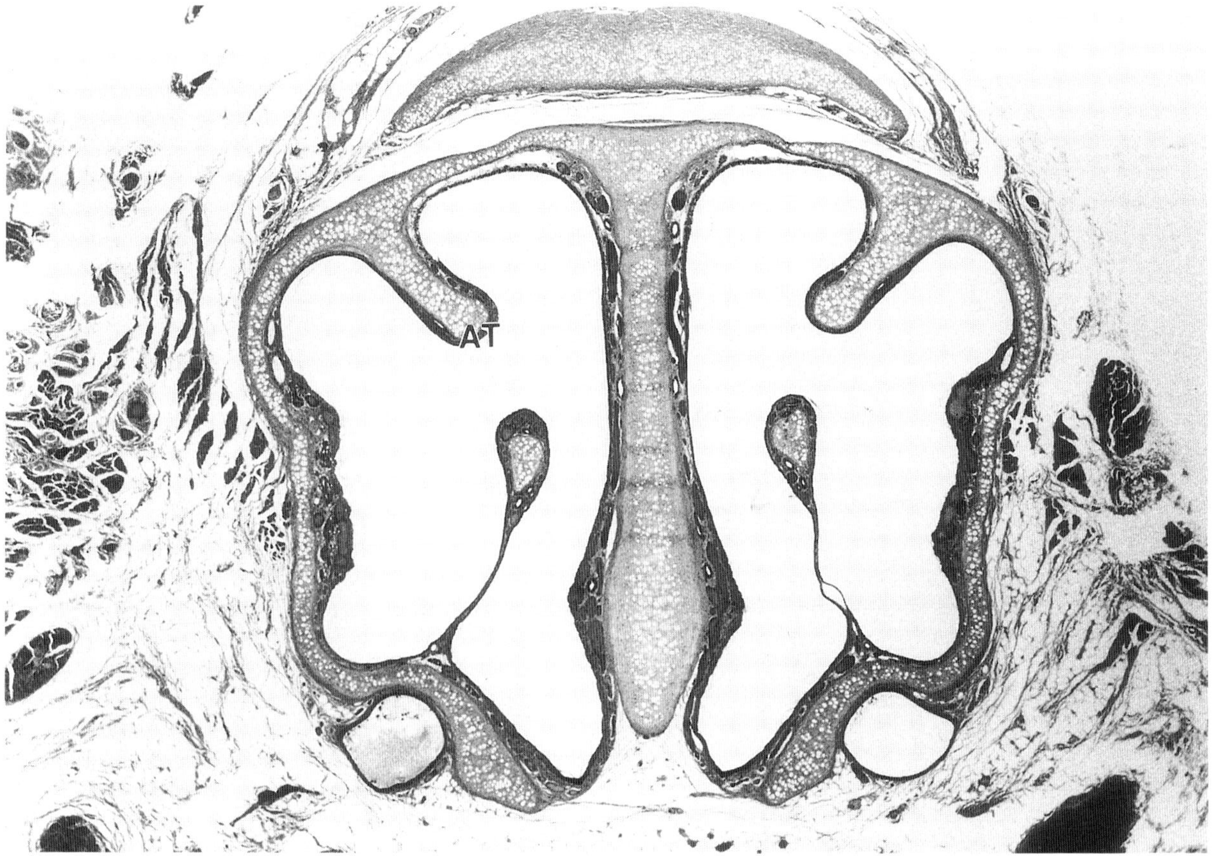

FIGURE 1. Photomicrograph of the nasal vestibule of a F344 rat. The nasal septum, atrioturbinates (AT), and lateral walls are lined by stratified squamous epithelium. Hematoxylin and eosin, original magnification ×5.

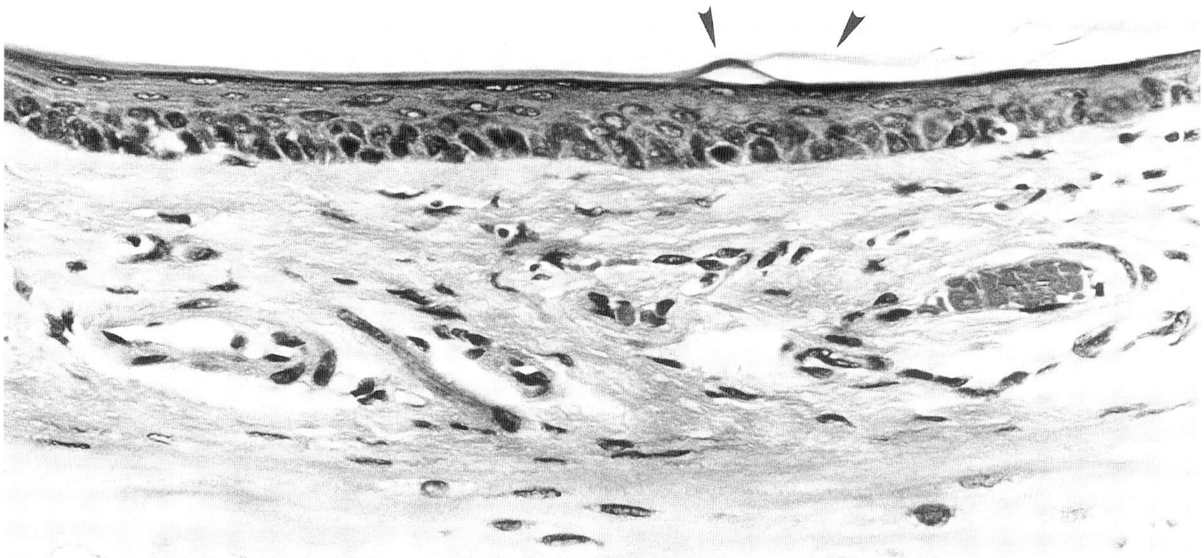

FIGURE 2. Photomicrograph of stratified squamous epithelium of the nasal vestibule of a F344 rat. Slight keratinization is present (arrowheads). Hematoxylin and eosin, original magnification ×100.

observed. By scanning electron microscopy (SEM), the surface of the nasal vestibule is characterized as an irregular mosaic of squamous cells separated by linear invaginations (Figure 3), the latter of which often appear as microridges by light microscopy (Figure 4). The underlying fibrovascular stroma of the squamous epithelium is devoid of sebaceous glands and hair follicles, in contrast to that in humans (Uriah & Maronpot, 1990).

The squamous epithelium of the vestibule in both rodents and nonhuman primates contains no histochemically detectable mucosubstances, unlike the other types of nasal epithelia (Harkema, 1991), and most likely functions as a protective barrier to the underlying tissues (Jiang et al., 1986). The nasal squamous epithelium has a fairly high rate of cell turnover. In the F344 rat, the nasal vestibular squamous epithelium has the greatest proportion of cells in DNA synthesis, as compared to the other epithelial types in the nasal passages (Johnson et al., 1990). This feature of efficient repair and regeneration of the squamous epithelium further supports its presumed role as a protective barrier to inhaled xenobiotics.

Spontaneous lesions of the squamous epithelium in the nasal vestibule are uncommon, and this epithelium is also fairly resistant to effects of inhaled xenobiotics (Jiang et al., 1986). Certain irritants, however, do induce squamous epithelial lesions (Table 1). Acute lesions include erosion and ulceration, often associated with an inflammatory cell infiltrate. The tips of the atrioturbinates are a common location for focal ulceration, resulting in consequent degeneration or atrophy of these structures.

Squamous-cell hyperplasia is usually observed with chronic insult and is often associated with inflammation and hyperkeratosis. With prolonged irritation, rete ridge formation may result, similar to that observed in the epidermis of the skin (Monticello et al., 1990). It is important to note, however, that the majority of squamous hyperplastic lesions in the rodent nose are located in areas of the nasal passages normally lined by respiratory epithelium. Respiratory squamous metaplasia with hyperplasia is a commonly encountered toxicant-induced lesion (Monticello et al., 1990). An infrequently observed change in the squamous epithelium is the presence of intracytoplasmic hyaline droplets (eosinophilic globules). These inclusions appear to consist of proteinaceous material of unknown origin and have been proposed to serve as an adaptive response to certain types of irritation (Monticello et al., 1990; St. Clair & Morgan, 1992).

Tumors originating from the region of the squamous epithelium of the nares and nasal vestibule are uncommon (Brown, 1990). Tumor types reported include squamous-cell carcinoma and papillomas. These tumors often form vegetative or erosive lesions that extend from the nasal orifices or onto the external nares, occasionally being classified as a primary skin tumor.

SUMMARY

Studies of the distribution of nasal lesions may provide insight into the mechanisms involved and the role of regional xenobiotic dosimetry in

FIGURE 3. Scanning electron micrograph (SEM) of the nasal vestibule of a Sprague-Dawley rat. Squamous epithelial cells are arranged in a mosaic pattern separated by interconnecting linear invaginations. Bar = 20 μm.

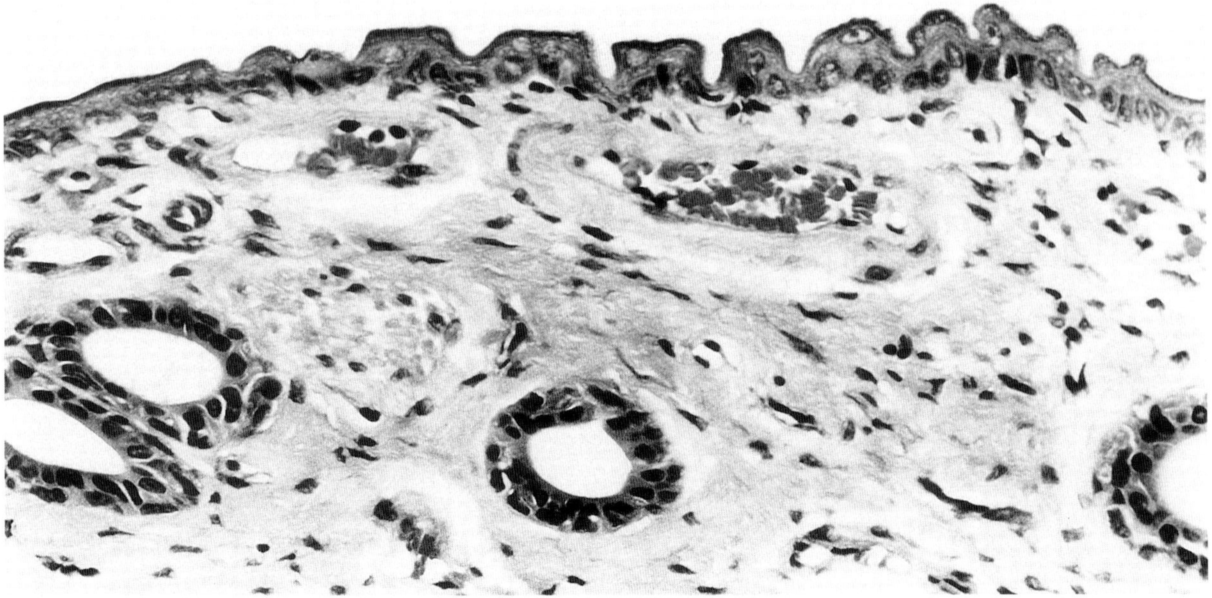

FIGURE 4. Photomicrograph of stratified squamous epithelium of the rat nasal vestibule. The undulating microridge appearance is the light microscopic correlate to the linear invaginations observed by SEM (see Figure 3). Hematoxylin and eosin, original magnification ×100.

TABLE 1. Selected Examples of Nasal Squamous Epithelial Lesions in Rodents Exposed to Inhaled Xenobiotics

Chemical	Squamous epithelial lesion[a]	Reference
Ammonia and other cage contaminants	Erosion, inflammation	Bolon et al. (1991)
Chlorine	Eosinophilic globules Erosion, ulceration Septal perforation	Gross et al. (1993)
Dimethylamine	Erosion, ulceration Septal perforation	Buckley et al. (1985)
Glutaraldehyde	Erosion, inflammation Hyperkeratosis, exfoliation	Gross et al. (1994)
Hydrogen chloride	Necrosis (ventral meatus)	Jiang et al. (1986)
Toluene diisocyanate	Inflammation, exfoliation	Buckley et al. (1984)

[a]Reported in the nasal vestibule unless otherwise noted.

lesion location. Toxicology studies, which can benefit from careful description and mapping of nasal lesions, reveal characteristic lesion distribution patterns that may be attributable to nasal airflow, regional metabolism, or other factors. Each of the epithelial types in the nasal passages exhibits a characteristic range of responses; however, the squamous region is generally the most refractory to chemical exposure. Only the nasal vestibule, and in rodents the ventral meatus and nasopalatine ducts, is lined by squamous epithelium. This epithelium resembles the epidermis of the skin in many respects. The small number of published examples of chemically induced squamous epithelial lesions in the nose of laboratory animals report specific patterns of responses. Further work is needed to characterize the role of regional nasal dosimetry in toxic responses of the squamous epithelial region of the nose and its relevance to human disease risk.

REFERENCES

Bolon, B., Bonnefoi, M. S., Roberts, K. C., Marshall, M. W., and Morgan, K. T. 1991. Toxic interactions in the rat nose: Pollutants from soiled bedding and methyl bromide. *Toxicol. Pathol.* 19:571–579.

Brown, H. R. 1990. Neoplastic and potentially preneoplastic changes in the upper respiratory tract of rats and mice. *Environ. Health Perspect.* 85:291–304.

Buckley, L. A., Jiang, X. Z., James, R. A., Morgan, K. T., and Barrow, C. S. 1984. Respiratory tract lesions induced by sensory irritants at the RD50 concentration. *Toxicol. Appl. Pharmacol.* 74:417–429.

Buckley, L. A., Morgan, K. T., Swenberg, J. A., James, R. A., Hamm, T. E., and Barrow, C. S. 1985. The toxicity of dimethylamine in F344 rats and B6C3F1 mice following a 1-year inhalation exposure. *Fundam. Appl. Toxicol.* 5:341–352.

Casanova, M., Morgan, K. T., Steinhagen, W. H., Everitt, J. I., Popp, J. A., and Heck, H. d'A. 1991. Covalent binding of inhaled formaldehyde to DNA in the respiratory tract of rhesus monkeys: Pharmacokinetics, rat-to-monkey interspecies scaling, and extrapolation to man. *Fundam. Appl. Toxicol.* 17:409–428.

Gross, E. A., Swenberg, J. A., Fields, S., and Popp, J. A. 1982. Comparative morphometry of the nasal cavity in rats and mice. *J. Anat.* 135:83–88.

Gross, E. A., Wolf, D. C., Barrow, C. S., Morgan, K. T., Moss, O. R., and Popp, J. A. 1993. Chronic inhalation study of chlorine in B6C3F1 mice and F344 rats. *Toxicologist* 13:150.

Gross, E. A., Mellick, P. W., Kari, F. W., Miller, F. J., and Morgan, K. T. 1994. Histopathology and cell replication responses in the respiratory tract of rats and mice exposed by inhalation to glutaraldehyde for up to thirteen weeks. *Fundam. Appl. Toxicol.*, in press.

Harkema, J. R. 1991. Comparative aspects of nasal airway anatomy: Relevance to inhalation toxicology *Toxicol. Pathol.* 19:321–336.

Harkema, J. R., Plopper, C. G., Hyde, D. M., St. George, J. A., Wilson, D. W., and Dungworth, D. L. 1987. Response of the macaque nasal epithelium to ambient levels of ozone. *Am. J. Pathol.* 128:29–44.

Hurtt, M. E., Thomas, D. A., Working, P. K., Monticello, T. M., and Morgan, K. T. 1988. Degeneration and regeneration of the olfactory epithelium following inhalation exposure to methyl bromide: pathology, cell kinetics, and olfactory function. *Toxicol. Appl. Pharmacol.* 94:311–328.

Jiang, X. Z., Morgan, K. T., and Beauchamp, R. O. 1986. Histopathology of acute and subacute nasal toxicity. In *Toxicology of the nasal passages*, ed. C. S. Barrow, pp. 51–65. New York: Hemisphere.

Johnson, N. F., Hotchkiss, J. R., Harkema, J. R., and Henderson, R. F. 1990. Proliferative responses of rat nasal epithelia to ozone. *Toxicol. Appl. Pharmacol.* 103:143–155.

Kimbell, J. S. 1994. Issues in modeling dosimetry in rats and primates. *Inhal. Toxicol.* 6(suppl):73–83.

Kimbell, J. S., Gross, E. A., Joyner, D. R., Godo, M. N., and Morgan, K. T. 1993. Application of computational fluid dynamics to regional dosimetry of inhaled chemicals in the upper respiratory tract of the rat. *Toxicol. Appl. Pharmacol.* 121:253–263.

Monticello, T. M., Morgan, K. T., Everitt, J. I., and Popp, J. A. 1989. Effects of formaldehyde gas on the respiratory tract of rhesus monkey: Pathology and cell proliferation. *Am. J. Pathol.* 134:515–527.

Monticello, T. M., Morgan, K. T., and Uriah, L. 1990. Nonneoplastic nasal lesions in rats and mice. *Environ. Health. Perspect.* 85:249–274.

Morgan, K. T. 1991. Approaches to the identification and recording of nasal lesions in toxicology studies. *Toxicol. Pathol.* 19:337–351.

Morgan, K. T. and Monticello, T. M. 1990. Airflow, gas deposition, and lesion distribution in the nasal passages. *Environ. Health. Perspect.* 88:209–218.

Morgan, K. T., Kimbell, J. S., Monticello, T. M., Patra, A. L., and Fleishman, A. 1991. Studies of inspiratory airflow patterns in the nasal passages of the F344 rat and rhesus monkey using nasal molds: Relevance to formaldehyde toxicity. *Toxicol. Appl. Pharmacol.* 110:223–240.

Negus, V. E., ed. 1958. *The comparative anatomy and physiology of the nose and paranasal sinuses.* Edinburgh: Livingstone.

St. Clair, M. B. G., and Morgan, K. T. 1992. Age-related changes in the upper respiratory tract of the rat. In *Pathobiology of the aging rat,* Vol. 1, eds. U. Mohr, D. L. Dungworth, and C. C. Capen, pp. 111–127. Washington, DC: ILSI.

Sorekin, S. P. 1977. The respiratory system. In *Histology,* 4th ed., eds. L. Weiss and R. O. Greep, Chapter 21. New York: McGraw-Hill.

Uriah, L. C., and Maronpot, R. R. 1990. Normal histology of the nasal cavity and application of special techniques. *Environ. Health Perspect.* 85:187–208.

Young, J. T. 1981. Histopathologic examination of the rat nasal cavity. *Fundam. Appl. Toxicol.* 1:309–312.

OZONE-INDUCED PROLIFERATIVE AND METAPLASTIC LESIONS IN NASAL TRANSITIONAL AND RESPIRATORY EPITHELIUM: COMPARATIVE PATHOLOGY

Jack R. Harkema, Jon A. Hotchkiss

Inhalation Toxicology Research Institute, Albuquerque, New Mexico, USA

This article reviews the proliferative and metaplastic responses of two distinct nonolfactory nasal epithelia (i.e., transitional and respiratory) in laboratory animals after repeated exposures to a common urban air pollutant, ozone. Alterations in these epithelia after either acute or chronic exposures are described. Ozone exposures induce lesions predominantly in the anterior aspects of the nasal airways in laboratory rats and monkeys. The principal epithelial alteration induced by this potent oxidant pollutant is a marked mucous cell hyperplasia/metaplasia in the transitional epithelium. Attenuation and loss of cilia are common ozone-induced lesions in the respiratory epithelium. In this article, we also compare the nasal epithelial lesions induced by ozone exposures in laboratory animals to nasal histopathology reported in human subjects living in ozone-polluted regions of Mexico City.

The nasal cavity is lined by surface epithelial cell populations that have specific roles in conducting and maintaining normal functions of the nose. For example, the olfactory epithelium (OE) contains neuroepithelial cells that are essential for the detection of odorants and sustentacular cells that are thought to play crucial roles in the metabolism of inhaled xenobiotics. The nasal respiratory epithelium (RE), which lines over 80% of the nasal airway in humans, is a pseudostratified, ciliated epithelium with numerous mucus-secreting cells that provide important nonolfactory functions. The ciliated cells with their motile cilia, along with the mucus-secreting cells in this specific epithelium, are the primary cellular components in the dynamic mucociliary apparatus. This apparatus is an important respiratory defense mechanism of the upper and lower respiratory tracts that protects the body from potentially injurious, inhaled, inorganic and organic particles and gases. These inhaled foreign agents deposit on the sticky mucous layer that is constantly being renewed by the underlying secretory cells and are propelled by the fast and synchronized beating of the cilia from the airway to the digestive tract, where the agents are eliminated from the body. Because of this highly efficient defense mechanism, the nasal airways have been

This review was sponsored in part by the U.S. Department of Energy, Office of Health and Environmental Research, under Contract DE-AC04-76EV01013 and by the National Heart, Lung, and Blood Institute under Grant RD1-HL51712.

Address correspondence to Dr. Jack R. Harkema, Department of Pathology, A46 VMC, College of Veterinary Medicine, Michigan State University, East Lansing, MI 48824.

described as a "scrubbing tower" that removes inhaled chemicals potentially harmful to the lower tracheobronchial airways and the delicate alveolar tissues in the lung where gas exchange occurs (Brain, 1970). Although the nasal airways have important protective roles in the defense of the lower respiratory tract, their surface epithelial cells, nonetheless, are still vulnerable to injury from inhaled toxicants. The proximity of these surface epithelial cell populations to the inhaled airflow makes them prime targets for toxicant-induced damage. How these surface epithelial cells respond to these agents is predominantly dependent on the chemical makeup of the toxicant, the inherent sensitivity of the cell, its in vivo environment, and the dose of the toxicant to the cells.

The purpose of this article is to provide some examples of how two specific surface epithelia in the nose, the transitional epithelium (TE) and RE, can respond to inhaled ozone. The discussion focuses on the structural (morphologic) alterations that have been shown to occur in the nasal epithelia of laboratory animals after irritant exposure. Though several inhaled toxicants have been reported to induce injury to the nasal mucosa (Buckley et al., 1984), structural alterations induced by exposures to ozone, the principal oxidant air pollutant in urban smog, is the focus of this manuscript. The discussion is also restricted to noncarcinogenic responses (i.e., hyperplasia, metaplasia) of these two distinct nasal epithelial cell populations.

COMPARATIVE MORPHOLOGY OF NASAL TRANSITIONAL AND RESPIRATORY EPITHELIUM

Before the extent and nature of toxicant-induced injury to the nasal TE or RE can be assessed, the normal tissue and cellular morphology and the location of these epithelia in the nasal cavity must be understood. Although the gross anatomy of the nasal airways in several animal species has been known for a long time, only recently have studies been conducted to systematically and thoroughly characterize the cells lining this important upper airway. Current knowledge of nasal cellular morphology is primarily the result of recent animal studies that were designed to quantitatively examine the response of the nasal epithelium to various toxicants. There has been no equivalent morphological study of the human nasal epithelium. Therefore, our knowledge of human nasal epithelial cells is limited, for the most part, to morphology of biopsied tissue from only a particular location in the nasal cavity (i.e., anterior tip of the inferior or middle nasal turbinate).

Though the epithelial lining of the nasal airways in laboratory animals is now recognized as a major site of injury from exposure to environmental toxicants, extrapolation from animals to humans is complicated by the complexity and diversity among species of this upper airway. Responsible estimates of the risks of nasal toxicants to human health must be made with a working knowledge of the differences and similarities of the structural components in the human and animal nasal airways.

Besides differences in the gross size and architecture of the nose among species, there are also species differences in the surface epithelial populations lining the nasal passages. These differences among species are found in the distribution of nasal epithelial populations and the types of nasal cells within these populations. There are, however, four distinct nasal epithelial populations in most laboratory animal species. These include a stratified squamous epithelium, which is primarily restricted to the nasal vestibule; ciliated, pseudostratified RE in the main nasal chamber and nasopharynx; nonciliated or poorly ciliated TE, lying between squamous epithelium and RE in the anterior or proximal aspect of the main chamber; and the OE located in the dorsal or dorsoposterior aspect of the nasal airway. Figure 1 illustrates the distribution of these distinct epithelial cell populations in the nasal lateral wall of the bonnet monkey and Fischer (F344/N) rat. As mentioned earlier, this discussion focuses on the responses of two of these nasal epithelial populations, the TE and RE.

Transitional Epithelium

The TE is caudal or posterior to the squamous epithelium and proximal or anterior to ciliated RE. It is a narrow zone of nonciliated, microvilli-covered

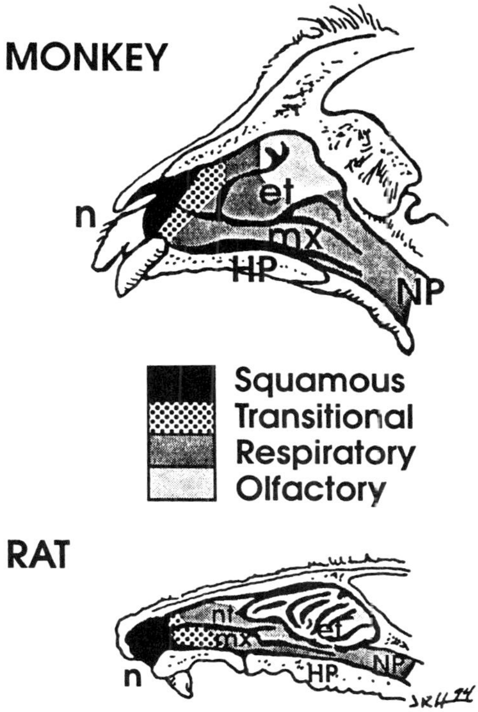

FIGURE 1. Diagrammatic representation of the exposed lateral wall of one nasal passage of the bonnet monkey and F344 rat with the nasal septum removed. Distribution of the different epithelial populations lining the luminal surfaces are designated; nt, nasoturbinate; mx, maxilloturbinate; NP, nasopharynx; n, naris; et, ethmoturbinates; HP, hard palate.

surface epithelium, which has been referred to as nonciliated RE in rats and mice by some authors (Matulionis & Parks, 1972; Montiero-Riviere & Popp, 1984; Popp & Martin, 1984) and as nasal TE in rats, dogs, and humans by others (Adams & Hotchkiss, 1983; Harkema et al., 1987a, 1989; Mygrind et al., 1982). Common distinctive features of this nasal epithelium in all species include (1) anatomical location in the anterior aspect of the nasal cavity between squamous epithelium and RE; (2) the presence of nonciliated cuboidal or columnar surface cells and basal cells; (3) scarcity of mucous cells and paucity of intraepithelial mucosubstances; and (4) an abrupt morphological border with squamous epithelium, but a less abrupt border with RE.

In rats and mice, this epithelium predominantly lines the lateral wall and lateral aspects of the nasoturbinates and maxilloturbinates in the anterior aspect of the nasal cavity. Recent morphometric analyses have found that the TE covers approximately 10–12% of the surface of the nasal cavity in the F344/N rat (E. Gross & J. Kimbell, personal communication). In addition, most of the TE in rodents is thin (i.e., one to two cells thick; Figure 2A), pseudostratified, and composed of three distinct cell types (basal, cuboidal, and columnar). In contrast, TE in monkeys is thick (i.e., four to five cells; Figure 2B), stratified, and composed of at least five different cell types [basal, nonciliated cells with few secretory granules, mucous (goblet) cells, small mucous granule cells, and cells with intracytoplasmic lumina] (Harkema et al., 1987a).

The TE in monkeys does contain luminal mucous cells that contribute to the overlying airway mucous layer. These cells contain both acidic and neutral mucosubstances (Harkema et al., 1987d). In contrast to the TE of the monkey, there are few, if any, mucous cells with intracytoplasmic mucosubstances in the normal TE of rats (Harkema et al., 1989).

Respiratory Epithelium

The majority of the nonolfactory nasal epithelium of laboratory animals and humans is ciliated RE. Approximately 46% of the nasal cavity in a F344 rat is lined by RE along with TE (Gross et al., 1982). Although this pseudostratified nasal epithelium is similar to ciliated epithelium lining other proximal airways (i.e., trachea and bronchi), it also has unique features. The RE in the rat is composed of six morphologically distinct cell types: mucous (goblet) cells, ciliated cells, nonciliated columnar cells, cuboidal cells, brush cells, and basal cells (Montiero-Riviere & Popp, 1984). These cells are unevenly distributed along the rat mucosal surface. Using scanning electron microscopy, Popp and Martin (1984) demonstrated an anterioposterior increase in ciliated cells along the lateral walls of the rat. The nasal septum of the rat, however, contains a uniform density of ciliated cells from anterior to posterior.

Like the RE of rats, the RE of monkeys contains ciliated, mucous, and basal cells (Harkema et al., 1987a). In addition, the monkey RE contains

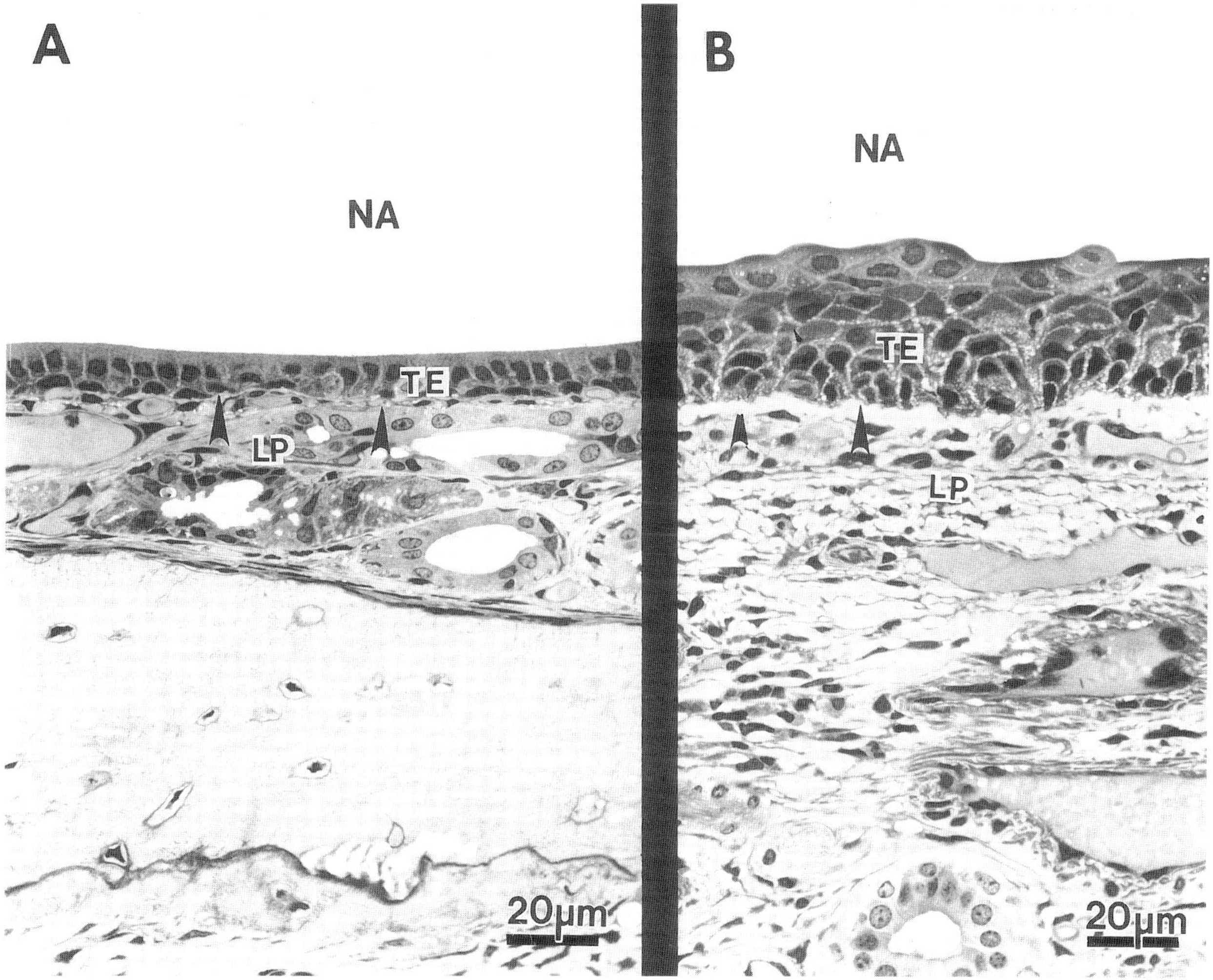

FIGURE 2. Photomicrographs of the nasal transitional epithelium (TE) in (A) an F344 rat and (B) a bonnet monkey. NA, nasal airway; arrowheads, basal lamina; LP, lamina propria. The rat TE is thin, one to two cells in thickness and pseudostratified, while the monkey TE is thick, four to six cells in thickness and stratified. Adapted from an illustration in Harkema (1991).

small mucous granule cells and cells with intracytoplasmic lumina. No brush cells have been observed in monkey RE.

RESPONSES OF NASAL TRANSITIONAL AND RESPIRATORY EPITHELIUM TO INHALED OZONE

As mentioned earlier, nasal epithelium can be the target for inhaled toxicants and can undergo considerable alteration after acute or chronic exposures to noxious agents. Several inhaled chemical irritants have been demonstrated to cause significant injury to the nasal surface epithelium (Buckley et al., 1984; Monticello et al., 1990). Often acute lesions induced by inhaled irritants are restricted to the anterior nasal passages, with an anterior to posterior distributional decrease in severity of the lesions. These alterations to the nasal mucosa can include acute inflammation with a conspicuous influx of neutrophils into the lamina propria and surface epithelium from the blood vessels; degeneration and necrosis of surface epithelium; epithelial ulceration; subsequent increases in the numbers of resistant epithelial cells (i.e., hyperplasia); or even transformation of the normal epithelium to a morphologically different, adaptive epithelium (e.g., squamous metaplasia, mucous cell metaplasia). These responses are highly dependent on the chemical nature of the toxicant. For example, repeated exposures of rats to formaldehyde have been shown to induce marked squamous metaplasia in the TE and RE of the anterior nasal passages (Kerns et al., 1983). In contrast, TE in similar regions of the rat nose will undergo mucous cell metaplasia when repeatedly exposed to ozone (Harkema et al., 1989, 1992; Hotchkiss et al., 1989). These examples illustrate the interesting fact that the nasal surface epithelium can respond to irritant insults in different ways. Both the nasal TE and RE can respond quickly to injurious toxic exposures, not only to repair themselves from sudden induced injury, but also to transform themselves (i.e., metaplasia) to adapt to continuous, long-term exposures.

In the remainder of this article, we describe in detail the known responses that are initiated in TE and RE of laboratory animals after short- and long-term exposures to a common urban, air pollutant, ozone. Ozone is an irritating oxidant gas and the principal deleterious agent in photochemical smog. Currently, ozone is the most pervasive problem of the major air pollutants for which National Ambient Air Quality Standards (NAAQS) have been designated under the Clean Air Act (Steinfeld, 1991). The U.S. Environmental Protection Agency has estimated that 13 million healthy adults are exposed to ozone in excess of the NAAQS concentration (0.12 ppm) for at least 1 h each week during the summer (Paul et al., 1987). The episodic high concentrations of ozone in large metropolitan areas like Los Angeles and Mexico City pose significant threats to the health of their inhabitants.

Ozone-Induced Alterations in Monkey Nasal Epithelium

Although there have been numerous studies characterizing ozone-induced lesions in the lungs of several laboratory animal species, only recently have studies been designed specifically to examine the effects of ozone on the nose. The first study that demonstrated that ozone could be a nasal toxicant used bonnet monkeys (*Macaca radiata*) as the experimental animals (Harkema et al., 1987b, 1987c). In this study, monkeys were exposed to 0.15 or 0.30 ppm ozone for 6 or 90 days (8 h/day). Lesions to the airway mucosa were restricted to the very anterior aspects of the nasal passages and involved the TE and the adjacent RE. The principal ozone-induced lesions included acute rhinitis characterized by a conspicuous intraepithelial influx of neutrophils (6 days only), attenuation of cilia along with individual ciliated cell degeneration and necrosis (6 and 90 days), an increase in the number of epithelial cells (i.e., epithelial hyperplasia; 6 days only), and a marked increase in mucus-secreting cells [i.e., mucous (goblet) cell hyperplasia; 6 and 90 days; Figure 3]. The magnitude of this change was not dependent on the ozone concentration (i.e., 0.15 vs. 0.30 ppm), but on the duration of the exposure (i.e., more mucous cells after 90 days of exposure than after 6 days of exposure). In the TE of ozone-exposed

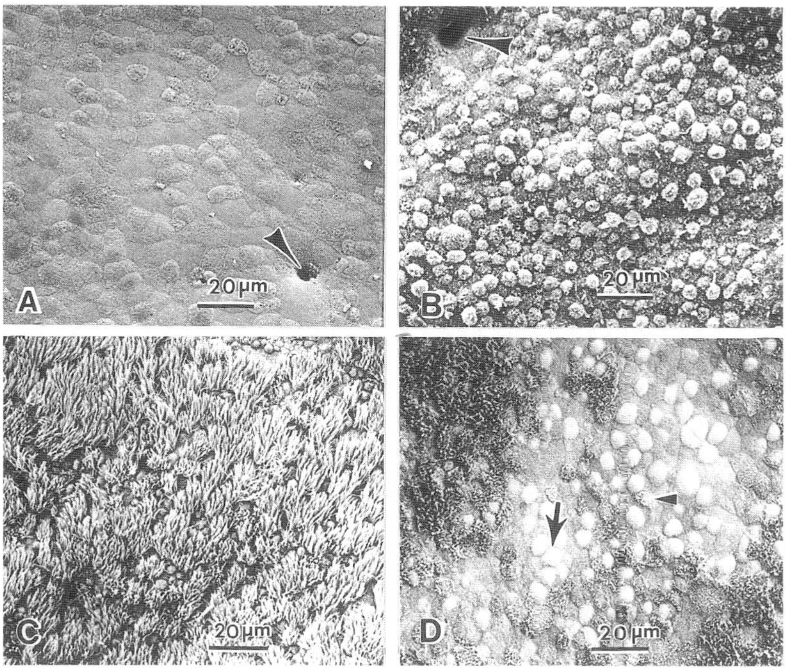

FIGURE 3. Scanning electron photomicrographs of luminal surfaces of nasal transitional (A and B) and respiratory epithelium (C and D) from the anterior nasal septum of bonnet monkeys exposed to 0 ppm (A and C) or 0.30 ppm (B and D) ozone for 90 days. In (B) and (D), there are numerous ozone-induced mucous cells with domelike, apical surfaces projecting above the general luminal surface (arrow in D). Also in (D), there are ozone-induced losses of cilia and cells with shortened cilia (small arrowhead). The large arrowheads in (A) and (B) point to openings of ducts from subepithelial glands. Adapted from illustrations in Harkema et al. (1987c).

monkeys, the numbers of nonciliated cells with mucus-secretory granules were 15–20% greater than in air-exposed (0 ppm ozone) controls. In addition, the amount of mucous material in the TE was 300% greater in monkeys exposed to 0.15 ppm ozone for 6 days (8 h/day), compared to air-exposed controls (Harkema et al., 1987b). These nasal epithelial alterations induced by ozone may indicate alterations in normal physiologic functions (for example, mucociliary clearance) that are important upper respiratory tract mechanisms to defend the lungs from excessive burdens of harmful agents. Interestingly, ozone-induced epithelial alterations (i.e., hyperplasia of cuboidal bronchiolar cells) were also evident in the respiratory bronchioles within the lungs of these same monkeys (Harkema et al., 1993). This type of nasal damage could also be a factor in increasing susceptibility to acute infections of the upper respiratory tract. The findings of this study have added significance because of their presence in a species of nonhuman primate whose nasal airways resemble, at a gross and microscopic level, those of humans (Harkema, 1991).

Ozone-Induced Alterations in Rat Nasal Epithelium

In 1989, we reported that Fischer rats (F344/N) exposed for 7 days to 0.8 ppm ozone, 6 h/day, developed mucous cell metaplasia in the TE lining the maxilloturbinates, lateral wall, and lateral aspects of the nasoturbinates in the proximal nasal passages (Harkema et al., 1989). The ozone-induced lesions in the F344/N rat resembled those that were experimentally induced in the nasal cavity of bonnet monkeys. There were, however, some noticeable differences. The increase in mucous cells in the rat nasal airway was restricted to the TE that is normally devoid of mucous cells. Therefore, the ozone-induced nasal epithelial alteration in the rat was a true metaplastic response (transformation of an epithelium with no mucous cells into an epithelium with numerous mucous cells), which differed from the hyperplastic response (increase in the number of mucous cells in an epithelium that normally contains some mucous cells) induced in monkeys after ozone exposure. In addition, no epithelial alterations were evident in the RE in the anterior nasal airways of the rat, unlike the monkey. The rats exposed to 0.8 ppm ozone did, however, have attenuation of cilia in the lateral walls of the nasopharynx. Finally, no metaplastic alterations were evident in the TE of rats exposed to 0.12 ppm, 6 h/day for 7 days. This was in contrast to the conspicuous mucous cell hyperplasia that was induced in monkeys after a 6-day (8 h/day) exposure to 0.15 ppm ozone. This suggests that the rat TE may not be as sensitive to ozone-induced injury as monkey TE.

The amount of histochemically stained mucosubstances within the TE (i.e., stored secretory product within mucous cells) has been determined to be a sensitive and reproducible quantitative indicator of the severity of the ozone-induced metaplasia (Harkema et al., 1987b, 1989). To estimate the amount of intraepithelial mucosubstances in the nasal epithelium, transverse tissue sections from specifically designated locations throughout the nasal airway were processed for light microscopy and stained with alcian blue (pH 2.5)/periodic acid-Schiff's sequence (AB/PAS) for demonstration of acidic and neutral mucosubstances. The volume density of acidic and neutral mucosubstances within the TE was quantitated by using a computerized image analysis system. Figure 4 photomicroscopically illustrates the differences in amounts of AB/PAS-stained mucosubstances in the TE of the maxilloturbinate in F344/N rats that were exposed to 0 or 0.8 ppm ozone for 7 days, 6 h/day and killed immediately after the end of the exposure. The mean amount of intraepithelial mucosubstances in the TE of the maxilloturbinates immediately after the 7-day ozone exposure was 0.9 ± 0.1 nl/mm^2 (mean ± standard error of the mean; n = 5), compared to 0 nl/mm^2 in controls exposed to filtered air (0 ppm ozone; n = 8). Interestingly, the amounts of intraepithelial mucosubstances in the TE do not decrease 7 days after the end of the exposure, but instead markedly increase. Further, the volume densities of the mucosubstances in the TE of rats killed 7 days after the end of exposure was approximately 3 times greater than 0.8 ppm ozone-exposed rats killed immediately after exposure (2.0 ± 0.5 nl/mm^2 compared to 0.9 ± 0.1 nl/mm^2; n = 5/group). The persistence of this ozone-induced change in intraepithelial mucosubstances, however, has not been adequately examined for time periods greater than 7 days after the end of exposure.

In this issue of the journal, a brief report by Harkema et al. describes a study designed to determine if ozone-induced, mucous cell metaplasia in TE was different among inbred strains/outbred stocks of rats. We found that the severity of the mucous cell metaplasia significantly varied among the ozone-exposed rat strains/stocks. The results of this study indicated that Wistar rats and F344/NTac rats were more sensitive to ozone exposure compared to F344/N rats from the Inhalation Toxicology Research Institute's colony and Sprague-Dawley rats. The reason for the strain/stock-related variability in the severity of this ozone-induced lesion was not determined.

In an attempt to better understand the pathogenesis of the ozone-induced mucous cell metaplasia, we have characterized the ozone-induced events in the nasal mucosa of rats that occur prior to the onset of the mucous cell metaplasia. We reported that an influx of neutrophils (Hotchkiss et al., 1989) and an increase in epithelial DNA synthesis (Hotchkiss & Harkema, 1992; Johnson et al., 1990) always precede the mucous cell metaplasia induced by ozone, and these inflammatory and proliferative responses decrease as the mucous cell metaplasia develops in the TE. Six hours after the start of an exposure to 0.8 ppm ozone, there is an acute rhinitis, with an influx of neutrophils into the nasal TE. With repeated exposures of ozone (6 h/day, for 3 or 7 days), there is a gradual attenuation of the inflammatory response as measured by the number of intraepithelial neutrophils (Figure 5). The reason for the attenuation of this inflammatory response is unknown, but it may be due to an induction of mucosal antioxidants that protect the tissue from further ozone-induced injury. Interestingly, as the ozone-induced influx of neutrophils decrease,

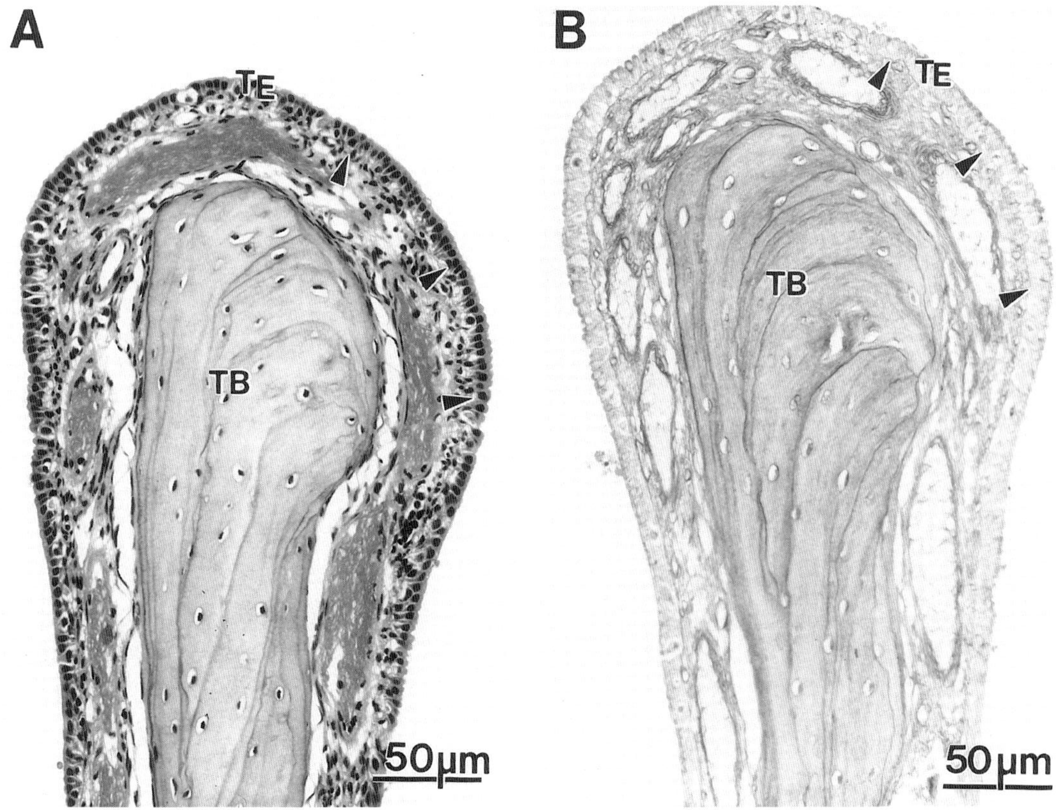

FIGURE 4. Light photomicrographs of maxilloturbinates from rats exposed to 0 ppm (control; A and B) for 7 days (6 h/day). TE, transitional epithelium; TB, turbinate bone; arrowheads, basal lamina between surface epithelium and underlying lamina propria. Tissues A and C were stained with hematoxylin and eosin and AB (pH = 2.5)/PAS. Tissues B and D were stained with AB (pH = 2.5)/PAS for mucosubstances.

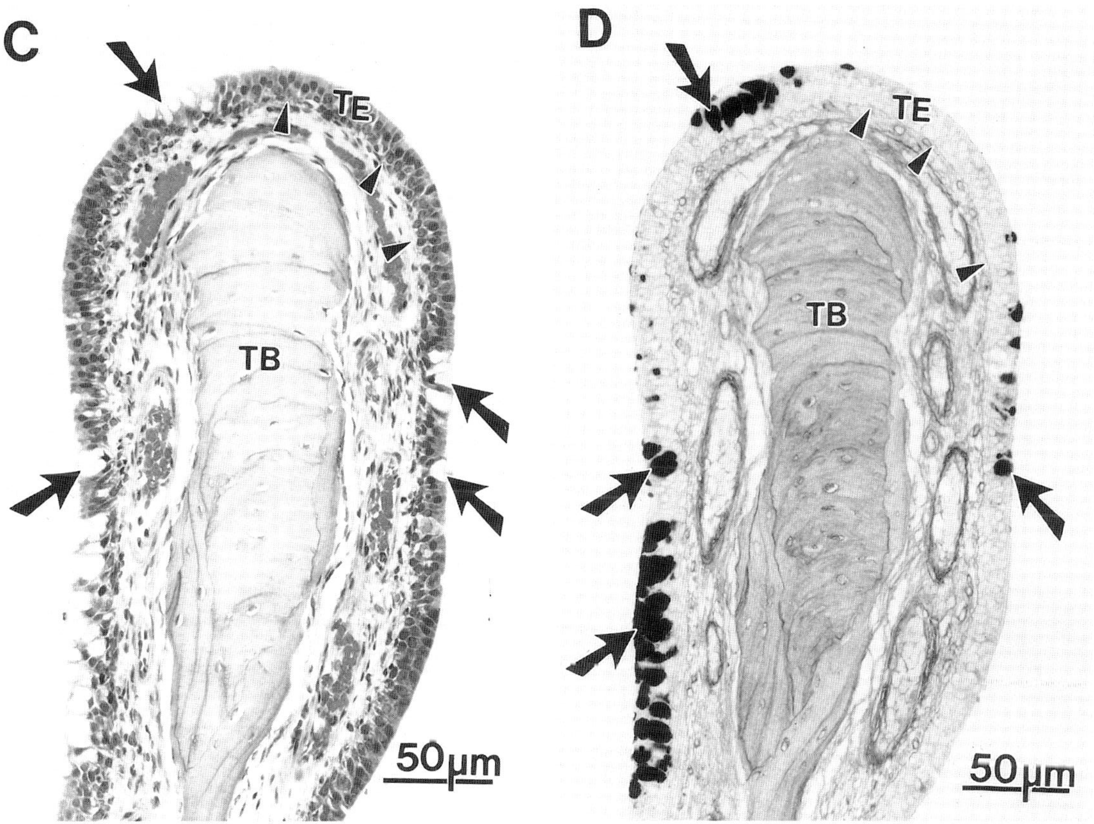

FIGURE 4. (*Continued*) Light photomicrographs of maxilloturbinates from rats exposed to 0.8 ppm ozone (C and D) for 7 days (6 h/day). Ozone-induced epithelial hyperplasia and mucous cell metaplasia (arrows point to mucous cells) are present in the surface epithelium of (C) with conspicuous amounts of AB/PAS-stained intraepithelial mucosubstances (D; arrows point to stained mucosubstances). TE, transitional epithelium; TB, turbinate bone; arrowheads, basal lamina between surface epithelium and underlying lamina propria. Tissues A and C were stained with hematoxylin and eosin and AB (pH = 2.5)/PAS. Tissues B and D were stained with AB (pH = 2.5)/PAS for mucosubstances.

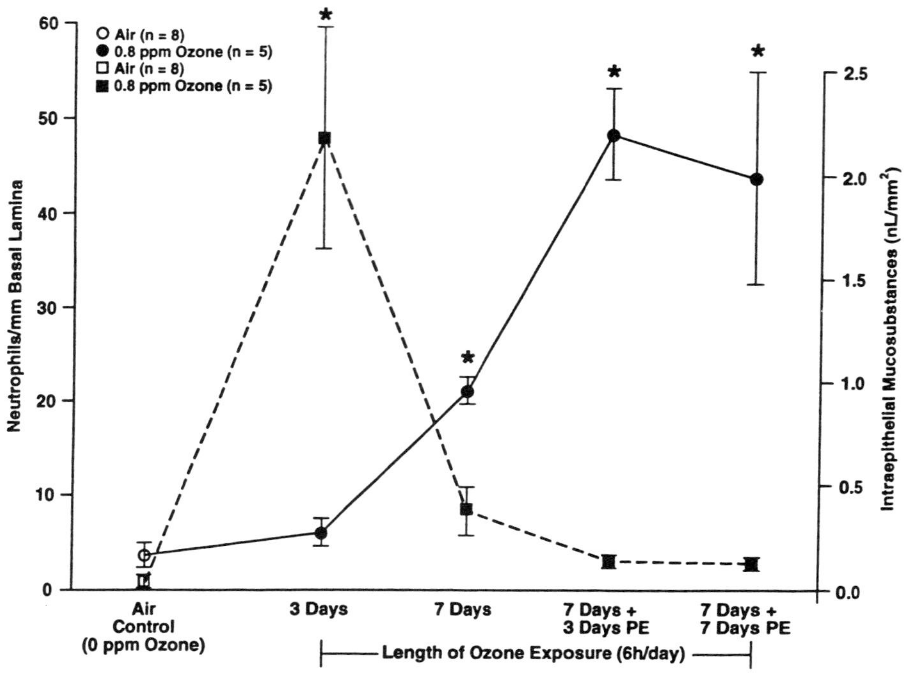

FIGURE 5. Mean numeric densities of neutrophils (squares connected by broken lines) and the mean volume densities of intraepithelial mucosubstances (circles connected by solid lines) in the transitional epithelium lining the maxilloturbinate from the anterior noses of rats exposed to 0.8 ppm ozone (6 h/day) for 3 or 7 days and sacrificed immediately after exposure, or 3 or 7 days postexposure (PE). All points and bars represent the mean and standard error of the mean, respectively, for each experimental group. Asterisk indicates significantly different from air-control group, $p < .05$.

there is a concomitant increase in mucosubstances within the surface TE (Figure 5). Since airway mucus is known to be a good antioxidant (Cross et al., 1984), it is plausible that these newly formed intraepithelial mucosubstances, with a corresponding increase in secreted mucins, act as a protective shield for the nasal epithelium and underlying lamina propria, preventing further injury by the continued ozone exposure. Therefore, the reduction in surface injury would lead to the subsequent attenuation of the inflammatory response. Although this is a plausible explanation, more studies must be conducted to verify this hypothesis.

In addition to the acute inflammatory response to ozone exposure in the rat TE, there is also marked induction of DNA synthesis in the TE prior to the onset of the mucous cell metaplasia. In a study by Johnson et al. (1990), the effects of ozone exposure (3- or 7-day exposure, 6 h/day, to 0, 0.12, or 0.8 ppm ozone) on DNA synthesis in all four types of rat nasal epithelium (squamous, TE, RE, and OE) were examined in vivo. DNA synthesis was monitored by immunohistochemically staining epithelial cells that had incorporated bromodeoxyuridine into their nuclei (BrdU; an analog of thymidine). RE, TE, and OE in control rats (exposed to 0 ppm ozone) were found to have low levels of DNA synthesis (1 BrdU-labeled cell/mm basal lamina). In contrast, SE had a high level of DNA synthesis (40 BrdU-labeled cells/mm basal lamina). Only rats exposed to 0.8 ppm ozone had significantly greater DNA synthesis in nasal epithelia compared to the air-exposed (0 ppm) controls. Following exposure to 0.8 ppm ozone, a transient, but marked, increase in DNA synthesis (30 times more than controls) was evident in the TE, while in the RE and OE the transient increase was less marked (five and four times greater than controls) (Figure 6). The number of S-phase cells in the squamous epithelium was similar to controls after 3 or 7 days of ozone exposure. The ozone-induced increase in DNA synthesis was prominent after 3 days of exposure and absent by 7 days of exposure in the RE and OE. The ozone-induced increase in BrdU-labeled cells in the TE was, however, markedly less at 7 days of exposure compared to 3 days of exposure (10- and 30-fold greater than controls, respectively). The transient nature of the DNA response in the TE and RE to the week-long, repeated ozone exposure was similar to the transient nature of

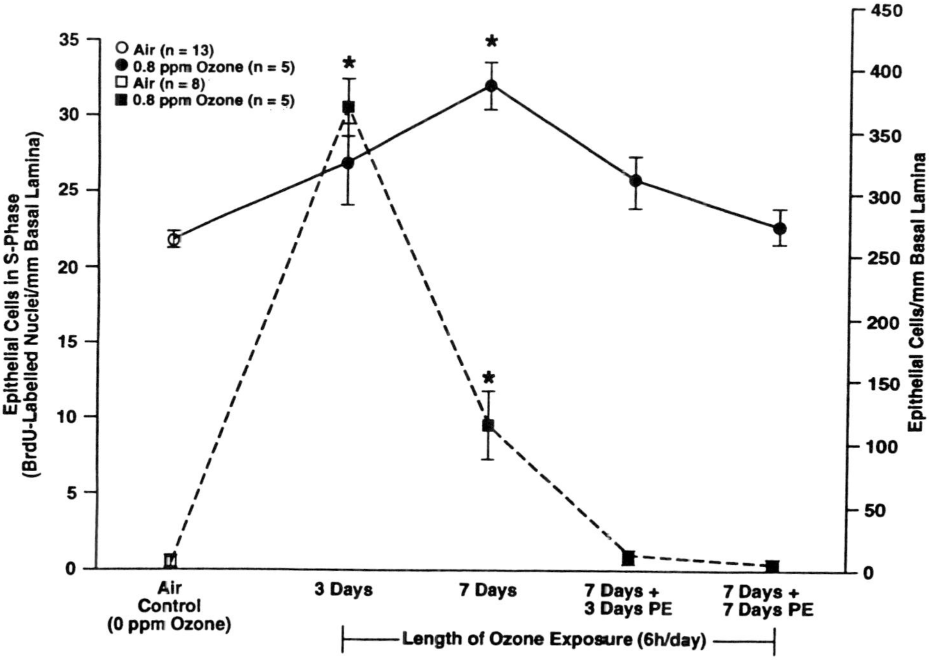

FIGURE 6. Mean numeric densities of epithelial cells in the S-phase of the cell cycle (undergoing DNA synthesis; squares connected by broken lines) and total epithelial cells (circles connected by solid lines) in the nasal transitional epithelium of F344 rats exposed to 0.8 ppm ozone (6 h/day) for 3 or 7 days and sacrificed immediately after exposure, or 3 or 7 days postexposure (PE). All points and bars represent the mean and standard error of the mean, respectively, for each experimental group. Asterisk indicates significantly different from air-controls, $p < .05$.

the inflammatory response in the nasal mucosa brought on by similar exposures to ozone that were mentioned earlier.

The reader is also directed to another study by Hotchkiss et al. (this volume) that compares the regional differences in ozone-induced DNA synthesis in nasal epithelial of the anterior nasal airways of the rat with predicted ozone dosimetry based on computational airflow-driven mass flux patterns. The results of this study suggest that while dosimetry may play some role in the observed pattern of ozone-induced DNA synthesis, much of the observed response must be due to inherent differences in epithelial susceptibility to ozone-induced injury.

Henderson et al. (1993) determined the lowest ozone exposure level that would induce DNA synthesis (uptake of BrdU) in the TE in the nose of rats and to determine the role of cumulative exposure (concentration × time) versus concentration of ozone in the response of TE to ozone-induced injury. Female F344/N rats, 11–13 wk of age, were exposed to a matrix of equal concentration × time values that included exposures to 0, 0.12, 0.24, and 0.48 ppm ozone for 3, 6, 12, or 24 h. No increased DNA synthesis was observed in rats exposed to 0.12 ppm ozone for any time periods. For exposures higher than 0.12 ppm ozone, the response of the TE was similar for equal cumulative exposures. The responses, however, were not linearly related to the cumulative (concentration × time) exposures. There was a general linear increase in DNA synthesis with increasing concentration, but not time, with concentrations above 0.12 ppm. Interestingly, no frank toxicity or cellular necrosis was observed, indicating that sublethal cell damage was sufficient to induce DNA synthesis above the 0.12 ppm concentration.

The early, but transient, increase in epithelial DNA synthesis after a 3-day exposure to 0.8 ppm ozone in the TE of F344/N rats results in conspicuous hyperplasia of this nasal epithelium (increase in the numeric density of epithelial cells; Figure 6). We have demonstrated that ozone-induced epithelial hyperplasia and mucous cell metaplasia in the TE of the F344/N rat can be induced with only a 3-day exposure of 0.8 ppm, 6 h/day, and the amount of intraepithelial mucosubstances at 7 days after the start of the exposures is similar to the amount found in the TE after a 7-day exposure to the same concentration of ozone (Hotchkiss et al., 1991). The fact that the type and magnitude of the alterations induced by either 3 or 7 days of ozone exposure were similar suggests that, once initiated, the development of ozone-induced phenotypic changes within the epithelium was not dependent on further ozone exposure. Four additional days of exposure to 0.8 ppm ozone had no significant effect on the location or magnitude of the ozone-induced proliferative and metaplastic changes that were observed in rats exposed to 0.8 ppm ozone for 3 days followed by a 4-day exposure to air.

Further research must be designed to determine if the ozone-induced mucous cell metaplasia is dependent on neutrophil influx and/or increased DNA synthesis and epithelial proliferation. The basic question is whether ozone can directly stimulate the TE cells to undergo transformation to mucous cells, or is mucous cell metaplasia the secondary result of ozone-induced inflammatory or proliferative events? This question may be better answered by exposing TE cells in vitro.

Effects of Chronic Ozone Exposures on the Nasal Transitional and Respiratory Epithelium

Recently, the effects of chronic ozone exposures on the structure and function of the nasal epithelium of F344 rats exposed to, 0.12, 0.5, or 1 ppm ozone for 20 mo (6 h/day, 5 days/wk) have been examined (Harkema et al., 1992). Rats were killed 1 wk after the end of the exposures. Immediately after death, flow rates of mucus throughout the nasal passages were determined using in vitro video-motion analysis. Following assessment of mucociliary function, the nasal tissues were processed for light microscopy and stained with AB/PAS for detection of intraepithelial mucosubstances. Rats exposed to 0.5 and 1 ppm ozone had markedly slower flow rates of mucus over the lateral walls and turbinates of the anterior third of the nasal passages compared to rats exposed to 0 and 0.12 ppm ozone. Interestingly, these intranasal regions in the rats exposed to 0.5 and 1 ppm ozone also had marked mucous cell metaplasias and 200 and 400 times greater amounts of intraepithelial mucosubstances than controls (0 ppm ozone), respectively. There were no significant differences in the mucous flow rates between the 0 ppm and 0.12 ppm exposed rats, and no ozone-induced morphologic alterations were present in the nasal epithelium of rats exposed to 0.12 ppm ozone. This is the first study to demonstrate that the ozone-induced, mucous cell metaplasia in the rat nasal TE is associated with a functional alteration (i.e., decreased mucous flow rates). This alteration could significantly modify an important respiratory defense mechanism of the upper airways (i.e., mucociliary clearance) and leave the more distal pulmonary airways vulnerable to potentially injurious concentrations of inhaled xenobiotics of infectious agents.

Nasal Mucosa in Humans Living in Ozone-Polluted Atmospheres

Mexico City is one of the more air-polluted urban areas in the world. Recently Calderon-Garcidueñas and her colleagues have demonstrated that residents of the southwestern metropolitan area of Mexico City develop nasal mucosa dysplasias in less than 30 days of arrival from regions of Mexico with no significant air pollution (Calderon-Garcidueñas et al., 1992). They reported the development of nasal mucosa dysplasias in 17.6% and 78.7% of <30-day and >60-day residents of the southwestern metropolitan area of the city, respectively. Together with the loss of normal RE, there was prominent basal cell hyperplasia, squamous metaplasia, and submucosal vascular proliferation. These investigators have suggested that ozone, the most important photochemical air pollutant in this urban atmosphere, could play a synergistic role in combination with other toxicants.

It is interesting to compare the results of this human study with ozone toxicity studies in laboratory animals. Unlike most of the animals that demonstrate marked mucous cell metaplasia induced after repeated exposures to high ambient concentrations of ozone, Calderon-Garcidueñas and her colleagues have not identified an increase in mucous cells, but rather a marked decrease in secretory cells due to squamous metaplasia. Many factors could be contributing to this difference, including the difference in exposure regimens (possible mixture of toxicants in the human exposures compared to a single toxicant in animal studies; differences in the concentrations of ozone to which the human subjects and animals were exposed; and the lengths of exposures) and the location of tissue sampling (the entire nasal mucosa was examined in animals, while only the inferior turbinate lined by RE was examined in humans). In addition, it is possible that the inherent susceptibility of nasal epithelium to ozone-induced injury is different between humans and animals.

Better designed human and animal studies using similar exposure conditions and tissue sampling, processing, and analyzing techniques are needed to better understand the pathogenesis of ozone nasal toxicity, and to develop good methods for establishing the human risk to ozone based on animal data. Advances in this area could be made if methods were developed to determine the dosimetry of ozone to specific nasal epithelium within human and animal airways.

SUMMARY

In this article, we have reviewed the normal structure of the nasal TE and RE in laboratory animals and have provided examples of how these epithelia respond to short- and long-term exposures of ozone. Although there are some suggestions of species-specific differences in the location and magnitude of response(s) in the TE and RE after similar ozone exposure regimens, the principal morphologic features of ozone-induced nasal toxicity in rats and monkeys are similar. In both of these species, repeated ozone exposures can initiate inflammatory, proliferative, and metaplastic responses in the nasal mucosa. The TE in laboratory animals is particularly responsive to ozone-induced alteration and quickly undergoes a wave of DNA synthesis, followed by cellular proliferation and mucous cell metaplasia. Human subjects living in highly ozone-polluted urban atmospheres in Mexico City have significant pollution-associated nasal lesions such as chronic rhinitis and dysplasia of RE. This suggests that high ambient ozone exposures could be a significant health risk to the human population. In addition, recent studies of F344/N rats chronically exposed to ambient levels of ozone for much of their lifetime have demonstrated that these animals have functional impairments of the nasal mucociliary apparatus (e.g., mucostasis and ciliastasis) that are associated with the metaplastic lesions in the TE and RE. This suggests that the ozone-induced epithelial lesions recognized in the nose can have deleterious functional consequences.

REFERENCES

Adams, D. R., and Hotchkiss, D. K. 1983. The canine nasal mucosa. *Anat. Histol. Embryol.* 12:109–125.

Brain, J. D. 1970. The uptake of inhaled gases by the nose. *Ann. Otol. Rhinol. Laryngol.* 79:529–539.

Buckley, L. A., Jiang, X. Z., James, R. A., Morgan, R. T., and Barrow, C. S. 1984. Respiratory tract lesions induced by sensory irritants at the RD50 concentration. Toxicol. *Appl. Pharmacol.* 74:417–429.

Calderon-Garcidueñas, L., Osnoro-Velazquez, A., Bravo-Alvarez, H., Delgado-Chavez, R., and Barrios-Marquez, R. 1992. Histopathological changes of nasal mucosa in southwest metropolitan Mexico City inhabitants. *Am. J. Pathol.* 140:225–232.

Cross, C. E., Halliwell, B., and Allen, A. 1984. Antioxidant protection: A function of tracheobronchial and gastrointestinal mucus. *Lancet* 1:1328–1330.

Gross, E. A., Swenberg, J. A., Field, S., and Popp, J. A. 1982. Comparative morphometry of the nasal cavity in rats and mice. *J. Anat.* 135:83–88.

Harkema, J. R. 1991. Comparative aspects of nasal airway anatomy: Relevance to inhalation toxicology. *Toxicol. Pathol.* 19:321–336.

Harkema, J. R., Plopper, C. G., Hyde, D. M., Wilson, D. W., St. George, J. A., and Wong, V. J. 1987a. Nonolfactory surface epithelium of the nasal cavity of the bonnet monkey: A morphologic and morphometric study of the transitional and respiratory epithelium. *Am. J. Anat.* 180:266–279.

Harkema, J. R., Plopper, C. G., Hyde, D. M., St. George, J. A., and Dungworth, D. L. 1987b. Effects of an ambient level of ozone on primate nasal epithelial mucosubstances: Quantitative histochemistry. *Am. J. Pathol.* 127:90–96.

Harkema, J. R., Plopper, C. G., Hyde, D. M., St. George, J. A., Wilson, D. W., and Dungworth, D. L. 1987c. Response of the macaque nasal epithelium to ambient levels of ozone: A morphologic and morphometric study of the transitional and respiratory epithelium. *Am. J. Pathol.* 128:29–44.

Harkema, J. R., Plopper, C. G., Hyde, D. M., and St. George, J. A. 1987d. Regional differences in quantities of histochemically detectable mucosubstances in the nasal, paranasal, and nasopharyngeal epithelium of the bonnet monkey. *J. Histochem. Cytochem.* 35:279–286.

Harkema, J. R., Hotchkiss, J. A., and Henderson, R. F. 1989. Effects of 0.12 and 0.80 ppm ozone on rat nasal and nasopharyngeal epithelial mucosubstances: Quantitative histochemistry. *Toxicol. Pathol.* 17:525–535.

Harkema, J. R., Bermudez, E. G., Morgan, K. T., and Mellick, P. W. 1992. Effects of chronic ozone exposure on the nasal mucociliary apparatus in the rat. *Am. Rev. Respir. Dis.* 145:A98.

Harkema, J. R., Plopper, C. G., Hyde, D. M., St. George, J. A., Wilson, D. W., and Dungworth, D. L. 1993. Response of macaque bronchiolar epithelium to ambient concentrations of ozone. *Am. J. Pathol.* 143:857–866.

Henderson, R. F., Hotchkiss, J. A., Chang, I. Y., Scott, B. R., and Harkema, J. R. 1993. Effect of cumulative exposure on nasal response to ozone. *Toxicol. Appl. Pharmacol.* 119:59–65.

Hotchkiss, J. A., and Harkema, J. R. 1992. Endotoxin or cytokines attenuate ozone-induced DNA synthesis in rat nasal transitional epithelium. *Toxicol. Appl. Pharmacol.* 114:182–187.

Hotchkiss, J. A., Harkema, J. R., Sun, J. D., and Henderson, R. F. 1989. Comparison of acute ozone-induced nasal and pulmonary inflammatory responses in rats. *Toxicol. Appl. Pharmacol.* 98:289–302.

Hotchkiss, J. A., Harkema, J. R., and Henderson, R. F. 1991. Effect of cumulative ozone exposure on ozone-induced nasal epithelial hyperplasia and secretory metaplasia. *Exp. Lung Res.* 15:589–600.

Johnson, N. F., Hotchkiss, J. A., Harkema, J. R., and Henderson, R. F. 1990. Proliferative responses of rat epithelia to ozone. *Toxicol. Appl. Pharmacol.* 103:143–155.

Kerns, W. D., Pavkov, K. L., Donofrio, D. J., Gralla, E. J., and Swenberg, J. A. 1983. Carcinogenicity of formaldehyde in rats and mice after long-term inhalation exposure. *Cancer Res.* 43:4382–4392.

Matulionis, D. H., and Parks, H. R. 1972. Ultrastructural morphology of normal nasal respiratory epithelium of the mouse. *Anat. Res.* 176:65–84.

Monticello, T. M., Morgan, K. T., and Uriah, L. 1990. Nonneoplastic nasal lesions in rats and mice. *Environ. Health Perspect.* 85:249–274.

Montiero-Riviere, N. and Popp, J. A. 1984. Ultrastructural characterization of the nasal respiratory epithelium in the rat. *Am. J. Anat.* 169:31–43.

Mygrind, N., Pedersen, M., and Nielsen, M. H. 1982. Morphology of the upper airway epithelium. In *The nose: Upper airway physiology and the atmospheric environment,* eds. D. F. Proctor and I. Andersen, pp. 71–97. Amsterdam: Elsevier.

Paul, R. A., Biller, W. F., and McCurdy, T. 1987. National estimates of population exposure to ozone. In *Proceedings of the 80th Annual Meeting and Exhibition of the Air Pollution Control Association,* Paper 87-42.7, Pittsburgh, PA: American Pollution Control Association.

Popp, J. A., and Martin, J. T. 1984. Surface topography and distribution of cell types in the rat nasal respiratory epithelium: Scanning electron microscopic observations. *Am. J. Anat.* 169:425–436.

Steinfeld, J. H. 1991. *Rethinking the ozone problem in urban and regional air pollution.* p. 1. Washington, DC: National Academy Press.

OLFACTORY MUCOSAL TOXICITY. INTEGRATION OF MORPHOLOGICAL AND BIOCHEMICAL DATA IN MECHANISTIC STUDIES: DIBASIC ESTERS AS AN EXAMPLE

Matthew S. Bogdanffy, Steven R. Frame

Haskell Laboratory for Toxicology and Industrial Medicine,
E. I. du Pont de Nemours and Co., Inc., Newark, Delaware, USA

The olfactory mucosa of the common laboratory rat is a complicated organ comprised of a variety of cell types. Each cell type has specific biochemical and physiologic functions critical to the normal function of the tissue as the organ of olfaction. Damage to the olfactory epithelium leading to a dysfunctional state can result in anosmia, and since olfaction is an important defense mechanism against noxious chemicals, loss of olfaction could compromise the ability to avoid chemical exposure. The olfactory mucosa is organized into epithelial and lamina propria components, each separated by a continuous basement membrane. The epithelium is comprised of sensory cells, sustentacular cells, ducts of Bowman's glands, and basal cells, while the lamina propria contains acini of Bowman's glands, connective tissue, capillary endothelium, and nerve bundles. Histochemical techniques can be used to identify specific cell types within the mucosa, and these are often useful as early markers of cellular differentiation following atrophy and regenerative hyperplastic responses to chemical insult. Localization of biotransformation enzymes using histochemical techniques is a valuable contribution to mechanistic studies of olfactory toxicity. A variety of histochemical markers are reviewed. A range of nonneoplastic lesions of the olfactory epithelium is also reviewed, with emphasis on the importance of histopathology to the elucidation of mechanisms of olfactory toxicity. Finally, the mechanism of action of dibasic esters solvents is reviewed as an example of the role of histochemical and biochemical studies in olfactory toxicity research. Research on dibasic esters reveals a common mechanism of action for organic esters on olfactory epithelium. Organic esters absorbed by the olfactory mucosa are hydrolyzed via carboxylesterases in sustentacular cells and Bowman's glands to their corresponding acids, which are the ultimate toxicant. Organic acids ultimately induce toxicity through a mechanism that probably includes a decrease in intracellular pH. Studies with human nasal tissue demonstrate that carboxylesterase-mediated hydrolysis of dibasic esters is less pronounced than in rat tissue. Thus, humans are at greatly reduced risk of dibasic ester toxicity of the olfactory epithelium.

The olfactory mucosa of rodents is a complicated structure of cellular heterogeneity designed to carry out the functions of humidification and olfaction of inspired air. Since rodents are obligate nasal breathers (Proctor & Chang, 1983) and rely heavily on olfaction for survival, their olfactory system is highly developed. In contrast, the human olfactory system, discussed elsewhere in this volume, is poorly developed, although the basic

Address correspondence to Matthew S. Bogdanffy, Haskell Laboratory for Toxicology and Industrial Medicine, E. I. du Pont de Nemours and Co., Inc., P.O. Box 50. Newark, DE 19714, USA.

structure of the epithelium is similar. Understanding the spectrum of lesions induced by nasal toxicants in the context of mechanisms of action will facilitate our assessment of the relevance to humans of the health risks posed by these materials. This article describes a spectrum of nonneoplastic rodent olfactory nasal lesions and shows, by example, how mechanistic studies that utilize histology, histochemistry, and biochemistry are useful for investigating mechanisms of action.

GENERAL MORPHOLOGY OF THE OLFACTORY MUCOSA

The normal histology of the rodent olfactory mucosa has been described recently (Uriah & Maronpot, 1990) and is reviewed here only briefly. The olfactory mucosa covers most of the turbinates (endo- and ectoturbinates) in the posterior regions of the nasal cavity and approximately the posterior two-thirds of the dorsal meatus. Using the transverse sections described by Young (1981) as a point of reference, level IV of the nasal cavity is lined completely with olfactory mucosa, except for small patches of respiratory epithelium located on lateral recesses of the olfactory turbinates and lateral wall. The olfactory mucosa is comprised of two distinct layers; the epithelial layer, and the lamina propria. These are separated from each other by a basement membrane. Within the lamina propria are blood vessels, fibroblasts and connective tissue. Associated specificially with the olfactory system are acini and short ducts of Bowman's glands, the major secretory element in the olfactory mucosa, and nerve bundles comprised of unmyelinated olfactory nerves that pass down from the epithelial layer and are channeled to the olfactory bulbs of the brain.

The epithelial layer is comprised of basal cells, long ducts of Bowman's glands, sensory cells, and sustentacular, or support, cells. The basal cells lie juxtaposed against the basement membrane and are believed to be the progenitor cells for regenerating sensory and sustentacular cells (Graziadei & de Han, 1973; Graziadei & Graziadei, 1979; Harding et al., 1977; Moulton, 1974). However, it has been suggested that ducts of Bowman's glands, proliferating in response to toxic insult, may also differentiate into sensory cells (Mulvaney & Heist, 1971; Uriah et al., 1987). Ducts of Bowman's glands protrude from beneath the epithelial layer and extend to the surface of the airway and carry mucosubstances from the acini. The sensory cells, which detect odorants depositing on the surface of the epithelium, are composed of long dendritic processes that extend above a prominent nucleus and are capped at the surface by the olfactory vesicle. Extending from the vesicle into the mucus layer are long cilia that possess receptors for odorants. A long axonal process extends beneath the dense nucleus and penetrates the basesment membrane to become compiled with other axons into nerve bundles. The nuclei of the sensory cells, because of their tight arrangement, give the appearance of being stacked in columns. These columns are typically about six nuclei in height, depending on location. The sustentacular cells are shaped much like an elongated wine glass, with a large apical portion containing microvilli and cilia, a large nucleus that stains lightly (relative to sensory cells), and a long stem with a footlike projection that anchors the cell to the basement membrane. Nuclei of adjacent sustentacular cells are generally lined up in a continuous row above the columns of sensory cell nuclei. The sustentacular cells perform a critical support function for the sensory cells.

HISTOCHEMICAL MARKERS OF OLFACTORY MUCOSAL CELLS

Various histochemical markers have been applied to studies of the olfactory epithelium. Randall et al. (1987) applied several common enzyme histochemical stains to nasal tissue fixed in acetone and embedded in glycol methacrylate under cold conditions. The technique provided excellent morphology. In olfactory epithelium, ATPase, acid phosphatase, alkaline phosphatase, glucose 6-phosphatase, naphthylbutryate esterase, and γ-glutamyl transpeptidase all stained positively, although the cellular and intracellular distribution varied. For example, naphthylbutyrate esterase was found in sustentacular cells, but not sensory cells, whereas the opposite was noted for acid phosphatase. Alkaline phosphatase was found on the surface of basal cells and the luminal surface of sustentacular cells, whereas acid phosphatase was localized exclusively in the cell body of sensory cells. Thus, routine histochemical markers can be very useful for identifying specific cell types. Through use of these enzyme histochemical markers, Morgan et al. (1989) showed that inhalation exposure to methyl bromide induces complete degeneration of the olfactory epithelium but spares the basal cells from which sensory cells and sustentacular cells were regenerated (see later discussion).

Several xenobiotic-metabolizing enzymes and cofactors have been localized in olfactory mucosa. Aldehyde dehydrogenase, formaldehyde dehydrogenase, and glutathione are all found in olfactory mucosa, although at lower levels than in respiratory mucosa (Bogdanffy et al., 1986; Keller et al., 1990). Immunohistochemical techniques have been valuable for identifying xenobiotic metabolizing enzymes. Cytochromes P-450 have been identified, primarily in Bowman's glands and sustentacular cells, and are in good association with cytochrome P-450 reductase (Adams et al., 1991; Foster et al., 1986; Voigt et al., 1985, 1993). The distribution of P-450 and related enzymes has contributed to much of our understanding about the biochemical basis for olfactory lesions associated with P-450 substrates, such as the tobacco-specific nitrosamine, NNK (Belinsky et al., 1987). Carboxylesterase was localized immunohistochemically in the same structures using a rabbit antiserum directed against the rat microsomal 59-kD carboxylesterase (Olson et al., 1993). The distribution correlated well with the enzyme histochemical distribution of naphthylbutyrate esterase noted earlier and described in more detail later. Finally, the cyanide-metabolizing enzyme rhodanese has also been localized immunohistochemically in the olfactory mucosa of laboratory animals and in respiratory epithelium of humans (Lewis et al., 1991, 1992).

COMMON NONNEOPLASTIC LESIONS OF THE OLFACTORY EPITHELIUM

Because the nasal turbinates are anatomically complex, accurate recording of the lesion location is critical for clues into mechanism of action. Information on airflow pattern-related deposition of toxicants versus systemic delivery and metabolic activation can be gleaned from this detail (Morgan, 1991). An excellent review of common nonneoplastic responses of the olfactory epithelium has recently been published (Monticello et al., 1990). Following are a basic description and several examples of common lesions of the olfactory epithelium.

The response of olfactory mucosal epithelium to toxic insult is similar to that seen in other tissues—degeneration, necrosis, atrophy, hyperplasia, and metaplasia. Depending on the mechanism of action of the causative agent and the concentration and duration of exposure, the response of the oflactory epithelium to injury may be nonspecific, affecting all cell layers, or specific to particularly cell types. Nonspecific degeneration and necrosis affecting all olfactory mucosal cell types may occur following exposure to

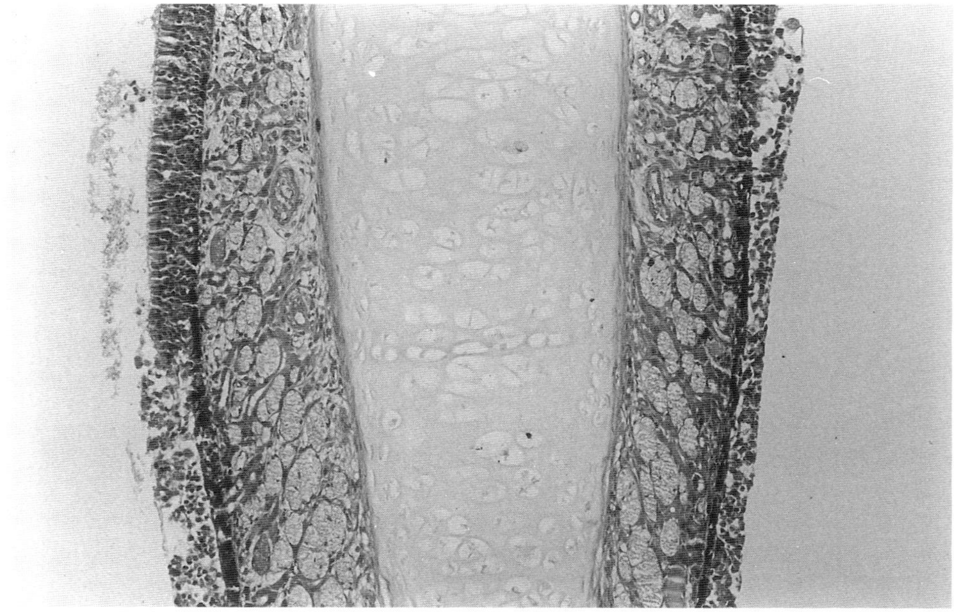

FIGURE 1. Dorsal nasal septum from a rat exposed to 1000 ppm of vinyl acetate for 1 day. The olfactory mucosa has acute necrosis and desquamation of olfactory epithelium (×144).

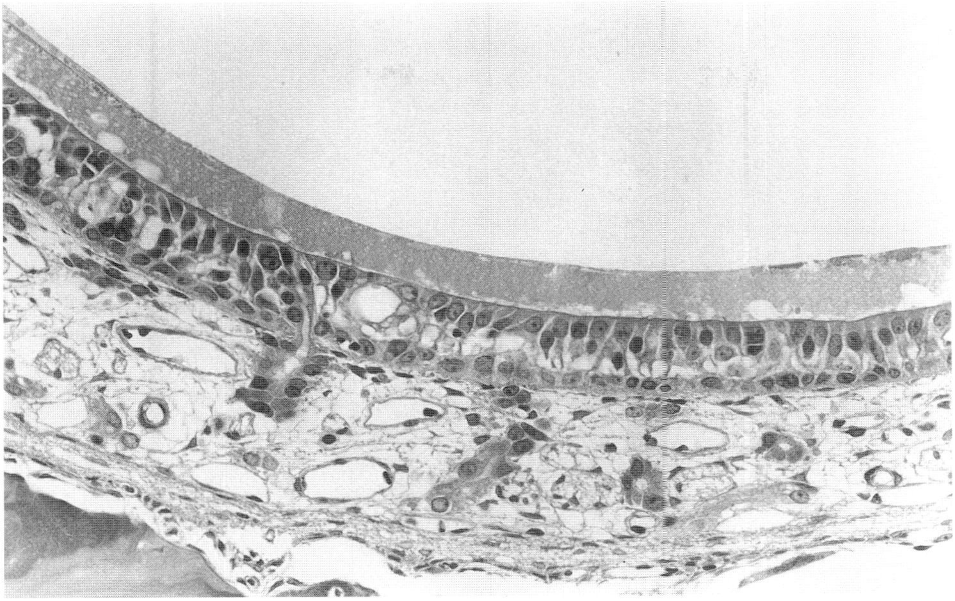

FIGURE 2. Olfactory mucosa from a rat exposed to a proprietary compound for 4 wk. There is loss of sensory cells but relative sparing of sustentacular cells (×360).

irritants. Early lesions are characterized by cell vacuolation and swelling and disorganization of the normal layered organization of the mucosa. With more severe insult, lesions may include necrosis, desquamation, and ulceration of olfactory mucosa (Figure 1). Cell-specific toxicity in the olfactory epithelium may occur, with primary effects on either sensory or sustentacular cells. Figure 2 illustrates an example of degeneration of the olfactory mucosa that, segmentally, was associated with loss of sensory cells but sparing of the sustentacular cells. This lesion may be similar to that produced by the neurotoxin 3,3′-imidodipropionitrile (St. Clair et al., 1991). In contrast, sequential studies with dibasic esters (DBE) induce a primary effect on sustentacular cells with sensory-cell degeneration and necrosis occurring subsequent to loss of the support cells (see later). Irrespective of the causative agent and the initial site of cell damage, severe or chronic olfactory injury may result in mucosal atrophy characterized by thinning and disorganization of the epithelium that is composed of a few layers of cuboidal or attenuated epithelial cells.

The olfactory mucosa is capable of a number of adaptive and regenerative responses following chronic insult. Adaptive responses include squamous metaplasia and respiratory metaplasia. With squamous metaplasia, the normal olfactory mucosa is replaced by a more resistant squamous epithelium of variable thickness. Respiratory metaplasia is characterized by replacement of the normal olfactory mucosa by ciliated columnar epithelium

(Figure 3). The histogenesis of respiratory metaplasia has not been determined, but it represents an adaptive response by sustentacular cells.

Olfactory sensory cells are unique among nerve cells in their ability to regenerate following cell damage or loss (Figure 4). Epithelial hyperplasia may occur in basal cells or sensory cells. Regenerative hyperplasia of sensory neurons may occur in a disorganized manner, with production of pseudoglandular or cystlike structures, or the regenerative response may be completely restorative. Expsoure of rats to 200 ppm methyl bromide for 1 day produced degeneration and atrophy of the olfactory epithelium (Hurtt et al., 1988). However, despite continued exposure, regenerative hyperplasia with restoration of the normal layered architecture occurred by day 5.

INTEGRATION OF MORPHOLOGY, HISTOCHEMISTRY, AND BIOCHEMISTRY INTO MECHANISMS OF OLFACTORY TOXICITY: DIBASIC ESTERS

Dibasic ester (DBE) is a solvent mixture of the dimethyl esters of adipic, glutaric, and succinic acids. Initial hypotheses regarding the mechanism of action of dibasic esters were based on the observations that (1) DBE-induced lesions were confined to the olfactory epithlium of rats, and (2) carboxylesterase activity is localized in high concentration in nasal tissue, most notably olfactory epithelium (Bogdanffy et al., 1987).

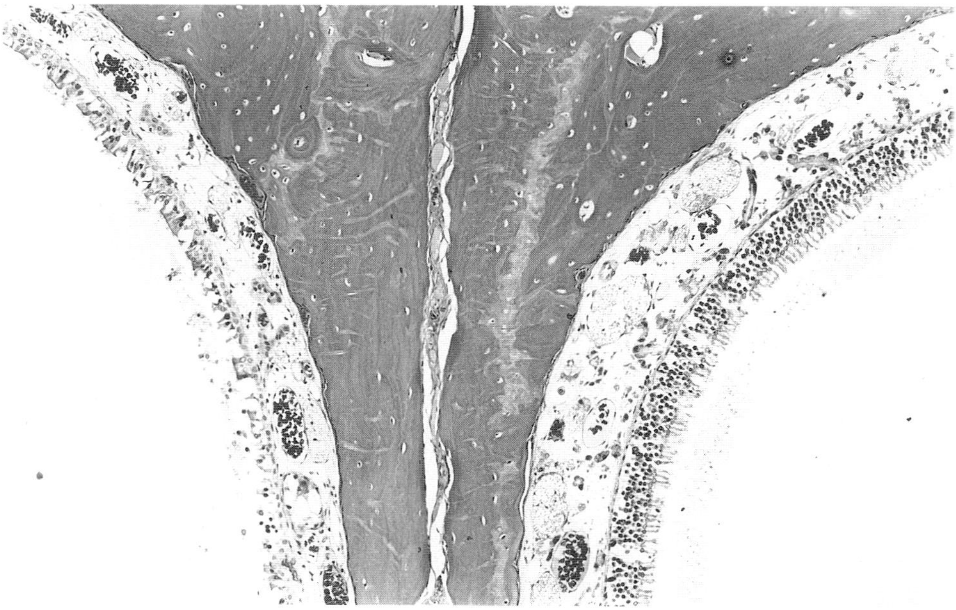

FIGURE 3. Dorsal region of the olfactory mucosa of a rat exposed to 390 mg/m^3 of dibasic esters mixture for 90 days (from Keenan et al., 1990). Respiratory metaplasia of olfactory mucosa is present on the right (×120).

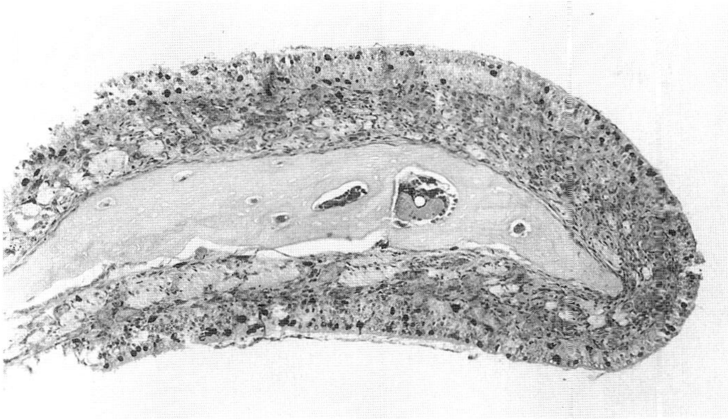

FIGURE 4. Section of an ethmoturbinate from a rat exposed to 1000 ppm vinyl acetate for 27 days. Rats were injected ip with 5-bromo-2′-deoxyuridine (BrdU) 2 h prior to sacrifice and the sections were stained with an anti-BrdU monoclonal antibody. The olfactory mucosa has increased cell labeling (indicated by dark nuclei) consistent with hyperplasia (×120).

DBE-induced lesions were confined to the olfactory epithelium of rats exposed to 20 to 390 mg/m^3 for 90 days and were characterized by a loss of sensory cells and sustentacular cells (Keenan et a ., 1990). Interestingly, a clear no-effect level was observed at 20 mg/m^3 in male rats, but 5/20 female rats of the same exposure group had minimal degeneration of the olfactory epithelium, suggesting a greater sensitivity of female rats. The lesions were reversible, and by 45 days postexposure, the lesion incidence in the 20-mg/m^3 females was reduced to 0/10. Attempts at repair were made evident by a shift in the morphology of the lesion from active necrosis to a disorganization of the sensory cell layer and occasional focal respiratory metaplasia.

Carboxylesterase activity has been localized histochemically within nasal respiratory and olfactory mucosa using α-naphthylbutyrate as the substrate (Bogdanffy et al., 1987). Although present in substantial quantities in both mucosae, the distribution of activity is markedly different. In respiratory mucosa, carboxylesterase activity is confined primarily within the epithelial cell layer. Seromucous glands of the lamina propria contain only marginally detectible activity. All epithelial cell types are positive, to varying extents, for esterase activity, but the most intensely stained are cuboidal cells and ciliated columnar cells. In olfactory epithelium, esterase activity appears only in specific cell populations. The most notable are the sustentacular cells and ducts of Bowman's glands of the epithelial cell layer and the acini of Bowman's glands in the lamina propria. The activity in Bowman's glands is particularly intense.

The observations of the histochemical localization of carboxylesterase influenced the hypothesis of the mechanism of action of DBE in the following ways. First, the intense staining suggested that in situ DBE metabolism in olfactory mucosa might be a predominant factor. However, the presence of carboxylesterase in respiratory epithelial cells suggested that respiratory epithelium could be at risk of DBE toxicity. Other factors, such as relatively lower metabolic capacity or elevated mechanisms of detoxication, might spare the respiratory epithelium from observable toxicity at the concentrations tested in the 90-day study. Second, the localization of carboxylesterase activity in sustentacular cells suggested that sensory cell degeneration might be secondary to sustentacular cytotoxicity. Therefore, research to investigate the mechanism of DBE toxicity should bear out the following hypotheses: (1) DBE is metabolized by nasal carboxylesterases to toxic acid metabolites; (2) respiratory epithelium must be susceptible to DBE-induced toxicity at very high exposure levels; and (3) the earliest signs of toxicity must be observed in sustentacular cells.

To test the mechanism of action of DBE, an in vitro assay was developed in which nasal turbinates were excised and placed in tissue culture medium containing DBE. Cytotoxicity was assayed by measuring the release into the medium of the intracellular enzyme acid phosphatase. As described earlier, acid phosphatase is one of few enzyme markers that was known from histochemical studies to be present in all potential target epithelial cells, most importantly in sensory cells. DBEs and their metabolites in media were also measured by high-performance liquid chromatography (HPLC). Using this system it was possible to test DBEs, and their metabolites, directly on nasal turbinates and probe the mechanism of action by pretreating animals with the carboxylesterase inhibitor bis(*p*-nitrophenyl)phosphate (BNPP).

DBEs induced increases in acid phosphatase release from turbinate explants over a concentration range of 10-100 mM (Trela & Bogdanffy, 1991a). These increases were accompanied by parallel increases in monomethyl ester metabolite production and are exemplified by studies with dimethyl glutarate (Figure 5, A and B). All turbinates were responsive to DBE in this system. However, turbinates lined with olfactory mucosa were more sensitive to DBE-induced cytotoxicity than those lined with respiratory mucosa. Similarly, a greater production of monomethyl ester metabolites was observed in incubates from olfactory turbinates relative to respiratory turbinates.

To test the hypothesis that carboxylesterase-mediated DBE metabolism was required for cytotoxicity, experiments were conducted on turbinates from rats pretreated with BNPP. BNPP pretreatment reduced the cytotoxic response in all turbinate types (Figure 5C). Similarly, reduced levels of DBE metabolites were observed in the culture media (Figure 5D).

Similar experiments with the monomethyl ester and diacid metabolites were conducted on isolated nasal turbinates (Trela & Bogdanffy, 1991b). The monomethyl esters of all three DBEs were found to induce cytotoxicity and, again, parallel increases in the formation of the diacids were observed. BNPP was able to inhibit the cytotoxicity and metabolism of the monomethyl

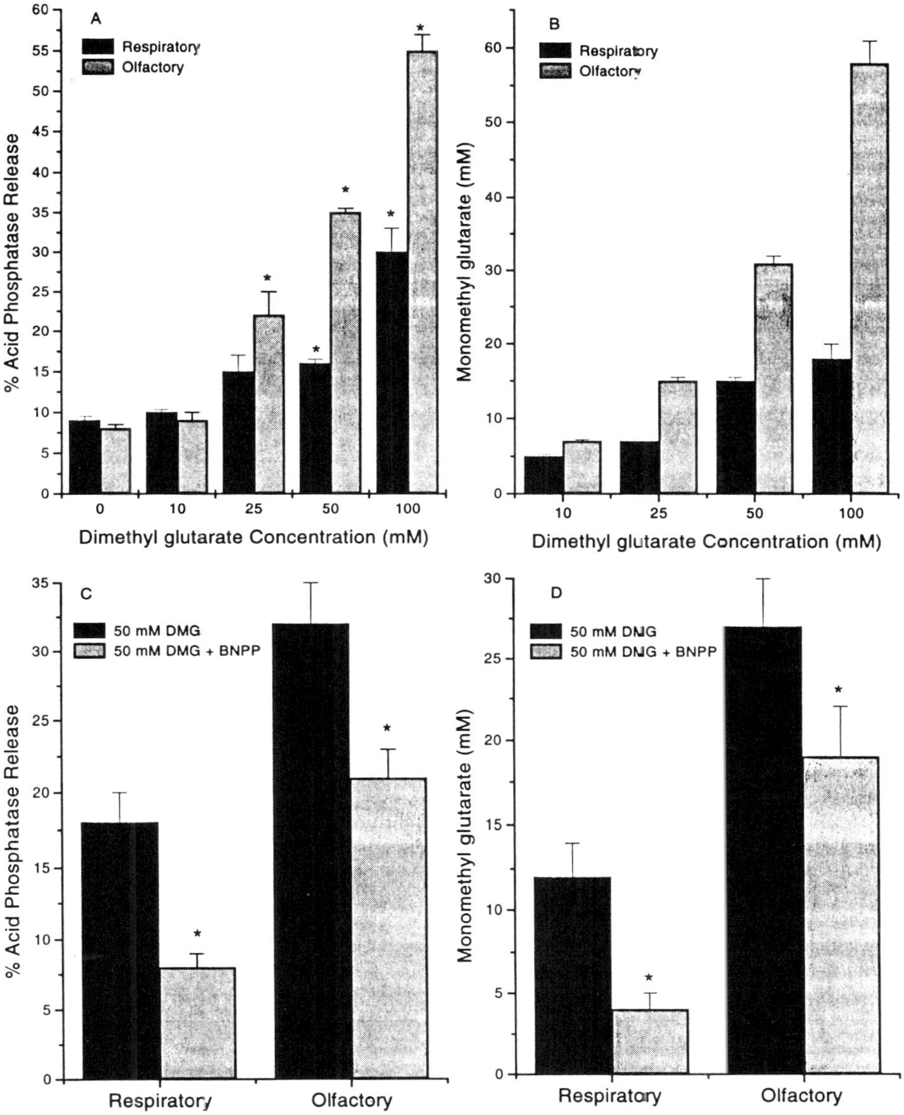

FIGURE 5. Cytotoxicity and metabolism of dimethyl glutarate (DMG) in rat nasal turbinate explants and inhibition by pretreatment of rats with a carboxylesterase inhibitor. (A) Cytotoxicity following in vitro exposure of explants for 2 h. (B) Monomethyl glutarate formation from DMG after 2 h. (C) Inhibition of cytotoxicity in turbinates from rats pretreated with bis(*p*-nitrophenyl)phosphate (BNPP). (D) Inhibition of monomethyl glutarate formation in turbinates from rats pretreated with ENPP. Adapted from Trela and Bogdanffy (1991a).

esters. The diacids were not significantly metabolized by nasal tissue, and pretreatment of rats with BNPP had no effect on diacid-induced cytotoxicity.

These results showed that, in fact, carboxylesterase-mediated metabolism of DBE is required for toxicity. Furthermore, the monomethyl ester and diacid metabolites of the DBEs contribute to the overall expression of cellular degeneration. Left unanswered is the extent to which monomethyl ester and diacid metabolites are formed in vivo.

The nasal deposition of DBEs were studied by Morris et al. (1991) and these studies provided evidence of in vivo metabolism. Nasal uptake of DBE vapor was measured in the surgically isolated upper respiratory tract using a unidirectional flow model. Individual DBE vapors deposited in nasal tissues with greater than 98% efficiency. Slightly lower deposition efficiencies (approximately 97%) were observed when rats were exposed to all three DBEs simultaneously. Finally, the deposition of ethyl acetate, a known substrate for nasal carboxylesterases, was inhibited by simultaneous exposure with dimethyl succinate (DMS).

DMS is metabolized least efficiently of the three DBEs. The kinetics of hydrolysis of the DBEs were measured in homogenates of rat nasal tissues (Bogdanffy et al., 1991). The value of V_{max}/K_m, a measure of the efficiency of hydrolysis at low substrate concentration, was four to six times higher for DMA relative to DMS (Table 1). Furthermore, the efficiency of hydrolysis in olfactory tissue was 4 to 11 times greater than that of respiratory tissue. The K_m values for a given substrate, when compared between respiratory and olfactory tissue, were fairly similar. Thus, the difference in efficiency is primarily due to capacity, that is, V_{max}. This conclusion is supported by the observation of the heavy staining pattern in Bowman's glands in olfactory mucosa. These glands must therefore contribute to the increased metabolic capacity of olfactory mucosa relative to respiratory mucosa. It follows that the increased capacity of the olfactory mucosa to transform DBEs to toxic acid metabolites is likely responsible for its increased sensitivity to the toxic effects of DBEs observed in the inhalation study.

Having linked DBE toxicity to carboxylesterase-mediated production of acidic metabolites, it was of interest to determine the metabolic capacity of

TABLE 1. V_{max}/K_m Values for Nasal Mucosal Carboxylesterase-Mediated Metabolism of Dibasic Esters

	Respiratory		Olfactory	
Dibasic Ester	Male	Female	Male	Female
Dimethyl succinate	61[a]	31	376	184
Dimethyl glutarate	93	133	1051	1162
Dimethyl adipate	378	250	1418	2373

Note. Values are nmol/min/mg/m*M* (adapted from Bogdanffy et al., 1991).

TABLE 2. Metabolism of Dibasic Esters to Monomethyl Esters by Human and Rat Nasal Mucosa

	Respiratory		Olfactory	
Dibasic Ester	Human	Rat	Human	Rat
Dimethyl succinate	nd–1.14	110	nd–2.73	1461
Dimethyl glutarate	nd	600	nd–10.25	3519
Dimethyl adipate	nd–1.66	385	nd–2.52	4199

Note. Tissue samples were obtained from autopsy cases and frozen in liquid nitrogen. Cases were aged 14–55 yr. Postmortem interval ranged from 3 to 7 h. Three samples of each mucosal type were selected from approximately 50 cases based on histopathological assessment. The selected tissues were histologically normal and at least moderately well preserved. Units are nmol/min/mg protein (nd, not detectable). Initial substrate concentration was 25 m*M*. Methods for measuring metabolism and the reported values for male rats were from Bogdanffy et al. (1991).

human nasal tissue. Table 2 shows that DBEs are metabolized in human nasal tissue at a rate that is 100 to more than 1000 times lower than that of rat tissue. This suggests that humans are at greatly reduced risk of DBE-induced nasal toxicity relative to rats.

Since respiratory epithelium has significant quantities of carboxylesterase activity, DBEs must have the potential to induce pathology in the epithelium, despite the lack of observation of such an effect in the 90-day inhalation study. The cytotoxic effects of DBE observed in the in vitro explant studies provide clear support for such potential. Perhaps the concentrations used in the 90-day inhalation study were not high enough to produce significant toxic metabolites or overcome endogenous, and as yet unknown, detoxication mechanisms. A study conducted with aerosols of DBE showed that, in fact, respiratory epithelium was susceptible to toxicity at high DBE concentrations (Lee et al., 1992). Exposure of rats to 5900 mg/m^3 DBE aerosol/vapor mixture produced nasal lesions that were distributed along the major inspiratory airflow routes, a result of aerosol deposition, and included necrosis and denudation of respiratory epithelial cells, in addition to the expected similar necrotic response in olfactory epithelium. This study also confirmed the reversibility of the DBE-induced nasal lesions and followed the repair at the ultrastructural level.

The primary target cell for DBE-induced olfactory toxicity was demonstrated through both in vitro and in vivo studies. The explant culture system provided a simple method for exposing nasal tissues to DBE and examining the early ultrastructural changes. Exposure of both respiratory and olfactory turbinates to DBE in culture media for up to 2 h resulted in mild degenerative changes characterized by cytoplasmic vacuolation and swelling of nuclei (Trela et al., 1992) (Figure 6). Olfactory epithelium was more sensitive than respiratory epithelium, and the earliest changes were associated with sustentacular cells. Ultrastructurally, sustentacular cells had variable,

FIGURE 6. Olfactory mucosa from a rat endoturbinate incubated with 10 m*M* dimethyl adipate. Sustentacular cells have vesiculation of the apical cytoplasm and nuclear swelling. Sensory cells are unaffected. From Trela et al. (1992) (×450).

sometimes severe, swelling of all cytoplasmic organelles, and the resultant swollen cytoplasm was occasionally seen extending above the neuronal olfactory vesicle. Nuclei of sustentacular cells were swollen and had dispersion and peripheral margination of chromatin. Ultrastructural changes in olfactory neurons, including cell bodies, axons, dentrites, and olfactory vesicles, were normal in most cells. Changes in sustentacular cell ultrastructure were observed following 2-h exposures to 10 m*M* dimethyl adipate. This treatment did not result in significant releases of acid phosphatase into the medium (Figure 5). Thus, since acid phosphatase is released primarily from sensory cells, the biochemical studies support the conclusion that the primary cell type affected by DBE is the sustentacular cell. Confirmation of this conclusion was derived from the in vivo high-exposure study. Twenty-four hours after exposure of rats to 5900 mg/m^3 DBE vapor/aerosol, olfactory neurons maintained a normal structure, whereas sustentacular cells had slight degenerative changes similar to those observed in the nasal explants exposed to dimethyl adipate (Lee et al., 1992). More severe sustentacular cell injury, accompanied by sloughing of sensory cells, was observed at subsequent time points. The pathogenesis of sensory-cell degeneration induced by DBE is therefore secondary to sustentacular cell degeneration. This conclusion is consistent with similar observations of early effects on sustentacular cell ultrastructure and secondary degeneration of sensory cells reported by Thomas and Morgan (1988). Although mitochondrial swelling was observed early in the spectrum of ultrastructural changes in sustentacular cells, and DBEs have been shown to inhibit mitochondrial ATP synthesis (Bogdanffy & Londergan, 1989), other nonspecific responses such as vesicular swelling of cytoplasmic organelles and nuclear swelling and dispersion of chromatin point toward a decrease in intracellular pH induced by accumulation of organic acids as the primary biochemical mechanism.

CONCLUSION

Toxic responses of the olfactory mucosa involve a complex interplay between sensory cells and the sustentacular cells supporting their viability, and the biochemical activities of all epithelial cells, especially the underlying Bowman's glands and their ducts that penetrate into the epithelial cell layer. Elucidation of the mechanisms of action of olfactory toxicants must therefore combine the techniques of light and electron microscopy, histochemistry, and biochemistry to synthesize cohesive mechanisms of action. Once these mechanisms are adequately defined in laboratory animal models, key biochemical pathways can be evaluated in human tissues to assess human risk. For dibasic esters, the apparent sensitivity of rats to DBE-induced nasal lesions of the olfactory mucosa appears to be related to the great efficiency with which DBEs are deposited in the mucosa and the high metabolic conversion of DBEs to their acid metabolites. Although deposition characteristics of DBEs in the human nasal cavity are not known, the metabolic capacity is orders of magnitude lower than rats, suggesting that humans are at reduced risk of DBE-induced nasal lesions.

REFERENCES

Adams, D. R., Jones, A. M., Plopper, C. G., Serabjit-Singh, C. J., and Philpot, R. M. 1991. Distribution of cytochrome P-450 monoxygenase enzymes in the nasal mucosa of the hamster and rat. *Am. J. Anat.* 190:291–298.

Belinsky, S. A., Walker, V. E., Maronpot, R. R., Swenberg, J. A., and Anderson, M. W. 1987. Molecular dosimetry and DNA adduct formation and cell toxicity in rat nasal mucosa following exposure to the tobacco specific nitrosamine 4-(*N*-methyl-*N*-nitrosamino)-1-(3-pyridyl)-1-butanone and their relationship to induction of neoplasia. *Cancer Res.* 47:6058–6065.

Bogdanffy, M. S., Randall, H. W., and Morgan, K. T. 1986. Histochemical localization of aldehyde dehydrogenase in the respiratory tract of the Fischer-344 rat. *Toxicol. Appl. Pharmacol.* 82:560–567.

Bogdanffy, M. S., Randall, H. W., and Morgan, K. T. 1987. Biochemical quantitation and histochemical localization of carboxylesterase in the nasal passages of the Fischer-344 rat and B6C3F1 mouse. *Toxicol. Appl. Pharmacol.* 88:183–194.

Bogdanffy, M. S., Kee, C. R., Hinchman, C. A., and Trela, B. A. 1991. Metabolism of dibasic esters by rat nasal mucosal carboxylesterase. *Drug Metab. Dispos.* 19,124–129.

Foster, J. R., Elcombe, C. R., Boobis, A. R., Davies, D. S., Sesardic, D., McQuade, J., Robson, R. T., Hayward, C., and Lock, E. A. 1986. Immunocytochemical localization of cytochrome P-450 in hepatic and extra-hepatic tissues of the rat with a monoclonal antibody against cytochrome P-450c. *Biochem. Pharmacol.* 35:4543–4554.

Graziadei, P. P. C., and de Han, R. S. 1973. Neuronal regeneration in frog olfactory system. *J. Cell Biol.* 59:525–530.

Graziadei, P. P. C., and Graziadei, G. A. M. 1979. Neurogenesis and neruon regeneration in the olfactory system of mammals. I. Morphological aspects of differentiation an structural organization of the olfactory sensory neurons. *J. Neurocytol.* 8:1–8.

Harding, J., Graziadei, P. P. C., Monti-Graziadei, G. A., and Margolis, F. L. 1977. Denervation in the primary pathway of mice. *Brain Res.* 132:11–28.

Hurtt, M. E., Thomas, D. A., Working, P. K., Monticello, T. M., and Morgan, K. T. 1988. Degeneration and regeneration of the olfactory epithelium following inhalation exposure to methyl bromide: Pathology, cell kinetics, and olfactory function. *Toxicol. Appl. Pharmacol.* 94:311–328.

Keenan, C. M., Kelly, D. P., and Bogdanffy, M. S. 1990. Degeneration and recovery of rat olfactory epithelium following inhalation of dibasic esters. *Fundam. Appl. Toxicol.* 15:381–393.

Keller, D. A., Heck, H. d'A., Randall, H. W., and Morgan, K. T. 1990. Histochemical localization of formaldehyde dehydrogenase in the rat. *Toxicol. Appl. Pharmacol.* 106:311–326.

Lee, K. P., Valentine, R., and Bogdanffy, M. S. 1992. Nasal lesion development and reversibility in rats exposed to aerosols of dibasic esters. *Toxicol. Pathol.* 20:376–393.

Lewis, J. L., Rhoades, C. E., Gervasi, P. G., Griffith, W. C., and Dahl, A. R. 1991. The cyanide-metabolizing enzyme rhodanese in human nasal respiratory mucosa. *Toxicol. Appl. Pharmacol.* 108:114–120.

Lewis, J. L., Rhoades, C. E., Bice, D. E., Harkema, J. R., Hotchkiss, J. A., Sylvester, D. M., and Dahl, A. R. 1992. Interspecies comparison of cellular localization of the cyanide metabolizing enzyme rhodanese within olfactory mucosa. *Anat. Rec.* 232:620–627.

Monticello, T. M., Morgan, K. T., and Uriah, L. 1990. Nonneoplastic nasal lesions in rats and mice. *Environ. Health Perspect.* 85:249–274.

Morgan, K. T. 1991. Approaches to the identification and recording of nasal lesions in toxicology studies. *Toxicol. Pathol.* 19:337–351.

Morgan, K. T., Thomas, D. A., and St. Clair, M. B. 1989. Enzyme markers for studies of olfactory epithelial regeneration in rats expsoed to methyl bromide. *Toxicologist* 9:147.

Morris, J. B., Clay, R. J., Trela, B. A., and Bogdanffy, M. S. 1991. Deposition of dibasic esters in the upper respiratory tract of the male and female Sprague-Dawley rat. *Toxicol. Appl. Pharmacol.* 108:538–546.

Moulton, D. G. 1974. Dynamics of cell population in the olfactory epithelium. *Ann. N.Y. Acad. Sci.* 237:52–61.

Mulvaney, B. D., and Heist, H. E. 1971. Regeneration of rabbit olfactory epithelium. *Am. J. Anat.* 131:241–252.

Olson, M. J., Maritn, J. L., LaRosa, A. C., Brady, A. N., and Pohl, L. R. 1993. Immunochemical localization of carboxylesterase in the nasal mucosa of rats. *J. Histochem. Cytochem.* 41:307–311.

Proctor, D. F., and Chang, J. C. F. 1983. Comparative anatomy and physiology of the nasal cavity. In *Nasal tumors in animals and man*, Vol. 1, eds. G. Reznick, S. F. Stinson, pp. 1–33, Boca Raton, FL: CRC Press.

Randall, H. W., Bogdanffy, M. S., and Morgan, K. T. 1987. Enzyme histochemistry of the rat nasasl mucosa embedded in cold glycol methacrylate. *Am. J. Anat.* 179:10–17.

St. Clair, M. B. G., Morgan, K. T., Peele, D. B., and Crofton, K. M. (1991). Olfactory toxicity of IDPN. *Neurotoxicology* 12:135.

Thomas, D. A., and Morgan, K. T. 1988. Olfactory toxicity: Studies of methyl bromide. *CIIT Activities* 8:1–7.

Trela, B. A., and Bogdanffy, M. S. 1991a. Carboxylesterase-dependent cytotoxicity of dibasic esters (DBE) in rat nasal explants. *Toxicol. Appl. Pharmacol.* 107:285–301.

Trela, B. A., and Bogdanffy, M. S. 1991b. Cytotoxicity of dibasic esters (DBE) metabolites in rat nasal explants. *Toxicol. Appl. Pharmacol.* 110:259–267.

Trela, B. A., Frame, S. R., and Bogdanffy, M. S. 1992. A microscopic and ultrastructural evaluation of dibasic esters (DBE) toxicity in rat nasal explants. *Exp. Mol. Pathol.* 56:208–218.

Uriah, L. C., and Maronpot, R. R. 1990. Normal histology of the nasal cavity and application of special techniques. *Enivron. Health Perspect.* 85:187–208.

Uriah, L. C., Talley, F. A., Mitumori, K., Gupta, B. N., Bucher, J. R., and Boorman, G. A. 1987.

Ultrastructural changes in the nasal mucosa of F-344 rats and B6C3F1 mice following an acute exposure to methyl isocyanate (MIC). *Environ. Health Perspect.* 72:77–88.

Voigt, J. M., Guengerich, F. P., and Baron, J. 1985. Localization of a cytochrome P-450 isozyme cytochrome P-450 Pb-B and NADPH cytochrome P-450 reductase EC-1.6.2.4 in rat nasal mucosa. *Cancer Lett.* 27:241–248.

Voigt, J. M., Guengerich, F. P., and Baron, J. 1993. Localization and induction of cytochrome P-450 1A1 and aryl hydrocarbon hydroxylase activity in rat nasal mucosa. *J. Histochem. Cytochem.* 41:877–885.

Young, J. T. 1981. Histologic examination of the rat nasal cavity. *Fundam. Appl. Toxicol.* 1:309–312.

ACCESSORY NASAL STRUCTURES IN TOXICOLOGY STUDIES

Robert C. Sills

National Institute of Environmental Health Sciences, Environmental Toxicology Program, Pathology Branch, Research Triangle Park, North Carolina, USA

Kevin T. Morgan

Chemical Industry Institute of Toxicology, Research Triangle Park, North Carolina, USA

Gary A. Boorman

National Institute of Environmental Health Sciences, Environmental Toxicology Program, Pathology Branch, Research Triangle Park, North Carolina, USA

In addition to air conditioning, the nose has a range of sensory functions that render this organ structurally complex due to the presence of diverse accessory structures, including the vomeronasal and septal olfactory organ, an array of mucoserous glands, and the trigeminal and other sensory nerve endings. The supporting structures of the nasal airways are beautifully "engineered" to permit integration of the nose and its accessory organs with adjacent structural components of the head, including the nasolacrimal and incisive ducts, and the bones of the maxilla and anterior cranial vault. A bilaterally symmetric architecture, comprised of muscle, bone, cartilage, nerves, blood vessels, and other connective tissue, is well adapted to the functional integrity of this organ. The purpose of this article is to consider the accessory and supporting nasal structures from the perspective of toxicologic pathology. Brief descriptions of selected spontaneous and treatment-related lesions are presented.

The nose serves multiple functions that are reflected in the complexity of its structure (Harkema, 1991; Negus, 1958; Proctor & Andersen, 1982; Reznik, 1983; Uraih et al., 1990). Accessory nasal structures, in this review, refer to a number of tissues or organs that lie within or in close proximity to the nasal passages, and that serve primary functions other than air transport or air conditioning. These accessory structures (Figures 1 and 2) are listed, along with their functions and selected references containing more detailed information, in Table 1. In addition, this article presents

We are grateful to a number of individuals whose work contributed to this article. Olfactory lesions of the vomeronasal organ were provided by Brian Short of SmithKlein Beecham. The ethmoid bone lesions were provided by Stéphane Méry of CIIT. Septal lesions were provided by Joel Leininger and Doug Wolf of CIIT. NALT slides were provided by Joelle Ibanes of the Chemical Industry Institute of Toxicology.

Address correspondence to Robert C. Sills, Environmental Toxicology Program, National Institute of Environmental Sciences, P.O. Box 12233, Research Triangle Park, NC 27709.

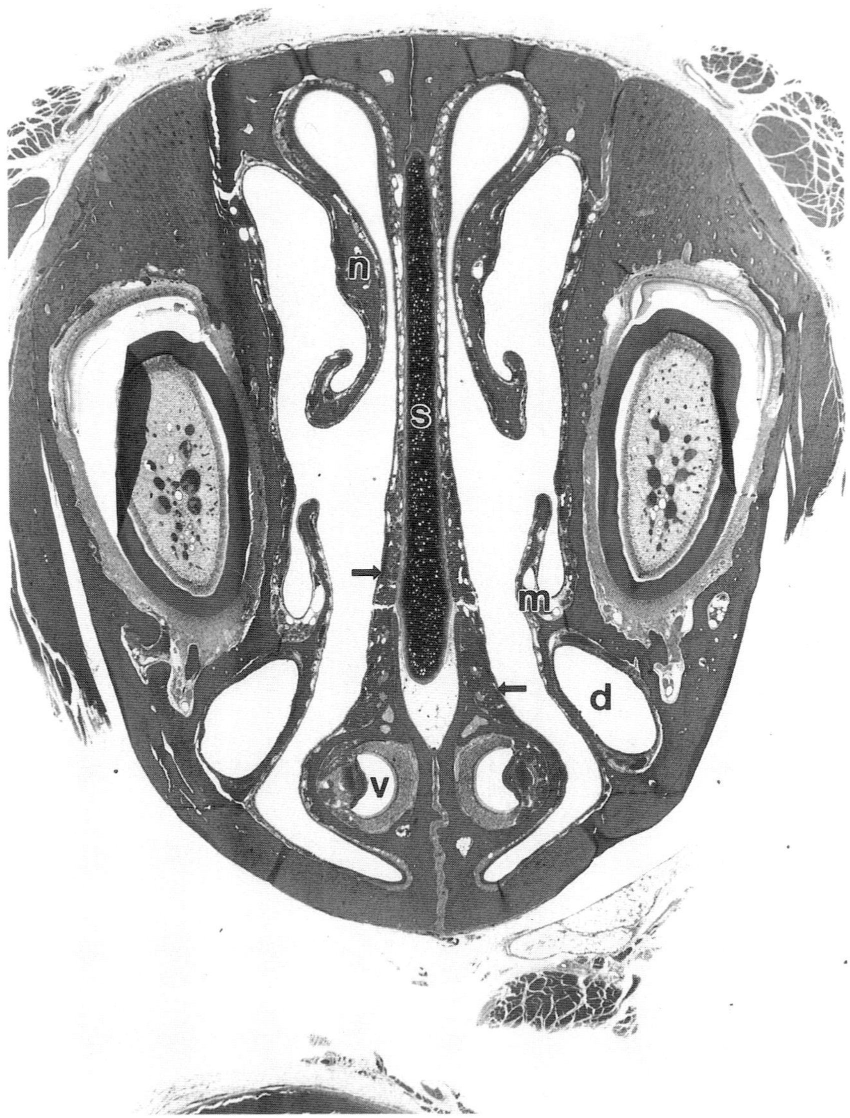

FIGURE 1. Anterior nasal cavity of an untreated control rat: vomeronasal organ (v), septal glands (arrows), nasolacrimal duct (d), septal cartilage (s), maxilloturbinate (m), and nasoturbinate (n).

selected lesions observed in nasal supporting structures, such as the septal cartilage, turbinate bones, and nasal vasculature.

The purpose of this article is therefore to focus attention away from the nasal mucosae, which have been addressed in detail elsewhere in this volume on other features of the nose. The objectives of this article are (1) to review the anatomy of accessory nasal structures and other selected nasal structures in rats and mice; (2) to present selected examples of lesions in these structures, found in toxicity and carcinogenicity studies; and (3) where applicable, to discuss the development of lesions as they relate to mechanisms of toxicity. Toxic lesions may be broadly categorized into degenerative, inflammatory, reparative, adaptive, or proliferative. This arti-

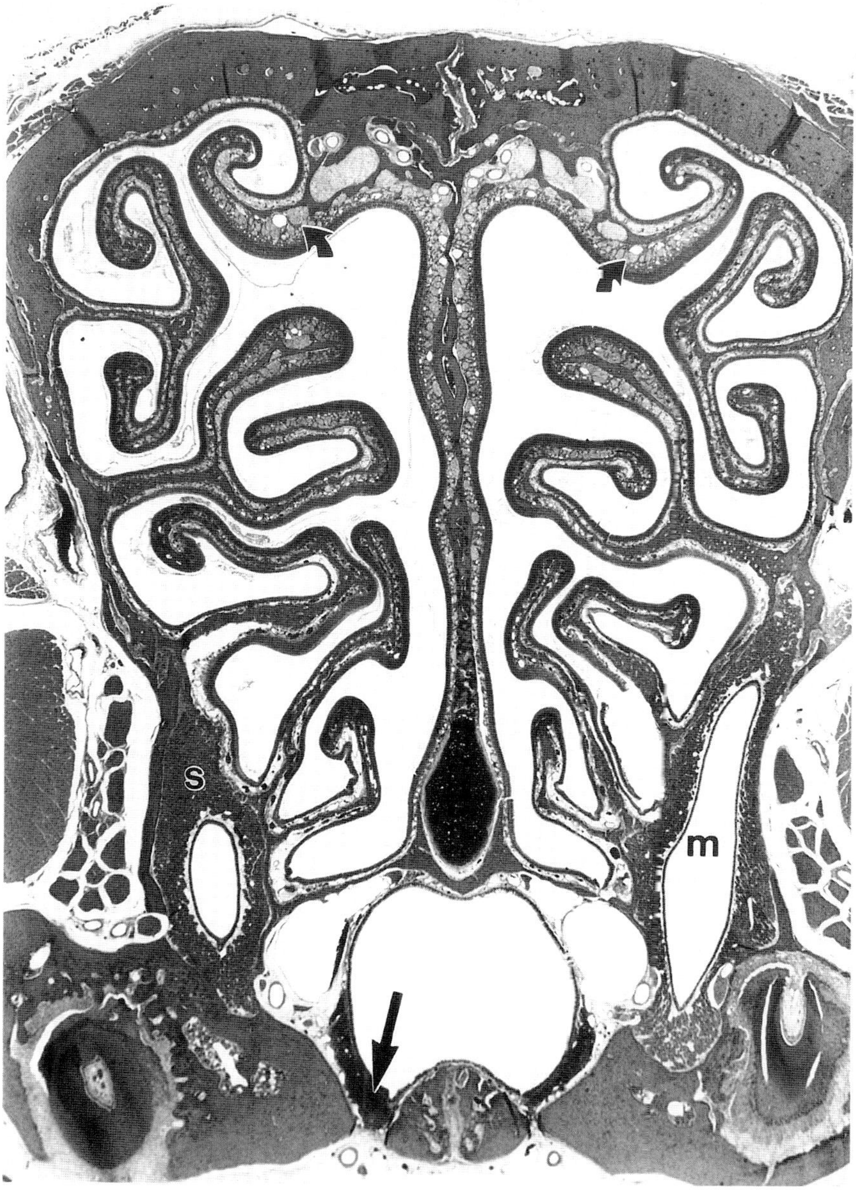

FIGURE 2. Posterior nasal cavity of an untreated control rat: Steno's glands (s), maxillary sinus (m), nose-associated lymphoid tissue (arrow), olfactory nerves (arrowhead).

TABLE 1. Accessory Nasal Structures

Accessory nasal structures	Function	Reference
Vomeronasal organ (Jacobson's organ)	Suckling reflex in young Pheromonal recognition and food flavor perception	Negus (1958) Vaccarezza et al. (1981)
Septal olfactory organ (organ of Masera, organ of Rodolfo-Masera)	Chemosensory odor stimulus assessment	Marshall and Maruniak (1986)
Bowman's glands	Seromucous gland Biotransformation	Bogdanffy (1990) Dahl and Hadley (1991) Reed (1993)
Steno's glands	Biotransformation enzymes	Belinsky et al. (1987)
Nasolacrimal duct	Transport lacrimal secretions to the nasal cavity	Hebel and Stromberg (1986)
Nasopalatine duct	Passage of materials from nasal to buccal cavities	Negus (1958)
Nose-associated lymphoid tissue (NALT)	Immune system	Kuper et al. (1989) Spit et al. (1989)

cle covers lesions that fall into these categories, with the exception of proliferative lesions considered to be features of neoplastic processes.

MATERIALS AND METHODS

Examples of normal control tissues and tissues exhibiting chemically induced, age-associated, background, or other lesions involving accessory and supporting nasal structures were obtained from the National Toxicology Program Archives and from previous inhalation studies conducted at the Chemical Industry Institute of Toxicology, the U.S. Environmental Protection Agency, and SmithKlein Beecham Pharmaceuticals. Histological sections, principally paraffin-embedded and stained with hematoxylin and eosin (HE), were examined and photographed by bright-field light microscopy.

RESULTS AND DISCUSSION

Vomeronasal (Jacobson's) Organ

The vomeronasal organ consists of paired tubular diverticula (Figure 3), located ventral to the nasal septum, in the vomer bone of the anterior nasal cavity (Uraih & Maronpot, 1990; Vaccarezza et al., 1981). These tubular diverticula, which are discrete chemoreceptor structures, are lined by tall columnar epithelium along one side while the opposite side is lined by vomeronasal (chemosensory) epithelium (Figure 4). The distribution of these

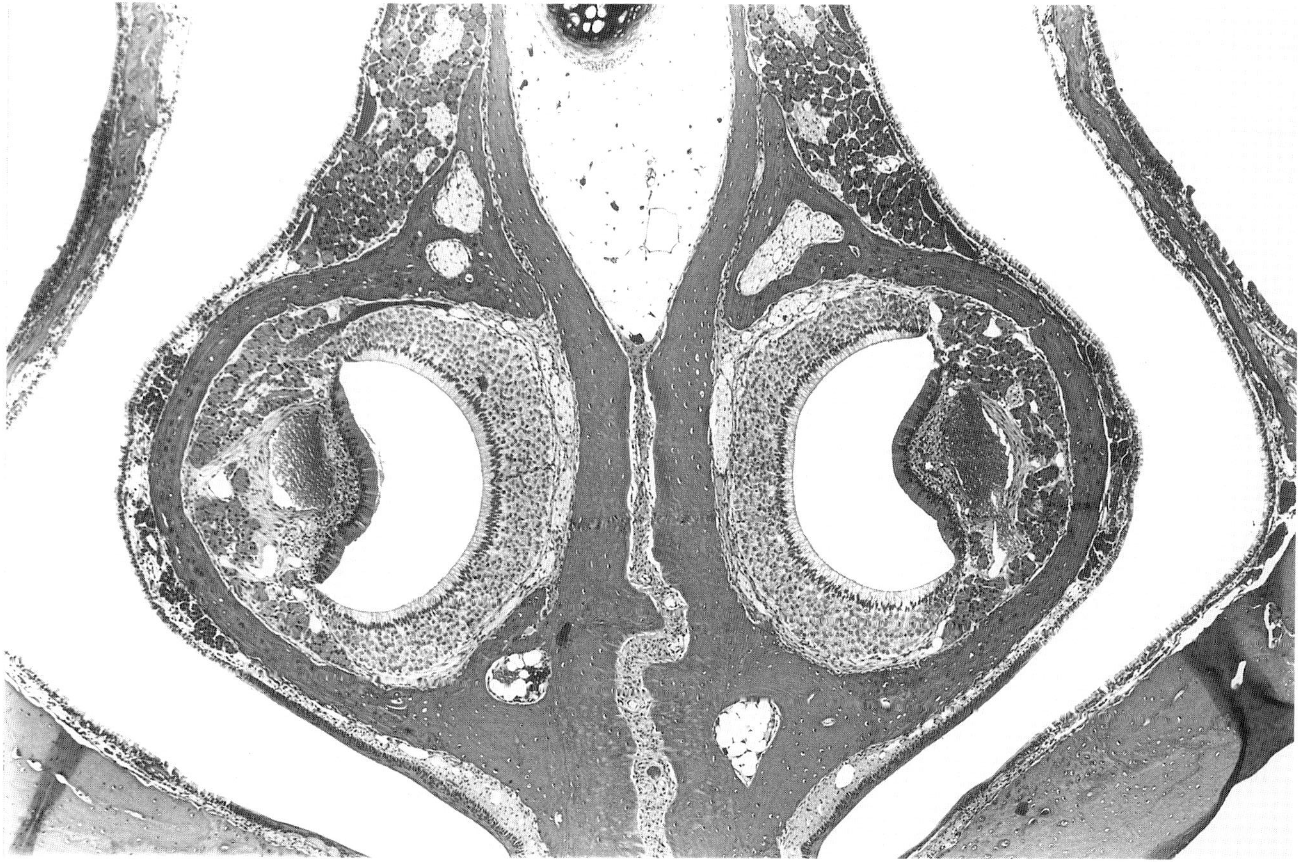

FIGURE 3. Vomeronasal organ of a control rat. Note paired tubular diverticula.

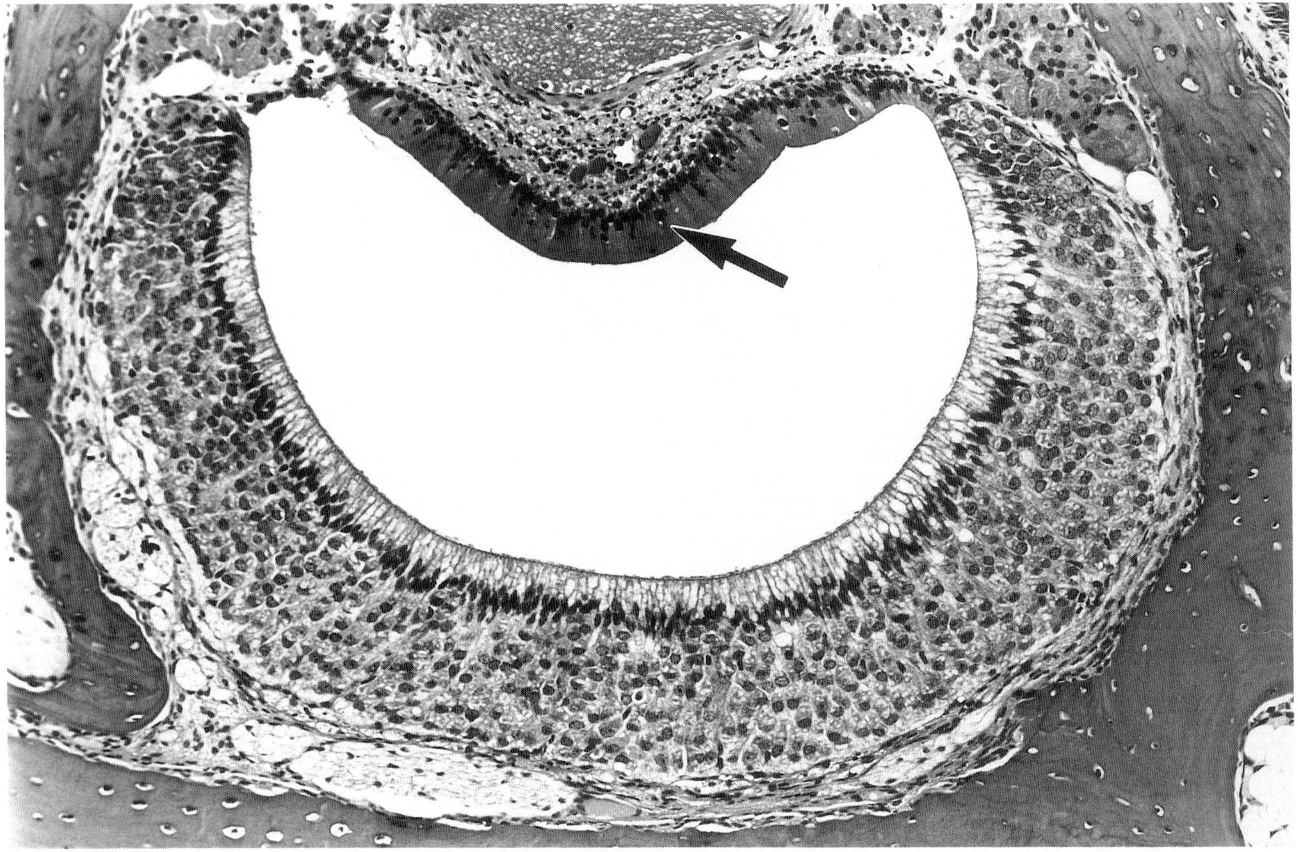

FIGURE 4. Tubular diverticulum vomeronasal organ. One side is lined by tall columnar epithelium (arrow) while the opposite side is lined by chemosensory epithelium, consisting of sustentacular and sensory cells.

epithelia exhibits a spiral pattern along the long axis of the nose (Uraih & Maronpot, 1990). The chemosensory epithelium resembles the olfactory epithelium, comprised of both supporting or sustentacular cells and sensory neurons. However, in contrast to the olfactory epithelium, these sensory cells do not possess cilia, and they are replaced laterally from a region at the junction of the columnar and chemosensory epithelia (Uraih & Maronpot, 1990). The vomeronasal organ of many animal species is considered responsible for the detection of pheromones (Negus, 1958), and thus plays a role in sexual behavior and "emotion." Recent studies indicate that the vomeronasal organ plays a similar role in humans (Taylor, 1994).

Selective toxicity of the vomeronasal chemosensory and columnar epithelia has been reported rarely in rodent toxicology studies. Recently, rats exposed to a growth hormone releasing peptide (GH-RP-6) by intranasal instillation exhibited atrophy and degeneration of the vomeronasal chemosensory epithelium with sparing of the columnar epithelium (Macia et al., 1993). In exposed animals, the affected epithelium was markedly decreased in thickness due to degeneration of both the sustentacular cells and sensory neurons (Figure 5).

In another recent study (National Toxicology Program, unpublished observations), mice were exposed to the volatile, iron-containing compound ferrocene, and toxic responses were observed specifically in the columnar epithelium of the vomeronasal organ, sparing the chemosensory region. Lesions were characterized by disorganization and degeneration of cells within the columnar epithelium and an associated accumulation of the iron-containing compound in affected areas (Figure 6). As for lesions elsewhere in the nose, site-specific epithelial damage noted for these two compounds may be a consequence of local dose, tissue-specific metabolism, or other factors. However, much remains to be learned about the toxicology of the vomeronasal organ.

Septal Olfactory Organ (Organ of Masera)

The septal olfactory organ is located on the nasal septum, posterior to the vomeronasal organ and dorsal to the nasopalatine duct (Adams & McFarland, 1971; Rodolfo-Masera, 1943). It consists of a patch of epithelium morphologically similar to olfactory epithelium elsewhere in the nose, and its function is chemosensory, primarily concerned with olfaction. The position of this structure adjacent to inspiratory airstreams (Morgan et al., 1991) indicates a specialized sensory function. Extensive lesions of the olfactory mucosa may also involve the septal olfactory organ, but this region does not appear to have been systematically examined for toxic responses.

Nasal Glands

Familiarity with the distribution of nasal glands in rodents is important when interpreting toxic lesions, such as degeneration, atrophy, and ductular ectasia (Figure 7). The nasal cavity contains four major glands or glandular complexes: Bowman's, Steno's, and the septal and turbinate glands

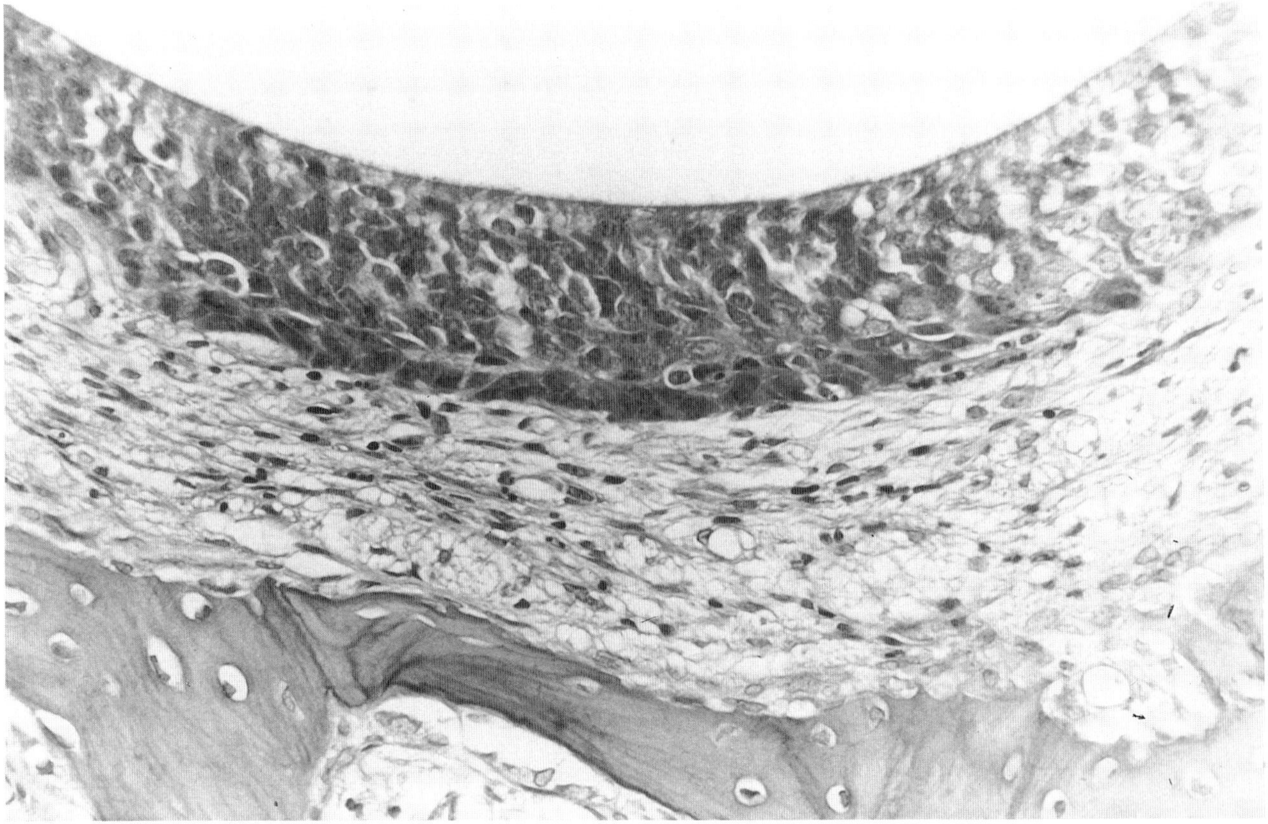

FIGURE 5. Vomeronasal organ from a rat exposed to a growth-hormone-releasing peptide. Note chemosensory epithelium is markedly reduced in thickness, with degeneration of sustentacular cells and sensory neurons. Due to atrophy, the supporting stroma are prominent.

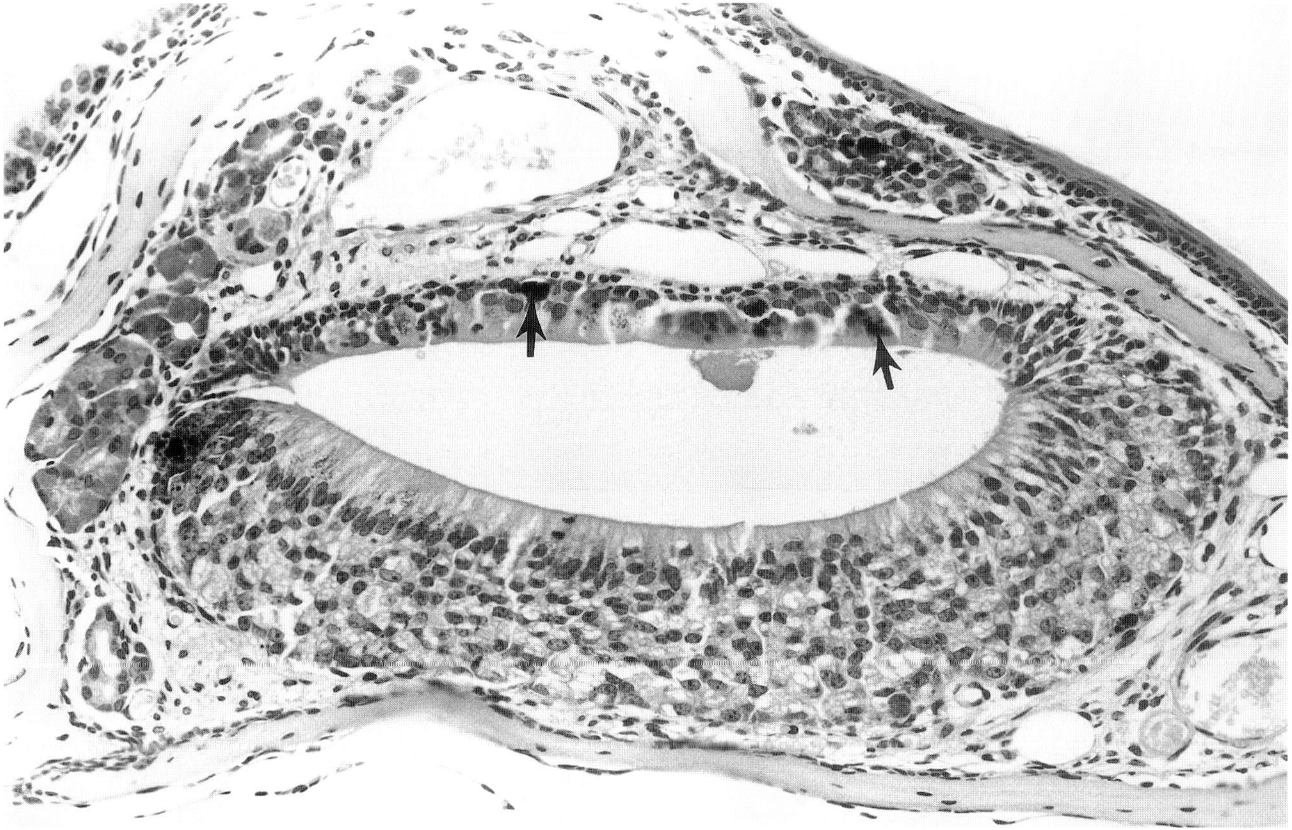

FIGURE 6. Vomeronasal organ from a mouse exposed to a volatile, iron-containing compound, ferrocene. Columnar epithelial cells are disorganized. The iron-containing compound (arrows) is closely associated with degenerate respiratory epithelial cells. Note the chemosensory epithelium is normal.

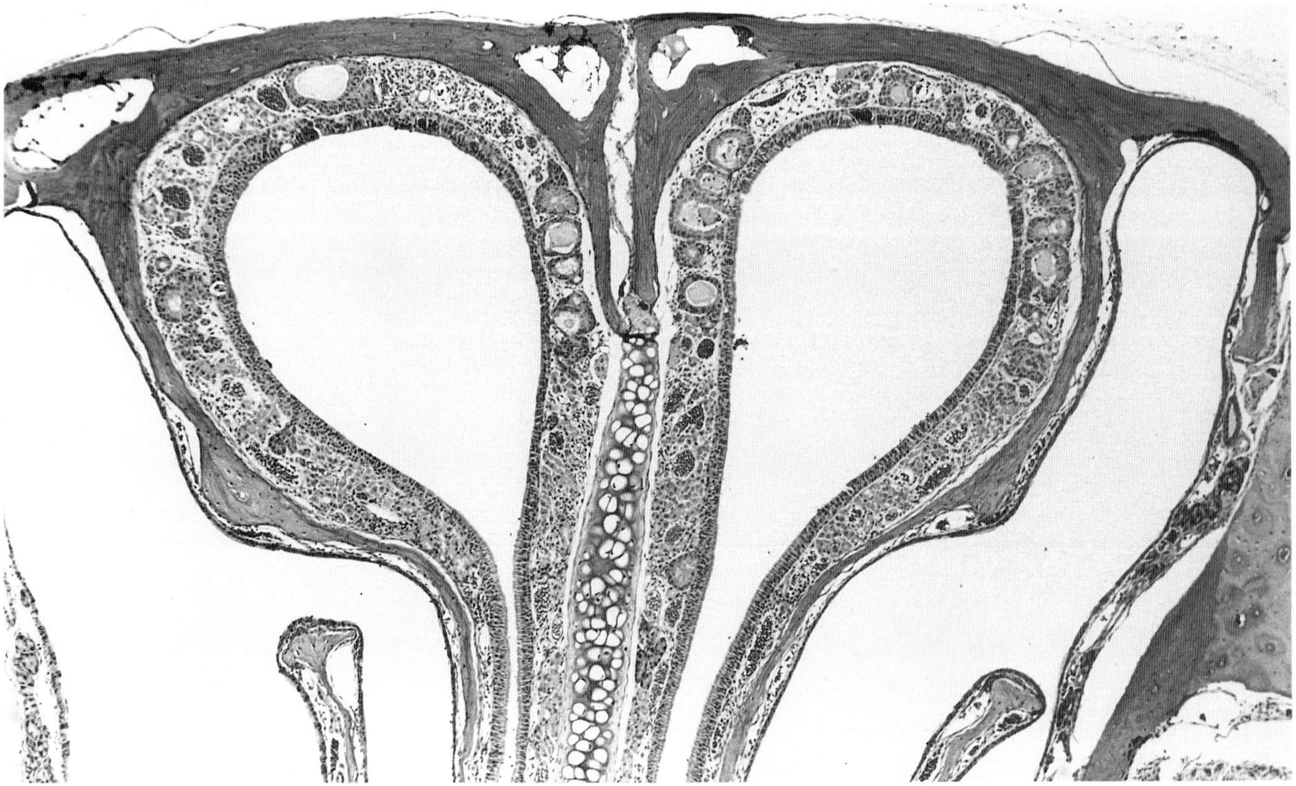

FIGURE 7. Bowman's glands located in the lamina propria of the olfactory epithelium. Note the bilateral ductular ectasia.

(Boorman et al., 1990). Bowman's and Steno's glands are known to have high metabolic capacity (Bogdanffy, 1990; Dahl & Hadley, 1991; Reed, 1993), whereas the septal and turbinate glands are primarily secretory in nature (Katz & Merzel, 1977; Klaassen et al., 1981).

The majority of nasal lesions that involve these glands have been reported in Bowman's glands. However, the structure, function, and toxicology of the Bowman's glands of the olfactory mucosa are addressed elsewhere (Gaskell, 1990; Morgan, 1994). Recent studies of the tobacco-specific nitrosamine 4-(*N*-methyl-*N*-nitrosamino)-1-(3-pyricyl)-1-butanone (NNK) have revealed toxic responses specifically targeting the Steno's (lateral nasal) glands of the rat (Belinsky et al., 1987). Steno's glands, paired glands that completely surround the maxillary sinus (Moe & Bojsen-Moller, 1971), have extensive metabolic capacity (Boorman et al., 1990; Monticello et al., 1990). Depending on the section of nasal cavity examined, the distribution of Steno's glands may appear to vary. In the anterior nasal cavity, Steno's glands are dorsal to the incisor teeth, whereas more posteriorly they appear somewhat lateral to the nose associated lymphoid tissue. Parental administration of NNK induced necrosis and atrophy of Steno's glands. Necrosis of Bowman's glands was also observed in the study. Bowman's glands damage by NNK was attributed to P-450-mediated metabolism, but the mechanism of damage to the Steno's glands is unknown.

The septal glands are located within the lamina propria of the respiratory mucosa. The anterior septal gland secretions are primarily serous (PAS^-), whereas the posterior septal gland secretions are mainly mucus (PAS^+) (Katz & Merzel, 1977). Spontaneous and chemically related lesions are infrequently observed in septal or turbinate glands. Incidental lesions include focal dilatation of septal glands or dilatation of multiple turbinate glands (Figure 8). Severe lesions in septal or turbinate glands may be sufficiently severe to alter the contour of the nasal airways.

Nasolacrimal Duct

Nasolacrimal ducts carry lacrimal secretions from the medial canthus of the eye to the nasal cavity. Depending on the section of nasal cavity examined, nasolacrimal ducts vary in location and size (Uraih & Maronpot, 1990). From anterior to posterior, the nasolacrimal ducts arise from a narrow orifice in the nasal vestibule to pass medial and then ventral medial to the incisor teeth. These ducts then pass to the lateral aspect of the large curved incisor teeth to sweep dorsally toward the eye. Beginning at the anterior nasal cavity, the nasolacrimal ducts are small, become large and saccular in their center, and taper down in size as they proceed toward the eye. Some common nasolacrimal duct lesions include inflammation (Figure 9), hyperplasia, and squamous metaplasia (Boorman et al., 1990). Acute inflammation, degeneration, and individual cell necrosis are sometimes seen.

During an inhalation study of vinyl chloride, marked toxic changes were found in the nasolacrimal duct of rats and mice. It was felt that

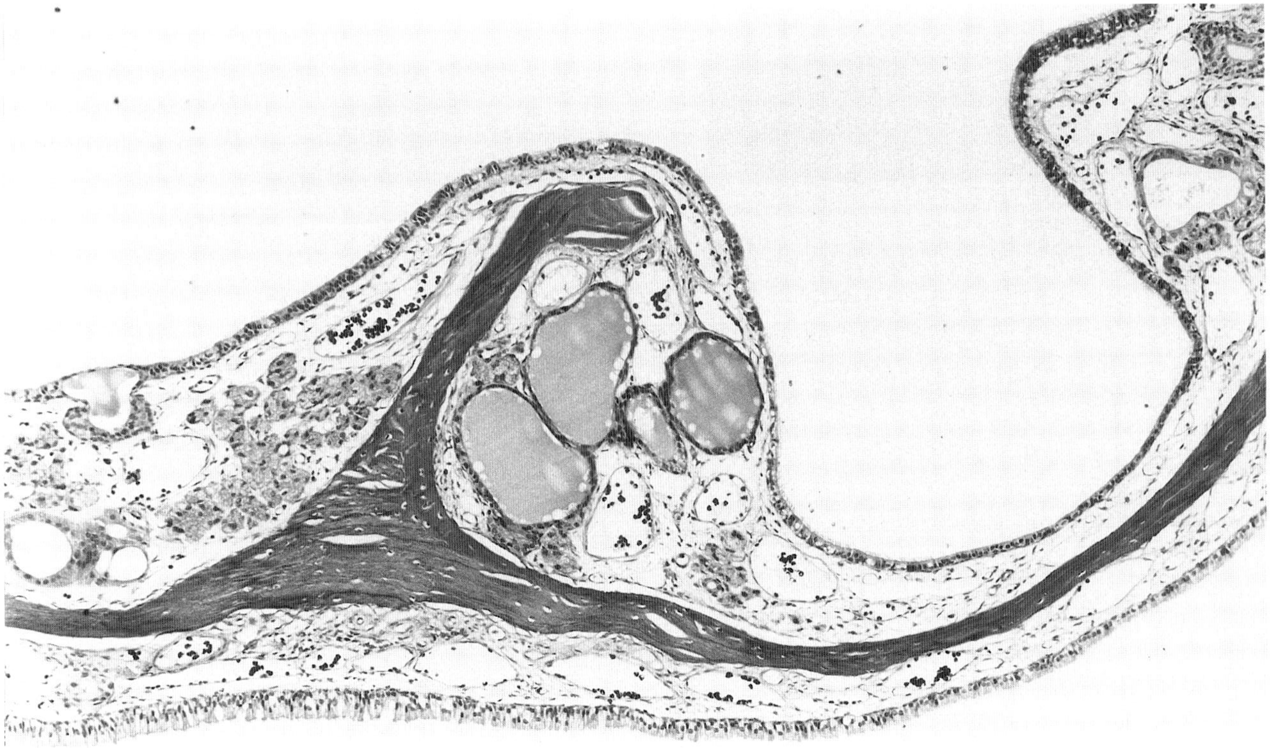

FIGURE 8. Turbinate gland of a rat. Dilatation of multiple glandular structures.

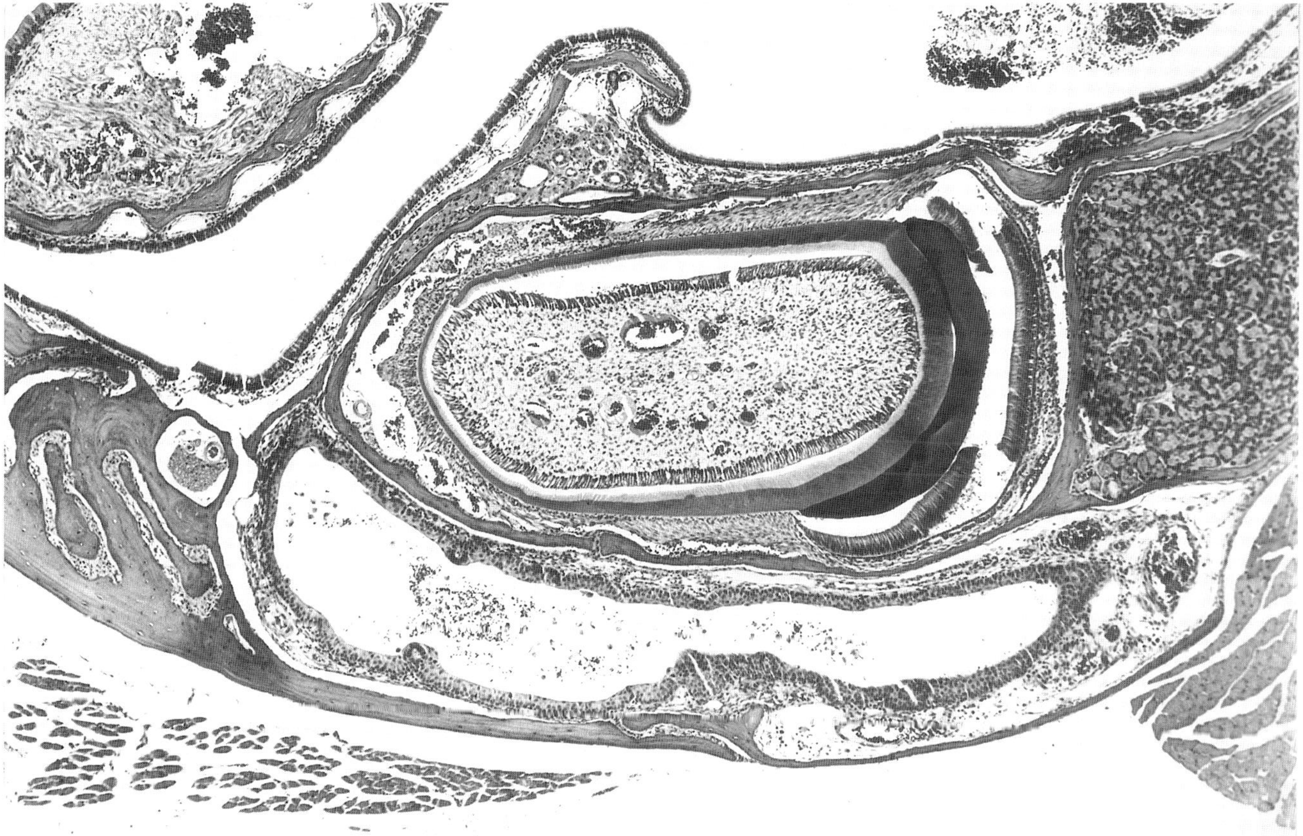

FIGURE 9. Nasolacrimal duct located lateral to the root of the incisor tooth at this level of the nose. Note the inflammatory exudate within the lumen and hyperplasia of the epithelial lining.

because of its water solubility, vinyl chloride in lacrimal secretions caused direct toxicity (Drew et al., 1983). In disease-free rats, focal minimal inflammation consisting of neutrophils, mononuclear cells, or lymphocytes of the nasolacrimal duct is common. In some cases, squamous metaplasia of the nasolacrimal duct epithelium may be associated with keratinization (Figures 10 and 11). Squamous metaplasia is an adaptive response often associated with prolonged inflammation or irritant effects. This metaplastic change is characterized by the replacement of the columnar epithelium by squamous epithelium, which is more resistant to injury. Hyperplastic lesions are also seen in the nasolacrimal duct. Hyperplastic lesions, when present, should be differentiated from neoplastic lesions of the nasolacrimal duct. Neoplastic lesions are usually larger and consist of atypical neoplastic cells arranged in a disorderly manner (Figure 12). In aging rats, inflammation of incisors and molars may result in abnormal tooth growth, termed dysplasia, which can cause obliteration of structures such as the nasolacrimal duct (Figure 13).

Nasopalatine Duct

The nasopalatine duct provides paired narrow passageways between the nasal and oral cavities, with a valvelike arrangement on the buccal side in the form of the incisive papilla. This valve appears to permit only unidirectional flow of material from the nasal to the buccal cavities (Negus, 1958). The function of this duct is unclear, but it may play a role in olfaction or gustation. This duct, which passes through the incisive canal of the hard palate, is lined by squamous epithelium and has rarely been reported as a site of treatment-induced lesions. Erosion of the squamous epithelium of the ventral meatus in mice exposed by inhalation to 50 ppm hydrogen chloride extended posteriorly to involve the lining of the nasopalatine duct (Jiang et al., 1986). Food material may also become lodged in these ducts to induce local inflammatory responses (Monticello et al., 1990).

Nose-Associated Lymphoid Tissue

The nose-associated lymphoid tissue (NALT) of rodents is the equivalent of the Waldeyer ring of humans (Koornstra et al., 1991). NALT is located in the ventral floor of the posterior nose (Figure 14) in close proximity to the septal window at the anterior opening of the nasopharyngeal meatus (Kuper et al., 1989). The NALT consists of lymphoid aggregates with a specialized epithelial covering termed the lymphoepithelium. The NALT resembles the bronchus-associated lymphoid tissue (BALT) in many respects, including the presence of specialized antigen-processing cells, called M cells (Wolf & Bye, 1984). For a detailed comparison of NALT, BALT, and Peyer's patches of the gastrointestinal tract, see Kuper et al. (1992). Responses of the NALT to xenobiotics and other foreign materials include lymphoid hyperplasia and the formation of lymphoid follicles. There is, however, a paucity of information on responses of the NALT to inhaled chemicals. Recent studies of a rat-adapted influenza virus reported the sequence of changes in the NALT of infected rats (Ibanes et al., 1993).

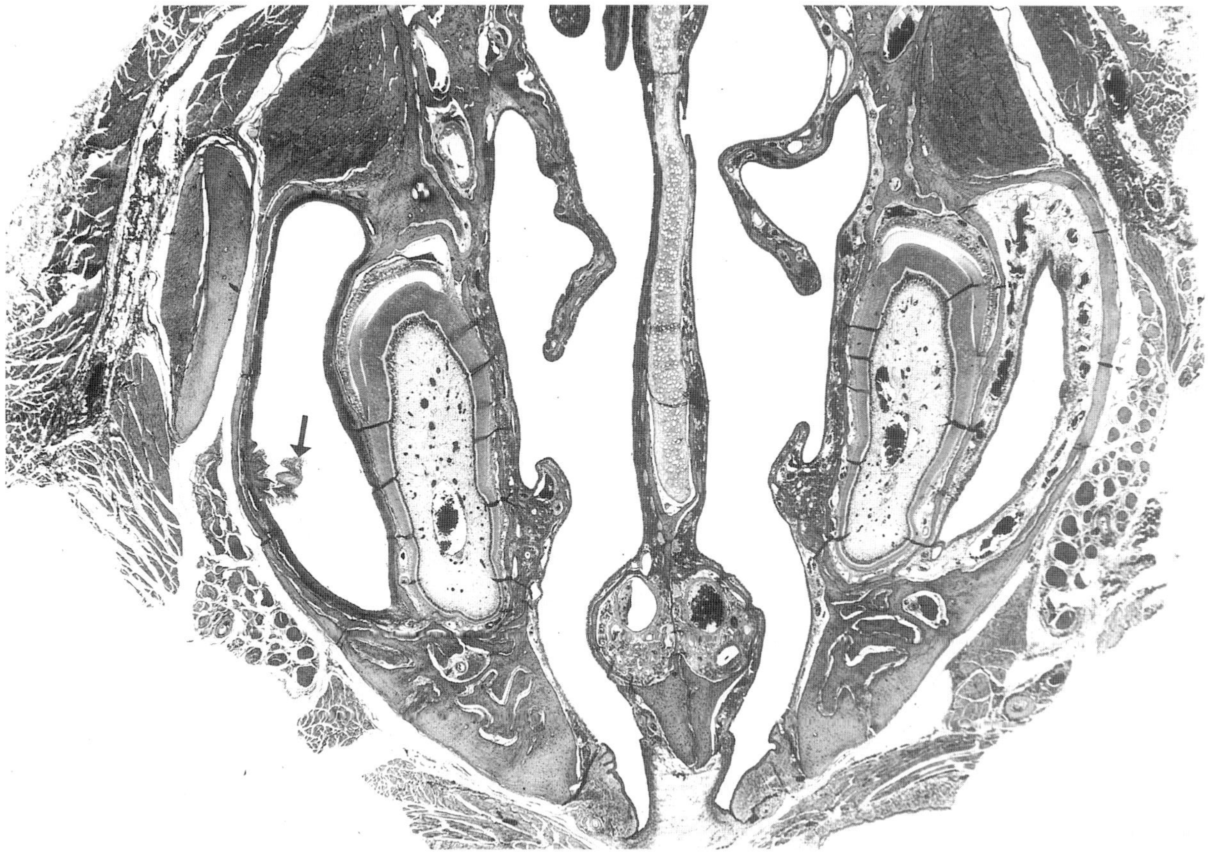

FIGURE 10. Nasolacrimal duct. Focal squamous metaplasia of lining epithelium associated with keratinization (arrow).

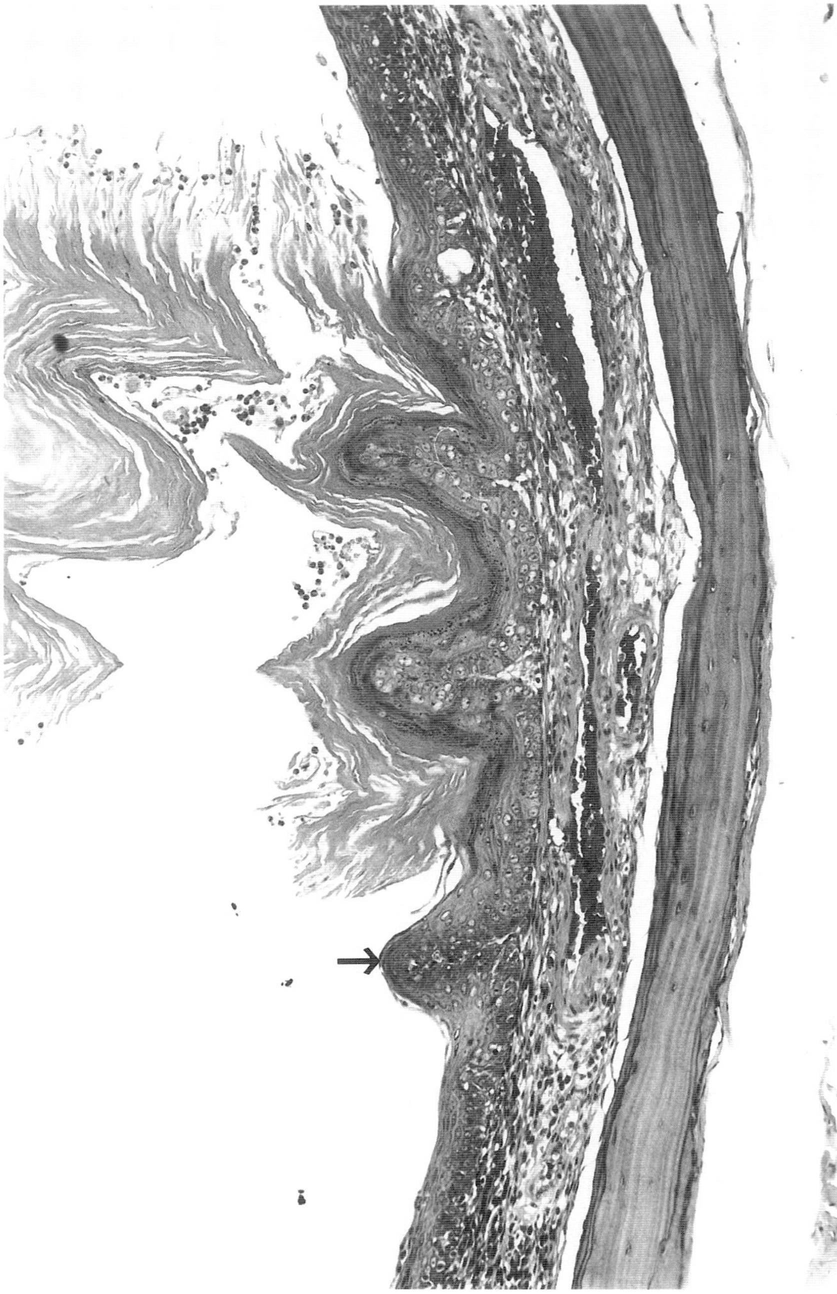

FIGURE 11. High magnification of Figure 10. Note the prominent squamous metaplasia (arrow) and keratinization.

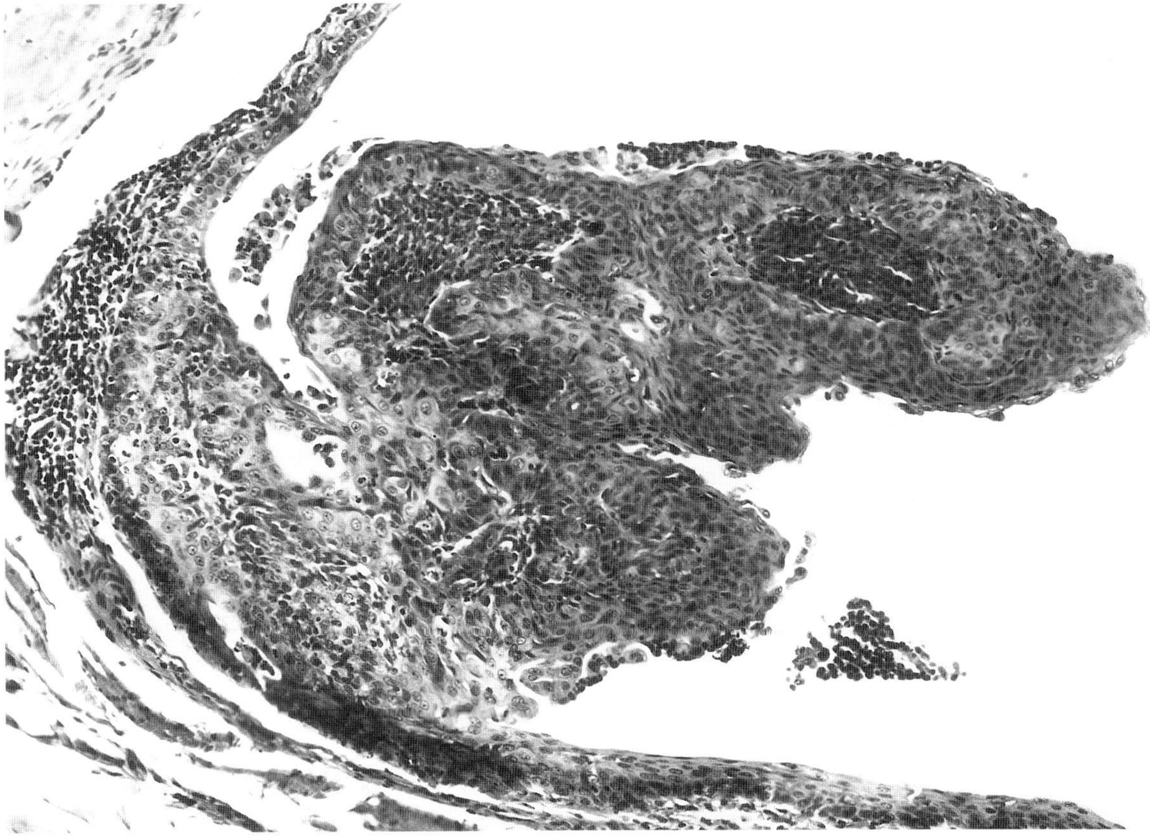

FIGURE 12. Adenoma of nasolacrimal duct epithelium consist of atypical neoplastic cells with a disorderly arrangement.

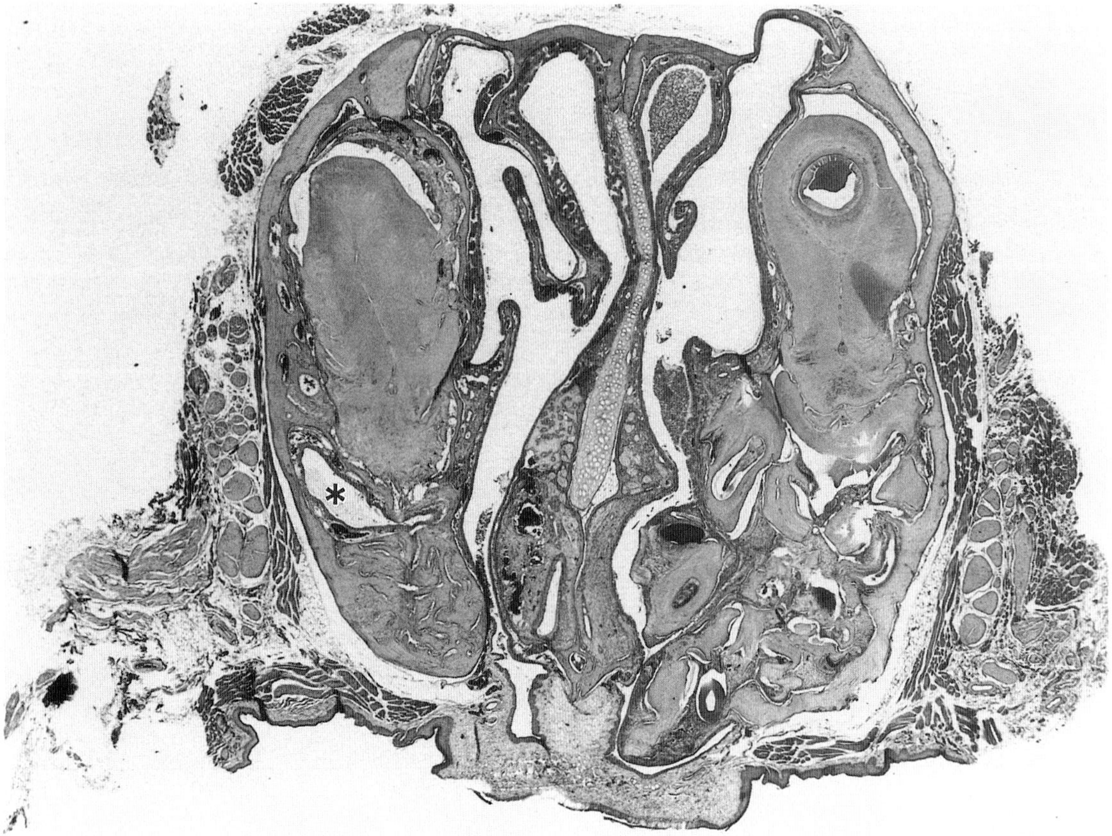

FIGURE 13. Dysplastic tooth obliterating the nasolacrimal duct. Note a remnant of the nasolacrimal duct (asterisk).

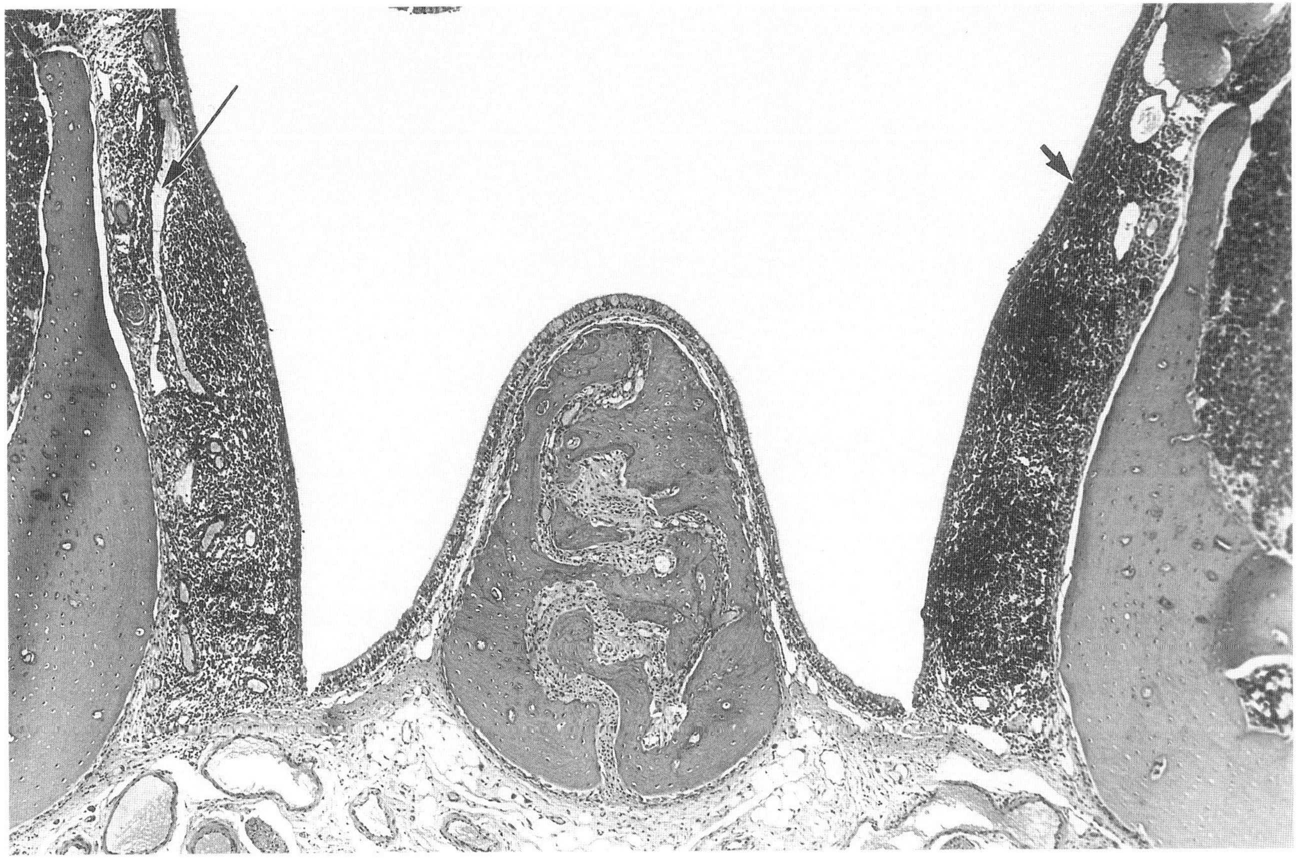

FIGURE 14. Nose-associated lymphoid tissue (NALT) from a rat. Note lymphoid follicles, lymphoepithelium (arrowhead), and high endothelial venules (arrow).

Cartilage and Bone

Generally, toxic lesions of the nasal septal cartilage or nasal turbinate bones are infrequently reported. Most turbinate and nasal bone lesions described are usually related to systemic changes, such as osteopetrosis (Marks, 1973), or may be secondary to advanced renal disease (Boorman et al., 1990). In a recent study, mice exposed to chlorine gas exhibited perforation (fenestration) of the septum of the nasal vestibule, with involvement of the anterior septal cartilage in some cases (Chemical Industry Institute of Toxicology, 1993). This lesion involved the anterior septum, beginning with focal necrosis of the septal cartilage, followed by partial to complete re-epithelialization of the margins of the perforated region (Figure 15). Exposure of rats and mice to cage contaminants has also been associated with septal perforations (Bolon et al., 1991).

Remodeling of bone is often a secondary response to local respiratory or olfactory epithelial ulceration, inflammation, necrosis, or degeneration. In rats exposed to chloroform, remodeling of bone was observed in the posterior nasal cavity, was most severe peripherally, and generally spared regions adjacent to the central airways (Méry et al., 1994). In this study, there was no necrosis or degeneration of the olfactory epithelium and superficial Bowman's glands; however, deep Bowman's glands were targets of toxicity. Remodeling consisted of extensive new bone formation and periosteal cell proliferation (Figure 16). Associated with the extensive new bone formation was a high proliferative index for osteogenic cells. The mechanism for the new bone formation is unknown. It was proposed that the osseous lesions induced by chloroform may be secondary to lesions of the adjacent Bowman's glands. A more extensive remodeling of the ethmoid bones was reported for rats exposed to 3-methyl furan (Morse et al., 1984) or 3-methyl indole (Peele et al., 1990), demonstrating the "plasticity" of these delicate boney scrolls. Ozone exposure has also induced lesions of nasal turbinate bones in rats, which were characterized by marked atrophy (Harkema et al., 1993).

Blood Vessels and Nerves

The lamina propria of the nasal passages have an extensive blood supply, with separate blood vessels supplying each of the major airway epithelia (Hebel & Stromberg, 1976; Warwick & Williams, 1973; Weiss, 1983). Dilated blood vessels (angiectasis) may be associated with chemical exposure, with mononuclear cell leukemia in F344 rats, or is sometimes observed in debilitated moribund animals (Figure 17) (Boorman et al., 1990). An uncommon vascular lesion encountered in the rodent nasal cavity is thrombosis, with specific involvement of arteries in the anterior nose (Monticello et al., 1990). The mechanisms responsible for this lesion remain to be determined, and there is a need for systematic study of nasal vascular responses to toxic xenobiotics. The most common lesion of nerve bundles in the nose is loss of unmyelinated olfactory nerves in the lamina

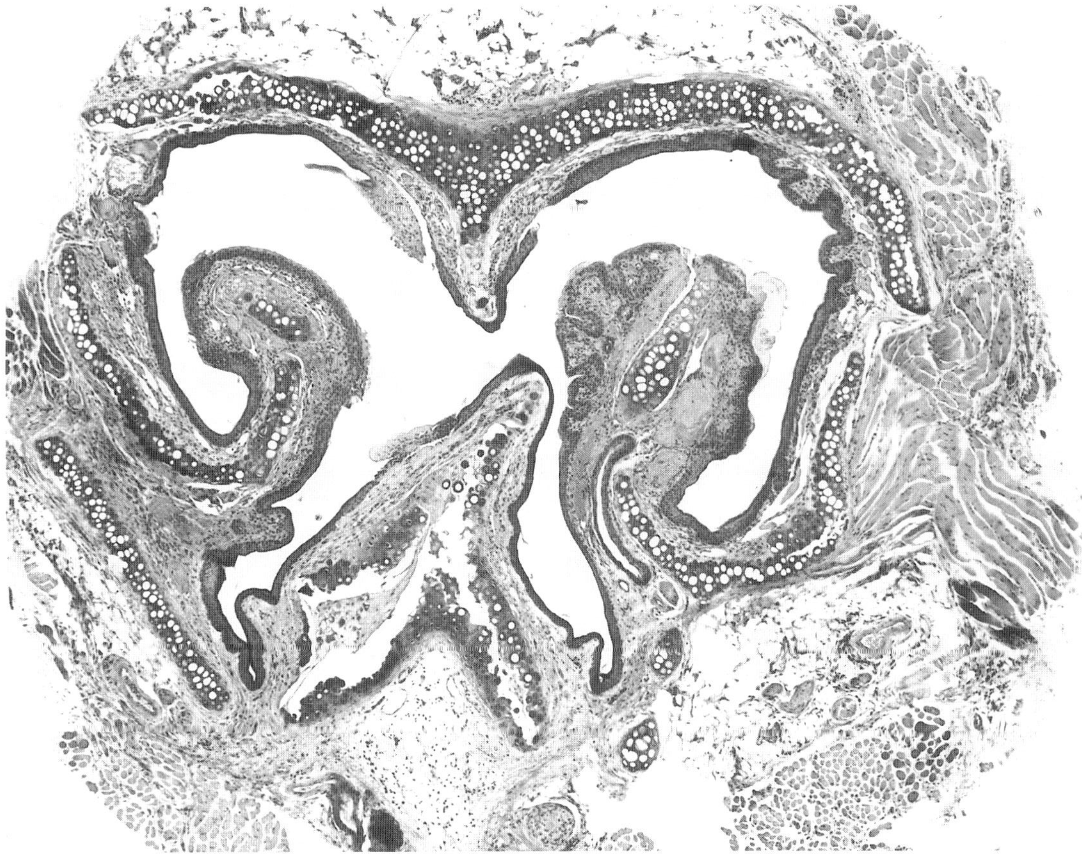

FIGURE 15. Perforation of septal cartilage with reepithelialization of the remaining cartilage in a mouse exposed to chlorine gas. Note the marked epithelial hyperplasia of the respiratory epithelium lining the maxilloturbinates.

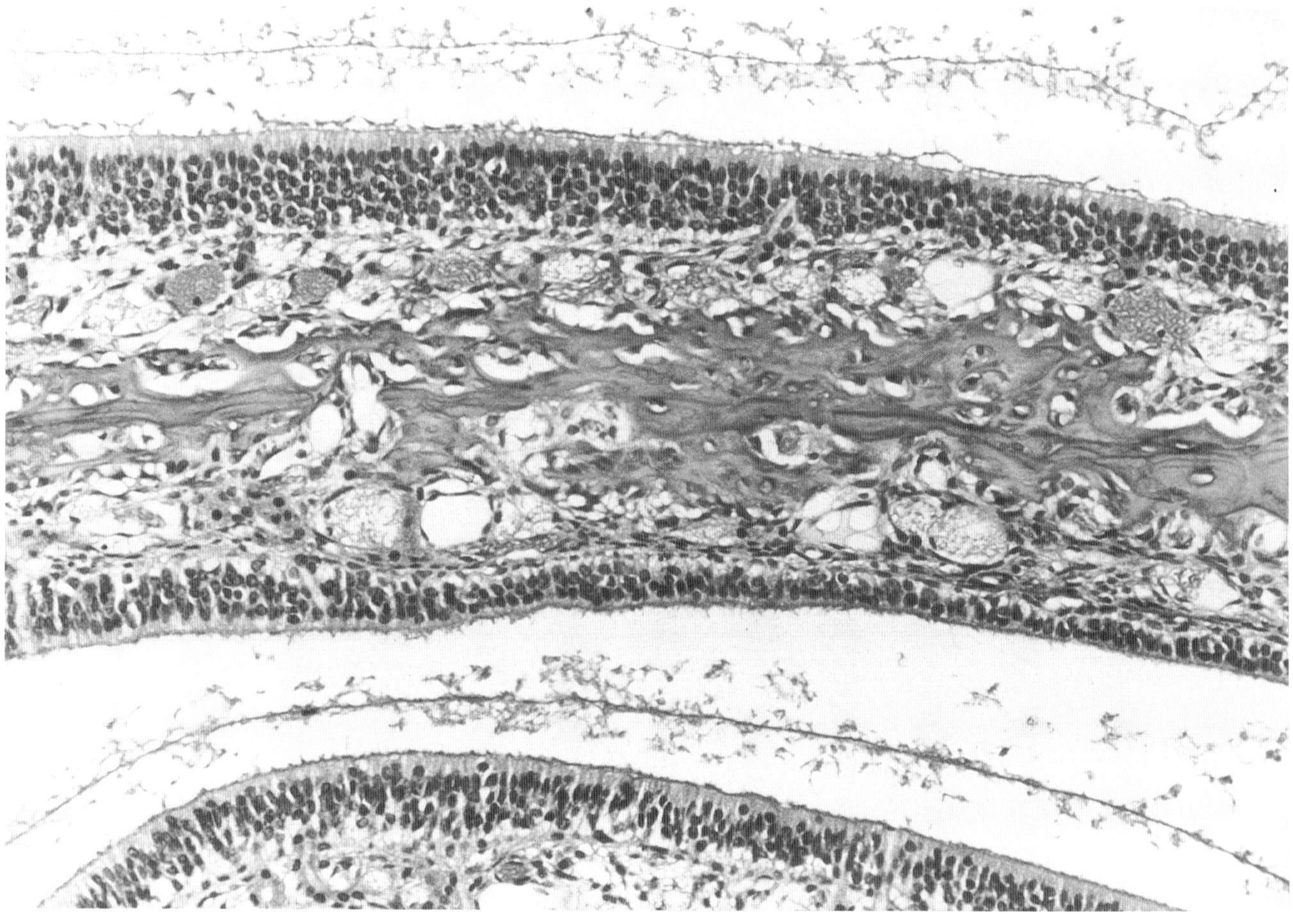

FIGURE 16. Sections of the first endoturbinate from a rat exposed to chloroform. Note extensive new bone formation, marked periosteal cell proliferation, and intact olfactory epithelium.

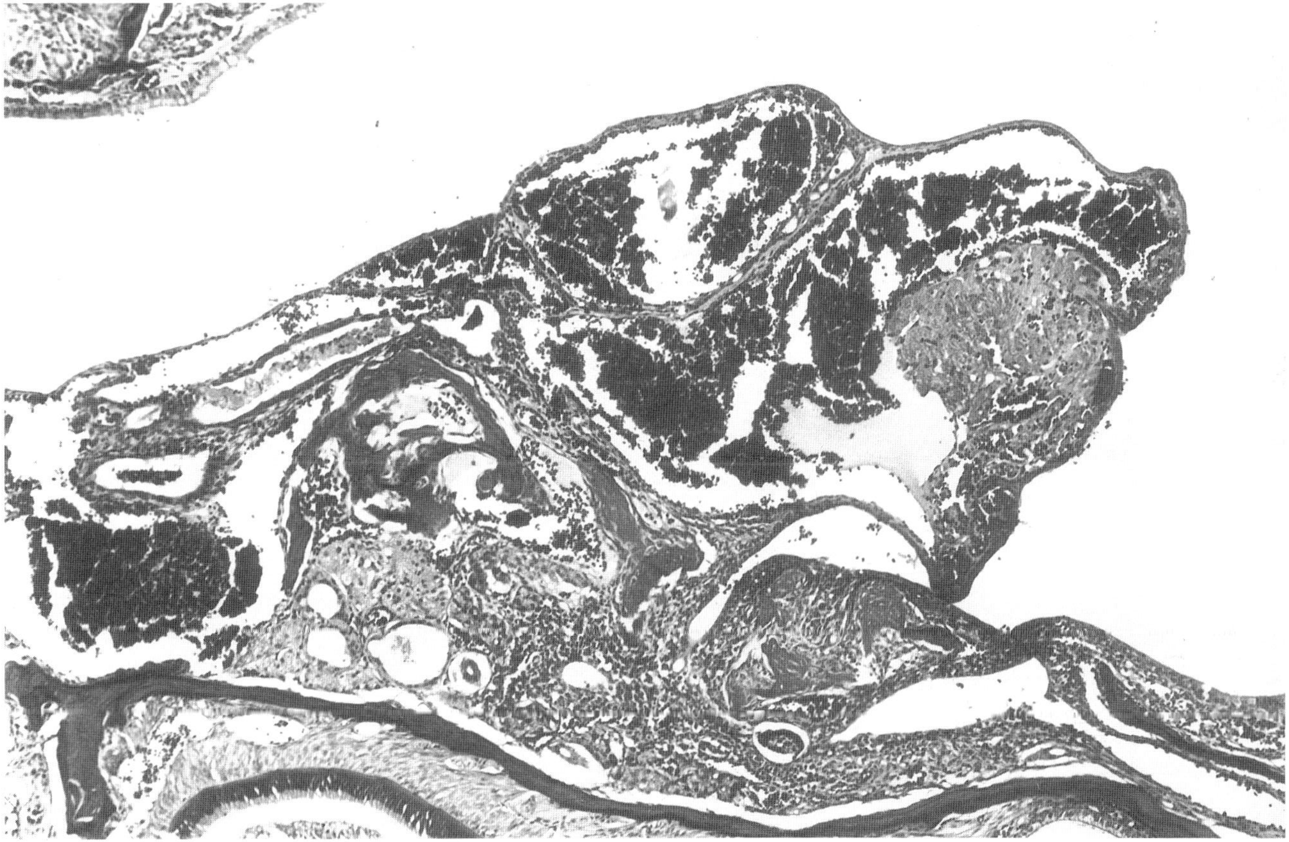

FIGURE 17. Lamina propria of the nasal passages with angiectasis. Note the markedly dilated blood vessels containing numerous red blood cells.

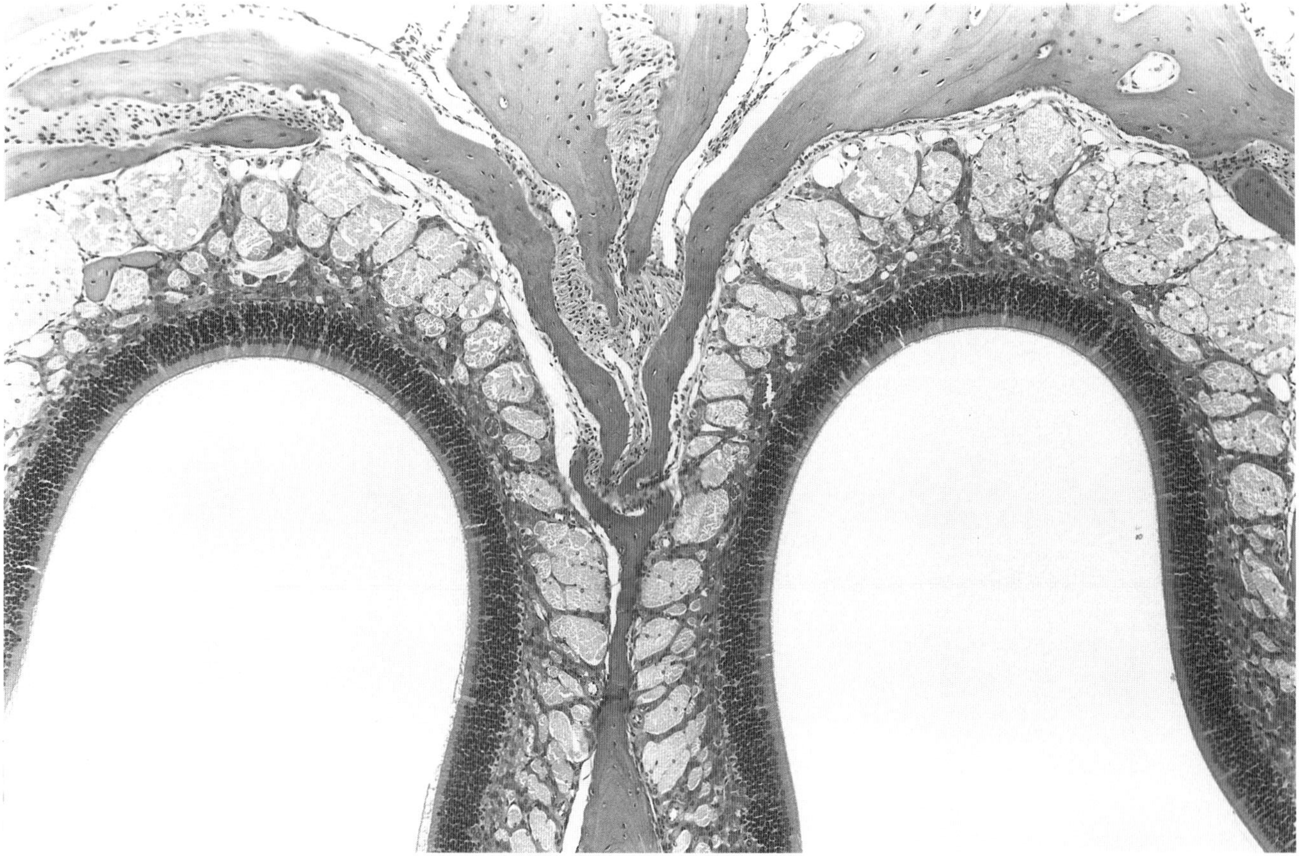

FIGURE 18. (a) Lamina propria of the olfactory region of an aging untreated control rat showing normal compliment of large unmyelinated nerve bundles.

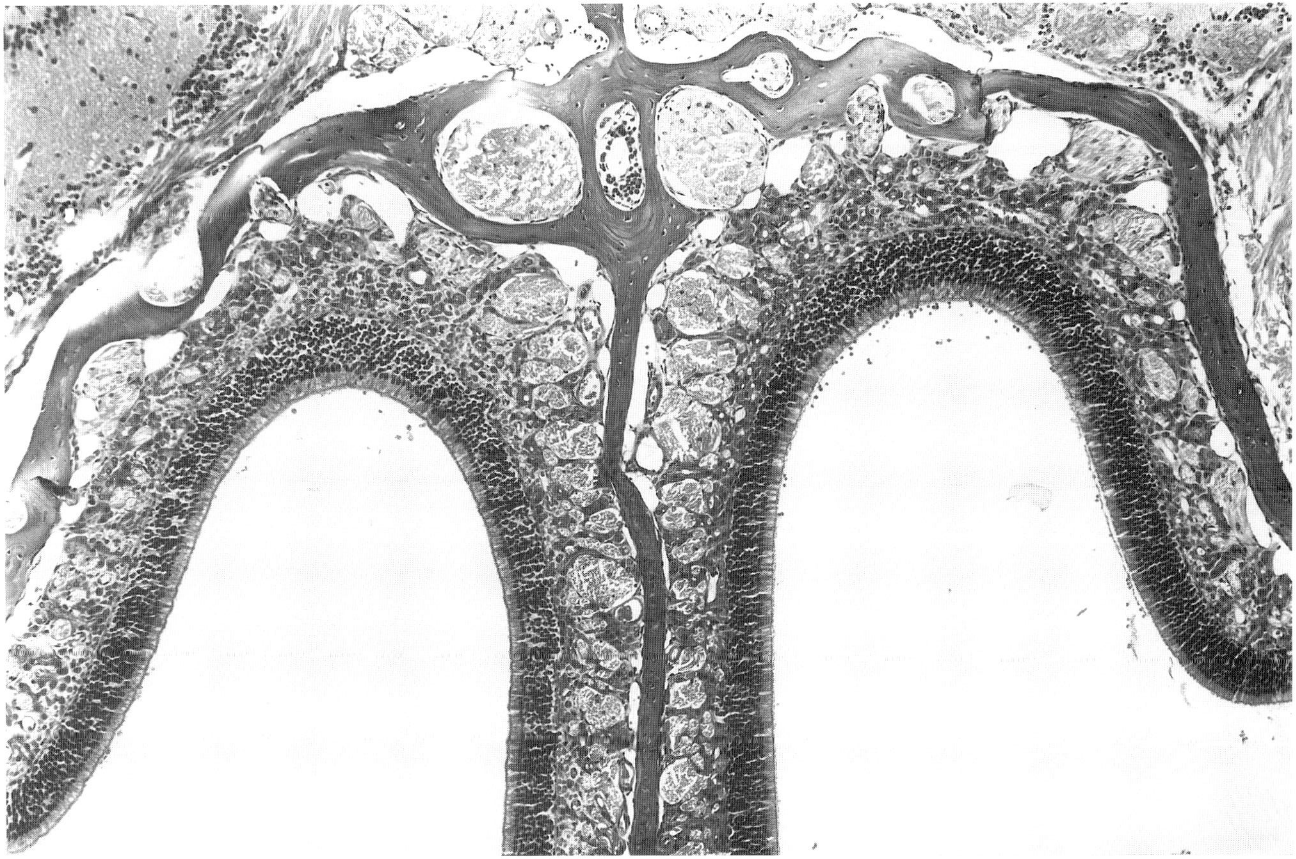

FIGURE 18. (*Continued*) (b) Bilateral atrophy and degeneration of the olfactory nerve bundles of an untreated control aging rat. Note the marked reduction in the size of the olfactory nerve bundles in the lamina propria of the dorsal region of the dorsal medial meatus. Compare with (a).

propria of the ethmoid region in association with olfactory epithelial damage (Buckley et al., 1985; Genter et al., 1992). Loss of these nerve bundles may also occur in aging animals (Figure 18).

CONCLUSION

This overview provided insights on the complexity and potential importance of a careful assessment of toxic responses in the accessory nasal structures in toxicology studies. It is evident that these structures are potentially important target sites and that local xenobiotic dosimetry should be considered when attempting to assess the underlying mechanisms. Multiple factors may influence the distribution of lesions in accessory nasal structures, such as tissue-specific toxicity of the columnar versus the chemosensory epithelium in the vomeronasal organ. A number of interesting questions have been raised by detection of lesions in the accessory and supporting structures of the nose. For instance, do metabolic properties, such as differences in enzyme distribution patterns between the chemosensory epithelium of the vomeronasal organ and the olfactory epithelium lining the nasal airways (Randall et al., 1987), account for lesion site specificity? Does the olfactory epithelium of the vomeronasal organ have high levels of cytochrome P-450, considered important in the metabolism and removal of odorants (Dahl, 1988)? Depending on airflow patterns, regional deposition or metabolism, accessory nasal structures in the anterior versus the posterior nasal cavity may be more prone to develop toxic lesions. The relevance of many toxic lesions in the accessory nasal structures of rodents as related to human risk assessment is unclear but remains a challenge to scientists in toxicology research.

REFERENCES

Adams, D. R., and McFarland, L. Z. 1971. Septal olfactory organ in *Peromyscus*. *Comp. Biochem. Physiol.* 40:971–974.

Belinsky, S. A., Walker, V. E., Maronpot, R. R., Swenberg, J. A., and Anderson, M. W. 1987. Molecular dosimetry of DNA adduct formation and cell toxicity in rat nasal mucosa following exposure to the tobacco specific nitrosamine 4-(*N*-methyl-*N*-nitrosamino)-1-(3-pyridyl)-1-butanone and their relationship to induction of neoplasia. *Cancer Res.* 47:6058–6065.

Bogdanffy, M. S. 1990. Biotransformation enzymes in the rodent nasal mucosa: The value of a histochemical approach. *Environ. Health Perspect.* 85:177–186.

Bolon, B., Bonnefoi, M. S., Roberts, K. C., Marshall, M. W., and Morgan, K. T. 1991. Toxic interactions in the rat nose: Pollutants from soiled bedding and methyl bromide. *Toxicol. Pathol.* 19:571–579.

Boorman, G. A., Morgan, K. T., and Uraih, L. C. 1990. Nose, larynx, and trachea. In *Pathology of the Fischer 344 rat*, eds. G. A. Boorman, S. L. Eustis, M. R. Elwell, C. A. Montgomery, and W. F. MacKenzie, pp. 315–337. New York: Academic Press.

Buckley, L. A., Morgan, K. T., Swenberg, J. A., James, R. A., Hamm, T. E. J., and Barrow, C. S. 1985. The toxicity of dimethylamine in F-344 rats and B6C3F1 mice following a 1 year inhalation exposure. *Fundam. Appl. Toxicol.* 5:341–352.

Chemical Industry Institute of Toxicology. 1993. A Chronic Inhalation Toxicity Study of Chlorine in Female and Male B6C3F1 Mice and Fischer 344 Rats. Chemical Industry Institute of Toxicology, Research Triangle Park, NC.

Dahl, A. R. 1988. The effect of cytochrome P-450-dependent metabolism and other enzyme activities on olfaction. In *Molecular neurobiology of the olfactory system, molecular, membranous, and cytological studies*, eds. F. L. Margolis and T. V. Getchell, pp. 51–70. New York: Plenum Press.

Dahl, A. R., and Hadley, W. H. 1991. Nasal cavity enzymes involved in xenobiotic metabolism: effects on the toxicity of inhalants. *Crit. Rev. Toxicol.* 21:345–372.

Drew, R. T., Boorman, G. A., McConnell, E. E., Haseman, J. K., Busey, W., and Moore, J. A. 1983. The effect of age and exposure duration on cancer induction by a known carcinogen in rats, mice and hamsters. *Toxicol. Appl. Pharmacol.* 68:120–130.

Gaskell, B. A. 1990. Nonneoplastic changes in the olfactory epithelium—Experimental studies. *Environ. Health Perspect.* 85:275–289.

Genter, M. B., Llorens, J., O'Callaghan, J. P., Peele, D. B., Morgan, K. T., and Crofton, K. M. 1992. Olfactory toxicity of β,β'-iminodiproprionitrile (IDPN) in the rat. *J. Pharmacol. Exp. Ther.* 263:1432–1439.

Harkema, J. R. 1991. Comparative aspects of nasal airway anatomy: Relevance to inhalation toxicology. *Toxicol. Pathol.* 19:321–336.

Harkema, J. R., Herrera, L., and Mellick, P. 1993. Ozone-induced chronic rhinitis with boney atrophy of the turbinates in male rats. *Vet. Pathol.* 30:441.

Hebel, R., and Stromberg, M. W. 1976. *Anatomy of the laboratory rat.* Baltimore: Williams & Wilkins.

Hebel, R., and Stromberg, M. W. 1986. *Anatomy and embryology of the laboratory rat.* Baltimore: Williams & Wilkins.

Ibanes, J. D., Morgan, K. T., and Burleson, G. R. 1993. Histopathological and flow cytometric analyses in nasal epithelium and nasal associated lymphoid tissue as a result of virus infection and pollutant exposure. *Toxicologist* 13:91.

Jiang, X. Z., Morgan, K. T., and Beauchamp, R. O. 1986. Histopathology of acute and subacute nasal toxicity. In *Toxicology of the nasal passages*, ed. C. S. Barrow, pp. 51–66. New York: Hemisphere.

Katz, S., and Merzel, J. 1977. Distribution of epithelia and glands of the nasal septum mucosa in the rat. *Acta Anat.* 99:58–66.

Klaassen, A. B. M., Kuijpers, W., and Jap, P. H. K. 1981. Secretory characteristics of the nasal glands. *Rhinol. Suppl.* 1 ASIAN:93–98.

Koornstra, P. J., deJong, F. I. C. R. S., Vlek, L. F. M., Marres, E. H. M. A., and Vriesman, P. J. C. V. B. 1991. The Waldeyer ring equivalent in the rat. *Acta Otolaryngol.* (Stockh.) 111:591–599.

Kuper, C. F., Spit, B. J., Bruijntjes, J. P., Hendriksen, E. G. J., and Hameleers, D. M. H. 1989. Nasal lymphoid tissue, its specialized position in the upper respiratory tract. In *Nasal carcinogenesis in rodents: Relevance to human health risk*, eds. V. J. Feron and M. C. Bosland, pp. 30–35. Wageningen, The Netherlands: Pudoc.

Kuper, C. F., Koornstra, P. J., Hameleers, D. M. H., Biewenga, J., Spit, B. J., Duijvestijn, A. M., Vriesman, P. J. C. B., and Sminia, T. 1992. The role of nasopharyngeal lymphoid tissue. *Immunol. Today* 13:122–128.

Macia, R. A., Short, B. G., and Kerns, W. D. 1993. Selective damage of the olfactory epithelium (OE) of the vomeronasal organ following intranasal instillation of growth hormone releasing peptide (GH-RP-6) in rats. *Toxicologist* 13:262.

Marks, S. C. 1973. Pathogenesis of osteopetrosis in the ia rat: Reduced bone resorption due to reduced osteoclast function. *Am. J. Anat.* 138:165–190.

Marshall, D. A., and Maruniak, J. A. 1986. Masera's organ responds to odorants. *Brain Res.* 366:329–332.

Méry, S., Larson, J. L., Butterworth, B. E., Wolf, D. C., Harden, R., and Morgan, K. T. 1994. Nasal toxicity of chloroform in male F344 rats and female B6C3F1 mice following one week inhalation exposure. *Toxicol. Appl. Pharmacol.*, in press.

Moe, H., and Bojsen-Moller, F. 1971. The fine structure of the lateral nasal gland (Steno's gland) of the rat. *J. Ultrastruct. Res.* 36:127–148.

Monticello, T. M., Morgan, K. T., and Uraih, L. 1990. Nonneoplastic nasal lesions in rats and mice. *Environ. Health Perspect.* 85:249–274.

Morgan, K. T. 1994. Nasal dosimetry, lesion distribution, and the toxicologic pathologist: A brief review. *Inhal. Toxicol.* 6(suppl):41–57.

Morgan, K. T., Kimbell, J. S., Monticello, T. M., Patra, A. L., and Fleishman, A. 1991. Studies of inspiratory airflow patterns in the nasal passages of the F344 rat and rhesus monkey using nasal molds: Relevance to formaldehyde toxicity. *Toxicol. Appl. Pharmacol.* 110:223–240.

Morse, C. C., Boyd, M., and Witschi, H. 1984. The effect of 3-methylfuran inhalation exposure on the rat nasal cavity. *Toxicology* 30:195–204.

Negus, V. 1958. *The comparative anatomy and physiology of the nose and paranasal sinuses*. London: Livingstone.

Peele, D. B., Allison, S. D., Bolon, B., Prah, J. D., Jensen, K. F., and Morgan, K. T. 1990. Functional deficits produced by 3-methylindole-induced olfactory mucosal damage revealed by a simple olfactory learning task. *Toxicol. Appl. Pharmacol.* 107:191–202.

Proctor, D. F., and Andersen, I., ed. 1982. *The nose, upper airway physiology and the atmospheric environment*. New York: Elsevier Biomedical Press.

Randall, H. W., Bogdanffy, M. S., and Morgan, K. T. 1987. Enzyme histochemistry of the rat nasal mucosa embedded in cold glycol methacrylate. *Am. J. Anat.* 179:10–17.

Reed, C. J. 1993. Drug metabolism in the nasal cavity: Relevance to toxicology. *Drug Metab. Rev.* 25:173–205.

Reznik, G. 1983. Comparative anatomy and histomorphology of the nasal and paranasal cavities in rodents. In *Nasal tumors in animals and man*, eds. G. Reznik and S. F. Stinson, pp. 35–44. Boca Raton, FL: CRC Press.

Rodolfo-Masera, D. T. 1943. Su l'esistenza di un particolare organo olfaltivo nel setto nasale della civia e di altri roditori. *Arch. Ital. Anat. Embriol.* 48:157–212.

Spit, B. J., Hendriksen, E. G. J., Bruijntjes, J. P., and Kuper, C. F. 1989. Nasal lymphoid tissue in the rat. *Cell Tissue Res.* 255:193–198.

Taylor, R. 1994. Brave new nose: Sniffing out human sexual chemistry. *J. NIH Res.* 6:47–51.

Uraih, L. C., and Maronpot, R. R. 1990. Normal histology of the nasal cavity and application of special techniques. *Environ. Health Perspect.* 85:187–208.

Uraih, L. C., Morgan, K. T., and Maronpot, R. R. 1990. Symposium on toxicologic pathology of the upper respiratory system. *Environ. Health Perspect.* 85:161–352.

Vaccarezza, O. L., Sepich, L. N., and Tramezzani, J. F. 1981. The vomeronasal organ of the rat. *J. Anat.* 132:167–185.

Warwick, R., and Williams, P. L., eds. 1973. *Gray's anatomy*. Philadelphia: W. B. Saunders.

Weiss, L., ed. 1983. *Histology, cell and tissue biology*. New York: Elsevier.

Wolf, J. L., and Bye, W. A. 1984. The membraneous epithelial (M) cell and the mucosal immune system. *Annu. Rev. Med.* 35:95–112.

HUMAN OLFACTORY EPITHELIUM: NORMAL PATTERNS AND TYPES OF LESIONS FOUND IN THE GENERAL POPULATION

Barbara R. Talamo, Wen-Hui Feng, Margo Stockmayer

Neuroscience Program, Departments of Neurology and Physiology, Tufts University School of Medicine, Boston, Massachusetts, USA

Immunocytochemical studies of tissue sections and whole-mount preparations have been used to analyze human olfactory epithelium in autopsies obtained from cases ranging from 2 days to 91 yr of age. Olfactory epithelium in humans is very uniform in the neonate and young child, but sensory areas and scattered neurons become interspersed with nonsensory areas in the adult. A few olfactory neurons may remain in areas that include respiratory cells, goblet cells, or secretory glands; conversion to squamous epithelium occurs in some cases. Islands of densely packed olfactory neurons may remain even when most of the region has been converted to respiratory epithelium, suggesting that neuronal stem cells cannot effectively migrate laterally to reconstitute the sensory epithelium. Deep pits lined with olfactory neurons are seen at all ages and have not been observed in other animals, except for monkeys. Their function is unknown. Large invaginations of the epithelium into the lamina propria are common in older individuals. In some cases, including Alzheimer's disease patients, misdirected neurites accumulate within the epithelium in masses, and these neurites, as well as some swollen and degenerate cells, show abnormal expression of neurofilament proteins and tau. This phenotype of ectopic neurite outgrowth and altered expression of cytoskeletal proteins may represent a necrotic neurodegenerative pathway that differs from the more frequent route of neuronal degeneration triggered by most damaging events, which may be through programmed cell death. Elucidation of the cellular pathways and signals that regulate differentiation of the assorted cell types that repopulate the epithelium will help identify the targets of various xenobiotic and pathogenic agents and the mechanisms that lead to an altered epithelial phenotype.

PROPERTIES OF NORMAL EPITHELIUM

Odorant detection is carried out by neurons in the most dorsal and posterior region of the human nose, which is lined with sensory epithelium. Inhaled air, of course, carries not only chemicals detected as odorants, but also a variety of xenobiotic agents, microorganisms, and particulate matter. While the epithelium has evolved to deal with this normal condition, various lesions develop with time that probably reflect the inability of endogenous mechanisms to deal with these challenges. Analysis of the normal structure of the human olfactory epithelium using immunocytochemical probes indicates that aberrant structures sometimes arise, disrupting the normal organization. These studies describe the appearance of young healthy

Address correspondence to Dr. Barbara R. Talamo, Neuroscience Laboratories, Tufts University School of Medicine, 136 Harrison Ave., Boston, MA 02111, USA.

human olfactory epithelium and also the variety of structures that has been observed in many clinical specimens obtained at autopsy.

The olfactory epithelium is specialized to detect volatile chemicals, and both the cellular composition and organization of the epithelial layers maximize the accessibility to air of neuronal structures bearing odorant detectors. Four types of cells have been generally recognized in the sensory epithelium. These are the olfactory receptor neurons themselves, nonneuronal support or sustentacular cells, and two types of basal cells that lie on or just above the basement membrane of the epithelium. Sustentacular cell bodies occupy the outermost layer of the epithelium; neuronal cell bodies are situated in the upper and middle layers of the epithelium, and the precursor or stem cells lie in the deeper layers. Beneath the basement membrane is the connective tissue-rich lamina propria, which contains blood vessels, olfactory axon bundles, trigeminal and autonomic nerve fibers, and Bowman's glands.

The various cell types are similar in most vertebrates. Sustentacular cells have a microvillar surface in contact with the mucous layer and are filled with vesicular profiles. The function of these cells is largely unknown, although a role as a support cell for neurons or a source of xenobiotic metabolizing enzymes has been suggested (Harkema, 1991). The most mature neurons lie just below the sustentacular cells, with immature neurons below them. In all species thus far examined, neurogenesis continues into adulthood. Neurons arise from basal stem cells within the epithelium. There are two types of basal cells. Flattened horizontal basal cells are in direct contact with the basement membrane, and globose or "light" basal cells lie directly above them. The direct, mitotically active precursors of the olfactory receptor neurons are thought to be the globose basal cells (Levey et al., 1991). In the human nose, as in virtually all other mammals, the receptor neuron sends a single dendrite to the epithelial surface, where it swells to form a knob from which modified cilia emerge. Chemical detectors, presumably receptor protein molecules, are located in the plasma membrane of these specialized nonmotile cilia that float at the tissue/mucus/air interface. At the opposite pole of the neuronal cell body, a fine, unmyelinated axon emerges, traverses the basement membrane, and fasciculates with other olfactory axons to form large axon bundles in the lamina propria. The unbranched axons are ensheathed by glial cells that have characteristics of both peripheral and central glia. The axons cross into the cranial cavity through openings in the cribriform plate of the ethmoid bone and travel along the surface of the olfactory bulb, eventually synapsing with mitral cell dendrites and other neuronal targets in the glomeruli of the bulb. In a variety of species, including humans, smaller urn-shaped microvillar cells that appear to be a second type of sensory neuron have been described in the upper region of the olfactory epithelium (Moran et al., 1982; Morrison & Costanzo, 1990; Rowley et al., 1989).

This anatomical organization of the epithelium optimizes direct contact between odorants and receptor molecules in the surface membrane of the cilia by making the receptor elements accessible to volatile odorants, but it also means that these structures are exposed to detrimental airborne substances, including chemicals and viral or bacterial pathogens. Nowhere else in the body are neurons so vulnerable to external agents. The mucus layer secreted by glands and by secretory cells in the epithelium forms a barrier and provides protection from dryness and heat, as well as acting as a solvent and carrier for antibodies, carrier proteins, and enzymes that may chemically modify dissolved molecules. A wealth of enzymes and binding proteins participate in the conversion of compounds from active to inactive states, and vice versa. New compounds formed by these processes may be benign or harmful. Many different conversion enzymes have been identified that are capable of detoxifying potentially damaging reactive species, including oxidants and other toxic or pathogenic molecules. Some of these enzymes, including those from the cytochrome P-450 class (Dahl & Hadley, 1991), are highly enriched in selected areas of the nasal epithelium. A network of scavenging and regulatory cells secretes antibodies, cytokines, enzymes, and carrier proteins that mediate challenges to epithelial integrity.

In fact, the sensory neurons appear to be vulnerable to agents in the environment, and evidence gathered from studies of various vertebrate and primate models indicates that neuronal death is a feature of the normal olfactory epithelium. Dying receptor cells are replaced by new sensory neurons; this continuous renewal is an unusual feature of the olfactory epithelium not commonly found elsewhere in the adult nervous system. The lifetime of olfactory receptor neurons has been estimated at 30–90 days in mice (Graziadei & Monti-Graziadei, 1979; Mackay-Sim & Kittel, 1991a, 1991b). Experimental manipulations dramatically demonstrate the capacity for neurogenesis. Olfactory nerve transection or olfactory bulb removal causes massive neuronal death in the epithelium, followed by generation of a new population of olfactory receptor neurons. This phenomenon has been replicated in many species, including primates (Graziadei et al., 1980). However, the reason for cell turnover in the absence of damage to the olfactory receptor cell axon is unclear. It has been postulated that receptor neuron turnover is due to rapid cellular aging (Graziadei & Monti-Graziadei, 1979), or that differentiated receptor neurons die if they fail to make a synapse in the olfactory bulb (Hinds et al., 1984; Farbman, 1990, Schwob et al., 1992). Under normal circumstances, the only other cause of neuronal cell death may be direct damage to receptor neurons caused by trauma (at the surface of the epithelium or to the axon through head injury and shearing at the cribriform plate) or by disease or exposure to pathogens and toxins in the environment.

In the human nose, the sensory epithelium described earlier is often found to be intermixed with non-sensory epithelium (Morrison & Costanzo, 1990; Paik et al., 1992) composed of different cell types, including respiratory cells with motile cilia and nonciliated cells and goblet cells (Harkema, 1991). The presence of both receptor neurons and epithelium containing different varieties of cells within the olfactory region of the human nose suggests that the normal pathways for generation of sensory epithelium are altered under certain conditions, as yet not understood. Other specialized epithelia have been identified in the normal lining of different regions of the nose in both animals and in normal human tissue. Squamous epithelium, transitional epithelium (a nonciliated cuboidal epithelium), respiratory epithelium containing ciliated as well as secretory cells, and olfactory epithelium have been characterized (Harkema, 1991). Identification of the precursor cells for the wide variety of cell types that populate the various epithelial regions and identification of the factors that direct appropriate differentiation are challenging and important problems. While the basal cells that lie along the basement membrane of these characteristic epithelia all appear similar, it is apparent that differences in the stem cells or in local signals in each region direct the course of differentiation to produce the particular cell types specific to those epithelia. Knowledge of these lineages, signals, and mechanisms will help us to identify the targets of toxic, traumatic, and pathogenic events that alter the normal composition and function of the epithelium and lead to the appearance of metaplastic tissue within a formerly homogeneous area.

The purpose of the study reported here was to characterize the sensory epithelium in human subjects of various ages with antibodies that identify particular cell types, using both conventional tissue sections and whole-mount procedures to survey the extent of tissue variability.

METHODS

All tissue was obtained at autopsy from human subjects. Cases were obtained with permission through the Neuropathology Department at New England Medical Center, from referrals, and by courtesy of Dr. John Growden and Dr. Tessa Hedley-White at Massachusetts General Hospital and Dr. Jim Hamos at the University of Massachusetts Medical Center. The cribriform plate and associated nasal tissues, including as much of the septal and lateral walls of the superior ethmoturbinate as possible, were removed en bloc and immersion-fixed in 2% paraformaldehyde, 10% saturated picric acid, in the cold, for 4 h. (Monkeys were perfused with 4% paraformaldehyde and the nasal block was then placed in paraformaldehyde/picric acid and treated similarly to human tissue.) Tissue was rinsed in phosphate buffer and then carefully dissected from the underlying bony and cartilaginous structures. The tissue was then cryoprotected in 30% sucrose and frozen. Two types of preparation were examined by immunocytochemical methods: (1) whole-mount preparations in which the entire tissue was treated, and (2) frozen sections. Staining was carried out with antibodies that were selected to identify particular cell types. Three antibodies are general neuronal markers that stain olfactory neurons as well as other types of nerve fibers that innervate the mucosa and structures in the lamina propria.

Both immature and highly differentiated neurons are labeled with antibodies to neuron-specific enolase (NSE; rabbit polyclonal from Polysciences), microtubule-associated protein 5 or 1B (MAP 5; mouse monoclonal from Sigma), and neuron-specific tubulin (Class III, β; mouse monoclonal Tu-J1, gift from A. Frankfurter; Moody et al., 1989). Antibody to olfactory marker protein (OMP; a goat polyclonal, gift of F. Margolis) was used because it appears to be selective only for mature, differentiated olfactory neurons (Farbman & Margolis, 1980). Other antibodies were selected because they stain abnormal neurons that appear to be undergoing transdifferentiation prior to or accompanying cell degeneration, as previously described for Alzheimer's disease patients (Talamo et al., 1989a). In this study, antibody to neurofilament proteins (Labsystems) was used. Antibody binding was visualized using standard methods with appropriate secondaries directly coupled to fluorescent tags or biotinylated secondary antibodies plus streptavidin-Texas red (BRL) or -Bodipy (Molecular Probes). In some cases the peroxidase-antiperoxidase method was used, with diaminobenzidine as a substrate. When nuclear staining was carried out, sections either were dipped in 50 μm bisbenzamide (Hoechst 33258, Aldrich Chemical) for fluorescence microscopy or were counterstained with hematoxylin for light microscopy. Tissue was dehydrated and mounted in Permount or Histoclear. The whole-mount methodology is described elsewhere (Feng et al., 1991; Feng and Talamo, unpublished data). Alcian blue staining (Clark, 1981) was combined with immunocytochemistry to visualize secretory cells.

RESULTS

Olfactory Epithelium in the Child

Whole-mount preparations of human olfactory tissue obtained from newborns or children were stained with antibodies to MAP 5 or Tu-J1 to tubulin to reveal neurons and then viewed from the epithelial surface. A continuous sheet of regularly spaced dendritic knobs of olfactory neurons (interspersed with unstained sustentacular cells) is seen within the olfactory region (Figure 1). Small circular areas (diameter 15–20 μm) that are free of dendritic knobs are easily seen at high magnification. Examination of sections of olfactory tissue cut perpendicular to the surface shows that these dendrite-free areas correspond to regions where the ducts of secretory glands cross the epithelium and emerge at the surface (see Figure 13, color plate). The sensory epithelium is thick and well organized, containing several layers of mature olfactory sensory neurons. Basal cells lie deep, just above the basement membrane; immature and mature neurons occupy the intermediate layers, and oval nuclei of sustentacular cells (which are not stained with any of the neuronal antibodies) are superficially located near the epithelial surface (see Figure 13). Antibody to OMP, a marker of mature neurons, always stains fewer neurons than antibodies to more generalized

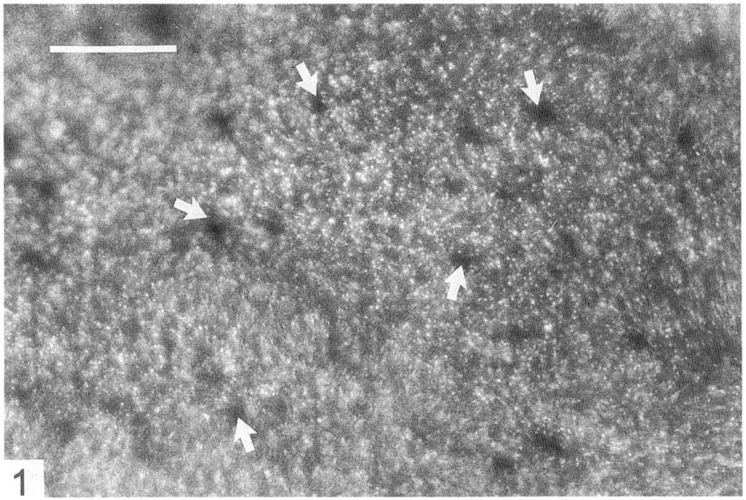

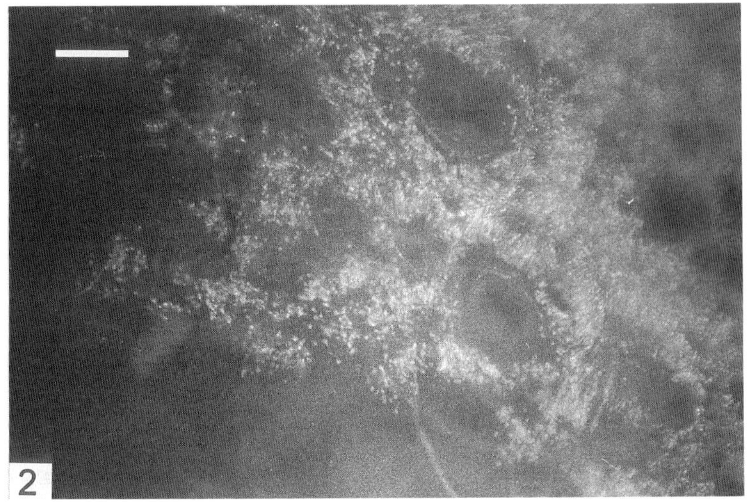

FIGURE 1. Uniform distribution of brightly stained knobs of olfactory receptor neurons across the surface of human olfactory epithelium. Whole mount of nasal epithelium stained with Tu-J1. Small unstained circular regions surround the openings of Bowman's gland ducts at the surface (arrows). Specimen A90-83; 2 mo/M, congenital heart disease. Scale bar = 50 μm. **FIGURE 2.** Olfactory and respiratory epithelium border. Whole mount of nasal epithelium stained with Tu-J1. Near the transition, some of the duct openings are surrounded by enlarged neuron-free regions. Dendritic knobs are less densely packed near the transition to respiratory epithelium. Specimen A90-83, 2 mo/M, congenital heart disease. Scale bar = 100 μm.

neuronal markers such as MAP 5 and Tu-J1, which stain both immature and mature neurons (not shown). In some cases, superficially located urn-shaped neurons are observed (see Figure 19A, arrow), probably corresponding to microvillar neurons recently described in humans. Horizontal basal cells may be identified by immunostaining with cytokeratin antibodies (data not shown; Yamagishi et al., 1989).

Olfactory neurons are all contained within the sensory area, which is clearly separated from the adjacent respiratory epithelium where there are no concentrated collections of sensory cells (Figure 2). At the interface of sensory and respiratory epithelium, there are localized nonsensory areas within the sensory epithelium, typically apparent as relatively large neuron-free "circles" of respiratory epithelium (diameter about 65–135 μm). These regions usually surround a Bowman's gland duct opening, but the neuron-free region around the opening is enlarged compared to that around ducts within predominantly olfactory areas. At high magnification, it can also be appreciated that the density of sensory cell knobs decreases near the respiratory border, and occasional neurons are actually located among respiratory cells, as also may be seen in cross section in Figure 3 (see also Yamagishi & Nakano, 1992; Morrison & Costanzo, 1990).

Another feature of the olfactory epithelium that appears to be unique to primates (Feng et al., 1992), are the large round or oval openings (diameter 50–70 μm) on the epithelial surface that are closely rimmed with olfactory neuron dendrites (Figure 4). These openings or pits extend into the lamina propria, giving rise to deep, narrow pockets lined completely with sensory epithelium (see Figure 14). At the surface, these are easily distinguished from the much narrower openings of Bowman's glands, which are not closely rimmed with ciliated receptor neurons. These neuron-lined crypts, which are detectable at all ages and are variable in number, tend to be located anteriorally and dorsally on both septal and lateral wall surfaces. In this regard, rhesus monkey (*Macaca mulatta*) olfactory tissue appears similar to human tissue, showing crypts, areas with a reduced number of neuronal layers (not shown) (Talamo et al., 1991b; Feng et al., 1992), and patches of respiratory metaplasia within the olfactory epithelium (discussed later) (Figure 5).

Mature Human Olfactory Epithelium

In older human subjects, the distribution of dendritic knobs across the surface (and of olfactory receptor neuron somata within the epithelium) becomes less regular, with the epithelium becoming thinner in some regions. In cross section (Figure 6), areas with fewer layers of receptor neurons are seen adjacent to many-layered regions, and fewer dendritic knobs are visible at the epithelial surface of whole-mount preparations (Figure 7, high magnification). Areas containing no neurons at all are interspersed among regions that have normal complements of sensory cells, in an erratic and highly variable pattern (Figure 8 whole mount and Figure 9 section).

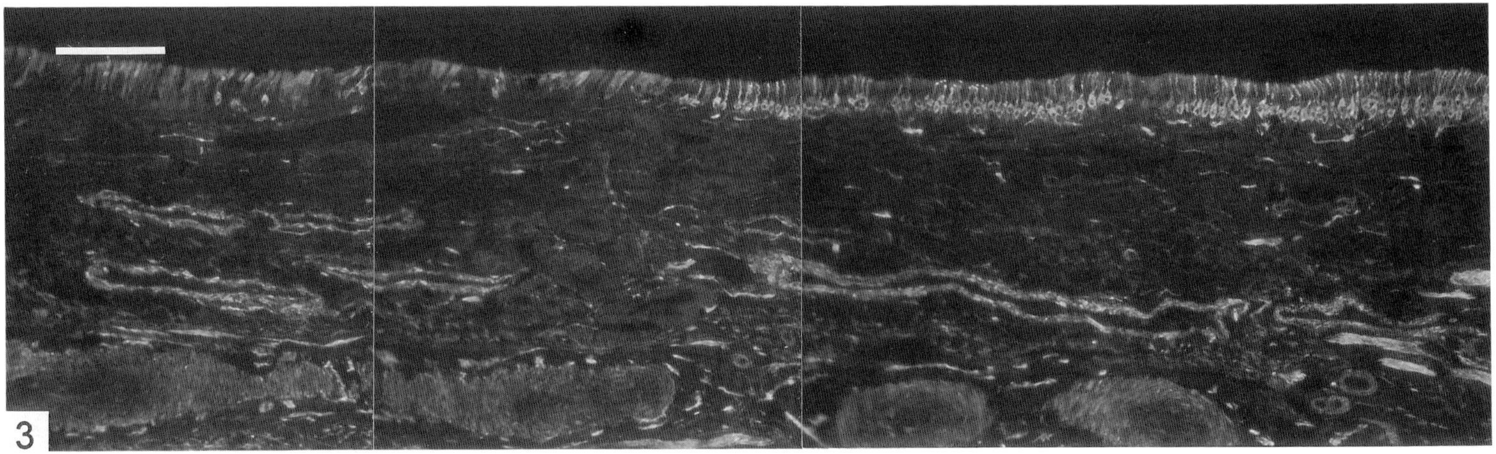

FIGURE 3. Olfactory/respiratory epithelium border. Cross section stained with Tu-J1. Olfactory receptor neuron density is reduced gradually at transition to respiratory epithelium. Specimen A90-88. 2 days/F, asphyxiation. Scale bar = 100 μm.

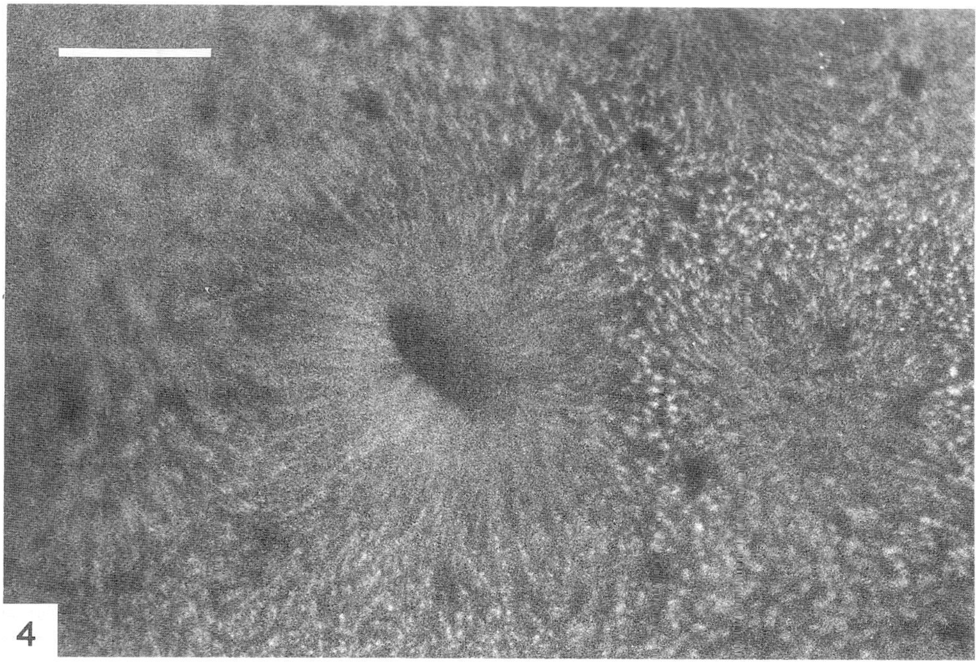

FIGURE 4. Olfactory pit. Whole mount of olfactory epithelium stained with Tu-J1. An enlarged opening of a pit rimmed with olfactory neuron dendrites is seen, as well as smaller dark circular areas that correspond to duct openings. Specimen A91-7, 2 days/M, congenital renal malformation. Scale bar = 100 μm.

The patchy distribution of olfactory and nonsensory or respiratory epithelium within the sensory region in humans also has been noted by others (Morrison & Costanzo, 1990, 1992; Nakashima et al., 1984, 1985, 1991; Paik et al., 1992; Doty & Snow, 1988; Naessen, 1971). Epithelial heterogeneity may be seen in both young and old, but is more prevalent in older subjects. Loss of sensory epithelium has been observed even in the very young (9 mo and 5 yr of age) (Figure 10). Islands of olfactory epithelium may remain within large areas of nonsensory epithelium (Figure 11, whole mount), while within other regions of the same specimen the olfactory epithelium appears normal (cf. with Figure 8). Loss of olfactory epithelium is common on the roof of the nasal cavity, as noted previously by others (Paik et al., 1992; Nakashima et al., 1984) (e.g. see Figure 17, color plate).

Types of Olfactory Epithelial Variation in Mature Tissue A variety of cell types is observed in nonsensory epithelium within degenerated human olfactory epithelium. Epithelium containing only one or a few layers of receptor somata is not uncommon (Figures 6 and 9). Occasional neurons, sometimes swollen and misshapen, are observed in these areas, and neuronal cell bodies are apparent even in epithelial layers that are only two cells thick (see Figure 15). Pits in these areas may retain some receptor neu-

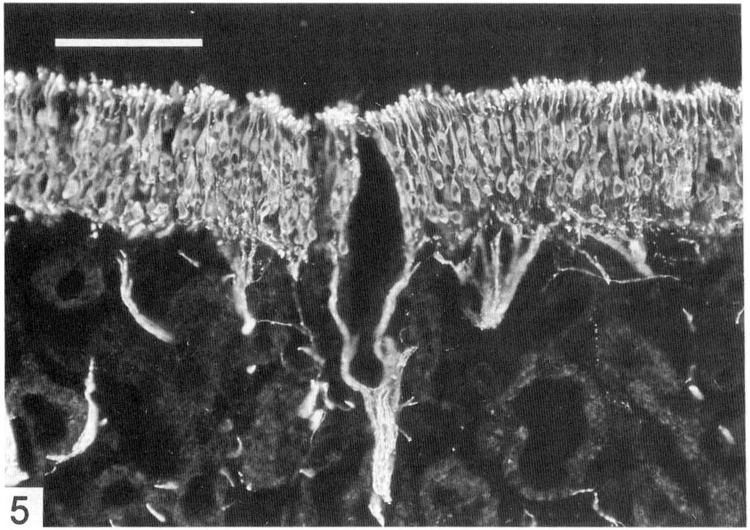

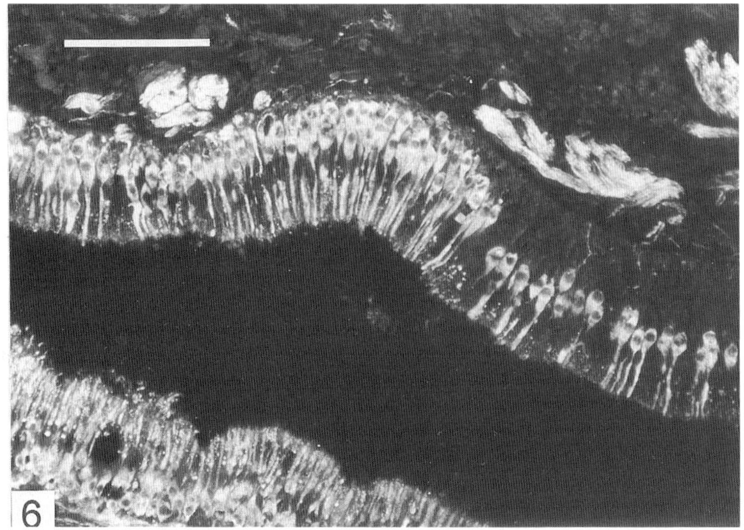

FIGURE 5. Monkey olfactory epithelium. Cross section stained with Tu-J1. A duct opening is visible at the surface. Olfactory axons bundle after they cross the basement membrane. *Macaca mulatta*, 24 yr old. Scale bar = 100 μm. **FIGURE 6.** Reduced number of olfactory neuron layers in older specimens. Cross section stained with antibody to MAP 5. Regions of olfactory epithelium in which olfactory neurons are stacked several cells deep lie adjacent to areas where receptor neurons are only one cell body thick or where they disappear altogether. Bright dots near the epithelial surface are autofluorescent pigment. Specimen A87-54, 45 yr/M, external hemorrhage. Scale bar = 100 μm.

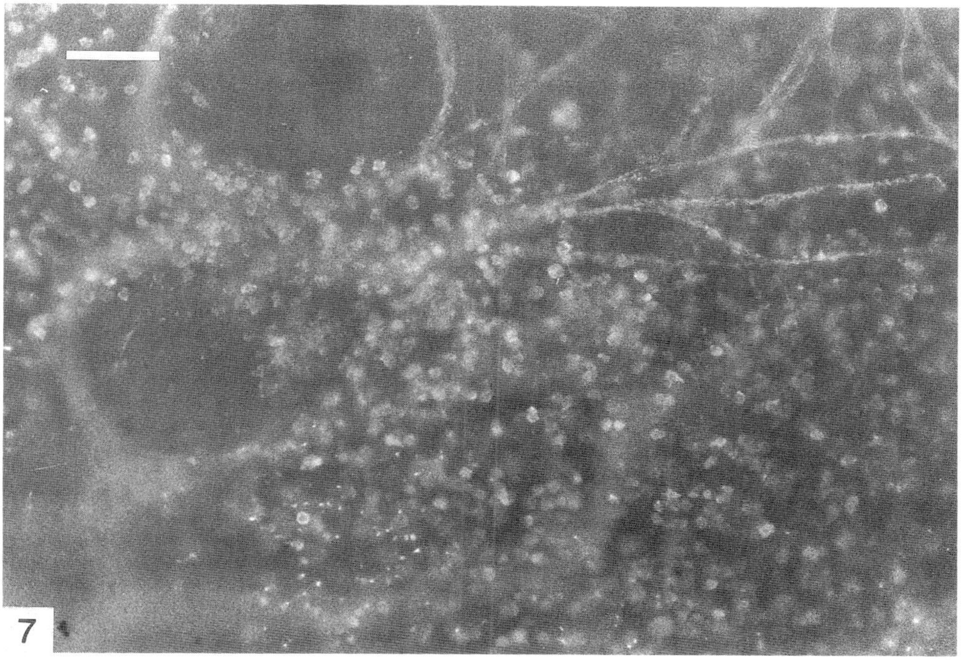

FIGURE 7. Sparse dendritic knobs and cell bodies below the surface. Whole mount stained with MAP 5. Specimen A90-26, 63 yr/M, alcoholic cirrhosis. Scale bar = 100 μm.

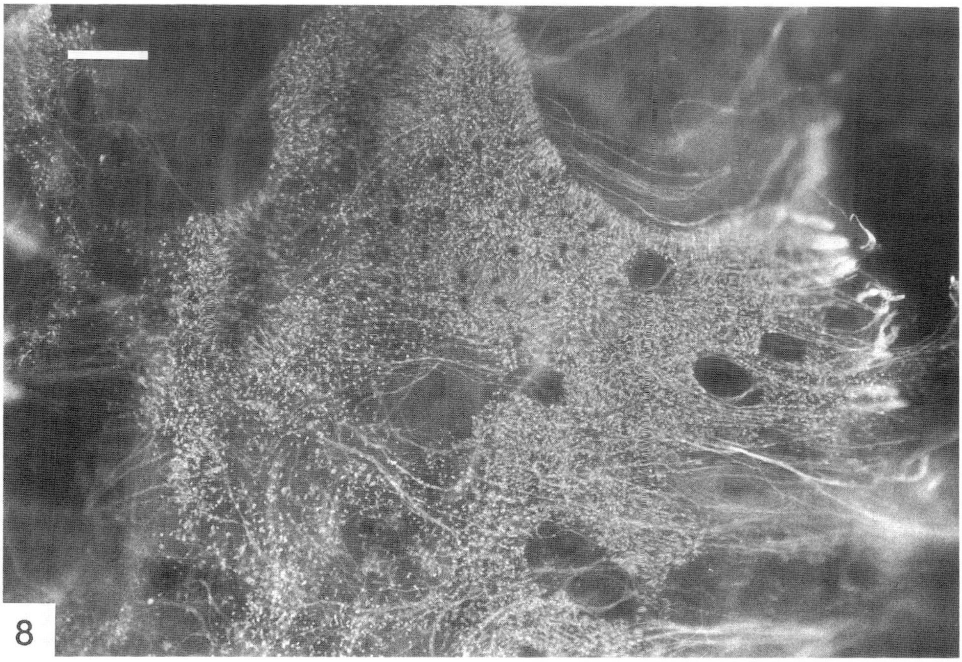

FIGURE 8. Intermixed sensory and non-sensory epithelium. Whole mount stained with Tu-J1. Relatively normal olfactory areas are intermixed with areas in which olfactory dendritic knobs are sparse or totally absent. Large nonsensory regions surround the olfactory area. Specimen A91-32, 64 yr/F, pneumonia, progressive supranuclear palsy. Scale bar = 100 μm.

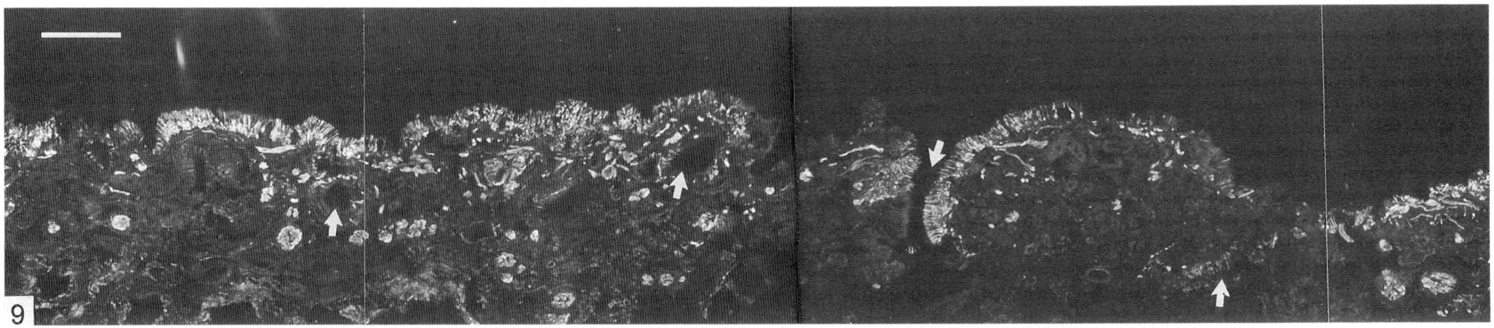

FIGURE 9. Intermixed olfactory and respiratory areas. Cross section stained with Tu-J1. Several crypts containing olfactory neurons are present in this montage (arrows). Specimen A91-25, 73 yr/F, cancer, amyotropic lateral sclerosis. Scale bar = 100 μm.

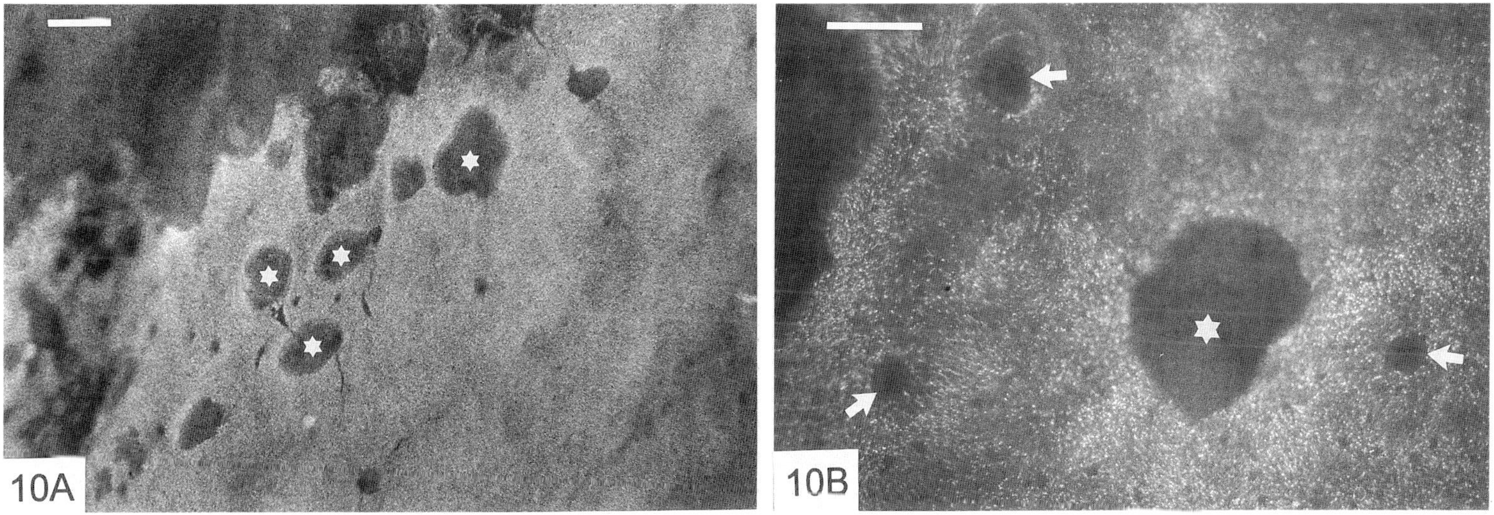

FIGURE 10. (A) Sensory epithelium loss in child. Whole mount stained with Tu-J1. Dendritic knobs are densely packed in some regions. Large metaplastic regions devoid of sensory neurons are visible. Specimen A 90-107, 9 yr/M; hemolytic anemia. Scale bar = 100 μm. (B) Same specimen as (A), higher magnification. The large nonsensory area is marked with a star; enlarged neuron-free regions around duct openings are indicated by arrows. Outside the large nonsensory area, it can be seen that dendritic knobs are less densely packed in some regions. Scale bar = 50 μm.

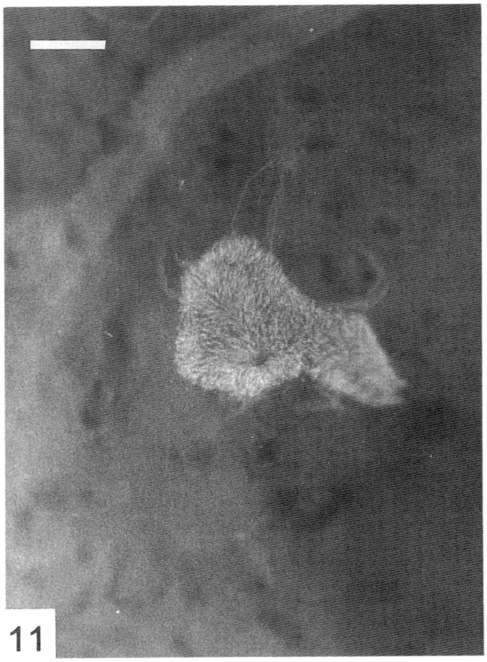

FIGURE 11. Isolated normal-looking olfactory patch with very large nonsensory surround. Whole mount stained with Tu-J1. Specimen A 91-32, 64 yr/F, pneumonia, progressive supranuclear palsy. Scale bar = 100 μm.

rons, even though most of the epithelial lining has become nonsensory. Less frequently, squamous epithelium also can be seen (Figure 12). Disrupted areas that contain a mixture of cells are frequent and cannot be clearly classified as to epithelial type. Tall, ciliated, respiratory epithelium may be observed within the sensory region (Figure 16, color plate). Alcian blue stains secretory cells and often reveals increased numbers of glandular cells and goblet cells (Figure 16) (Talamo et al., 1992).

Another striking feature is the invagination of the epithelium into broad lacunae that extend into the lamina propria (Figure 17), as also reported by Nakashima et al. (1991) and Paik et al. (1992). We have found that these lacunae frequently contain glandular duct openings (Feng et al., 1992) and Alcian blue-stained secretory cells, and may contain olfactory receptor neurons as well, a point not previously noted (see Figure 17).

Other Degenerate Structures A different cellular morphology and degenerative process has been observed in some human tissue; this pathology is particularly prominent in autopsy tissue taken from patients with confirmed Alzheimer's disease and also may be seen in some nondemented patients with less well-defined clinical conditions. Antibodies to cytoskeletal proteins that have been identified in dystrophic neurites, and neurofibril-

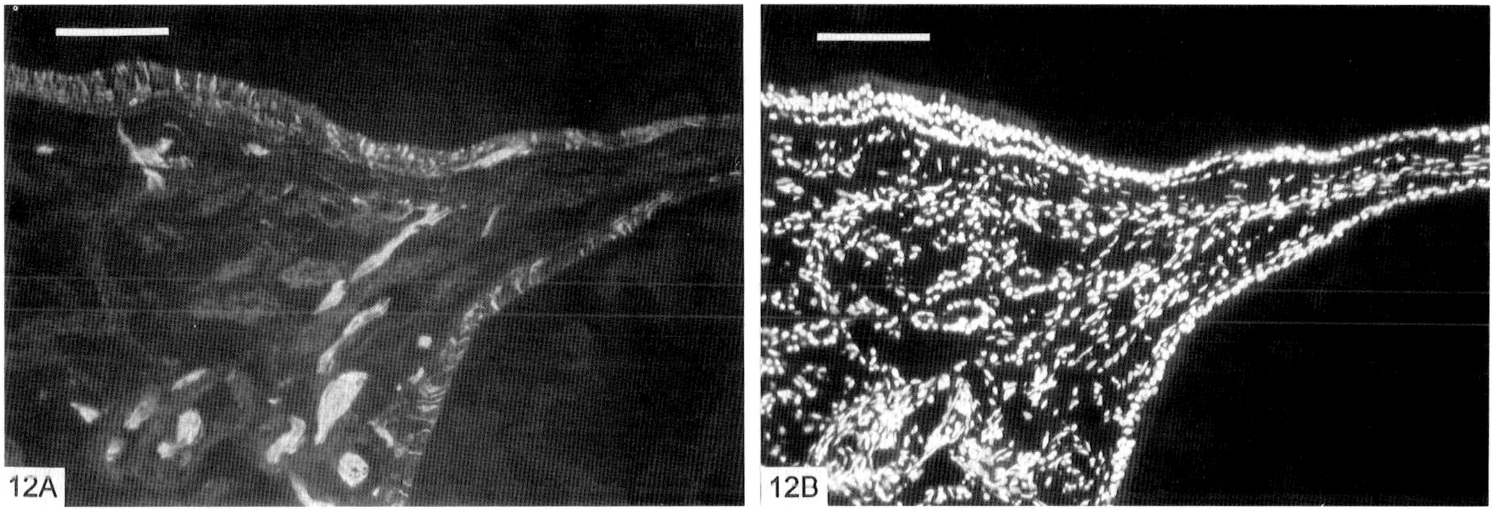

FIGURE 12. (A) Squamous epithelium adjacent to degenerate olfactory epithelium with scattered neurons. Cross section stained with Tu-J1. Specimen A90-50, 74 yr/M, cancer. Scale bar = 100 μm. (B) Same section photographed with filter to show nuclear fluorescence, illustrating thinning of epithelium to single celled squamous layer.

lary tangles of Alzheimer's brain stain many abnormal structures in the epithelium. These antibodies do not normally stain olfactory receptor neurons (Talamo et al., 1989a, 1989b; Schwob et al., 1986; Vollrath et al., 1985, but see Takahashi et al., 1984; Ophir, 1987; Bruch & Carr, 1991; Burton & Wentz, 1992; Yamagishi et al., 1989), but in some tissues they stain a subset of neurons in the epithelium and a subset of axons in bundles in the lamina propria. Abnormal structures can be observed even with antibodies to proteins found in all neurons (Figures 18 and 19). For example, with antibodies to neuron-specific enolase, some neurons appear relatively normal even in Alzheimer's tissue, with dendrites reaching to the epithelial surface and axons extending down through the basement membrane (Figure 19A). However, other neurons are swollen and misshapen, with stunted and distended dendrites (Figure 19B). Cytoskeletal antibodies to neurofilament protein and antibodies specific to medium-subunit neurofilament protein, phosphorylated heavy subunit neurofilament protein, tau, and the antibody ALZ50 (to heavily phosphorylated tau) detect ectopic nerve fibers in the epithelium as the most prominent feature (Talamo et al., 1989a) (not shown). Neurites accumulate in the epithelium as single fibers or in masses (Figure 20A) in which they mat or fasciculate to form a meshwork or basket-like structure. A subset of neurofilament-stained nerve fibers in the lamina propria also can be observed. These should not be confused with bundles of trigeminal or autonomic nerve fibers that normally stain for neurofilament and tau, which can be seen in deep layers of the lamina propria in some regions. Swollen cell bodies also may be stained (Figure 20B), while normal cell bodies in the same region are unstained. The abnormal neurites stain with antibodies to neuronal markers as well, such as neuron-specific tubulin, and may be located near relatively normal looking olfactory neurons or in areas rich in secretory cells (Figure 18). Neuritic structures of this type and staining for neurofilament and tau are not normally observed in young or healthy-looking olfactory epithelium and have never been observed in rodents or in young or old monkeys.

DISCUSSION

Human Epithelial Structure

The use of immunocytochemical techniques in whole-mount preparations, as well as in conventional cross sections, gives a broad picture of the distribution of normal sensory epithelium in the human nose. It also reveals the complexity of abnormal and altered features that appear in surveys of human olfactory tissue. The disappearance of neurons from epithelium that otherwise looks "sensory" and the occurrence of occasional neuronal cells or of fibers in an abnormal location are difficult to detect using conventional staining techniques, and the extent of alteration is not readily appreciated in the limited sample available for electron microscope (EM) examination.

In the newborn human, the epithelial types are well organized and con-

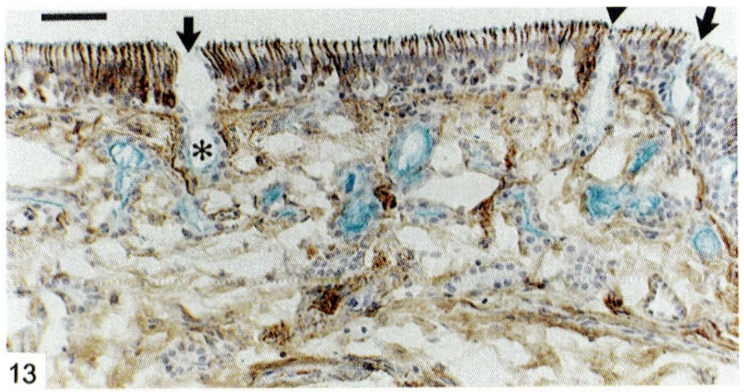

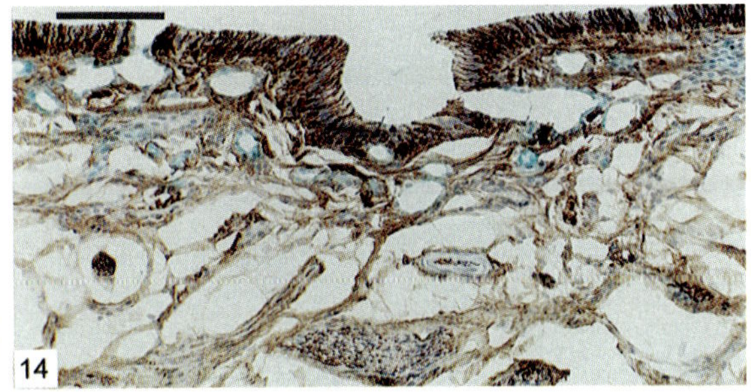

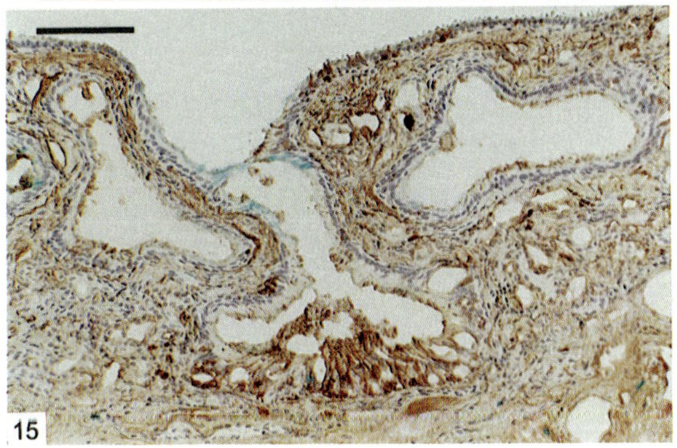

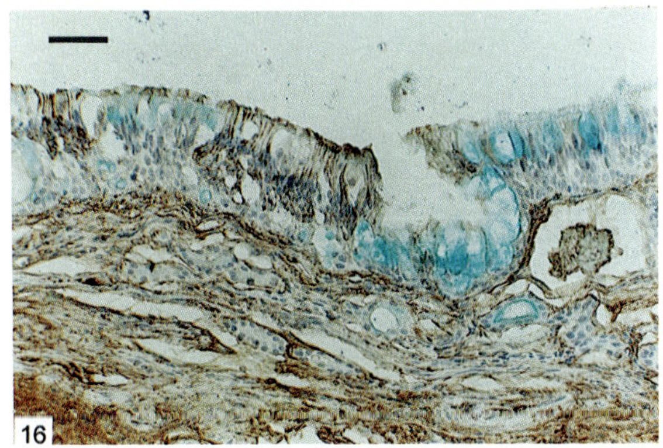

FIGURE 13. Cross sectional view of uniform olfactory epithelium interrupted with duct openings of Bowman's glands. Stained with Tu-J1, visualized with PAP and counterstained with Alcian blue and hematoxylin. Duct openings are indicated by arrows, an associated gland by asterisk. Specimen A91-7, 2 d/M, congenital renal malformation. Scale bar = 100 μm. **FIGURE 14.** Olfactory pit, cross section from same specimen as Figure 5A. Stained with Tu-J1 and Alcian blue. A shallow crypt, fully lined with olfactory receptor neurons is shown. A Bowmans's gland opens into the crypt. Another ductal opening of a gland to the surface of the epithelium is also visible. Scale bar = 100 μm. **FIGURE 15.** Degenerate olfactory epithelium. Cross section stained with Tu-J1 and Alcian blue. A thin sensory epithelium contains scattered olfactory neurons. A crypt is lined primarily with nonsensory epithelium, but contains olfactory neurons at its base. Specimen MGH91-327 85 yr/F, cancer. Scale bar = 100 μm. **FIGURE 16.** Intraepithelial gland in area of intermixed olfactory and respiratory epithelium. Cross section stained with Tu-J1 and Alcian blue. Specimen UM90-24, 75 yr/F, Alzheimer's disease, pneumonia. Scale bar = 100 μm.

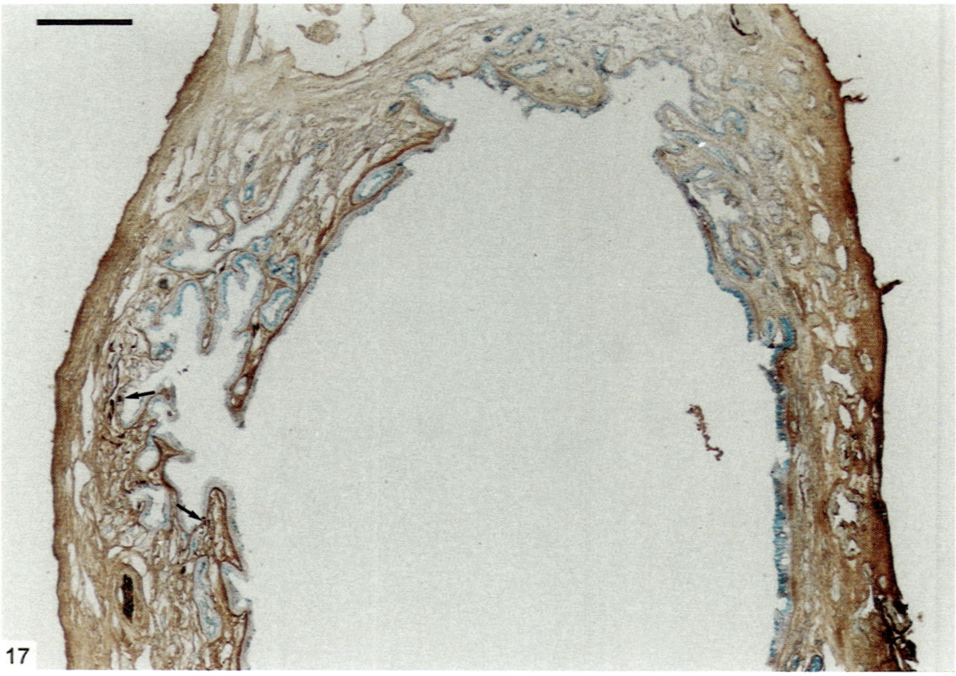

17

FIGURE 17. Large lacunae near roof of nasal cavity. Cross section stained with Tu-J1 and Alcian blue, low power. A few olfactory receptor neurons remain within lacunae (arrows). Most of the olfactory epithelium has been replaced with nonsensory epithelium. Many secretory cells are present. Specimen A 90-75, 67 yr/M, upper gastrointestinal hemorrhage. Scale bar = 100 μm.

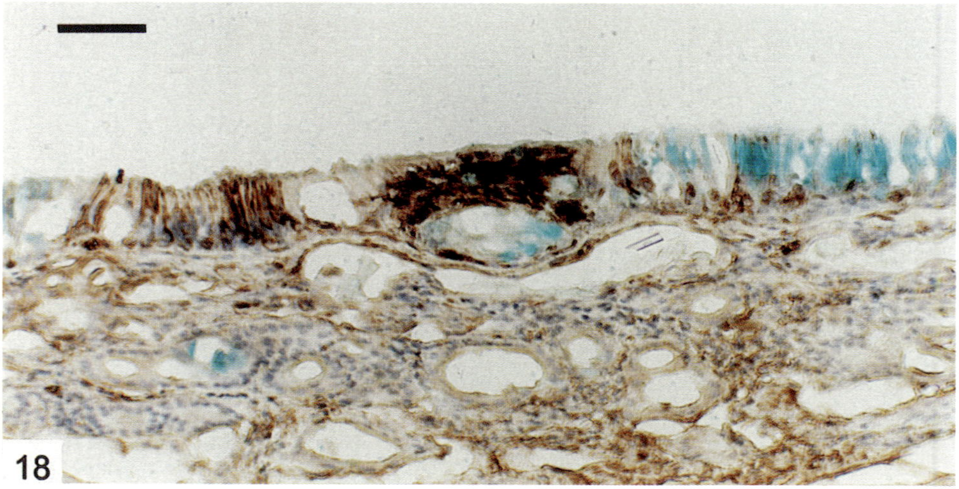

18

FIGURE 18. Relationship of normal looking neurons to secretory cells and neuritic mass. Cross section of olfactory epithelium from Alzheimer's patient, stained with Tu-J1 and Alcian blue. Specimen UM90-24, 75 yr/F, Alzheimer's disease, pneumonia. Scale bar = 100 μm.

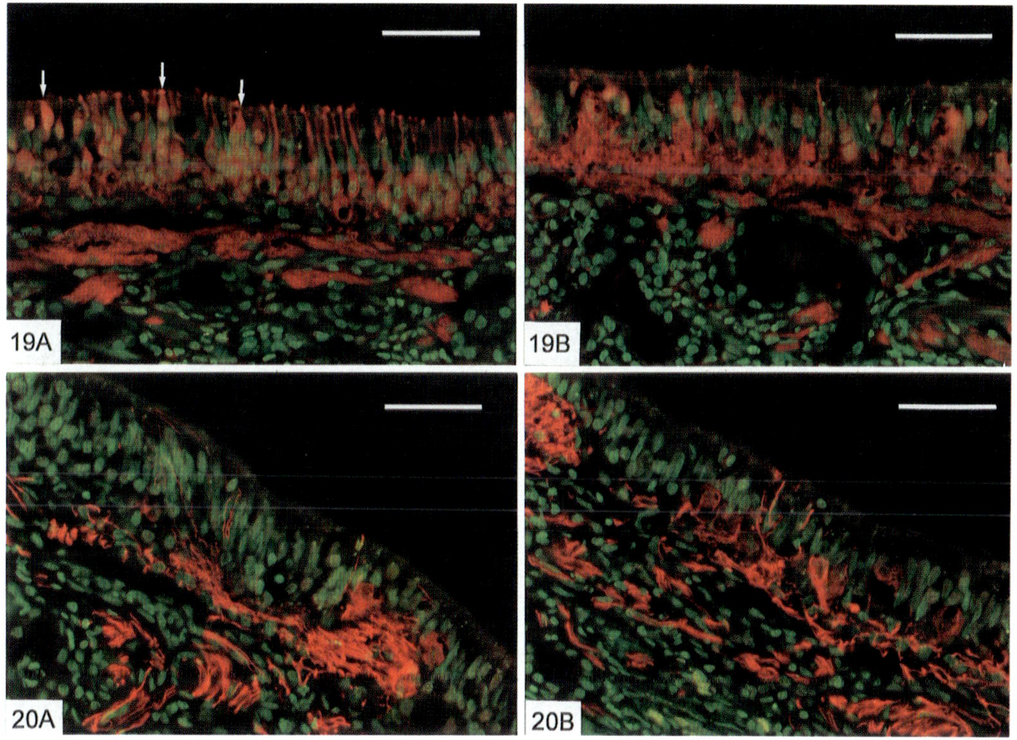

FIGURE 19. (A) Relatively normal area of olfactory epithelium in Alzheimer's patient. Cross section stained with antibody to NSE (red) and counterstained with DAPI (green). Double exposure. Oval nuclei of sustentacular cells are seen nearest the surface. Olfactory neurons (red) mostly lie deeper, but some near the surface are urn-shaped (arrows) and may correspond to microvillar-type neurons. Flattened nuclei along the basement membrane and round nuclei just above them correspond to horizontal basal and globose basal cells. Specimen A 87-30, 91 yr/F; Alzheimer's disease. Scale bar = 50 μm. (B) Abnormal region of the epithelium in the same patient as in (A). Cross section stained for NSE, counterstained with DAPI. Double exposure. This area of the section shows distorted neurons with distended dendrites. Scale bar = 50 μm. **FIGURE 20.** (A) Pathological neuritic structures stained for neurofilament. Cross section stained with antibody to neurofilament (Labsystems), showing single fibers and matted neurites in the epithelium. Specimen A87-30, 91 yr/F, Alzheimer's disease. Scale bar = 50 μm. (B) Swollen cell bodies filled with neurofilament protein. Another region from same patient as in (A). Scale bar = 50 μm.

fined to distinct regions, with sharp boundaries. Olfactory epithelium is contiguous and separated from respiratory epithelium. In older subjects, there is an increased intermixing of epithelial types; patches of respiratory epithelium are found within the olfactory sheet (also Naessen, 1971; Nakashima et al., 1984; Paik et al., 1992). At later times, there may be a good deal of microheterogeneity that is appreciated in immunocytochemical studies. The uniformity of epithelial organization and of cell types gives way to a juxtaposition of cells that are not normally found in close proximity to one another. In these circumstances, epithelial patches may be difficult to classify as either purely olfactory, respiratory, or transitional because cell types are intermixed.

It has been suggested that these changes may be due to the effects of age or to the effects of exposure to viral, bacterial, or environmental agents or to trauma (e.g., Naessen, 1971), but in most cases it is difficult to be certain of the cause due to the inadequacy of the clinical histories that are available. Efforts to classify the varieties of epithelial types that are observed in biopsy samples (Yamagishi & Nakano, 1992) are limited by difficulties in identifying particular cell types; it then becomes difficult, if not impossible, to identify relatively undifferentiated or transdifferentiated cells as elements of squamous, transitional, respiratory, or olfactory epithelium. It is therefore important to characterize particular types of cells to gain insight into the kinds of damage and transformation that may have occurred.

Sensory Cell Loss

In humans, the reasons for loss of olfactory neurons and for the immense variability across the epithelium of a single subject and among subjects are unknown. The presence of islands of apparently normal tissue in a sea of nonsensory epithelium and the presence of large, well-defined nonsensory areas within olfactory epithelium suggest that events have occurred that have altered the ability of stem cells to repopulate the area in a normal fashion. These events may be highly localized, not affecting adjacent areas. Such local effects may arise from the loss of a partially committed stem cell population that provides a particular lineage or from the loss of a particular cell type that secretes instructive signals (e.g., growth factors or cytokines) for precursor cell differentiation. The occurrence of nonsensory epithelium in areas that are primarily sensory suggests that sensory stem cells are not able to migrate laterally to the degree necessary to repopulate the area.

Epithelial Cell Specification

Two features of locally specialized organization in neonatal tissue may give insight into the disordered organization observed in older cases. Within the sensory epithelium, but only near the border with respiratory epithelium, enlarged neuron-free regions surround Bowman's gland openings, suggesting an influence of ductal or glandular cells or secretions on the expression of phenotype in the surrounding tissue. The enlarged neuron-free region at the transitional zone may indicate that competing local signals determine the differentiation pattern. At the respiratory/sensory border, the nonsensory signals may outweigh sensory differentiation signals.

Sensory Pits and Epithelial Lacunae

The function of the pits or crypts that are fully lined with receptor neurons in the neonate is unknown. Although their structure is somewhat reminiscent of the specialized vomeronasal invaginations in other species, their anatomical location is distinct from that of the vomeronasal organs, which are anteriorally and ventrally placed on the nasal septum, apart from the olfactory epithelium. Possibly these sensory pits contain subsets of neurons specialized to recognize particular cues such as pheromones, as suggested for vomeronasal neurons, but no selective staining has been recognized with immunocytochemical reagents that we have used to date. Alternatively, the pits may represent an adaptation to increase surface area or to alter airflow, resulting in areas of reduced air movement. The relationship of the well-defined crypts seen in specimens from young subjects to the very large epithelial invaginations extending into the lamina propria observed in older subjects is unclear, but the presence of a few olfactory receptor neurons remaining in the deeper inner recesses of invaginations (e.g., Figures 9, 15, and 17) suggests that this site may be more protected than the surface.

Alternative Pathways for Neurodegeneration

Both along the surface and in the pits, goblet cells, secretory cells and glandlike structures containing material stained by Alcian blue are common. A broad variety of epithelial configurations is observed, ranging from tall ciliated respiratorylike epithelium to low epithelium only one to two or three cells thick. Even in this flattened epithelium, olfactory receptor neurons are sometimes observed, and although they may appear short, they otherwise look normal. However, highly degenerate-looking cells also may be seen. These are often swollen, with misshapen dendrites, or are surrounded by amorphous material that stains positively for neuron-specific antigens. When these abnormal cells and structures are seen, they can stain for epitopes of neurofilament or phosphorylated cytoskeletal proteins, which are frequently seen in Alzheimer's disease patients. The rare appearance of one or several cells stained for phosphorylated neurofilament protein in young specimens (not shown) may indicate an infrequent neurodegenerative event that differs from the pathway that normally mediates neuronal turnover in the epithelium (see below).

Animal Models

Many of the features observed in human tissue appear analogous to changes reported in experimental animal models exposed to various xenobiotic agents. Secretory-cell hyperplasia has been reported in airway epithelia after exposure to damaging inhalants in various animal models, including the monkey (Harkema et al., 1987) and the rat (Walker et al., 1978; Lamb & Reid, 1968). Rats exposed to 3-methyl indole (Peele et al., 1991) sometimes also show fistulae with olfactory and respiratory epithelium invaginating into the lamina propria, as well as hypertrophy of Bowman's glands. Under other protocols such as inhalation of 3-trifluoromethyl pyridine, features reminiscent of changes in the human epithelium also were observed, including glandlike cells trapped in the epithelium (Gaskell et al., 1988, 1990). Intermittent replacement of olfactory epithelium with respiratory epithelium and local breakdown of olfactory epithelium organization have been reported in aging rats and rats exposed to toxic chemicals (Keenan et al., 1990; Morgan et al., 1993).

Degeneration and Cell Death of Olfactory Receptor Neurons

The nature of olfactory neuron cell death is poorly understood, and future studies should benefit from the current intense interest in mechanisms of programmed cell death and necrosis. At any given time the epithelium contains a mixture of neurons in various stages of development, including precursor cells for which there are no characteristic markers, immature neurons (characterized by their location deep in the epithelium and immunocytochemical staining for the growth-associated protein GAP 43 but not for OMP; Verhaagen et al., 1989, 1990), mature neurons that have extended dendrites to the mucosal surface and are extending axons through the lamina propria and into the olfactory bulb, and degenerating neurons. While neuronal cell death is common even in healthy tissue, swollen, degenerating cells are rarely seen. Neurons that are deprived of their target in the bulb are unstable, and accelerated neuronal birth and death are observed in the epithelium of bulbectomized animals (Levey et al., 1991; Schwob et al., 1992). In experimental animals, macrophages are observed within epithelium undergoing massive degeneration after bulbectomy (Monti-Graziadei & Graziadei, 1979), and it is likely that apoptotic rather than necrotic pathways account for cell death under these circumstances, although this has not been demonstrated directly. This pathway may be the one operating in response to neuronal damage caused by most environmental agents.

Swollen cells that stain for neurofilament protein and tau may represent a pathological, necrotic pathway for cell death and appear to be relatively rare even in areas that show high sensory cell loss in humans. Such cells have not been observed in laboratory animals, either under basal conditions or when cell turnover is induced experimentally.

The type of degeneration observed in humans with Alzheimer's disease and in other circumstances of unknown etiology has two striking features that differentiate it from the cell turnover that is a component of normal olfactory epithelium. These are ectopic neurite outgrowth and accumulation of atypical cytoskeletal proteins in both neurites and swollen cells. Intraepithelial accumulations of olfactory neurites also have been observed in humans with Kallmann's syndrome, where there is olfactory bulb dysgenesis and therefore no olfactory receptor cell target (Schwob et al., 1993), and in certain cases of anosmia (Jafek et al., 1989), but a change in cytoskeletal proteins has not been documented in these instances. Neurofilament and tau are produced in most mature neurons in the central nervous system (CNS), but are not characteristic of the more labile olfactory receptor neuron, which seems to continue to express neuronal proteins that are representative of juvenile cells (Talamo et al., 1989b; Schwob, 1992; Riederer et al., 1986).

Even when massive olfactory receptor neuron degeneration is triggered in the rat by olfactory bulbectomy, cell death is not accompanied by altered expression of neurofilament proteins, either in early stages or at later times (B. R. Talamo and W. H. Feng, unpublished data) when the epithelium has been repopulated with neurons that have an elevated turnover rate (Schwob et al., 1992). In animals, ectopic neurite outgrowth and accumulation of neuromalike masses in the epithelium can be elicited during recovery from olfactory receptor neuron death triggered either by olfactory bulbectomy (Hempstead & Morgan, 1985; Schwob et al., 1992; B. R. Talamo and R. Sutherland, unpublished data), by injection of IDPN (Peele et al., 1991) or by inhalation of 3-trifluoromethyl pyridine (Morgan et al., 1991). In bulbectomized rats and after IDPN, the ectopic neuritic masses stain with antibodies to OMP, confirming their origin from olfactory receptor neurons, but they do not stain with antibodies to neurofilament protein, unlike the cases observed in humans. Thus, abnormal neurite outgrowth and altered expression or modification of cytoskeletal components are not always coincident, and misdirected neurite growth may be caused by more than one mechanism. Necrotic cell death also may be induced when abnormal macromolecular expression pathways are activated, leading to production of novel neuron-specific products such as neurofilament proteins. Cells dying through a necrotic pathway release cytosolic components into the surrounding tissue, recruiting inflammatory cells that release cytokines, oxidative enzymes, and other potentially damaging agents that may exacerbate the cascade of cellular damage. Distinguishing toxic and pathogenic agents that activate the different pathways for cell destruction and their targets should lead to a better understanding of the mechanisms that give rise to the altered cellular phenotypes.

As noted earlier, the heterogeneity of cells in epithelium from older humans may indicate that regeneration and reconstitution of olfactory epithelium are not as well maintained in humans as in laboratory animals. Evidence that receptor neurons may be long-lived in protected circumstances was reported in a study of mice kept in the filtered air of laminar flow cages. In these mice, dividing progenitor cells labeled with [^{3}H]thymidine gave rise to labeled olfactory receptor neurons that were subsequently detected up to a year later (Hinds et al., 1984). This suggests that the shorter neuronal lifetime observed in laboratory animals may be due to harmful influences from the environment. A fairly healthy and well-organized olfactory epithelium, with some respiratory metaplasia, also has been observed in laboratory monkeys originally captured from the wild (and therefore presumably exposed to a variety of natural pathogens and xenobiotics) (Talamo & Feng, in preparation). Even in these animals, the sensory epithelium shows little sign of the cytoskeletal disruption seen in human olfactory neurons (Talamo et al., 1991a).

In experiments where laboratory animals have been deliberately exposed to various toxicants and xenobiotics, there is much more variability, encompassing most of the range of tissue changes seen in humans. Little information is available about the mechanism of this variation, about what signals might normally direct differentiation of the different epithelial cell types, or about how metaplastic regions develop. It is not known whether all basal cells are totipotent and able to give rise to a variety of epithelial cell types, each subject to specific local and diffusable signals, or whether there are partially determined stem cells with a limited capacity for differentiation. One could also ask whether there is a default pathway for differentiation in the absence of specific instructions. Questions most directly relevant to environmental studies include whether molecules derived from the environment or generated by (immune) surveillance cells in response to external (xenobiotic?) agents destroy or alter the normal molecular instructions to stem cells or eliminate sets of stem cells or sets of cells that provide necessary information for generation of "normal" epithelium in a particular area.

An experimental approach to answering some of these questions will require that the various cell types be clearly identified by markers, for example, with antibodies or through identification of specific mRNAs. An understanding of these pathways of differentiation and the role of early commitment versus extracellular instructional cues will be essential to interpreting how the diversity of epithelial structures is generated.

Etiology of Changes in Human Olfactory Epithelium

While the many changes in the epithelial organization in the olfactory area of the human nose can be documented, the lack of adequate clinical histories, as well as complications related to complex and unknown premortem and postmortem changes, makes it difficult to assign cause and effect relationships to these changes. Medical records of many cases that come to autopsy focus on the problem that brought the individual to the hospital, frequently unrelated to respiratory or olfactory problems. In addition to these problems, additional complications arise from the nature of the autopsy procedure. The amount of tissue acquired is variable and the entire olfactory region can rarely be obtained. The lack of landmarks in this region often makes it difficult to determine the precise anterior-posterior extent of the olfactory area, and it is therefore difficult to quantitate the amount of loss of olfactory epithelium. As noted earlier, the extreme patchiness and variability of the epithelium in the olfactory region must be kept in mind when tissue is sampled by biopsy (see also Paik et al., 1992). Identification and characterization of the various types of lesions may allow us to ascertain the type of insult that created the alteration and make up the first step in understanding the normal biology of these tissues. Such characterization will provide the framework for elucidating the signals and conversion pathways that trigger normal and abnormal cell death and recovery. Despite these problems, the accessibility, lability, and regenerative capacity of neurons in relationship to neighboring cells makes the olfactory epithelium an attractive system for examining questions about toxicity and recovery.

REFERENCES

Bruch, R. C., and Carr, V. M. 1991. Rat olfactory neurons express a 200 kDa neurofilament. *Brain Res.* 550:133–136.

Burton, P. R., and Wentz, M. A. 1992. Neurofilaments are prominent in bullfrog olfactory axons but are rarely seen in those of the tiger salamander *Ambystoma tigrinum*. *J. Comp. Neurol.* 317:396–406.

Clark, G. 1981. *Staining procedures*, 4th ed., pp. 201–202. Baltimore: Williams & Wilkins.

Dahl, A. R., and Hadley, W. M. 1991. Nasal cavity enzymes involved in xenobiotic metabolism: Effects on the toxicity of inhalants. *Crit. Rev. Toxicol.* 21:345–372.

Doty, R. L., and Snow, J. 1988. Age related alterations in olfactory structure and function. In *Molecular neurobiology of the olfactory system*, eds. F. L. Margolis and T. V. Getchell, pp. 355–374. New York: Plenum Press.

Farbman, A. I. 1990. Olfactory neurogenesis: Genetic or environmental controls? *TINS* 13:362–365.

Farbman, A. I., and Margolis, F. L. 1980. Olfactory marker protein during ontogeny: Immunohistochemical localization. *Dev. Biol.* 74:205–215.

Feng, W. H., Kauer, J. S., Stockmayer, M., and Talamo, B. 1992. A new structure, the "pit organ," in human and monkey olfactory mucosa. *Soc. Neurosci. Abstr.* 18:1198.

Feng, W. H., Kauer, J. S., and Talamo, B. R. 1991. Distribution of pathological neurites across the surface of Alzheimer's nasal epithelium: A new whole-mount immunocytochemical method. *Soc. Neurosci. Abstr.* 17:691.

Gaskell, B. A., Hext, P. M., Pigott, G. H., Hodge, M. D. H., and Tinston, D. J. 1988. Olfactory and hepatic changes following inhalation of 3-trifluoromethyl pyridine in rats. *Toxicology* 50:57–68.

Gaskell, B. A., Hext, P. M., Pigott, G. H., Doe, J. E., and Hodge, M. C. E. 1990. Olfactory and hepatic changes following a single inhalation exposure of 3-trifluoromethyl pyridine in rats: Concentration and temporal aspects. *Toxicology* 62:35–51.

Graziadei, P. P. C., and Monti-Graziadei, G. A. 1979. Neurogenesis and neuron regeneration in the olfactory system of mammals. I. Morphological aspects of differentiation and structural organization of the olfactory sensory neurons. *J. Neurocytol.* 8:1–18.

Graziadei, P. P. C., Karlan, M. S., Monti-Graziadei, G. A., and Bernstein, J. J. 1980. Neurogenesis of sensory neurons in the primate olfactory system after section of the fila olfactoria. *Brain Res.* 186:289–300.

Harkema, J. R. 1991. Comparative aspects of nasal airway anatomy: Relevance to inhalation toxicology. *Toxicol. Pathol.* 19:321–336.

Harkema, J. R., Plopper, C. G., Hyde, D. M., St. George, J. A., Wilson, D. W., and Dungworth, D. L. 1987. Response of the macaque nasal epithelium to ambient levels of ozone. A morphologic and morphometric study of the transitional and respiratory epithelium. *Am. J. Pathol.* 128:29–44.

Hempstead, J. L., and Morgan, J. I. 1985. Monoclonal antibodies reveal aspects of the biochemistry and organization of olfactory neurons following unilateral olfactory bulbectomy. *J. Neurosci.* 5:2382–2387.

Hinds, J. W., Hinds, P. L., and McNelly, N. A. 1984. An autoradiographic study of the mouse olfactory epithelium: Evidence for long-lived receptors. *Anat. Rec.* 210:375–383.

Jafek, B. W., Eller, P. M., Esses, B. A., and Moran, D. T. 1989. Post-traumatic anosmia. Ultrastructural correlates. *Arch. Neurol.* 46:300–304.

Keenan, C. M., Kelly, D. P., and Bogdanffy, M. S. 1990. Degeneration and recovery of rat olfactory epithelium following inhalation of dibasic esters. *Fundam. Appl. Toxicol.* 15:381–393.

Lamb, D., and Reid, L. 1968. Mitotic rates, goblet cell increase and histochemical changes in mucus in rat bronchial epithelium during exposure to sulphur dioxide. *J. Pathol. Bacteriol.* 96:97–111.

Levey, M. S., Chikaraishi, D. M., and Kauer, J. S. 1991. Characterization of potential precursor populations in the mouse olfactory epithelium using immunocytochemistry and autoradiography. *J. Neurosci* 11:3556–3564.

Mackay-Sim, A., and Kittel, P. 1991a. Cell dynamics in the adult mouse olfactory epithelium: A quantitative autoradiographic study. *J. Neurosci.* 11:979–984.

Mackay-Sim, A., and Kittel, P. W. 1991b. On the life span of olfactory receptor neurons. *Eur. J. Neurosci.* 3:209–215.

Monti-Graziadei, G. A., and Graziadei, P. P. C. 1979. Neurogenesis and neuron regeneration in the olfactory system of mammals. II. Degeneration and reconstitution of the olfactory sensory neurons after axotomy. *J. Neurocytol.* 8:197–213.

Moody, S. A., Quigg, M. S., and Frankfurter, A. 1989. Development of the peripheral trigeminal system in the chick revealed by an isotype-specific anti-beta-tubulin monoclonal antibody. *J. Comp. Neurol.* 279:567–580.

Moran, D. T., Rowley, J. C. I., and Jafek, B. W. 1982. Electron microscopy of human olfactory epithelium reveals a new cell type: The microvillar cell. *Brain Res.* 253:39–46.

Morgan, K. T., Kimbell, J. S., Monticello, T. M., Patra, A. L., and Fleishman, A. 1991. Studies of inspiratory airflow patterns in the nasal passages of the F344 rat and rhesus monkey using nasal molds: Relevance to formaldehyde toxicity. *Toxicol. Appl. Pharmacol.* 110:223–240.

Morgan, K. T., Gross, E. A., and Bonnefoi, M. 1993. Nasal structure, function, and toxicology. In *Toxicology of the lung*, eds. D. E. Gardner, J. D. Crapo, and R. O. McClellan, pp. 31–53. New York: Raven Press.

Morrison, E. E., and Costanzo, R. M. 1990. Morphology of the human olfactory epithelium. *J. Comp. Neurol.* 297:1–13.

Morrison, E. E., and Costanzo, R. M. 1992. Morphology of olfactory epithelium in humans and other vertebrates. *Microsc. and Res. Technique* 23:49–61.

Naessen, R. 1971. An enquiry on the morphological characteristics and possible changes with age in the olfactory region of man. *Acta Otolaryngol.* 71:49–62.

Nakashima, T., Kimmelman, C. P., and Snow, J. B. 1984. Structure of human fetal and adult olfactory neuroepithelium. *Arch. Otolaryngol.* 110:641–646.

Nakashima, T., Kimmelman, C. P., and Snow, J. 1985. Immunohistopathology of human olfactory epithelium, nerve and bulb. *Laryngoscope* 95:391–396.

Nakashima, T., Tanaka, M., Inamitsu, M., and Uemura, T. 1991. Immunohistopathology of variations of human olfactory mucosa. *Eur. Arch. Oto-Rhino-Laryngol.* 248:370–375.

Ophir, D. 1987. Intermediate filament expression in human fetal olfactory epithelium. *Arch. Otolaryngol. Head Neck Surg.* 113:155.

Paik, S. I., Lehman, M. N., Seiden, A. M., Duncan, H. J., and Smith, D. V. 1992. Human olfactory biopsy. The influence of age and receptor distribution. *Arch. Otolaryngol. Head Neck Surg.* 118:731–738.

Peele, D. B., Allison, S. D., Bolon, B., Prah, J. D., Jensen, K. F., and Morgan, K. T. 1991. Functional deficits produced by 3-methylindole-induced olfactory mucosal damage revealed by a simple olfactory learning task. *Toxicol. Appl. Pharmacol.* 107:191–202.

Riederer, B., Cohen, R., and Matus, A. 1986. MAP5: A novel brain microtubule-associated protein under strong developmental regulation. *J. Neurocytol.* 15:763–775.

Rowley, J. C., III, Moran, D. T., and Jafek, B. W. 1989. Peroxidase backfills suggest the mammalian olfactory epithelium contains a second morphologically distinct class of bipolar sensory neuron: The microvillar cell. *Brain Res.* 502:387–400.

Schwob, J. E. 1992. The biochemistry of olfactory neurons: Stages of differentiation and neuronal subsets. In *Science of olfaction*, eds. M. J. Serby and K. L. Chobor, pp. 80–125. New York: Springer-Verlag.

Schwob, J. E., Farber, N. B., and Gottleib, D. T. 1986. Neurons of the olfactory epithelium in adult rats contain vimentin. *J. Neurosci.* 6:208–217.

Schwob, J. E., Szumowski, K. E., Leopold, D. A., and Emko, P. 1993. Histopathology of olfactory mucosa in Kallmann's syndrome. *Ann. Otol. Rhino. Laryngol.* 102:117–122.

Schwob, J. E., Szumowski, K. E. M., and Stasky, A. A. 1992. Olfactory sensory neurons are trophically dependent on the olfactory bulb for their prolonged survival. *J. Neurosci.* 12 3896–3919.

Takahashi, S., Iwanaga, T., Takahashi, Y., Nakano, Y., and Fujita, T. 1984. Neuron-specific enoslase, neurofilament protein and S-100 protein in the olfactory mucosa of human fetuses. An immunohistochemical study. *Cell Tissue Res.* 238:231–234.

Talamo, B. R., Rudel, R. A., Kosik, K. S., Lee, V. M.-Y., Neff, S., Adelman, L., and Kauer, J. S. 1989a. Pathological changes in olfactory neurons in patients with Alzheimer's disease. *Nature* 337:736–739.

Talamo, B. R., Rudel, R. A., Lee, V. M.-Y., Kosik, K. S., and Kauer, J. S. 1989b. Neurofilaments and MAPS in human olfactory epithelium in Alzheimer's disease. *Soc. Neurosci Abstr.* 15:329.

Talamo, B. R., Feng, W. H., Perez-Cruet, M., Adelman, L., Kosik, K., Lee, V. M.-Y., Cork, L. C., and Kauer, J. S. 1991a. Pathological changes in olfactory neurons in Alzheimer's disease. In *Aging and Alzheimer's disease*, eds. J. H. Growden, S. Corkin, E. Ritter-Walker, and R. J. Wurtman, pp. 1–7. New York: New York Academy Science.

Talamo, B. R., Feng, W. H., Stockmayer, M., Cork, L., and Kauer, J. S. 1991b. Monkey and human olfactory epithelium: A comparative study. *Chem. Senses* 16:589.

Talamo, B. R., Feng, W. H., and Kauer, J. S. 1992. Development and degeneration of human olfactory epithelium. *Chem. Senses* 17:706.

Verhaagen, J., Oestreicher, A. B., Gispen, W. H., and Margolis, F. L. 1989. The expression of the growth associated protein B50/GAP43 in the olfactory system of neonatal and adult rats. *J. Neurosci.* 9:683–691.

Verhaagen, J., Greer, C. A., and Margolis, F. L. 1990. B-50/GAP43 gene expression in the rat olfactory system during postnatal development and aging. *Eur. J. Neurosci* 2:397–407.

Vollrath, M., Altmannsberger, M., Weber, K., and Osborn, M. 1985. An ultrastructural and immunohistological study of the rat olfactory epithelium: Unique properties of olfactory sensory cells. *Differentiation* 29:243–253.

Walker, D., Wilton, L. V., and Binns, R. 1978. Inhalation toxicity studies on cigarette smoke (VII) 6-week comparative experiments using modified flue-cured cigarettes: Histopathology of the conducting airways. *Toxicology* 10:241–259.

Yamagishi, M., and Nakano, Y. 1992. A re-evaluation of the classification of olfactory epithelia in patients with olfactory disorders. *Euro. Arch. Oto-Rhino-Laryngol.* 249:393–399.

Yamagishi, M., Nakamura, H., Takahashi, S., Nakano, Y., and Iwanaga, T. 1989. Olfactory receptor cells: Immunocytochemistry for nervous system-specific proteins and re-evaluation of their precursor cells. *Arch. Histol. Cytol.* 52(Suppl.):375–381.

TOXICITY OF NASAL RESPIRATORY MUCOSA IN HUMANS

Valerie J. Lund

Institute of Laryngology and Otology, University College London, London, UK

As the first portal of entry to the respiratory tract, the nasal cavity and associated sinuses are exposed to a significant range of inhaled xenobiotics. The available methodology to qualify and quantify physiological damage in populations at risk from potentially toxic agents is examined, including endoscopy, assessment of mucociliary function, assessment of nasal airway, olfaction, and relevant hematology. Inhaled xenobiotics may have an etiological role in allergy, infection, and the development of neoplasia. An overview of established occupational risks is examined with specific reference to malignant melanoma, and the results of recent clinical studies regarding association between this rare tumor and formaldehyde are presented.

Much interest and work have concentrated on the development and application of experimental animal models in the investigation of sino-nasal toxicology. At the other end of the investigative spectrum is the clinician faced with pathological change that may be the result of or be exacerbated by external agents, in particular inhaled xenobiotics. The purpose of this article is to examine the evidence in the clinical milieu for these interrelationships and the methodology available to assess pathophysiological change.

METHODS OF INVESTIGATION

A range of techniques is available to the technician to qualify and quantify physiological damage in populations at risk from potentially toxic agents.

Endoscopy

In addition to conventional anterior and posterior rhinoscopy, examination of the respiratory mucosa is now considerably facilitated by the use of rigid Hopkins rods and flexible fiber-optic nasendoscopes. Administration of topical intranasal anesthesia is recommended at initial assessment, and specific attention can be paid to macroscopic change of the mucosa and distortion of mucous drainage. Biopsy of any abnormal tissue may be effected with greater accuracy under direct vision using such instrumentation, as may specimens for bacteriology.

Mucociliary Function

Naso-Mucociliary Clearance (NMCC) This simple test of the mucociliary apparatus can be performed by placing a 0 5-mm-diameter piece of saccharine on the inferior turbinate, 1 cm from the anterior end, and the time

Address correspondence to Valerie J. Lund, MS, FRCS, FRCSEd, Reader in Rhinology, The Institute of Laryngology and Otology, University College London, 330 Gray's Inn Road, London WCIX 8EE, UK.

taken to taste something sweet in the mouth measured (Andersen et al., 1974). Normally this occurs within 30 min. If longer than an hour has elapsed, NMCC is impaired, though it is worth repeating the test in case the particle has fallen out and checking that the patient is capable of tasting saccharine.

Ciliary Beat Frequency (CBF) When the saccharine test is prolonged or a specific ciliary abnormality is suspected, it is possible to examine the cilia directly by taking a sample with a bronchoscopy cytology brush and observing cilial activity under a phase-contrast microscope with a photometric cell (Rutland et al., 1982). The frequency can be measured with a real time analyzer and expressed in hertz (Hz), with the normal range for the inferior turbinate being 12–15 Hz.

Nasal Airway Assessment

Nasal Expiratory or Inspiratory Peak Flow (NIPF) This technique, which uses a peak flowmeter, has the advantage of being inexpensive, quick, and easy to perform; it is useful for repeated examinations and compares well with rhinomanometry (Holmstrom et al., 1990). Of the two methods, forced inspiration is preferred, although this method can produce significant vestibular collapse. Forced expiration inflates the eustachian tubes, which the patient may find distressing, and can produce an unpleasant quantity of mucus in the mask.

Rhinomanometry Rhinomanometry attempts to measure nasal airway resistance by making quantitative measurements of nasal flow and pressure (Clement, 1984). This technique employs the principle that air will flow only through a tube when there is a pressure differential, passing from areas of high to low pressure, a differential created by respiratory effort. When the nasal mucosa is decongested, the reproducibility of rhinomanometric results is good, but it is time-consuming and is mostly a research tool. Active anterior rhinomanometry is most commonly used, as the posterior technique can not be used in 20–25% of individuals, due to an inability to relax the soft palate.

Acoustic Rhinometry In this technique, an audible sound pulse (150–10,000 Hz) generated by a spark is propagated in a sound tube and is passed into the nose, where it is altered by variations in cross-sectional area. The reflected signal is picked up by a microphone and analyzed. It is thus possible to determine the area within the nasal cavity as a function of distance, and from this, volumes may be derived (Hilberg et al., 1989). Acoustic rhinometry does not measure airflow. The technique is still being evaluated, but is likely to enter routine clinical practice, allowing accurate sequential quantification of abnormalities, pre- and posttreatment.

Olfaction

Quantitative Olfactory Thresholds Estimation of olfactory thresholds may be established by presentation of serial dilutions of pure odorant such as pm-carbinol (Amoore, 1992). The patient is presented with two bottles, one containing only the diluent solution as the control, the other, the odorant, in progressively increasing or decreasing concentrations. Each is sniffed in turn. The point is reached at which the patient cannot distinguish between the control and test bottle, which indicates the minimum detectable odor.

Quantitative Olfactory Testing "Scratch and sniff" tests using patches impregnated with microencapsulated odorants are available (Doty et al., 1984). The patient is forced to choose between a number of options after scratching the patch to release the odor. The results take into account answers guessed correctly and deliberately given incorrectly.

Hematology

Hematologic measurements may be of value in assessing the degree of active and passive smoking, particularly since deception by smokers about their smoking habits has been demonstrated (Kirk et al., 1981; Sillett et al., 1978). Of the various objective markers available, serum cotinine is presently regarded as the most reliable and can be readily used in the investigation of patients with pathology in the upper respiratory tract (Whittet et al., 1991).

PATHOLOGICAL RESPONSES OF UPPER RESPIRATORY MUCOSA TO INHALED XENOBIOTICS

Allergy

It is of considerable interest that while pollen counts have fallen, seasonal allergic rhinitis appears to be increasing in prevalence.

Hay fever appears to be more common in urban compared to rural areas. In Denmark, the prevalence of hay fever among patients attending general practitioners was 19% in Copenhagen, compared to 6–11% in rural areas (Pedersen & Weeke, 1981). In the United States, a community study indicated that up to 75% complained of probable allergic rhinitis lived in the city, compared to 25% who were living in the surrounding rural areas (Broder et al., 1974). The reasons for these differences are not known, but may include air pollution (Kim, 1991).

Medical examination of Swedish army recruits has shown that the prevalence of hay fever increased from 4.4% in 1971 to 8.4% in 1981, with a higher increase in the more northerly region of the country; this developing geographical difference argues against the increase being attributable to changes in doctors' labeling preference (Aberg, 1989).

It has been suggested that an increased susceptibility to environmental allergens is responsible for such a rise in manifestation of symptoms. It is known that interleukin 8 (IL-8) is released by exposure to ozone and that this increases the inflammatory response. Tumor necrosis factor α (TNFα) up-regulates IL-1, attracting neutrophils, and VCAM 1 increases the numbers of eosinophils. Considerable work has already taken place on this aspect in the lower respiratory tract, particularly in relation to asthma (Wardlaw, 1993; Wjst et al., 1993).

Infection

Although bacterial pathogens are not strictly within the definition of xenobiotics, the work of Cole and others has shown that common upper respiratory tract pathogens are capable of producing ciliotoxic substances that disrupt mucociliary clearance and result in a vicious circle of escalating tissue damage (Cole & Wilson, 1989; Feldman et al., 1992; Read et al., 1992). Thus, infection may play an important interactive role in populations exposed to other toxic agents.

Medications

A considerable range of pharmaceutical products are used intranasally, and, although their primary intention is to improve nasal symptoms, the presence of preservatives such as methyl-*p*-hydroxybenzoate, propyl-*p*-hydroxybenzoate chlorbutol, and chlorocresol is known to damage cilial activity (Batts et al., 1990).

Neoplasia

Since the discovery of an association between the woodworking trade and adenocarcinoma of the ethmoid (Acheson et al., 1968), interest has focused on the association of sinonasal malignancy with occupational risks. However, sinonasal malignancy is rare, with less than 1/100,000 of the population affected in most countries, though geographical variations do exist. Data from the International Agency for Research on Cancer show that the incidence is strikingly higher in Japan, where a rate of 2.6/100,000 men between 1969 and 1973 in the Osaka area has been observed, and rates of 3.6/100,000 have been reported in other parts of Japan (Waterhouse et al., 1976). In Nigeria and Jamaica, similar increases have been observed in the black population (2.5 and 2.0/100,000 males, respectively), and in Saskatchewan, Canada, and Oxford, UK, two areas with established local risk factors from the woodworking industry, slightly increased rates are seen (1.2 and 1.0/100,000 males) compared to other parts of the respective countries, where the overall rate is less than 1/100,000.

Examination of distribution by histological classification shows squamous cell carcinoma to be the commonest overall type (Muir & Nectoux, 1980), though tumors in this area manifest the greatest histological diversity in the body (Harrison & Lund, 1993).

Distribution by site within the sinonasal region again demonstrates geographical differences (Muir & Nectoux, 1980). In Japan, the maxilla is predominantly affected, and this is also the commonest site in other countries, although to a lesser extent. However, in figures from the United Kingdom, a comparatively larger proportion of nasal and ethmoid cancer is found.

The ICD classification does not separate ethmoid from frontal or sphenoid disease, but the former almost certainly contributes the major part of this group (9 : 1 ratio) (Lund, 1991). It is, however, noteworthy that the majority of tumors occur within the sinus cavity and not in the nose itself, which is the primary focus of attention in experimental work using animal models.

The role of cigarette smoking in the development of head and neck squamous-cell cancer is well established (Michaels, 1987; Sarracci, 1987; Guenel et al., 1988). However, its specific role in the sinonasal region has yet to be defined. A recent Japanese study exonerated both passive and active smoking (Shimizu et al., 1989), but another study from the Netherlands suggested a relative risk of 3.1/100,000 in the development of squamous cell carcinoma in this region (Hayes et al., 1987).

OCCUPATIONAL RISKS

With changing patterns of work practice, improved working conditions, and a mobile population, the effects of such occupational factors will become increasingly difficult to unravel despite our greater vigilance in this respect. Reports on the association between adenocarcinoma of the ethmoid and the woodworking trade, in particular hardwood exposure, began in 1965 from the United Kingdom and have subsequently appeared from many other countries (Macbeth, 1965; Acheson et al., 1968; Hadfield, 1970; Cecchi et al., 1980; Gignoux & Bernard, 1969; Debois, 1969; Delamare & Themans, 1971; Mosbech & Acheson, 1971; Ironside & Matthews, 1975; Brinton et al., 1977; Engzell, 1978; Tola et al., 1980; Elwood, 1981; Hayes et al., 1986a). In addition to the association between hardwoods and adenocarcinoma, the effects of softwood in the development of squamous and anaplastic carcinoma have been reported (Hernberg et al., 1983; Voss et al., 1985; Boysen et al., 1986). It is, however, interesting that Japan, which has a large traditional woodworking industry, has found no such association in a country where squamous-cell carcinoma of the maxilla predominates (Takasaka et al., 1987).

The leather and shoe industry has also been the subject of extensive epidemiological study for squamous and adenocarcinoma since Acheson's report in 1968. Other occupations associated with an increased risk of adenocarcinoma are the manufacture of chrome pigment (Heuper, 1966; Enterline, 1974; Levy & Verritt, 1975), isopropyl alcohol (Fraumeni, 1975; International Agency for Research on Cancer Working Group, 1976; National Institute for Occupational Safety and Health, 1976), and the textile and clothing trade, the latter also being associated with malignant melanoma (Acheson et al., 1972). The relative risk and suspected carcinogens, other associated tumors, and, where calculated, the latent period in years are shown in Table 1.

Three occupations are linked with squamous-cell carcinoma in the nose and sinuses. Of these, radium dial painting has largely disappeared and

TABLE 1. Occupational Agents Correlated with Sinonasal Cancer

Occupation	Relative risk	Suspected carcinogen	Latent period	Other associated: Histology	Other associated: Cancers
Woodworkers	70	Dust 5 μm diameter Tar Aldehyde Aflatoxin Chromium Tannins	35 yr	Adenocarcinoma (hardwood) Squamous (softwood)	Lung Testis Brain
Leather/shoe	87	Dust Tar Aldehydes Tannins	55 yr	Adenocarcinoma	Rectum Bladder
Chrome pigment manufacturers	>21	Calcium chromate Zinc potassium chromate	—	Adenocarcinoma	Lung
Isopropyl alcohol manufacturers	>21	Isopropyl oil	<20 yr	Adenocarcinoma	Larynx
Textile and clothing	5–8	Wool dust and dyes		Adenocarcinoma Malignant melanoma	Tongue Mouth Pharynx
Nickel refining	>100	Nickel subsulphide, oxide,carbonyl	24 yr (5–40)	Squamous cell Anaplastic	Lung Larynx
Dial painting	—	Radium	15 yr	Squamous	Osteosarcoma Ca mastoid
Mustard gas manufacture	>30	β,β-Dichloroethyl sulphide	25 yr	Squamous	Tongue Pharynx Larynx Lung

Note. After Roush (1979).

mustard gas manufacture only occurred in Japan between 1924 and 1945 (Wada et al., 1968). Sintering and roasting processes in nickel refining have been implicated in reports from Russia, Norway, Canada, and Wales (Hill, 1939; Morgan, 1958; Sutherland, 1959; Doll et al., 1970; Pederson et al., 1973). There is a high relative risk with a minimum exposure of 6 mo, though the majority of those affected had contact in excess of 9 yr. The risk increased with duration of exposure and the age at first exposure and was probably eliminated before the first epidemiological studies were done. The literature contains other reports of occupational risks for nose and sinus malignancy, such as asbestos exposure in two documented cases (Bogouski et al., 1973; Stell & McGill, 1973) and baking (six cases—three squamous cell and three adenocarcinomas, Roush, 1979).

A questionnaire and personal interview conducted on 50 of the author's consecutive patients with sinonasal malignancy revealed a considerable range of carcinogenic exposure, the commonest of which was exposure to hardwoods in 20% of cases, all of whom suffered from adenocarcinoma of the ethmoid (Lund, 1991). However, other agents such as carbon tetrachloride, isopropyl alcohol, and formaldehyde were also revealed, and formaldehyde exposure was a significant occupational factor in three of eight patients with malignant melanoma of the nasal mucosa (Holmstrom & Lund, 1991).

As a result of this finding, a survey was conducted among 450 ear, nose, and throat (ENT) consultants in the United Kingdom. Information was requested on the numbers of sinonasal malignant melanoma treated and the period of practice covered. This revealed a further 181 cases in cumulative experience of 1820 yr (approximately 1 case per 10 years). Further examination of these figures suggested a clustering of cases in the Midlands area of Great Britain. Information from the West Midlands Cancer Register also suggested a higher incidence of malignant melanoma than published studies would suggest for the United Kingdom as a whole. Twenty-nine individuals were available for completion of a questionnaire and personal interview regarding possible environmental exposure to toxins. Three individuals were revealed who had worked predominantly within the tire manufacturing industry. The range of chemical exposure in the manufacture of tires is extensive (e.g., methyl isobutyl ketone, perchloroethylene, xylene, lead, petroleum, formaldehyde, zinc, stearate), and many of the substances have been implicated in previous toxicology studies, so it is difficult on such anecdotal case reports to unravel the exact role of a single substance, such as formaldehyde, which is just one of many possible causative agents. However, with a tumor of such rarity, an apparent association warrants documentation and further investigation.

Formaldehyde has been considered for the past few years as a risk factor for nasal cancer, since it was shown that it is a potent carcinogen in rats exposed long term to high doses (Kerns et al., 1983). Formaldehyde is a very reactive substance, causing single-strand breaks with DNA and DNA-protein cross-links (Morgan et al., 1986), and mutations in human cells in vitro (Goldmacher & Thilly, 1983).

Formaldehyde is highly irritant to the upper respiratory tract in humans, and nasal symptoms are common in occupationally exposed workers, as are physiological disturbances of olfaction and nasomucociliary clearance (Holmstrom & Wilhelmsson, 1988). In workers occupationally exposed to formaldehyde, histological changes in nasal mucosa, such as loss of cilia and metaplasia, have been reported (Holmstrom et al., 1989); in one study dysplasia was also noted (Boysen et al., 1990). However, no convincing evidence has accrued to show that formaldehyde is carcinogenic in humans, despite several epidemiological studies. Two studies reported a correlation between formaldehyde and the occurrence of nasal cancer (Olsen et al., 1984; Hayes et al., 1986b), but others, in particular the mortality study of Blair et al. on more than 26,000 workers exposed to formaldehyde, did not

(Blair et al., 1986). In the study of Walrath and Fraumeni (1983) on embalmers, persons with probably the highest occupational exposure, no correlation between formaldehyde exposure and nasal cancer was found although mortality was significantly raised for cancer of the colon and of the skin [8 cases of skin cancer (4 malignant melanomas) were found compared with the 3.6 expected; Walrath & Fraumeni, 1983]. However, in a recent study from France (Luce et al., 1993) looking at the relationship between sinonasal cancer and occupational exposure to formaldehyde and other substances, it appeared that exposure to both formaldehyde and wood dust may increase the risk of nasal adenocarcinoma by comparison with the risk due to wood dust alone.

DISCUSSION

While experimental toxicologists pursue the "Holy Grail" of the perfect animal model, the clinician is faced with the symptomatic patient suffering from conditions posing management problems of differing complexity. Populations may be increasingly identified where external environmental factors place individuals at risk of pathological change. Recognition of these factors and attempts to measure their effects may allow us to minimize their impact. Unfortunately, the present economic situation imposes financial constraints that may compromise commitment to the large-scale epidemiological studies necessary to confirm our suspicions.

REFERENCES

Aberg, N. 1989. Asthma and allergic rhinitis in Swedish conscripts. *Clin. Exp. Allergy.* 19:59–63.

Acheson, E. D., Cowdell, R. H., Hadfield, E. H., and Macbeth, R. G. 1968. Nasal cancer in woodworkers in the furniture industry. *Br. Med. J.* 2:587–596.

Acheson, E. D., Cowdell, R. H., and Rang, E. 1972. Adenocarcinoma of the nasal cavity and sinuses in England and Wales. *Br. J. Ind. Med.* 29:21–30.

Amoore, J. E. 1992. Odor standards in squeeze bottle kits for matching quality and intensity. *Water Sci. Technol.* 25:1–9.

Andersen, I., Camner, P. Jensen, P. L., Philipson, K., and Proctor, D. 1974. Nasal clearance of monozygotic twins. *Am. Rev. Respir. Dis.* 100:301–305.

Batts, A. H., Marriott, C., Martin, G. P., Wood, C. F., and Bond, S. W. 1990. The effect of some preservatives used in nasal preparations on the mucus and ciliary components of mucociliary clearance. *J. Pharm. Pharmacol.* 42:145–151.

Blair, A., Stewart, P., and O'Berg, M. 1986. Mortality among industrial workers exposed to formaldehyde. *JNCI* 76:1071–1084.

Bogouski, P., Gilson, J. C., and Timorell, V. 1973. *Biological effects of asbestos.* Lyon: International Agency for Research on Cancer (IARC).

Boysen, M., Voss, R., and Solberg, L. A. 1986. The nasal mucosa in softwood exposed furniture workers. *Acta Otolaryngol. (Stockh.)* 101:501–508.

Boysen, M., Zadig, E., Digernes, V., Abeler, V., and Reith, A. 1990. Nasal mucosa in workers exposed to formaldehyde: a pilot study. *Br. J. Ind. Med.* 47:116–121.

Brinton, L. A., Blot, W. J., Stone, B. J., and Fraumeni, J. F., Jr. 1977. A death certificate study of nasal cancer among furniture workers in North Carolina. *Cancer Res.* 37:3473–3474.

Broder, I., Higgins, M. W., Matthews, K. P., and Keller, J. B. 1974. Epidemiology of asthma and allergic rhinitis in a total community. *J. Allergy Clin. Immunol.* 53:127–138.

Cecchi, F., Buiatti, E., Kriebel, D., Nastasi, L., and Santucci, M. 1980. Adenocarcinoma of the nose and paranasal sinuses in shoemakers and woodworkers in the province of Florence, Italy (1963–77). *Brit. J. Ind. Med.* 37:222–225.
Clement, P. A. R. 1984. Committee report on standardization of rhinomanometry. *Rhinology* 22:151–155.
Cole, P. J., and Wilson, R. 1989. Host-microbial interrelationships in respiratory infection. *Chest* 95:2175–2215.
Debois, J. M. 1969. Tumoren van de neusholte bij houtbewerkers. *Neder. Tijdschr. Geneesk.* 25:92–93.
Delamare, J. F. M., and Themans, H. H. 1971. Het adenocarcinoom van de neusholten. *Neder. Tijdsch. Geneesk.* 115:688–690.
Doll, R., Morgan, L. G., and Speizer, F. E. 1970. Cancer of the lung and nasal sinuses in nickel workers. *Br. J. Cancer.* 24:623–632.
Doty, R. L., Shaman, P., and Dann, M. 1984. Development of the University of Pennsylvania Smell Identification Test: A standardized microencapsulated test of olfactory function. *Physiol. Behav.* 32:489–502.
Elwood, J. M. 1981. Wood exposure and smoking: Association with cancer of the nasal cavity and paranasal sinuses in British Columbia. *Can. Med. Assoc. J.* 124:1573–1577.
Engzell, U. 1978. Occupational etiology and nasal cancer. *Acta Otolaryngol. (Stockh.)* 86(Suppl. 360):126–128.
Enterline, P. E. 1974. Respiratory cancer among chromate workers. *J. Occup. Med.* 16:523–526.
Feldman, C., Read., R., Rutman, A., Jeffrey, P. K., Brain, A., Lund, V. J., Mitchell, T. J., Andrew, P. W., Boulnois, G. J., Todd, H. C., Cole, P. J., and Wilson, R. 1992. The interaction of *Streptococcus pneumoniae* with intact human respiratory mucosa in vitro. *Eur. Respir. J.* 5:576–583.
Fraumeni, J. F., Jr. 1975. Respiratory carcinogenesis: An epidemiologic appraisal. *J. Natl. Cancer Inst.* 55:1039–1046.
Gignoux, M., and Bernard, P. 1969. Tumeurs malignes de l'ethmoide chez les travailleurs du bois. *J. Med. Lyon* 50:731–736.
Goldmacher, V. S., and Thilly, W. G. 1983. Formaldehyde is mutagenic for cultured human cells. *Mutat. Res.* 116:417–422.
Guenel, P., Chastang, J. F., Luce, D., Leclerc, A., and Brugere, J. 1988. A study of the interaction of alcohol drinking and tobacco smoking among French cases of laryngeal cancer. *J. Epidemiol. Community Health* 42:350–354.
Hadfield, E. H. 1970. A study of adenocarcinoma of the paranasal sinuses in woodworkers in the furniture industry. *Ann. R. Coll. Surg. Engl.* 46:301–309.
Harrison, D. F. N., and Lund, V. J. 1993. *Tumours of the upper jaw.* London: Churchill Livingstone.
Hayes, R. B., Gerin, M., Raatgever, J. W., and De Bruyn, A. 1986a. Wood-related occupations, wood dust exposure and sinonasal cancer. *Am. J. Epidemiol.* 124:569–577.
Hayes, R. B., Raatgever, J. W., and De Bruyn, A. 1986b. Cancer of the nasal cavity and paranasal sinuses and formaldehyde exposure. *Int. J. Cancer* 37:487–492.
Hayes, R. B., Kardaun, J. W. P. F., and De Bruyn, A. 1987. Tobacco use and sinonasal cancer. *Br. J. Cancer* 56:843–846.
Hernberg, S., Westerholm, P., and Schultz Larsen, K. 1983. Nasal and sinonasal cancer. Connection with occupational exposures in Denmark, Finland and Sweden. *Scand. J. Work Environ. Health* 9:315–326.
Heuper, W. C. 1966. Occupational and environmental cancers of the respiratory system. In *Recent results in cancer research*, ed. P. Rentchinck, Vol. 3. New York: Springer Verlag.
Hilberg, O., Jackson, A. C., and Swift, D. L. 1989. Acoustic rhinometry. Evaluation of nasal cavity geometry by acoustic deflections. *J. Appl. Physiol.* 66:295–303.
Hill, A. B. 1939. Statistical report to the Mond Nickel Company relating to the incidence of carcinoma of the respiratory system at the Clydach Works, Wales. Mond Nickel Company.
Holmstrom, M., and Lund, V. J. 1991. Malignant melanomas of the nasal cavity after occupational exposure to formaldehyde. *Br. J. Ind. Med.* 48:9–11.
Holmstrom, M., and Wilhelmsson, B. 1988. Respiratory symptoms and pathophysiological effects of occupational exposure to formaldehyde and wood dust. *Scand. J. Work Environ. Health* 14:306–311.

Holmstrom, M., Scadding, G. K., and Lund, V. J. 1990. The assessment of nasal obstructiona comparison between rhinomanometry and nasal inspiratory peak flow. *Rhinology* 28:191–196.

Holmstrom, M., Wilhelmsson, B., Hellquist, H., and Rosen, G. 1989. Histological changes in the nasal mucosa in persons occupationally exposed to formaldehyde alone and in combination with wood dust. *Acta Otolaryngol. (Stockh.)* 107:120–129.

International Agency for Research on Cancer Working Group. 1976. Isopropyl alcohol and isopropyl oils. In *Evaluation of carcinogenic risk of chemicals to man*, pp. 223–243. Geneva: World Health Organization.

Ironside, P., and Matthews, J. 1975. Carcinoma of the nose and paranasal sinuses in woodworkers in the state of Victoria, Australia. *Cancer* 36:1115–1121.

Kerns, W. D., Pavkov, K. I., Donofrio, D. J., Gralla, E. J., and Swenberg, J. A. 1983. Carcinogenicity of formaldehyde in rats and mice after long-term inhalation exposure. *Cancer Res.* 43:4382–4392.

Kim, K. H. 1991. Allergic rhinitis and house dust mite at urban district Seoul, Korea. *Proc. 7th Asia-Oceania Congress of ORL Societies,* Hong Kong.

Kirk, C. J. C., Woolcock, N. E., and Lund, V. J. 1981. The effect of advice to stop smoking on arterial patients by serum thiocyanate levels. In *Smoking and arterial disease*, ed. R. M. Greenhalgh. Bath: Pitman Medical.

Levy, L. S., and Verritt, S. 1975. Carcinogenic and mutagenic activities of chromium-containing materials. *Br. J. Cancer.* 32:262–263.

Luce, D., Gerin, M., Leclerc, A., Morcet, J. F., Brugere, J., and Goldberg, M. 1993. Sinonasal cancer and occupational exposure to formaldehyde and other substances. *Int. J. Cancer* 53:224–231.

Lund, V. J. 1991. Malignancy of the nose and sinuses. Epidemiological and aetiological considerations. *Rhinology* 29:57–68.

Macbeth, R. G. 1965. Malignant disease of the paranasal sinuses. *J. Laryngol. Otol.* 79:592–612.

Michaels, L. 1987. *Ear nose and throat histopathology*, pp. 371–373. Berlin: Springer Verlag.

Morgan, J. G. 1958. Some observations on the incidence of respiratory cancer in nickel workers. *Br. J. Ind. Med.* 15:224–242.

Morgan, K. T., Jiang, X-Z., Starr, T. B., and Kerns, W. D. 1986. More precise localization of nasal tumors associated with chronic exposure of F-344 rats to formaldehyde gas. *Toxicol. Appl. Pharmacol.* 82:264–271.

Mosbech, B. J., and Acheson, E. D. 1971. Nasal cancer in furniture-makers in Denmark. *Dan. Med. Bull.* 18:34–35.

Muir, C. S., and Nectoux, J. 1980. Descriptive epidemiology of malignant neoplasms of nose, nasal cavities, middle ear and accessory sinuses. *Clin. Otolaryngol.* 5:195–211.

National Institute for Occupational Safety and Health. 1976. Criteria for a Recommended Standard Occupational Exposure to Isopropyl Alcohol. Washington, DC: DHEW Publication 76.

Olsen, J. H., Plough Jensen, S., Hink, M., Faurbo, K., Breum, N. O., and Jensen, O. M. 1984. Occupational formaldehyde exposure and increased cancer risk in man. *Int. J. Cancer* 34:639–644.

Pedersen, P. A., and Weeke, E. R. 1981. Allergic rhinitis in Danish general practice. *Allergy* 36:375–379.

Pederson, E., Hogetvit, A. C., and Andersen, A. 1973. Cancer of the respiratory organs among workers as a nickel refinery in Norway. *Int. J. Cancer* 12:32–41.

Read, R. C., Roberts, P., Munro, N., Rutman, A., Hastie, A., Shryock, T., Hall, McDonald-Gibson, W., Lund, V. J., Taylor, G., Cole, P. J., and Wilson, R. 1992. Effect of *Pseudomonas aeruginosa* rhamnolipids on mucociliary transport and ciliary beating. *J. Appl. Physiol.* 72:2271–2277.

Roush, G. C. 1979. Epidemiology of cancer of the nose and paranasal sinuses: Current concepts. *Head Neck Surg.* 2:3–11.

Rutland, J., Dewar, A., Cox, T., and Cole, P. 1982. Nasal brushing for the study of ciliary ultrastructure. *J. Clin. Pathol.* 35:357–359.

Saracci, R. 1987. The interactions of tobacco smoking and other agents in cancer aetiology. *Epidemiol. Rev.* 9:175–191.

Shimizu, H., Hozawa, J., Saito, H., Murai, K., Hirata, H., Takasaka, T., Togawa, K., Konno, A., Kimura, Y., and Kukuchi, A. 1989. Chronic sinusitis and woodworking as risk factors for cancer of the maxillary sinus in northeast Japan. *Laryngoscope* 99:58–61.

Sillett, R. W., Wilson, M. B., Malcolm, R. E., and Ball, K. P. 1978. Deception among smokers. *Br. Med. J.* 2:1185–1186.

Stell, P. M., and McGill, T. 1973. Asbestos and cancer of the head and neck. *Lancet* 1:678.

Sutherland, R. B. 1959. Respiratory cancer mortality in workers employed in an Ontario nickel refinery 1930–1957. Report of the Industrial Division of Hygiene, Ontario, Canada. Ontario Department of Health.

Takasaka, T., Kawamoto, K., and Nakamura, K. 1987. A case-control study of nasal cancers. *Acta Otolaryngol. (Stockh.)* Suppl. 435:136–142.

Tola, S., Hernberg, S., Collan, Y., Linderborg, H., and Korkala, M. L. 1980. A case-control study of the etiology of nasal cancer in Finland. *Int. Arch. Occup. Environ. Health* 46:79–85.

Voss, R., Stenersen, T., Oppedal, B. R., and Boysen, M. 1985. Sinonasal cancer and exposure to softwood. *Acta Otolaryngol. (Stockh.)* 99:172–178.

Wada, S., Miyanishi, M., Nishimoto, Y., Kambe, S., and Miller, R. W. 1968. Mustard gas as a cause of neoplasia in man. *Lancet* 1:1161–1163.

Walrath, J., and Fraumeni, J. F, Jr. 1983. Mortality patterns among embalmers. *Int. J. Cancer* 31:497–511.

Wardlaw, A. J. 1993. The role of air pollution in asthma. *Clin. Exp. Allergy* 23:81–96.

Waterhouse, J., Muir, C., Correa, P., and Powell, J. eds. 1976. *Cancer incidence in five continents III*, p. 584. Lyon: IARC Scientific Publications 15.

Whittet, H., Lund, V. J., Brockbank, M., and Feyerbend, C. 1991. Serum cotinine as an objective marker for smoking habit in head and neck malignancy. *J. Laryngol. Otol.* 105:1036–1039.

Wjst, M., Reitmeir, P., Dold, S., Wulff, A., Nicolai, T., Freifrau von Loeffelholz-Colberg, E., and von Mutius, E. 1993. Road traffic and adverse effects on respiratory health in children. *Br. Med. J.* 307:596–600.

Section 4. Implications for Human Health and Exposure Limits

THRESHOLD LIMIT VALUES: HOW THEY ARE ESTABLISHED AND THEIR ROLE IN THE WORKPLACE

John Doull

University of Kansas Medical Center, Kansas City, Kansas, USA

It is now almost 50 years since a subcommittee of the American Conference of Governmental Industrial Hygienists (ACGIH) proposed setting limits for 63 workplace air contaminants. Today there are nearly 700 chemical substances and mineral dusts for which TLVs have been adopted by the ACGIH and there are over 90 more substances under consideration. These values were originally called maximum allowable concentrations (MACs), but after 2 yr, the name was changed to threshold limit values (TLVs). When the TLVs were adopted by OSHA, they were called permissible exposure limits (PELs), and NIOSH uses the term recommended exposure limits (RELs) for their values. Both the International Labor Organization and WHO have adopted the designation of occupational exposure limit (OEL), which I like for two reasons. First, this designation clearly indicates that these are worker exposure values, and second, it avoids the term "threshold," which has recently become somewhat controversial.

The ACGIH TLVs are defined as airborne concentrations of substances to which nearly all workers may be repeatedly exposed without adverse effects. We emphasize that there are not fine lines between safe and dangerous concentrations, and we acknowledge that they may not be effective in protecting all segments of the worker population. They are intended to serve only as guidelines for use in the practice of industrial hygiene and not as a relative index of the toxicity of any substance or group of substances. Our TLVs are based on industrial experience and on information from experimental human and animal studies. The basis for each recommendation is available in the documentation that is reviewed annually, along with all proposals for additions and changes. The TLV committee of the ACGIH consists of 22 voting and 2 nonvoting members. Voting members must be employed by a government agency (including the military) or a university, and nonvoting members come from labor unions or profit-making corporations. Recommendations for new TLVs or revisions come from committee members, professional colleagues throughout the world, members of the staff and ACGIH, manufacturers, labor unions, and others. These recommendations go first to the appropriate subcommittee for creation of the documentation and TLV recommendation. Subcommittees cover dusts and inorganics (D&I); hydrogen, oxygen, carbon compounds (HOC); hydrogen,

Address correspondence to John Doull, University of Kansas Medical Center, Kansas City, KS 66103.

oxygen, carbon or other chemical element compounds (MISCO); and carcinogens. The products of the subcommittees then go to the full committee for review and voting, and finally to the board and the ACGIH membership. If approved, they are placed on the Notice of Intended Changes for 1 yr to permit comment by interested parties, after which they are added to the TLV listings. Since the ACGIH is a professional society, the TLV recommendations have no legal status, although they are widely used in the United States and throughout the world.

Three categories of TLVs are currently specified: the threshold limit value time-weighted average (TLV-TWA), the threshold limit value short-term exposure limit (TLV-STEL), and the threshold limit value ceiling (TLV-C). The TLV listing also identifies substances for which the TLV is higher than the OSHA-PEL or NIOSH-REL, those substances for which there are also biological exposure indices (BEIs) for monitoring, substances for which the cutaneous route may significantly contribute to the overall exposure (SKIN NOTATION), substances that have been identified as being either confirmed or suspected human carcinogens, and suggestions for dealing with additive and independent effects from mixtures. The major criteria for setting the ACGIH's TLV is the toxicity of the inhaled substance, although levels have also been established based on the ability of substances to produce irritation, narcosis, or sensitization. Ceiling values are used primarily with substances that evoke an acute toxic effect, and short-term exposure limits (STELs) are basically ceilings for exposures of 15 min or less.

Since the TLV is a time-weighted average based on an 8-h workday, excursions are permitted for substances that do not have a ceiling value, provided that (1) the excursions above the TLV are compensated for during the workday by equivalent excursions below the TLV, and (2) the magnitude of the excursion does not exceed the limits that are recommended in the documentation provided with the TLV listings.

The process used by ACGIH to establish TLVs is closely related to the process that we refer to in toxicology as hazard evaluation. Toxicologists have traditionally considered hazard evaluation to be a judgmental or weight-of-evidence approach based on information relating to the agent, the subject, and the exposure. The first step in this process is to characterize the toxicity of the agent. This is done by identifying all of the potential adverse effects that can result from either acute or chronic exposure to the agent and by establishing dose-response relationships for each of these adverse effects. This information, together with the physical chemical data plus all of the gene-tox, toxicokinetics, and mechanistic studies, constitutes what is referred to as the tox database for the agent. The next step is to determine whether the tox database obtained with the test species is appropriate or relevant for the subject or target species and also for the actual exposure conditions. In those cases where the tox database is adequate and relevant and where there is a threshold or no-effect-level for the adverse effect of interest, the standard (TLV, ADI, tolerance, etc.) is established by simply dividing the NOEL by an appropriate safety factor. The final step in this process is to evaluate the data provided by the industrial hygiene, occupational medicine, and epidemiology studies to ensure consistency with the proposed TLV or standard.

The goal of this approach is to predict a "safe" dose for exposure of the subject to the agent; except for nonthreshold adverse effects, this approach continues to be the most practical and widely used method for protecting the health of the worker, the public, and the environment. Alternative methods, such as the extrapolation or model approach, that focus on the risk rather than on safety have been developed for the risk assessment of nonthreshold carcinogens and mutagens. The selection of the safety factor in the threshold approach is a judgmental decision that not only reflects the confidence of the evaluator in the quality of the data but also takes into account differences in susceptibility and kinetics between the test and target species and also between individual members of the exposed population. In the safety factors associated with the ACGIH TLVs, we may use lower safety factors with irritant substances than with substances producing more serious toxic effects, and since we recognize the priority of target species data over test species data, we also generally use lower safety factors for substances with adequate human data based on long industrial use and medical experience.

My assignment for this presentation was to cover the organization, procedures, and challenges of the ACGIH TLV committee. As one of the newer members of this committee, my concerns about the challenges facing our committee may not reflect the concerns of other members, but I suspect that we would all agree that one of our major challenges is to ensure that our documentation contains all of the current relevant information. With over 600 agents and the additional job of preparing documentation for several new agents each year, this is a major task for the committee members, who are all volunteers, and the staff, who are dedicated but overworked. Our committee welcomes input from all sources, and we are doing two things to facilitate this. First, we are publishing the documentation for all agents on the Notice of Intended Change (NIC) list in *Applied Occupational and Environmental Hygiene* to encourage comments, and second, we are holding public meetings to encourage open interaction between the committee members and scientists from other interested groups.

Another challenge that our committee has addressed during the past few years is the classification of occupational carcinogens. We have developed and adopted a five-category system that will accommodate those animal carcinogens that are unlikely to cause cancer in workers, agents for which the data are inadequate, and agents where there is reliable evidence that they do not produce a significant risk of cancer in humans. Our committee is also exploring the issue of reproductive or fetotoxic hazards to male and female workers and the need for guidelines similar to those in the German or MAK listing. We also have committees that are reevaluating our guidelines for STELs and the SKIN NOTATION, and we have recommended a more flexible comment period for agents on the NIC listing. My final challenge is one that affects not only our TLV committee but all professional and trade associations, and that is our involvement and role in the European Community developments. ACGIH participated in the Hague Workshop in which guidelines were developed for the harmonization of health-based recommendations for occupational exposure limits within the European Community, and we look forward to continuing this kind of interaction to achieve our common goal of improving health in the workplace.

MANAGEMENT OF OCCUPATIONAL EXPOSURE LEVELS IN THE UNITED STATES

Richard W. Niemeier

National Institute for Occupational Safety and Health, Cincinnati, Ohio, USA

The responsibilities of developing occupational exposure limits in the United States is described in terms of the related processes used by the National Institute for Occupational Safety and Health (NIOSH) and the Department of Labor (DOL). In the United States the processes are characteristically lengthy and litigious. Attempts are being made to expedite the promulgation of occupational exposure limits (OELs) by developing and applying more generic forms of decision logic and risk management policies to the process. Dialogue needs to begin on integrating these policies with the concepts of degree of adversity of occupational disease and injury and the acceptability of risk.

The Department of Labor (DOL) is primarily responsible for promulgating occupational safety and health regulations by the U.S. government. In addition to the establishment of the Occupational Safety and Health Administration (OSHA), the Occupational Safety and Health Act (OSHAct) of 1970 also directed the establishment of NIOSH within the Department of Health and Human Services. The specific paragraph of the OSHAct [Public Law 91-596, Section 20 (a) (3)], which directs NIOSH to make recommendations is, "The Secretary of Health, Education and Welfare . . . shall develop criteria dealing with toxic materials and harmful physical agents and substances which will describe exposure levels that are safe for various periods of employment, including but not limited to the exposure levels at which no employee will suffer impaired health or functional capacities or diminished life expectancy as a result of his work experience." Congress in 1977 reformulated the Mine Safety and Health legislation of 1969, which created the Mine Safety and Health Administration (MSHA) within the DOL. Responsibilities for NIOSH in these acts are listed in Table 1.

The Department of Labor (OSHA and MSHA) has established a scheme which is routinely used to establish new and to revise existing regulations. The sources of topics for regulations are varied and may result from DOL initiatives, NIOSH recommendations, Environmental Protection Agency-Toxic Substances Control Act (EPA-TSCA) Section 6 referrals, DOL ad hoc Advisory Committee Recommendations, legal challenges and other sources. This scheme usually follows the procedure listed in Figure 1. An Advance Notice of Proposed Rulemaking (ANPR) or a Request for Information is pub-

This article originally appeared in *Occupational Exposure Limits—Health Based Values or Administrative Norms? Proceedings of the First International Course on OELs*, eds. B. Beije and P. Lundberg. *Arbete ach Hälsa* 15:33–40. This article is not subject to U.S. copyright laws.

Address correspondence to Richard W. Niemeier, PhD, National Institute for Occupational Safety and Health, 4676 Columbia Parkway, Cincinnati, OH 45226-1998, USA.

TABLE 1. NIOSH Responsibilities

OSHAct
Conduct research (direct, grants, contracts)
Develop innovative methods for occupational safety and health problems
Recommend criteria
List toxic substances
Conduct industry-wide studies
Assess manpower needs and develop training programs
Develop reporting procedures in occupational safety and health
Consult in occupational safety and health
MHSAct
Survey toxic substances and physical agents and make recommendations
Administer x-ray surveillance program for coal miners
Test and certify respirators and hazard measuring devices
Conduct mine health research

lished in the *Federal Register*, which allows various times for written responses from interested parties such as labor, government, industry and academia. The usual time period varies from 30 to 90 days. During that period the interested parties assemble and submit comments, advice, and data to the docket which is the official recordkeeping function of the DOL.

After an indeterminate period during which the DOL examines the available information, it publishes a Notice of Proposed Rulemaking (NPRM) in the *Federal Register*, which adheres to standard practices of pre-

Topic Sources

OSHA Initiatives	Advisory Committee
NIOSH Recommendations	Legal Challenges
EPA-TSCA Section 6 Referrals	Other

↓

Advanced Notice of Proposed Rulemaking (ANPR)

↓

Comments, Advice, Data Submissions

↓

Notice of Proposed Rulemaking (NPRM)

↓

Comments, Advice, Data Submissions

↓

Informal Public Hearing (Optional)

↓

Final Rule

FIGURE 1. OSH regulatory scheme in United States.

senting the collected information in a preamble, justification of the standard, and the proposed language of the new regulation. The notice invites written comments, testimony, and new data to be submitted within a certain time to the docket. An informal public hearing is usually scheduled at the time of the publication of the NPRM and interested parties are encouraged to participate. Following this hearing a final rule is published, which becomes the new enforceable regulation after a certain waiting period. NIOSH fully participates in all phases of this rulemaking process.

NIOSH makes formal recommendations in the various forms of documents that it publishes. These include (1) the more formal and exhaustive scientific review in the form of Criteria Documents; (2) shorter and more focused reviews entitled Current Intelligence Bulletins; (3) abbreviated reports based on a collection of new case studies in the form of Alerts; (4) written testimony presented at DOL hearings; and (5) miscellaneous documents such as the NIOSH Pocket Guide to Chemical Hazards, Occupational Guidelines for Chemical Hazards, Hazard Reviews, and Advisories. These miscellaneous documents usually do not include new formal policy recommendations.

NIOSH recognizes that the process of developing criteria documents has historically been long and tedious and that more efficient ways are needed to address the long list of workplace chemicals that are not currently regulated and other hazardous working conditions. This list constitutes approximately 4000 chemical hazards. The bases for recommended standards have been routinely published by other countries and are recognized as very useful in the standard-setting process. NIOSH for about the past 6 years has been collaborating with Sweden and the Nordic Expert Group as well as the World Health Organization/International Programme on Chemical Safety in joint authorship or peer review of criteria documents. It is clearly evident that this collaboration is essential on a world-wide basis, but even this international effort will not seriously reduce the backlog of unregulated hazards (Niemeier, 1992).

The process that NIOSH currently uses to develop criteria documents is outlined in Figure 2. High regard is given to the contributions of the public, as noted at various steps in the process. Various sources of nominations from outside the institute, as well as those topics determined internally, are forwarded to the priority review committee. These nominations are examined and preliminary prioritizations are made based on various selection criteria, which are listed as follows:

1. Frequency of use (occupations, processes, number of industries)
2. Extent of exposure and background exposures
3. Number of workers exposed
4. Severity and type of adverse health effects
5. Technical feasibility of controls
6. Availability of sampling and analytical methods

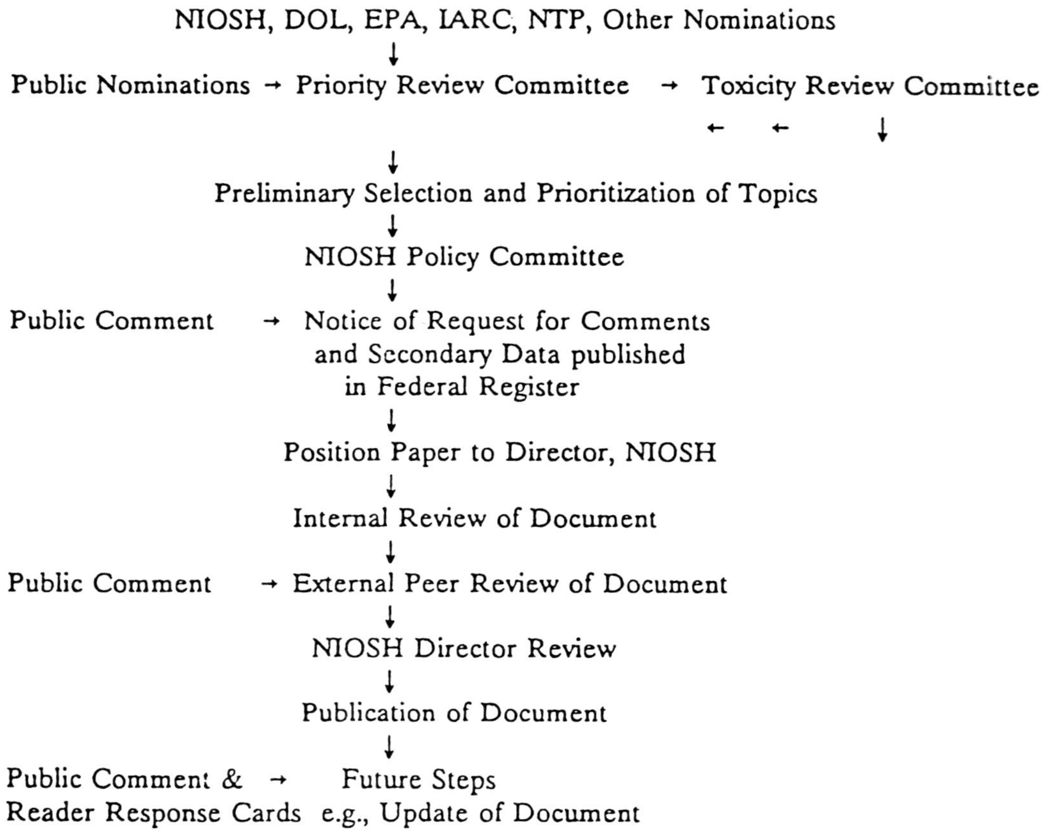

FIGURE 2. NIOSH REL and Criteria Document process.

7. Quality of available data
8. Substitution
9. Interactions—synergistic, multiplicative
10. Existing standards
11. Other considerations

The preliminary selection of topics is presented to the NIOSH Policy Committee (which has a representative from each NIOSH division), and the final list is derived after comment from the committee. The presentation to the policy committee not only helps assess priorities but also enables other divisions to begin to plan research on identified gaps in information that might otherwise prevent full development of a criteria document.

The next step is to publish in the *Federal Register* a request for comments and data from interested parties. This request usually lists data already obtained by NIOSH literature searches but does not necessarily signify that a set of recommendations will follow, since in some cases there is very limited information. The internal process continues with assembly of the position paper, which contains critical data, policy choices and implications of new recommendations, gaps in information, and research recommendations; this paper is presented to the Director of NIOSH. Following discussions, a draft document is developed and submitted for internal and external peer review. After addressing all peer review comments, the final document is presented to the Director of NIOSH for his or her, approval, published, and submitted to the DOL. The DOL also receives a full set of the peer review comments and the responses to these comments.

A final opportunity for public involvement is allowed through the mechanism of attaching reader response cards to the document to permit an open dialogue for future updates of the document.

NIOSH routinely publishes the criteria document in a standard format, for which a detailed outline is presented in Table 2. These chapters not only identify the hazard and review the physical/chemical, toxicological, industrial hygiene, and epidemiological information but also include recommendations for exposure, work practices, protective equipment, labeling, training, engineering and work practice controls, and medical and environmental monitoring and recordkeeping. NIOSH considers these elements of the criteria document to be essential in leading to the prevention of future occupational disease, injury, and death.

TABLE 2. Contents of NIOSH Criteria Documents

I. Recommendations
  - A. Definitions/glossary/abbreviations
  - B. Recommended exposure limit (REL)
  - C. Medical monitoring
  - D. Labeling and posting
  - E. Protective clothing and equipment
  - F. Worker information and training
  - G. Work practices and engineering
  - H. Sanitation
  - I. Exposure monitoring
  - J. Recordkeeping

II. Introduction
  - A. Scope of document
  - B. Industry or process description
  - C. Chemical and physical properties
  - D. Production and use
  - E. Number of workers
  - F. Exposure characteristics
  - G. Existing health standards

III. Effects of exposure
  - A. Health Hazards
    - 1. Animal toxicity
    - 2. Human toxicity
  - B. Safety hazards
  - C. Summary and conclusions

IV. Basis for recommended standard
  - A. Assessment of effects
  - B. Risk assessment
  - C. Technical feasibility
  - D. Recommendations

V. Hazard identification
  - A. Sampling and analytical methods
  - B. Sampling strategy
  - C. Medical monitoring
    - 1. Preplacement
    - 2. Periodic
    - 3. Biological
    - 4. Recordkeeping

VI. Methods of protecting workers
  - A. Informing workers (information and training)
  - B. Work practices
  - C. Engineering
  - D. Personal protective equipment
  - E. Availability of substitutes
  - F. Sanitation
  - G. Labeling posting
  - H. Exposure monitoring
  - I. Recordkeeping and surveillance

VII. Research needs

VIII. Reference

IX. Appendices

NIOSH is currently exploring and attempting to develop more efficient processes to address the majority of unregulated hazards in an expedited manner. This activity has been tentatively labeled the I-REL (Interim-Recommended Exposure Limit) process. Based on the extensive work that was performed for the 1989 OSHA PEL regulation, some additional insights have guided these new efforts to help lessen the burdens of the process. It was evident from the 1989 PEL exercise that very limited data are available to permit complete evaluation of each chemical or hazard. In the near future computer modeling of analogous or missing information to estimate toxicity using structure activity relationships will be necessary to fill the information gaps. NIOSH has also begun an analysis, using the past 22 years of experience in making recommendations, to define the critical elements and identify which of these data must be collected to make scientifically valid and meaningful recommendations. This may lead to the development of an expert system using artificial intelligence techniques to arrive at preliminary recommendations similar to those being developed in other fields (Harber et al., 1991; Harber & McCoy, 1989; Jelovsek et al., 1990; Tait, 1992).

Collaboration has also continued with the ONE Committee (OSHA, NIOSH, EPA, and MSHA) to coordinate the process. A new activity has been initiated to determine the criteria, data needs, and a decision logic for the determination of the "skin" designation. As progress is made in addressing this and other critical issues it is anticipated that better use of the broad authority allowed under EPA-TSCA and more thorough coordination with DOL will help to acquire a prioritized minimum set of data necessary for these policy decisions.

Finally, as part of this new approach NIOSH has begun to formulate policies for application of quantitative risk assessment results in developing RELs and for controlling risks to specific hazardous diseases. The disease end points include carcinogenesis, mutagenesis, reproduction, skin effects, and sensitive populations. Generic policies are also being formulated to guide future recommendations on risk management decisions, which may be dependent on disease and injury type and outcome and on severity of the risk. The risk management areas include hazard communication and training, recordkeeping, medical surveillance, environmental monitoring and sampling strategy, personal protective equipment, and ventilation. These policies are being integrated to fit the preliminary concept of a sliding, but overlapping, scale of adverse risks suggested in Table 3. Continual attention needs to be given during the development of each of these policies, centering on the issues of degree of adversity of disease and injury and acceptability of risk. Other agencies have begun addressing these same issues in terms of negligible, tolerable, and unacceptable risks, either on a voluntary or involuntary basis, and the risk management issues that accompany these terms (Illing, 1991a, 1991b; Koundakjian & Illing, 1991).

TABLE 3. Scale of Adverse Risks

Category	Characteristics of health effects	Example	Significant lifetime risk
High	Life shortening Severe effect on quality of life Irreversible Major societal burden	Lung cancer Birth defects Spontaneous abortion Germinal mutations	$\geq 10^{-6}$ to $\leq 10^{-5}$
Moderate	Variable—Mild to fatal Reversible depending on extent of disease	Pulmonary function loss Musculoskeletal disorders Reversible semen quality Reversible hormone imbalances with secondary effects	$\geq 10^{-4}$ to $\leq 10^{-5}$
Low	Fully reversible Short duration Generally not disabling	Dermatitis Nasal irritation Hormone imbalance without secondary effects	$\geq 10^{-4}$ to $\leq 10^{-3}$

REFERENCES

Harber, P., and McCoy, J. 1989. Predicate calculus, artificial intelligence, and workers' compensation. *J. Occup. Med.* 31:484–489.

Harber, P., McCoy, J., Shimozaki, S. Coffman, P., and Bailey, K. 1991. The structure of expert diagnostic knowledge in occupational medicine. *Am. J. Ind. Med.* 19:109–120.

Illing, H. P. 1991a. Extrapolating from toxicity data to occupational exposure limits: some considerations. *Ann. Occup. Hyg.* 35:569–580.

Illing, H. P. 1991b. Possible risk considerations for toxic risk assessment. *Hum. Exp. Toxicol.* 10:215–219.

Jelovsek, F., Mattison, D., and Young, J. 1990. Eliciting principles of hazard identification from experts. *Tera* 42:521–533.

Koundakjian, P., and Illing, H. P. 1991. Introduction. In *Risk management of chemicals*, pp. 1–13. London: Crown.

Niemeier, R. 1992. Setting occupational exposure standards—Past, present and future practice. *IOHA Conf.*, Brussels, Belgium, December 7.

Tait, K. 1992. The workplace exposure assessment expert system (Workspert). *Am. Ind. Hyg. Assoc. J.* 53:84–98.

INHALATION RfC METHODOLOGY: DOSIMETRIC ADJUSTMENTS AND DOSE-RESPONSE ESTIMATION OF NONCANCER TOXICITY IN THE UPPER RESPIRATORY TRACT

Annie M. Jarabek

Environmental Criteria and Assessment Office, U.S. Environmental Protection Agency, Research Triangle Park, North Carolina, USA

Evaluation of noncancer toxicity involves the evaluation of an array of different adverse end points across various target tissues. The U.S. Environmental Protection Agency (EPA) inhalation reference concentration (RfC) methodology represents an approach for derivation of dose-response estimates for noncancer toxicity. The RfC methodology differs from other methodologies described at this symposium in that it incorporates dosimetric adjustments to account for species-specific relationships of exposure concentration to deposited/delivered dose in order to extrapolate across species. The physicochemical characteristics of the inhaled chemical are considered as key determinants to its interaction with respiratory tract components and ultimate deposition. This article discusses the RfC approach to noncancer toxicity with emphasis on the evaluation of toxicity in the upper respiratory tract (URT). The derivation of default dosimetric adjustments for extrapolation of URT delivered dose across species is also presented. Areas of controversial interpretation or those requiring additional research are highlighted.

This article outlines the U.S. EPA's inhalation reference concentration (RfC)[1] methodology and highlights issues of interpretation and extrapolation unique to the upper respiratory tract (URT) that would benefit from directed laboratory research. The RfC methodology is an approach for the derivation of dose-response estimates for noncancer toxicity of inhalation exposures and has been described elsewhere (Jarabek et al., 1990). Some of the forthcoming revisions to the methodology are also discussed.

Dose-response[2] analysis is a component of risk assessment and is separate from risk management as defined by the framework proposed by the

[1]Inhalation reference concentration (RfC) was originally called inhalation reference dose (RfD_i).

[2]The NAS recommended that the scientific aspects of risk assessment should be explicitly separated from the policy aspects of risk management. Risk assessment was defined as the characterization of the potential adverse health effects of exposures to environmental hazards, and consists of four steps: (1) hazard identification: the determination of whether a chemical is or is not causally linked to a particular health effect; (2) dose-response assessment: the estimation of the relation between the magnitude of exposure and the occurrence of the health effects in question; (3) exposure assessment: the determination of the extent of human exposure; and (4) risk characterization: the description of the nature and often the magnitude of human risk, including attendant uncertainty.

Address correspondence to Annie M. Jarabek, Environmental Criteria and Assessment Office, MD-52, U.S. Environmental Protection Agency, Research Triangle Park, NC 27711.

U.S. National Academy of Sciences (U.S. NAS, 1983). Unlike evaluation of carcinogenicity, which is usually based on tumor outcome, evaluation of noncancer toxicity requires determination of the critical toxic effect from an array of different target tissues and end points of various severities within specific disease continuums. The RfC guidance on how to evaluate these toxicity data is described with emphasis on interpretation issues for nasal lesions.

The RfC methodology differs from other approaches described in this volume in that it incorporates dosimetric adjustments to account for the species-specific relationships of exposure concentrations to deposited/delivered doses. The physicochemical characteristics of the inhaled agent are considered as key determinants to its interaction with respiratory tract components and ultimate disposition. The dosimetric adjustments used to calculate doses to the upper respiratory tract (URT) for particles versus gases are presented. Data needed to provide parameters and refine the default modeling are discussed.

DEFINITION AND ASSUMPTIONS

The inhalation reference concentration (RfC) is defined as an estimate (with uncertainty spanning perhaps an order of magnitude) of a continuous inhalation exposure to the human population (including sensitive subgroups) that is likely to be without appreciable risk of deleterious noncancer health effects during a lifetime. The RfC approach described here represents one component of the risk assessment process, the dose-response[3] assessment component, and the RfC estimate must be compared against an exposure estimate in order to characterize risk. The attendant uncertainties and default assumptions of the dose-response estimate should be evaluated in context with those of the exposure estimate (e.g., assumptions of fate and transport modeling or type of sampling and averaging time of a measured exposure) to ascertain whether the two are appropriate to integrate. Further, the assumptions and uncertainties of the risk characterization components (dose-response and exposure assessments) must be explicitly communicated to the risk management arena for application to intended management scenarios. Often RfC values are compared inappropriately with risk management numbers that are intended for different exposure scenarios and populations. The intended use of a dose-response or risk management estimate influences its derivation (Jarabek & Segal, 1993). Table 1 provides a comparison for the different assumptions and derivation methods inherent with the risk assessment and risk management estimates discussed at this symposium.

[3]Although the strict definitions of "dose," "response," and "effect" are recognized, the conventions of the NAS paradigm are used, with the RfC being synonymous with a "dose-response" assessment. Therefore, in the qualitative sense, the term "dose" may encompass administered dose (i.e., exposure concentration), delivered dose, or target tissue dose. Likewise, "response" in the qualitative sense is an indication of an adverse influence regardless of whether the data were measured as quantal, count, continuous, or ordered categorical.

TABLE 1. Comparison of Risk Assessment and Risk Management Estimates

Estimate	NAS paradigm	Exposure scenario	Effect level	Population	Database	Dosimetry	SF or UF[a]
ACGIH TLV-TWA	Management	8 h/day 40 h/wk 40 yr	Impairment of health or freedom from irritation Narcosis Nuisance Stress	Healthy worker	Industrial experience Experimental human and animal	No	SF
NIOSH REL	Characterization	10 h/day 40 h/wk "Working lifetime"	Impairment of health or functional capacity and technical feasibility	Healthy worker	Medical Biological Chemical Trade	No	SF
OSHA PEL	Management	8 h/day 40 h/wk 45 yr	Impairment of health or functional capacity and technical feasibility	Healthy worker	Medical Biological Chemical Trade	No	SF
EPA RfC	Dose-response	24 h/day 70 yr	NOAEL	General population including susceptible	Occupational Experimental human and animal	Yes	UF
ATSDR MRL	Dose-response	24 h/day 70 yr	NOAEL	General population including susceptible	Occupational Experimental human and animal	No	UF

[a]SF = safety factor; UF = uncertainty factor for explicit extrapolations applied to data.

Noncancer toxicity is defined as adverse health effects or toxic end points, other than cancer and gene mutations, that are due to the effects of environmental agents on the structure or function of various organ systems. In the case of inhalation exposures, these effects include those in the respiratory tract as the portal-of-entry as well as effects that follow absorption and distribution of the toxicant to sites remote to its entry location. Most chemicals that produce noncancer toxicity do not cause a similar degree of toxicity in all organs, but usually demonstrate major toxicity to one or two organs referred to as the target organs (Klaassen, 1986). Generally, based on current understanding of homeostatic and adaptive mechanisms, dose-response assessment procedures approach noncancer health effects as though there is an identifiable threshold below which adverse effects are not observable. For an individual, the threshold concept presumes that a range of exposures from zero to some finite value can be tolerated by the organism without adverse effects. This threshold will vary from one individual to another, so that there will be a distribution of thresholds in an exposed population. The identification of a threshold distinguishes noncancer assessment approaches from those for carcinogenic end points that are typically approached as resulting from nonthreshold processes. Identification of a threshold is a function of the available data and current understanding of the exposure-dose-response continuum,[4] so that the identification may be revised as more information, such as data from studies encompassing additional end points or more sensitive indicators of toxicity (e.g., mechanistic determinants such as cell proliferation), is developed and evaluated. As more mechanistic insights are gained into relationships between exposure, dose, and response, however, this distinction between approaches for noncancer and cancer response assessment will blur for some chemicals.

Another distinction between approaches for cancer and noncancer risk assessment is that the definition of response for the latter may be defined as either an increase in incidence or in severity of an end point, whereas tumors are always counted as quantal (dichotomous). For dichotomous responses, dose-response model estimates describe probabilities of events in individuals. These probabilities can also be thought of as the fraction of a population that will show the response. Since continuous effects are graded and measured (e.g., percent change in a parameter), models estimate expected changes in individuals. These expected changes can be expressed as shifts in population means. For practical and sound conceptual reasons, responses and effects can be considered to be identical (Klaassen, 1986). That is, in a qualitative sense when trying to ascertain if a toxic agent exerts

[4]Because the toxic moiety for a given response is not always proportional to the applied dose of a compound, emphasis has been placed on the need to clearly distinguish between exposure concentration and dose to critical target tissues. The term "exposure-dose-response assessment" has been recommended as more accurate and comprehensive (Andersen et al., 1992). This expression refers not only to the determination of the quantitative relationship between exposure concentrations and target tissue dose, but also to the relationship between tissue dose and the observed/expected responses in laboratory animals and humans.

an adverse influence, the distinction is unimportant. The distinction between effects and responses must be carefully applied when employing mathematical models to calculate estimates and development of approaches that adequately address this issue, representing a future challenge for noncancer assessment.

DERIVATION

This section provides an outline of the steps to calculate an inhalation RfC. Aspects of the derivation that are peculiar or problematic when applied to toxicity in the URT are highlighted.

Database Evaluation

By definition, a database for derivation of a dose-response estimate for noncancer toxicity should ensure that both appropriate and adequate numbers of end points have been evaluated. Table 2 shows that the minimum

TABLE 2. Minimum Animal Bioassay Database for Various Levels of Confidence in the Inhalation Reference Concentration (RfC)

Mammalian database[a]	Confidence	Comments
1. A. Two inhalation bioassays[b] in different species B. One two-generation reproductive study C. Two developmental toxicity studies in different species	High	Minimum database for high confidence
2. 1A and 1B, as above	Medium to high	
3. Two of three studies, as above in 1A and 1B; one or two developmental toxicity studies	Medium to high	
4. Two of three studies, as above in 1A and 1B	Medium	
5. One of three studies, as above in 1A and 1B; one or two developmental toxicity studies	Medium to low	
6. One inhalation bioassay[c]	Low	Minimum database for estimation of an RfC

[a]Composed of studies published in refereed journals, final quality assured/quality checked and approved contract laboratory studies, or core minimum Office of Pesticide Programs rated studies. It is understood that adequate toxicity data in humans can form the basis of an RfC and yield high confidence in the RfC without this database. Pharmacokinetic data that indicate insignificant distribution occurs remote to the respiratory tract may decrease requirements for reproductive and developmental data.

[b]Chronic data.

[c]Chronic data preferred but subchronic acceptable.

requirement for derivation of an RfC with low confidence is a well-conducted subchronic inhalation bioassay that evaluated a comprehensive array of end points, including an adequate evaluation of respiratory tract effects, and established an unequivocal no-observed-adverse-effect level (NOAEL) and lowest-observed- adverse-effect level (LOAEL). Chronic inhalation bioassay data in two different mammalian species, developmental studies in two different mammalian species, and a two-generation reproductive study may be required to establish high confidence. The rationale supporting these database requirements is that, since the objective of the RfC is to serve as a lifetime estimate, all potential end points at various critical life stages must be evaluated. Well-defined and -conducted subchronic toxicity studies are considered to be reliable predictors of many forms of chronic toxicity, with the notable exceptions of carcinogenic, teratogenic, and reproductive effects. The specific requirement for adequate respiratory tract evaluation arises from the increased potential for the portal-of-entry tissue to interact intimately with chemicals. Consideration of the location and number of section levels and the level of detail in reporting nasal lesions can enter into the determination of database adequacy, especially when the chemical is expected to manifest respiratory tract toxicity (e.g., an irritant). Dosimetry data that indicate distribution to extrarespiratory tract sites is insignificant may obviate the requirement for reproductive and developmental data. If these minimum database requirements are not met, an RfC is not derived.

Designation of Effect Levels

Designation of effect levels can be defined as the association of adversity[5] with exposure concentration and is one of the most difficult procedures of any dose-response analysis for noncancer toxicity. The NOAEL is the highest level tested in a study at which the specified adverse effect is not produced; that is, by definition the NOAEL is a subthreshold level (Klaassen, 1986). The LOAEL is an exposure in a study at which the specified adverse effect is produced and is therefore above threshold. Table 3 presents the four types of effect levels that may be applicable when evaluating an individual study. The information in this table was predominantly based on systemic toxicity and would benefit from evaluation for application as guidance for the lower respiratory tract and URT. In particular, integration of histopathologic versus functional end points and evaluation of the proximal to distal distribution of lesions, given the differences across species, require consideration.

The distinction between adverse and nonadverse effects remains prob-

[5]Adverse effects are considered to be functional impairments or pathological lesions that may affect the performance of the whole organism or that reduce an organism's ability to respond to an additional challenge (*Federal Register*, 1980). One of the major problems encountered with this concept is the reporting of "observed effect levels" as contrasted to "observed adverse effect levels." The terms "adverse" and "not adverse" are at times satisfactorily defined, but because more subtle responses continue to be identified due to increasingly sophisticated testing protocols, scientific judgement is needed regarding the exact definition of adversity.

TABLE 3. Effect Levels Considered in Deriving RfCs in Relationship to Empirical Severity Rating Values

Effect or No-Effect Level	Rank	General effect
NOEL	0	No observed effects
NOAEL	1	Enzyme induction or other biochemical change, consistent with possible mechanism of action, with no pathologic changes and no change in organ weights
NOAEL	2	Enzyme induction and subcellular proliferation or other changes in organelles, consistent with possible mechanism of action, but no other apparent effects
NOAEL	3	Hyperplasia, hypertrophy, or atrophy, but no change in organ weights
NOAEL/LOAEL	4	Hyperplasia, hypertrophy, or atrophy, with changes in organ weights
LOAEL	5	Reversible cellular changes including cloudy swelling, hydropic change, or fatty changes
(LO)AEL[a]	6	Degenerative or necrotic tissue changes with no apparent decrement in organ function
(LO)AEL/FEL	7	Reversible slight changes in organ function
FEL	8	Pathological changes with definite organ dysfunction that are unlikely to be fully reversible
FEL	9	Pronounced pathologic changes with severe organ dysfunction with long-term sequelae
FEL	10	Death or pronounced life-shortening

Note. Adapted from DeRosa et al. (1985) and Hartung (1986). Ranks are from lowest to highest severity.

[a]The parentheses around the "LO" in the acronym "LOAEL" refer to the fact that any study may have a series of doses that evoke toxic effects of rank 5 through 7. All such doses are referred to as adverse effect levels (AELs). The lowest AEL is the (LO)AEL.

lematic. Although diseases are dynamic processes (e.g., injury, adaptation, dysfunction, or healing), a pathologist records a morphologic change at a single point in time (e.g., scheduled sacrifice), and these "freeze-frame" data are used to determine the probable cause and pathogenesis (past) and probable progression or outcome (future). Determining the degree of adversity in order to designate an effect level as a NOAEL or LOAEL requires interpretation of the data based on an ability to deduce the preceding events that have led to the observed change and to predict the outcome or progression. The relationship between structural alterations and altered function is not usually straightforward. Determining whether altered morphology is an adaptive response or truly an expression of toxicity (functional impairment) can be extremely difficult and even controversial (Burger et al., 1989; Ruben & Rousseaux, 1991). Structural alterations may occur while normal function remains unimpaired in target tissues with functional reserve, such as the respiratory tract, liver, and kidney. Not all tissues demonstrate this high reserve—for example, the central nervous system (CNS) can compensate to only a limited degree so that "focal" damage in this system may be critical. The lack of observed functional change may be due to failure to detect subtle or unknown functional changes rather than to their absence. Temporal-spatial patterns are particularly challenging when evaluating toxicologic pathology. Problems concerning time include reversibility, adaptation versus toxicity, progression versus regression, and peracute lethal toxicity. Problems concerning space are limited to missing the lesion completely or missing a relevant area because of sampling method.

Morgan (1991) has recommended a systematic but flexible approach to evaluation of lesions in the URT that considers selection of section level in context with the physicochemical characteristics of the inhaled gas (e.g., water solubility and reactivity), the role of factors that may account for lesion distribution (e.g., dosimetry and tissue susceptibility), and development of a pathogenesis profile or a chronological order of events (e.g., degenerative, adaptive, and adaptive/regenerative changes versus time). The nasal diagrams proposed by Mery et al. (1994) and used by Lomax et al. (1994) in this volume offer an approach to recording data and mapping lesions that aids this type of interpretation strategy. This approach is also likely the best to compile the data and precludes the restraint to interpretation and mathematical modeling presented by data scored categorically for severity (e.g., + = mild, ++ = moderate; and +++ = severe) and/or without sufficient section detail with respect to lesion location.

It should be emphasized that considerations of assigning adversity are not unique to the NOAEL/LOAEL paradigm. The limitations of the NOAEL/LOAEL approach have been discussed (Crump, 1984; U.S. EPA, 1990) and include the property that designation of effect levels is a function of spacing of the exposure levels and sample sizes used in the experimental design. Because concentration-response modeling, sometimes referred to as the "benchmark dose"[6] (BMD) approach, corrects these limitations, it has been proposed to replace the NOAEL/LOAEL approach. Regardless of the approach, however, designation of the specified health effect remains a key assessment decision. That is, the toxicological definition of adversity (i.e., the decision that a specified effect is adverse and what the associated severity is), particularly across end points, remains as a critical determination regardless of the mathematical models applied. For example, using the BMD approach, the concentration at the lower 95% confidence limit on a maximum likelihood estimate at 10% or 5% incidence (BMDL10 or BMDL05, respectively) for dichotomous responses (e.g., presence/absence of a particular nasal lesion) has been proposed as a surrogate NOAEL for use in derivation of an RfC. It is presumed that this surrogate NOAEL would be derived from a study most representative of the threshold region for the data array (see choice of principal study). Due to dosimetry considerations, it is thus

[6]The benchmark dose approach involves the use of a specific mathematical model (e.g., Weibull, logistic, polynomial) to determine a concentration (applied dose) and its lower confidence bound that is associated with a predefined effect measure (e.g., 10% response of a dichotomous measure) as the "benchmark." Application of this approach (Barnes et al., 1994) has been proposed for developmental end points, which are generally dichotomous (quantal) in nature, but it has yet to be applied widely to other noncancer outcomes.

likely that for many compounds the BMDL10 or BMDL05 for nasal lesions will be considered as a critical effect, unless research suggests otherwise.

Duration Adjustment

Many inhalation toxicity studies of laboratory animals use discontinuous exposure regimens, such as, exposures of 6–8 h/day and 5 days/wk. By definition, the RfC is derived to estimate a benchmark level for continuous exposure (and by extension, the RfC level is assumed to be protective for discontinuous exposures). A normalization to that exposure (e.g., 24 h/day for a lifetime of 70 yr) is needed to adjust for the wide variety of experimental exposures to permit comparisons between studies. Attempts are made to derive RfCs based on experimental exposure duration and frequency data that most closely approximate the assumed human exposure scenario (e.g., chronic bioassay data or 24 h/day exposures in Thomas domes). The default convention is to perform a linear prorate adjustment. The rationale is that the resultant human exposure concentration should be the concentration times time ($C \times T$) equivalent of the experimental animal level. The adjustments of laboratory animal data and for human occupational data are as follows.

To calculate duration-adjusted exposure levels in milligrams per cubic meter for experimental animals, the equation is:

$$\text{NOAEL}^{*}_{[\text{ADJ}]}\ (\text{mg/m}^3) = E\ (\text{mg/m}^3) \times D\ (\text{h/24 h}) \times W\ (\text{days/7 days}) \qquad (1)$$

where the $\text{NOAEL}^{*}_{[\text{ADJ}]}$ is the NOAEL or analogous effect level obtained with an alternate approach such as the BMD, adjusted for duration of experimental regimen; E is the experimental exposure level; D is the number of hours exposed per 24 h; and W is the number of days of exposure/7 days. This duration adjustment is also applied to LOAELs but is not applied when optimal physiologically based pharmacokinetic (PBPK) models are used.

The default approach for human occupational exposure scenarios is to adjust by the default occupational ventilation rate and for the intermittent work week schedule:

$$\text{NOAEL}_{[\text{HEC}]} = \text{NOAEL}\ (\text{mg/m}^3) \times (V_{\text{Eho}}/V_{\text{Eh}}) \times (5\ \text{days/7 days}) \qquad (2)$$

where the $\text{NOAEL}_{[\text{HEC}]}$ is the human equivalent concentration for the assumed ambient scenario, the NOAEL is the occupational exposure level (time-weighted average), V_{Eho} is the human occupational default minute volume (10 m^3/day), and V_{Eh} is the average human default minute volume (20 m^3/day).

This linear $C \times T$ adjustment is acknowledged as tenuous since toxicity of a given chemical depends on the concentration, duration, and frequency of exposure (e.g., an effect may be mediated by peak tissue concentration vs. cumulative tissue dose). Mechanistic data are needed on the relationship of noncancer nasal lesions to concentration, duration, and frequency of exposures.

Dosimetric Adjustments

The various species used in inhalation toxicology studies do not receive identical doses in comparable respiratory tract regions (extrathoracic, ET; tracheobronchial, TB; and pulmonary, PU) when exposed to the same aerosol or gas (Brain & Mensah, 1983). The adverse toxic effect may be more directly related to the quantitative pattern of deposition within the respiratory tract than to the exposure concentration, because the regional deposition pattern determines not only the initial lung tissue dose but also the specific pathways and rates by which the inhaled material is cleared and redistributed (Schlesinger, 1985). The physicochemical characteristics of inhaled particles or gases also influence the disposition (includes deposition, absorption, distribution, metabolism, and elimination) of chemicals (Jarabek et al., 1990).

Mathematical dosimetry models that incorporate mechanistic determinants of disposition have been useful in describing relationships between exposure concentration and target tissue dose and to applications of risk assessment (Andersen et al., 1992; Overton, 1984). However, because much information on the mechanistic determinants of disposition is lacking for most chemicals to which the RfC methodology will be applied, default dosimetry adjustments are derived from models that incorporate only major determinants. The default adjustments are for particles versus gases, and within gases, for three categories based on reactivity (including both dissociation and local metabolism) and water solubility. It is recognized that these are default dosimetry models, so that the use of more detailed models that incorporate a more comprehensive structure to include information on the determinants (parameters) of the exposure-dose-response continuum is considered optimal for this extrapolation. For example, the model of Yu and Yoon (1990) was used to perform interspecies dose extrapolation for derivation of the RfC for diesel exhaust, since this model incorporates mechanisms of particle clearance and thus allows estimation of retained dose. This model was judged to be more appropriate than the default deposition model described later, since the observed toxicity is related to particle overload (U.S. EPA, 1992).

Use of a PBPK model obviates the duration adjustment as just described. In order to extrapolate laboratory animal data using a PBPK model, the animal experimental regimen (e.g., 6 h/day, 5 days/wk) is simulated and the resultant appropriate dose metric (e.g., metabolite tissue concentration) calculated. The simulation model is then used with the human parameters for the human exposure scenario (e.g., 24 h/day) to ascertain the exposure concentration that results in an equivalent dose metric. This exposure concentration is the human equivalent concentration (HEC).

The HEC is calculated with the default dosimetric adjustment factors as:

$$\text{NOAEL}^*_{[\text{HEC}]}\ (\text{mg/m}^3) = \text{NOAEL}^*_{[\text{ADJ}]}\ (\text{mg/m}^3) \times \text{DAF}_r \tag{3}$$

where $NOAEL^*_{[HEC]}$ is the NOAEL or analogous effect level obtained with an alternate approach such as the BMD, dosimetrically adjusted to an HEC; $NOAEL^*_{[ADJ]}$ is defined in Eq. 1; and DAF_r is a dosimetric adjustment factor for respiratory tract region r (ET, TB, PU, or TOTAL), either the regional deposited dose ratio ($RDDR_r$) for particles or the regional gas dose ratio ($RGDR_r$) for gases (described later). The DAF represents a multiplicative factor to adjust an observed exposure concentration in a particular laboratory species to an exposure concentration for humans that would be associated with the same delivered dose.

The default adjustment factors for the extrapolation of URT effects are described below for particle and gas exposures, respectively.

DAF for Particle Exposures

Fractional deposition of particles in each region of the respiratory tract is calculated using a default empirical model (Ménache et al., 1994). An empirical model is a system of equations fitted to experimental deposition data. This type of model was chosen because it does not require the detailed information needed for theoretical models (e.g., exact airflow patterns, complete measurements of the branching structure of the respiratory tract, pulmonary region mechanics). An empirical model of regional fractional deposition has been used to calculate deposition in humans (Miller et al., 1988).

Measurement techniques for deposition are such that only general regions can be defined, so that localized deposition (e.g., respiratory vs. olfactory epithelium) is not estimated. Nevertheless, deposition data were available under the same nose-only exposure conditions for five laboratory animal species (unanesthetized) for a wide range of particle sizes (Raabe et al., 1988). Although other insoluble particulate deposition data for laboratory animals exist, these data were chosen due to the experimental design and reporting detail.

The logistic function has mathematical properties that are consistent with the shape of the function for deposition efficiency, so that for the URT, deposition efficiency was fitted as a function of an impaction parameter, d_{ae}^2Q, where d_{ae} is the aerodynamic diameter of the particle and Q is the volumetric flow rate (Ménache et al., 1994). Figure 1 shows the fit of the logistic model of extrathoracic deposition efficiency versus the impaction parameter for the rat. These exposures were essentially to monodisperse particles, so that for polydisperse particle exposures ($\sigma_g > 1.3$), deposition fractions are calculated by integrating the product of the monodisperse deposition probabilities and the truncated log-normal distribution over the inspirable particle domain (Ménache et al., 1994).

The fractional depositions in the ET region calculated for the laboratory species of interest (e.g., the species used to identify a NOAEL) and humans by the respective models are then used in the following equation to com-

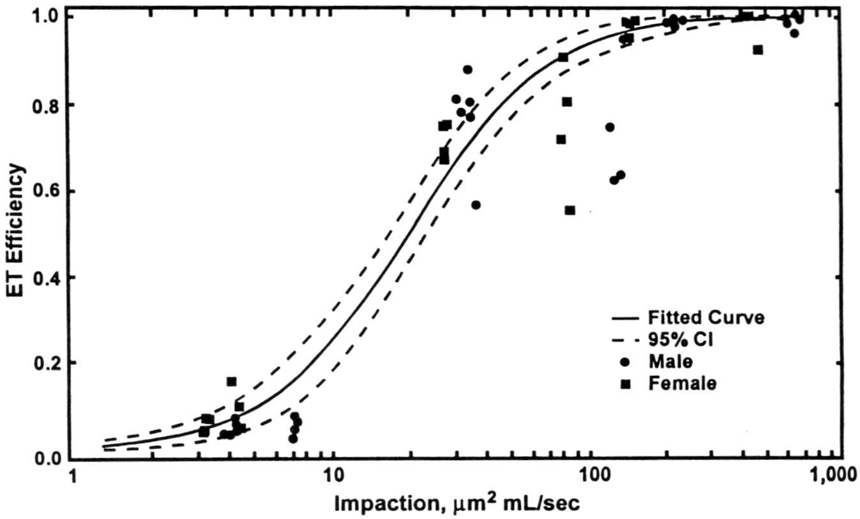

FIGURE 1. Extrathoracic deposition efficiency modeled as a logistic function of impaction parameter, d_{ae}^2Q, where d_{ae} is the particle aerodynamic diameter (μm) and Q is the ventilatory flow rate (ml/s). The empiric model is described in detail elsewhere (Ménache et al., 1994). Data shown for rats are those of Raabe et al. (1988).

pute a regional deposited dose ratio ($RDDR_{ET}$) as the product of a series of four ratios:

$$RDDR_r = \frac{(C_i)_A}{(C_i)_H} \times \frac{(\text{Normalizing factor})_H}{(\text{Normalizing factor})_A} \times \frac{(\dot{V}_E)_A}{(\dot{V}_E)_H} \times \frac{(F_r)_A}{(F_r)_H} \quad (4)$$

where $\dot{V}_E$ is the ventilation rate and F_r is the fractional deposition for a particular respiratory tract region r. For the purposes of calculating the $RDDR_r$, the exposure concentrations (C_i) for the laboratory animal, A, and human, H, are assumed to be the same, that is, as though the human had been in the exposure chamber rather than the test species. Therefore, the $RDDR_r$ provides a factor to adjust for the difference in dose delivered to the target tissue under the same exposure scenario. The first term thus equals 1. The second term is the ratio of the normalizing factors for the human and laboratory animal of interest. For effects in any of the three regions of the respiratory tract, surface area is the recommended normalizing factor. Surface area data to be used as default values for five laboratory animal species are available elsewhere (M. G. Ménache, personal communication, 1994). Default ventilation rates are calculated allometrically from default body weights for studies of subchronic or chronic duration (U.S. EPA, 1988). This $RDDR_{ET}$ would be used in Eq. 1 to calculate an $NOAEL_{[HEC]}$ for an ET effect of particle exposures.

DAF for Gas Exposures

Numerous model structures have been used to describe toxicant uptake in the respiratory tract. The type of model often reflects the physicochemical characteristics of the gases to which they are applied. For example, the model of Miller et al. (1985) for the respiratory tract uptake of ozone (highly reactive and poorly water soluble) is a detailed, distributed parameter model, whereas respiratory uptake for styrene (nonreactive, insoluble) can be described by a single ventilation-perfusion model compartment (Ramsey & Andersen, 1984). The chemical-specific or class-specific nature of these models has been dictated by the physicochemical characteristics of the subject gases, and therefore any single model is not applicable to the broad range of gases that the RfC methodology must address. A gas categorization scheme was constructed based on physicochemical characteristics as determinants of gas uptake. This scheme is shown in Figure 2. The dominant determinants are used to construct a conceptual framework that directs development of the default dosimetry model structures. These models are reduced further by simplifying assumptions to forms requiring a minimal number of parameters in order to derive a DAF for each category that is commensurate with the amount of data typically available.

Gases in category 3 are relatively water-insoluble and unreactive in the ET and TB surface liquid and tissue and thus result in a relatively small dose to these regions. The uptake of category 3 gases is predominantly in the pulmonary region and is perfusion limited. The site of toxicity of these

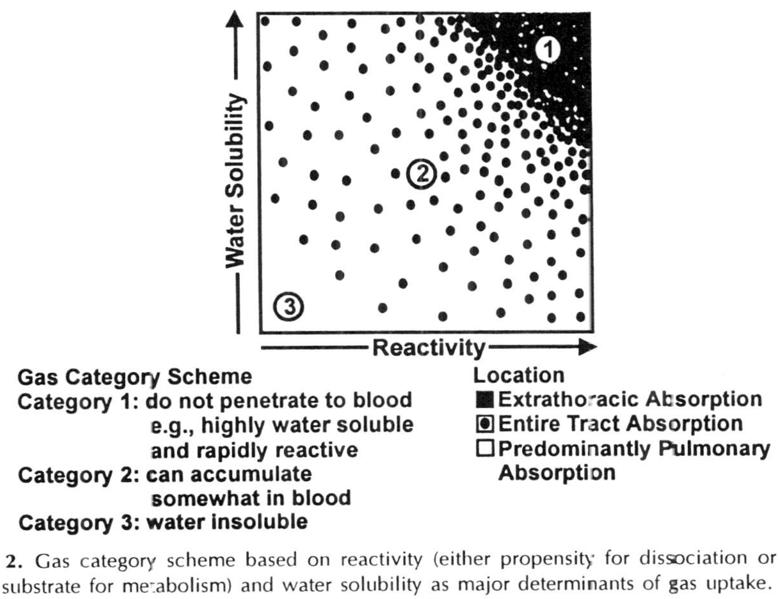

FIGURE 2. Gas category scheme based on reactivity (either propensity for dissociation or ability to serve as substrate for metabolism) and water solubility as major determinants of gas uptake. Definitive characteristic of each category and anticipated location for uptake are shown.

gases is generally at sites remote to the respiratory tract and requires a compartmental approach to describe distribution to various systemic tissues. Gases falling in category 3 are modeled using a PBPK model structure similar to Ramsey and Andersen (1984), and the derivation of the DAF for this category has been described previously (Overton & Jarabek, 1989).

The two categories of gases with the greatest potential for respiratory tract effects are (1) highly water-soluble and/or rapidly irreversibly reactive, and (2) water-soluble gases that may also be rapidly reversibly reactive or moderately to slowly irreversibly metabolized in respiratory tract tissue. The objective of the default modeling approach is to describe the effective dose to the three regions of the respiratory tract by addressing the absorption or "scrubbing" of a relatively water-soluble and/or reactive gas from the inspired airstream as it travels from the ET to PU region. That is, the dose to the peripheral regions (TB and PU) is affected by the dose to the region immediately proximal. The appropriateness of assessing proximal to distal dose representative of the scrubbing (uptake) pattern is supported by the often observed proximal to distal progression pattern of respiratory tract toxicity with increasing concentration. At low concentrations of relatively water-soluble and/or reactive gases, observed effects are largely isolated to the ET region. At higher concentrations, more severe effects occur in the ET region and toxicity is also observed to progress to the peripheral regions. The severity of toxicity also progresses peripherally with increased exposure concentrations. Respiratory tract uptake is not described in detail to the level of local airflow distribution (e.g., respiratory vs. olfactory epithelium), but the default model does adequately describe the scrubbing of the chemical from the inhaled airstream.

The approach used to model uptake of gases in categories 1 and 2 is based on the concept of an overall mass transport coefficient, K_g. The concept of the K_g is based on a concentration gradient analysis similar to Fick's law of diffusion and is utilized to describe transport through several different phases such as air and liquid. Two-phase mass transport resistance models using K_g have been used to describe absorption in the respiratory tract (Hanna et al., 1989). To simplify uptake by the respiratory tract as a two-phase resistance model, it must be assumed that the blood concentration is constant. For the gases in category 1, highly water-soluble and/or rapidly irreversibly reactive, the blood concentration can further be assumed to be zero. The overall mass transport resistance is defined by the reciprocal of the mass transport coefficient, $1/K_g$, composed of the resistance to lateral movement of the absorbing gas/vapor through the air, liquid lining fluids, and tissue, as shown in Figure 3. The resistance in series is given by

$$\frac{1}{K_g} = \frac{1}{k_g} + \frac{1}{H_T k_l} \tag{5}$$

where k_g is the transport coefficient in the gas phase, H_T is the partition coefficient between tissue and air, and k_l is the transport coefficient in the surface–liquid/tissue phase.

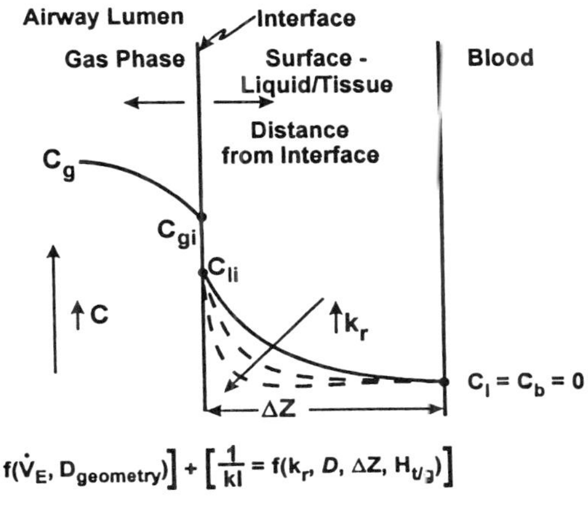

FIGURE 3. Schematic of two-phase mass transport resistance model used to describe respiratory tract uptake. This can be depicted as a resistance in a series where $1/K_g$ is the reciprocal of the overall mass transport coefficient, $1/k_g$ is the gas-phase resistance, and $1/k_l$ is the surface–liquid/tissue phase resistance. Parameter symbols and definitions are provided in the Appendix. Factors that influence flux are shown and described in the text. The definitive characteristic for category 1 gases, that the concentration in the blood (C_b) is zero, is illustrated.

The gas-phase term, k_g, is dependent on airflow rate (ventilation rate), flow geometry (local or regional airway geometry), and gas-phase diffusivity. The dependence of k_g on these parameters is discussed in greater detail by Hanna et al. (1989). The surface–liquid/tissue phase term, k_l, is determined by the phase thickness (Δz), the surface–liquid/tissue phase diffusivity, and the reactivity (i.e., the propensity for ionic dissociation or the ability to serve as substrate for metabolism) in the surface–liquid/tissue. The dependence of k_l on these parameters is discussed in greater detail by Ultman (1988).

Because category 1 gases are defined as highly water-soluble and/or rapidly irreversibly reactive in the surface–liquid/tissue phase of the respiratory tract, they are distinguished by the property that a significant backpressure from the surface–liquid/tissue phase during exhalation does not develop. Category 1 gases are also distinguished by the property that the gas does not significantly accumulate in the blood, which would reduce the concentration driving force and hence reduce the absorption rate. Examples of gases classified as category 1 are ozone, hydrogen fluoride, and formaldehyde.

The two-phase resistance approach is thus appropriate for category 1 gases, and a fractional penetration model of the type first proposed by Aharonson et al. (1974) and later utilized by Ultman (1988) and Hanna et al. (1989) is used to determine the fraction of the inhaled concentration absorbed in the ET region, the concentration entering and uptake in the TB region, and the remaining concentration entering the PU region. The uptake in the ET and the output to the TB (fractional penetration, fp_{ET}) is dependent on K_g, so that uptake in the ET is defined as $1 - fp_{ET}$. A ventilation–perfusion model is used to estimate the uptake in the PU region by substituting the concentration of the air exiting the TB region in place of the inhaled concentration. The overall schematic for the model for category 1 gases is shown in Figure 4.

The rate of mass absorbed at the gas–surface interface of the airway in a region is simply the product of the absorbed fraction, $(1 - fp_{ET})$, and the total mass inhaled during a single breath, $\dot{V}_E C_i$. The species-specific minute volume, $\dot{V}_E$, is used as the default volumetric flow rate since it approximates the flow rate at which the animal was breathing during the experimental exposure. Use of minute ventilation is also justified on the basis that by definition of category 1 gases, relatively little desorption would occur during exhalation, so that the uptake can be averaged over the entire cycle. If dose were to be defined on a mass per volume basis, it would implicitly assume that the outcome would be determined by concentration (i.e., mass/volume). This assumption would therefore argue that the most appropriate definition of dose should be one defined on the basis of surface area, that is, the mass flux (or dose) defined as mass per surface area-time. The mass flux implies a concentration gradient in the tissue such that the localized concentration would be highest at the surface. The mass flux is thereby a more accurate predictor of the peak localized concentration to use in defining dose for this application. Thus the dose, defined as the mass absorbed per surface area per minute (mg/cm²-min), to the ET region is given by

$$\text{Dose}_{ET}\ (\text{mg/cm}^2\text{–min}) = (1 - fp_{ET})\ \frac{C_i\dot{V}_E}{SA_{ET}} \tag{6}$$

Using the relationship of fp_{ET} to K_g, the dose to the ET region may also be expressed as

$$\text{Dose}_{ET} = \frac{C_i\dot{V}_E}{SA_{ET}}\left(1 - e^{\frac{-K_{g_{ET}}SA_{ET}}{\dot{V}_E}}\right) \tag{7}$$

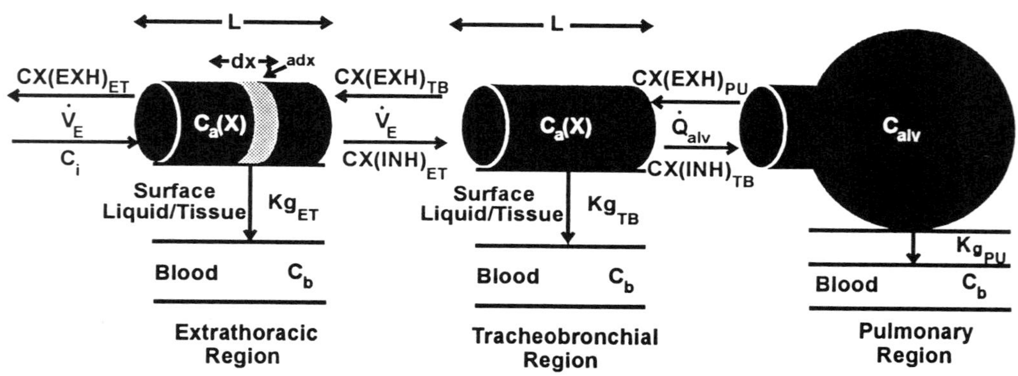

FIGURE 4. Schematic of model to estimate RGDR of gases in category 1. Parameter symbols and definitions are provided in the Appendix.

The ratio of the dose of differing species used to calculate the $RGDR_{ET}$ to serve as the DAF in Eq. 3 is then calculated as

$$\frac{(Dose_{ET})_A}{(Dose_{ET})_H} = \frac{\left(\frac{C_i\dot{V}_E}{SA_{ET}}\right)_A \left(1 - e^{\frac{-K_{g_{ET}} SA_{ET}}{\dot{V}_E}}\right)_A}{\left(\frac{C_i\dot{V}_E}{SA_{ET}}\right)_H \left(1 - e^{\frac{-K_{g_{ET}} SA_{ET}}{\dot{V}_E}}\right)_H} \tag{8}$$

Equation 8 represents the most general form of the ratio of extrathoracic regional dose between laboratory test species and humans for category 1 gases. This equation therefore serves as the basis for the default dosimetric adjustment. To evaluate the ratio, each term needs to be determined for the species of interest, so that values of extrathoracic uptake ($K_{g_{ET}}$) are needed for different animal species and gases. Values of K_g obtained in a single animal species may be scaled within a species for a different gas in the same category by decomposing K_g to the individual transport resistances (Hanna et al., 1989).

By definition, gases in category 1 would be associated with large K_g values due to high k_g and low k_l terms (Ultman, 1988). Thus, in these cases, the exponent is greater than or equal to 1. Under these circumstances, when the exponential term (the penetration fraction) approaches zero (less than 5% error when $K_{g_{ET}}$ is 5), the dose ratio simplifies to

$$\frac{(Dose_{ET})_A}{(Dose_{ET})_H} \cong \frac{\left(\frac{\dot{V}_E}{SA_{ET}}\right)_A}{\left(\frac{\dot{V}_E}{SA_{ET}}\right)_H} \tag{9}$$

where the ratio of the ventilation rates and surface areas determines the $RGDR_{ET}$. Equation 9 is the DAF used when K_g values are not available. Assuming that the penetration fraction reduces to zero is equivalent to assuming the gas is absorbed entirely in the ET region. Furthermore, based on Eq. 7, the absorption is assumed to be distributed equally. Studies currently in progress are anticipated to provide more localized measures of dose to the nasal cavity (Morgan et al., 1991; Kimbell et al., 1993).

The majority of gases with URT toxicity evaluated to date have been category 1 gases, so that the derivation of the default adjustments has been provided. Gases in category 2 (water soluble and rapidly reversibly reactive or moderately to slowly irreversibly metabolized in respiratory tract tissue) are those that have the potential for significant accumulation in the blood and thus have the potential for both respiratory and remote toxicity. The derivation of these defaults is beyond the scope of this paper, but the major issues underlying the model development are outlined.

Accumulation in the blood by category 2 gases will reduce the concentration driving force during inspiration and thereby reduce the absorption rate or dose upon inhalation. Thus, category 2 gases also have the potential for significant desorption during exhalation. A backpressure, that is, reversal of the concentration gradient at the air–liquid interface, may occur during exhalation when the exhaled air concentration is less than the concentration of the surface liquid established during inhalation. Thus, uptake for these gases can not be assumed to be modeled by a two-phase resistance model, and a hybrid model structure between that for category 1 and that for category 3 is proposed in order to account for uptake in the respiratory tract and the additional considerations of the potential for backpressure and for desorption on exhalation. The model structure is shown in Figure 5 and is structurally similar to that proposed by Overton and Graham (1994). The PBPK component is necessary to evaluate the steady-state blood concentration, which is required to calculate both the absorption flux on inhalation and the desorption flux during exhalation.

DATA ARRAY EVALUATION: CHOICE OF CRITICAL EFFECT

Although it is preferable to use human studies as the basis for RfC derivation, adequate human data are not usually available, often forcing reliance on laboratory animal data. The entire toxicity profile or data array is examined with the objective of selecting a prominent toxic effect, the critical effect, that is pertinent to the chemical's mechanism of action and that is at or just below the threshold for more serious effects. The critical effect chosen is generally characterized by the lowest $NOAEL_{[HEC]}$ that is representative of the threshold region for the entire data array. A challenging aspect of this evaluation is that across the data array, different effects

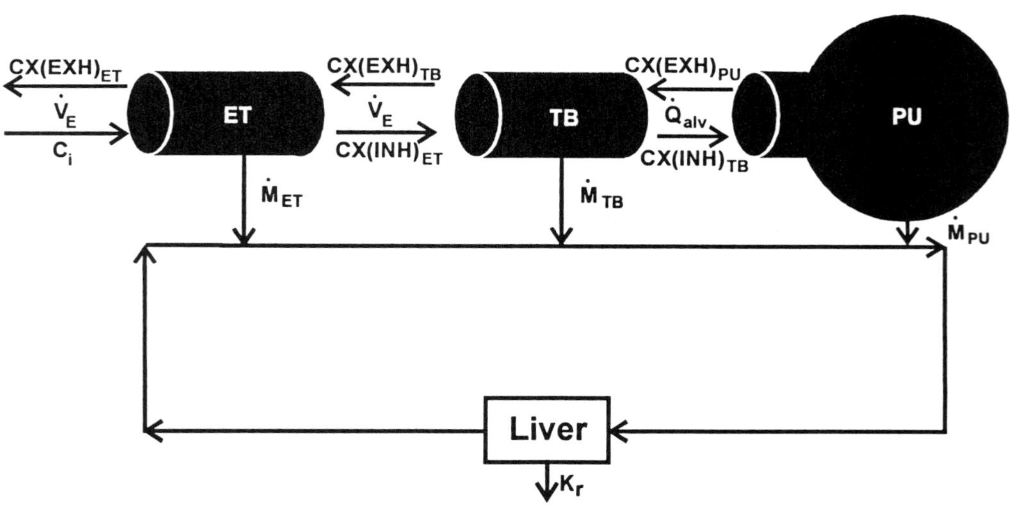

FIGURE 5. Schematic of model to estimate RGDR of gases in category 2. Parameter symbols and definitions are provided in the Appendix.

may be measured in the same tissue; different species are used in various studies; and each investigation may or may not be performed at similar exposure concentrations. The effects measured may or may not represent different and/or equivocal degrees of severity or adversity within disease continuums. Choice of the critical effect for a chemical represents a synthesis of the entire data array. If a relevant laboratory animal model exists, data on the critical effect in that species would be chosen. In the absence of that knowledge, data from the most sensitive species (i.e., the one exhibiting the lowest $NOAEL_{[HEC]}/LOAEL_{[HEC]}$ bracket) are chosen. The study that best characterizes the critical effect is called the principal study.

When disparity in dose–response patterns is apparent, studies must be evaluated to determine, if possible, whether the differences are due to (1) differences in the monitored end points or procedures across studies, (2) species differences in dose–response curves, or (3) choice of dose spacing if the BMD approach is not used. Differences in dose–response curves could be due to differences in concentration of the compound or metabolite reaching the receptor (dosimetry) or to differences in target receptor sensitivity (pharmacodynamics). Current controversy, with respect to the URT in this arena, involves the relevance of nasal lesions in laboratory rodents versus humans or other primates (DeSesso, 1993) and whether nasal lesions in rodents are somehow sentinel for effects in the lower respiratory tract of primates. A debate exists over whether nasal lesions in rodents should be used as the critical effect when functional, but less sensitive, tests, such as screening spirometry, are negative in humans. Physicochemical characteristics (water solubility, reactivity) and dosimetry may be used to ascertain the relative contribution of differences in inspiratory flow patterns (Morgan et al., 1991). Another question is whether or not primate data should be used without determination of whether the lack of an effect in that species versus rodents is due to small sample size. If the chemical is reactive and soluble but the URT was not adequately addressed, should an RfC be based on lower respiratory tract toxicity? Again, guidance on relevance of various nasal lesions in laboratory species to humans would greatly enhance the risk assessor's ability to address these issues and perform an analysis.

APPLICATION OF UNCERTAINTY FACTORS

The RfC is derived from the $NOAEL_{[HEC]}$ for the critical effect by consistent application of uncertainty factors (UFs). The uncertainty factors are applied to account for recognized uncertainties in the extrapolations from experimental data conditions to an estimate appropriate to the default human scenario. A modifying factor (MF) may also be applied when scientific uncertainties in the principal study are not explicitly addressed by the standard UFs. Table 4 provides guidance on how to apply the standard UFs and MF.

Thus, notationally, the RfC is defined as

$$RfC = NOAEL^{*}_{[HEC]}/(UF \times MF) \tag{10}$$

Application of UFs rather than safety factors represents an important conceptual difference of the RfC methodology when compared to other approaches. Uncertainty factors are applied in a systematic fashion for defined areas of extrapolation on a default basis, thereby creating a framework for data directed at those defined areas to decrease the magnitude of the UF applied. For example, in contrast to the oral reference dose methodology, which does not incorporate dosimetric adjustments, the UF for interspecies extrapolation in the RfC methods is 3. The usual factor of 10 for an area of uncertainty was reduced by one-half due to the default dosimetric adjustments. To the extent that more sophisticated models account for dosimetry and pharmacodynamic determinants between species, reduction of the UF for interspecies and intraspecies extrapolations can be anticipated.

The UF applied for extrapolating from studies of subchronic duration also identifies an area requiring additional research. Since lesions are usually evaluated only at scheduled sacrifice times, the risk assessor is forced to intuit the relationship of the lesion, with respect to concentration and duration, based on the incidence and/or severity of a given lesion at those points. Low statistical power often confounds interpretation at the interim sacrifices. As the bioassay progresses, changes with age begin to influence background rates and as well as potentially confounding the indices of the exposed groups. Figure 6 shows a schematic of the limits of that consideration. The UF for extrapolation of subchronic data would be obviated if data indicated the lesion was dependent on concentration as the major determinant. Research on reversibility of lesions after short-term exposures may not be particularly relevant, however. Data evaluated in the course of RfC derivations have indicated that nasal lesions in rats present after exposure to acrolein for 1 yr did not resolve after a recovery period of 52 wk. The LOAEL for nasal lesions in rats due to dimethylamine exposure decreased by five-fold from the 18-mo to 24-mo sacrifice. Other examples of the complexity of this evaluation are provided in this volume (Foureman et al., 1994; Guth & Jarabek, 1994).

CONFIDENCE STATEMENTS

The RfC as a quantitative dose-response estimate is not numeric alone. A confidence level of high, medium, or low is assigned to the principal study, to the overall database, and to the RfC itself. Confidence ascribed to the RfC estimate is a function of the confidence both in the quality of the study and in the completeness of the supporting database together. Confidence in the database takes precedence over that assigned to the study. High confidence in the RfC is an indication that the database included investigation of a comprehensive (or especially relevant) array of noncancer toxicity end points, established from studies of chronic duration in various mammalian species, and that the principal study (or studies) established an unequivocal $NOAEL_{[HEC]}$. A high confidence RfC is not likely to change as

TABLE 4. Guidelines for the Use of Uncertainty Factors in Deriving Inhalation Reference Concentration (RfC)

Standard Uncertainty Factors (UFs)	Processes considered in UF purview
H = Human to sensitive human Use a 10-fold factor when extrapolating from valid experimental results from studies using prolonged exposure to average healthy humans. This factor is intended to account for the variation in sensitivity among the members of the human population.	Pharmacokinetics/pharmacodynamics Sensitivity Differences in mass (children, obese) Concomitant exposures Activity pattern Does not account for idiosyncracies
A = Animal to human Use a 3-fold factor when extrapolating from valid results of long-term studies on experimental animals when results of studies of human exposure are not available or are inadequate. This factor is intended to account for the uncertainty in extrapolating animal data to the case of average healthy humans. Use of a 3 is recommended with default dosimetric adjustments. More rigorous adjustments may allow additional reduction. Conversely, judgment that the default may not be appropriate could result in an application of a 10-fold factor.	Pharmacokinetics/pharmacodynamics Relevance of laboratory animals model Species sensitivity
S = Subchronic to chronic Use a 10-fold factor when extrapolating from less than chronic results on experimental animals or humans when there are no useful long-term human data. This factor is intended to account for the uncertainty in extrapolating from less than chronic NOAELs to chronic NOAELs.	Accumulation/cumulative damage Pharmacokinetics/pharmacodynamics Severity of effect Recovery Duration of study Consistency of effect with duration
L = $LOAEL_{[HEC]}$ to $NOAEL_{[HEC]}$ Use a 10-fold factor when deriving an RfC from a $LOAEL_{[HEC]}$, instead of a $NOAEL_{[HEC]}$. This factor is intended to account for the uncertainty in extrapolating from $LOAEL_{[HEC]}$s to $NOAEL_{[HEC]}$s.	Severity Pharmacokinetics/pharmacodynamics Slope of dose-response curve Trend, consistency of effect Relationship of end points Functional vs. histopathological evidence Exposure uncertainties
D = Incomplete to complete database Use up to a 10-fold factor when extrapolating from valid results in experimental animals when the data are "incomplete." This factor is intended to account for the inability of any single animal study to adequately address all possible adverse outcomes in humans.	Quality of critical study Data gaps Power of critical study/supporting studies Exposure uncertainties
Modifying Factor (MF) Use professional judgment to determine whether another uncertainty factor (MF) that is ≤10 is needed. The magnitude of the MF depends upon the professional assessment of scientific uncertainties of the study and database not explicity treated above (e.g., the number of animals tested or quality of exposure characterization). The default value of the MF is 1.	

Note. Assuming the range of the UF is distributed log-normally, reduction of a standard 10-fold UF by half (i.e., 10^{-5}) results in a UF of 3. Composite UF for derivation involving 4 areas of uncertainty is 3000 in recognition of the lack of independence of these factors. Inhalation reference concentrations are not derived if all five areas of uncertainty are invoked.

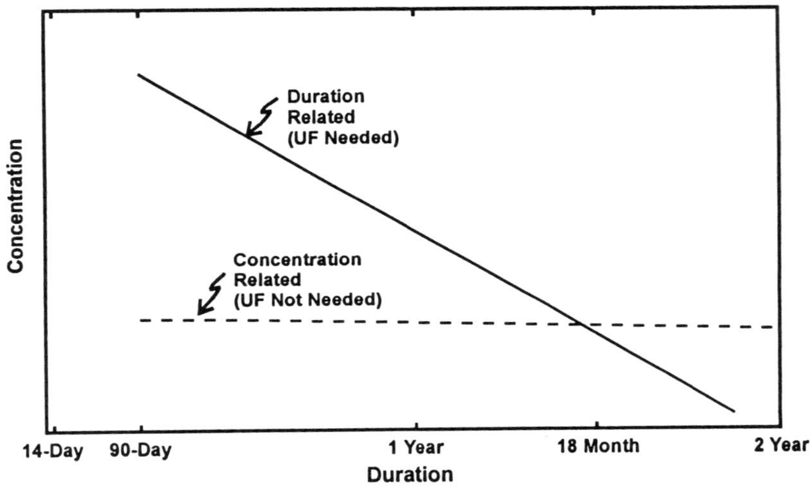

FIGURE 6. Schematic illustrating relationship of a fixed effect (toxicity indexed as either an increased incidence and/or an increase in severity) as a function of concentration and duration. Different lines indicate the relationship where toxicity is predominantly explained either by concentration or by duration.

more data become available, with the exception of additional mechanistic or sophisticated tests that may change the perspective of the evaluation. Low confidence in an RfC is usually applied to a derivation that is based on several extrapolations and indicates an estimate that may be especially vulnerable to change if new data become available.

A complete dose-response evaluation should include communication of the rationale for data selection, the strengths and weaknesses of the database, key assumptions, and resultant uncertainties. The rationale for the choice of the data from which the RfC is derived, a discussion of data gaps, rationale for UF application, and the confidence statements are all routinely provided on the summary of the RfC entered on the EPA's Integrated Risk Information System (IRIS).

SUMMARY

The U.S. EPA's inhalation RfC methodology for derivation of dose-response estimates of noncancer toxicity of inhalation has been described. Noncancer toxicity involves the evaluation of an array of different end points across various target tissues. The nature of lesions in the URT with respect to exposure duration, dosimetry, and disease outcomes needs to be characterized to facilitate accurate use of URT toxicity in this analysis. The RfC methodology differs from other methodologies in that it incorporates dosimetric adjustments to account for the species-specific relationships of exposure concentrations to deposited/delivered doses when extrapolating across species. Default dosimetric adjustments are derived from simplified models due to the lack of data required for more detailed models. Empiric data on deposition and uptake for particles and gases would enhance the accuracy of such adjustments. Data on local uptake mechanisms for the URT in other species used in risk assessment would provide a greater level of refinement.

REFERENCES

Aharonson, E. F., Menkes, H., Gurtner, G., Swift, D. L., and Proctor, D. F. 1974. Effect of respiratory airflow rate on removal of soluble vapors by the nose. *J. Appl. Physiol.* 37:654–657.

Andersen, M. E., Krishnan, K., Conolly, R. B., and McClellan, R. O. 1992. Mechanistic toxicology research and biologically-based modeling: Partners for improving quantitative risk assessments. *CIIT Activities* 12(1):1–7.

Barnes, D. G., Daston, G. P., Evans, J. S., Jarabek, A. M., Kavlok, R. J., Kimmel, L. A., Park, C., and Spitzer, H. L. 1994. Benchmark dose workshop: Criteria for use of benchmark dose to estimate a reference dose. *Regul. Toxicol. Pharmacol.* 20: in press.

Brain, J. D., and Mensah, G. A. 1983. Comparative toxicology of the respiratory tract. *Am. Rev. Respir. Dis.* 128:S87–S90.

Burger, G. T., Renne, R. A., Sagartz, J. W., Ayres, P. H., Coggins, C. R. E., Mosberg, A. T., and Hayes, A. W. 1989. Histologic changes in the respiratory tract induced by inhalation of xenobiotics: Physiologic adaptation or toxicity? *Toxicol. Appl. Pharmacol.* 101:521–542.

Crump, K. S. 1984. A new method for determining allowable daily intakes. *Fundam. Appl. Toxicol.* 4:854–871.

De Rosa, C. T., Stara, J. F., and Durkin, P. R. 1985. Ranking chemicals based on chronic toxicity data. *Toxicol. Ind. Health* 1:177–191.

DeSesso, J. 1993. The relevance to humans of animal models for inhalation studies in the nose and upper airways. *Qual. Assurance: Good Practice Reg. Law* 2(3):213–231.

Foureman, G. L., Greenberg, M. M., Sanga, G. K., Stuart, B. P., Shiotsuka, R. N., and Thyssen, J. H. 1994. Evaluation of nasal tract lesions in derivation of the inhalation reference concentration for hexamethylene diisocyanate. *Inhal. Toxicol.* 6(suppl):341–355.

Guth, D. J., and Jarabek, A. M. 1994. U. S. EPA inhalation reference concentration for bromomethane. *Inhal. Toxicol.* 6(suppl):327–339.

Hanna, L. M., Scherer, P. W., and Frank, R. 1989. Absorption of soluble gases and vapors in the respiratory tract. In *Respiratory physiology: An analytical approach*, eds. H. K. Chang and M. Paiva, pp. 277–316. New York: Marcel Dekker.

Hartung, R. 1986. Ranking the severity of toxic effects. In *Trace substances in environmental health–XX* (Proceedings of the University of Missouri's 20th annual conference), ed. D. D. Hemphill, pp. 204–211. Columbia, MO: University of Missouri.

Jarabek, A. M., and Segal, S. A. 1993. Noncancer toxicity of inhaled toxic air pollutants: Available approaches for risk assessment and risk management. In *Toxic air pollutant handbook*, ed. D. R. Patrick, pp. 100–127. New York: Van Nostrand Reinhold.

Jarabek, A. M., Menache, M. G., Overton, J. H., Jr., Dourson, M. L., and Miller, F. J. 1990. The U.S. Environmental Protection Agency's inhalation RfD methodology: Risk assessment for air toxics. *Toxicol. Ind. Health* 6(5):279–301.

Kimbell, J. S., Gross, E. A., Joyner, D. R., Godo, M. N., and Morgan, K. T. 1993. Application of computational fluid dynamics to regional dosimetry of inhaled chemicals in the upper respiratory tract of the rat. *Toxicol. Appl. Pharmacol.* 121:253–263.

Klaassen, C. D. 1986. Principles of toxicology. In *Casarett and Doull's toxicology: The basic science of poisons*, 3rd Ed., eds. C. D. Klaassen, J. Doull, and M. Amdur, pp. 11–32. New York: Macmillan.

Lomax et al. 1994. Regional histopathology of the mouse nasal cavity following two weeks of exposure to acrylic acid for either 6 or 22 hours per day. *Inhal. Toxicol.* 6(suppl):445–449.

Ménache, M. G., Raabe, O. G., and Miller, F. J. 1994. An empirical dosimetry model of aerodynamic particle deposition in the rat respiratory tract. *Reg. Toxicol.*

Mery, S., Gross, E. A., Joyner, D. R., Godo, M., and Morgan K. T. 1994. Nasal diagrams: A tool for recording the distribution of nasal lesions in rats and mice. *Toxicol. Pathol.* 22(4): in press.

Miller, F. J., Overton, J. H., Jr, Jaskot, R. H., and Menzel, D. B. 1985. A model of the regional uptake of gaseous pollutants in the lung: I. The sensitivity of the uptake of ozone in the human lung to lower respiratory tract secretions and exercise. *Toxicol. Appl. Pharmacol.* 79:11–27.

Miller, F. J., Mortonen, T. B., Menache, M. G., Graham, R. C., Spektor, D. M., and Lippmann, M. 1988. Influence of breathing mode and activity level on the regional deposition of inhaled particles and implications for regulatory standards. In *Inhaled particles VI: Proceedings of an international symposium and workshop on lung dosimetry*, eds. J. Dodgson, R. I. McCallum, M. R. Bailey, and D. R. Fisher. Cambridge, UK: British Occupational Hygiene Society.

Morgan, K. T. 1991. Approaches to the identification and recording of nasal lesions in toxicology studies *Toxicol. Pathol.* 19(4 Part 1): 337–351.

Morgan, K. T., Kimbell, J. S., Monticello, T. M., Patra, A. L., and Fleishman, A. 1991. Studies of inspiratory airflow patterns in the nasal passages of the F344 rat and rhesus monkey using nasal molds Relevance to formaldehyde toxicity. *Toxicol. Appl. Pharmacol.* 110:223–240.

Overton, J. H., Jr. 1984. Physicochemical processes and the formulation of dosimetry models. In *Fundamentals of extrapolation modeling of inhaled toxicants: Ozone and nitrogen dioxide*, eds F. J. Miller, and D. B. Menzel, pp. 93–114. New York: Hemisphere.

Overton, J. H., and Jarabek, A. M. 1989. Estimating human equivalent no observed adverse effects levels for VOCs based on minimal knowledge of physiological parameters. 82nd Annual meeting of the Air and Waste Management Association, Anaheim, CA, June. Also Air and Waste Management Association, Pittsburgh, PA, paper 89-91.8.

Overton, J. H., and Graham, R. C. 1994. Modeling the uptake of gases by the dog nasal-pharyngeal region: Effects of morphometric and physicochemical factors. *Inhal. Toxicol.* 6(suppl):113–124.

Raabe, O. G., Al-Bayati, M. A., Teague, S. V., and Rasolt, A. 1988. Regional deposition of inhaled monodisperse coarse and fine aerosol particles in small laboratory animals. In: *Inhaled particles VI. Proceedings of an international meeting and workshop on lung dosimetry*, eds. J. Dodgson, R. I. McCallum, M. R. Bailey, and D. R. Fisher. Cambridge, UK: British Occupational Hygiene Society.

Ramsey, J. C., and Andersen, M. E. 1984. A physiologically based description of the inhalation pharmacokinetics of styrene in rats and humans. *Toxicol. Appl. Pharmacol.* 73:159–175.

Ruben, Z., and Rousseaux, C. G. 1991. The limitations of toxicologic pathology. In *Handbook of toxicologic pathology*, eds. W. M. Haschek and C. G. Rousseaux, pp. 131–142. New York: Academic Press.

Schlesinger, R. B. 1985. Comparative deposition of inhaled aerosols in experimental animals and humans: A review. *J. Toxicol. Environ. Health* 15:197–214.

Ultman, J. S. 1988. Transport and uptake of inhaled gases. In *Air pollution, the automobile, and public health,* eds. A. Y. Watson, R. R. Bates, and D. Kennedy, pp. 323–366. Washington, DC: National Academy Press.

U.S. Environmental Protection Agency. 1988. Recommendations for and Documentation of Biological Values for Use in Risk Assessment. Cincinnati, OH: Office of Health and Environmental Assessment, Environmental Criteria and Assessment Office; EPA report EPA-600/6-87-008. Available from NTIS, Springfield, VA, PB88-179874.

U.S. Environmental Protection Agency. 1990. Science Advisory Board's Comments on the Use of Uncertainty and Modifying Factors in Establishing Reference Dose Levels. EPA/SAB/EHC/90-005. Letter from L. R. Loehr and A. Upton to W. Reilly, January 17.

U.S. Environmental Protection Agency. 1992. Integrated Risk Information System (IRIS). Reference concentration (RfC) for inhalation exposure for diesel engine emissions. Online. Office of Health and Environmental Assessment, Environmental Criteria and Assessment Office, Cincinnati, OH.

U.S. National Academy of Sciences. 1983. *Risk assessment in the federal government: Managing the process*. Washington, DC: National Academy Press.

Yu, C. P., and Yoon, K. J. 1990. Retention Modeling of Diesel Exhaust Particles in Rats and Humans. Amherst, NY: Health Effects Institute, Research Report 40.

APPENDIX. Definition of Parameter Symbols

a	Airway perimeter
C_i	Inhaled concentration
C_{alv}	Pulmonary region gas concentration
C_b	Blood concentration
C_g	Gas phase concentration in airway lumen
C_{gi}	Gas-phase concentration at the interface of the gas phase with the surface-liquid/tissue phase (mg/cm^3)
C_l	Surface–liquid/tissue phase concentration
$C_{l/g}$	Surface–liquid/tissue concentration in equilibrium with the gas phase
C_{li}	Surface–liquid/tissue concentration at the interface of the gas phase and the surface–liquid/tissue phase
$CX(EXH)_{TB}$	Concentration exiting from tracheobronchial region upon exhalation
$CX(EXH)_{PU}$	Concentration exiting from pulmonary region upon exhalation
$CX(INH)_{ET}$	Concentration exiting from extrathoracic region upon inhalation
$CX(INH)_{TB}$	Concentration exiting from tracheobronchial region upon inhalation
dx	Differential of axial distance into airway
ET	Extrathoracic respiratory region
fp_{ET}	Fractional penetration through the extrathoracic region
$H_{l/g}$	Surface–liquid/tissue: gas (air) partition coefficient
K_g	Overall mass transport coefficient
$K_{g_{ET}}$	Overall mass transport coefficient of the extrathoracic region
$K_{g_{TB}}$	Overall mass transport coefficient of the tracheobronchial region
$K_{g_{PU}}$	Overall mass transport coefficient of the pulmonary region
k_g	Transport coefficient in the airphase
k_l	Transport coefficient in the surface–liquid/tissue phase
k_r	Reaction rate constant in the blood
L	Airway length
$\dot{M}_r$	Mass flux for respiratory region r
PU	Pulmonary respiratory tract region
$\dot{Q}_{alv}$	Alveolar ventilation rate
SA_r	Surface area of unspecified respiratory region r
t	Time
t_{EXH}	Time (duration) of exhalation
TB	Tracheobronchial respiratory tract region
$\dot{V}_E$	Minute ventilation

U.S. EPA INHALATION REFERENCE CONCENTRATION FOR BROMOMETHANE

Daniel J. Guth, Annie M. Jarabek

U.S. Environmental Protection Agency, Environmental Criteria and Assessment Office, Hazardous Pollutant Assessment Branch, Research Triangle Park, North Carolina, USA

The EPA inhalation reference concentration (RfC) for bromomethane was derived based on lesions observed in rat nasal epithelium from a chronic study. Reuzel et al. (1991) exposed Wistar rats to 0, 3, 30, or 90 ppm bromomethane for 6 h/day, 5 days/wk for up to 29 mo. Hyperplastic changes in the basal cells accompanied by degeneration in the olfactory epithelium in the dorso-medial part of the nasal cavity were observed in all exposed groups of both sexes at 29 mo of exposure. These changes were concentration-related in both incidence and severity and were statistically significant at 29 mo. The effect was not statistically significant in smaller groups of animals examined after 12 or 24 mo of exposure. This study identifies a lowest-observed-adverse-effect level (LOAEL) at 3 ppm in the rat. In the chronic National Toxicology Program (NTP) study (NTP, 1992), B6C3F1 mice were exposed to 0, 10, 33, or 100 ppm bromomethane for 6 h/day, 5 days/wk, for either 6, 15, or 24 mo. A statistically significant increase in the incidence of olfactory epithelial necrosis and metaplasia was seen in the nasal cavities of male and female mice exposed to 100 ppm. Mice exposed to lower concentrations did not exhibit significant increases in any of the lesions. Based on the results of this study, a LOAEL of 100 ppm and a no-observed-adverse-effect level (NOAEL) of 33 ppm was identified in mice. The RfC was based on the more sensitive effect from the rat study. The apparent 30-fold difference in the effect level between the 24- and 29-mo durations is surprising, because comparison of the effects after exposures ranging from 1 day to 24 mo shows similar effect levels. The LOAEL of 3 mg/m^3 was adjusted to a continuous exposure scenario [LOAEL(ADJ) = 2.08 mg/m^3], and the human equivalent concentration was calculated to be 0.48 mg/m^3. This value was divided by an uncertainty factor of 100, which includes intraspecies uncertainty, use of a LOAEL for a mild effect, and interspecies extrapolation. The resulting RfC is 5 μg/m^3.

Bromomethane has been used widely as a fumigant and pesticide. The derivation of an inhalation reference concentration (RfC) for bromomethane was initiated because of its inclusion on the Clean Air Act Amendments (CAA), Section 112(b), list of hazardous air pollutants, as well as interest from other EPA programs. In implementation of the Clean Air Act, the RfC will be used in ranking of chemicals for the early reductions program, in evaluation of petitions to delist the chemical or to delist a source category that is an emitter of bromomethane, and in evaluation of residual risk after

The views expressed in this paper are those of the authors and do not necessarily reflect the views or policies of the U.S. Environmental Protection Agency.

Address correspondence to Dr. Daniel J. Guth, U.S. EPA/ECAO (MD-52), Research Triangle Park, NC 27711, USA.

technology-based emission standards are applied. Other areas of potential use of the RfC include the evaluation of urban air toxics, evaluation of area sources, and application in several specific studies mandated by the CAA.

The RfC is defined as an estimated exposure concentration that is likely to cause no adverse effects in the human population, including sensitive subgroups, under conditions of a continuous lifetime exposure, with approximately an order-of-magnitude uncertainty. The methodology for derivation of the inhalation RfC has been described (Jarabek et al., 1990). The RfC method is an approach to dose-response assessment that rests on a review of the entire toxicological literature on bromomethane. The RfC derivation begins with the determination of the lowest-observed-adverse-effect level (LOAEL) and the no-observed-adverse-effect level (NOAEL) from the exposures in available studies. The critical effect, defined as the adverse effect having the lowest experimental LOAEL or the highest experimental NOAEL, is selected as the basis for the RfC. If the effect levels are derived from animal studies or human occupational studies, the approaches discussed by Jarabek et al. (1990) are applied to estimate the human equivalent concentration for continuous exposure. The human equivalent concentration is then divided by uncertainty factors to account for areas of uncertainty that arise from extrapolation from the experimental conditions to the human exposure scenario of the RfC.

Critical aspects of the RfC derivation for bromomethane were the evaluation of the consistency of the lesions reported at various durations in different studies, the interpretation of the adversity of the nasal lesion observed in rats, and the selection of the study and toxicological end point to be used. These issues are discussed further and form the basis of this report.

STUDIES EVALUATED

The literature on the inhalation toxicity of bromomethane includes subchronic and chronic studies in rats and mice, as well as acute and short-term studies that evaluate effects on the respiratory, nervous, reproductive, and cardiovascular systems in rats, mice, guinea pigs, rabbits, and monkeys. Studies are also available that evaluate the developmental and reproductive toxicology of bromomethane. The database for this chemical is fairly complete with regard to laboratory animal studies because studies of long-term exposure in multiple species, as well as studies examining the specific end points of developmental and reproductive toxicology, are available. Limited information is available on the effects of exposures on humans from occupational or epidemiological studies. The most common signs of acute intoxication with bromomethane in humans are neurotoxic in nature and include headache, dizziness, fainting, apathy, weakness, loss of memory, speech impairment, visual effects, limb numbness, tremors, muscle twitching, paralysis, ataxia, seizures, convulsions, and unconsciousness (U.S. EPA, 1986; Alexeef & Kilgore, 1983). A cross-sectional occupational study (Anger et al., 1986) and a report of four cases of occupational poisoning (Herzstein & Cullen, 1990) have focused on the neurotoxic effects in humans. These human studies cannot serve as the basis for the derivation of an RfC for bromomethane because of design deficits, notably concurrent exposures to other chemicals, or inadequate quantitation of exposure levels and/or durations.

Although the human literature discusses primarily central and peripheral nervous system effects resulting from acute high-level or long-term occupational exposure, the laboratory animal literature shows the extrathoracic region of the respiratory tract to be a sensitive target of methyl bromide. Two chronic inhalation studies in rodents are available and are described next.

A series of inhalation toxicity studies of bromomethane was conducted under the sponsorship of the National Institute of Public Health and Environmental Hygiene of the Netherlands. Reuzel et al. (1987, 1991), exposed Wistar rats to 0, 3, 30, or 90 ppm of 98.8% pure bromomethane for 6 h/day, 5 days/wk for up to 29 mo. The test atmosphere was measured by gas chromatography every 30 min during exposure. Each exposure level consisted of 50 male and 60 female animals with three satellite groups of 10 animals/sex/exposure level. Animals in the satellite groups were sacrificed at 14, 53, and 105 weeks of exposure. Eleven organs were weighed at necropsy, and 36 tissues were examined histopathologically, including the lungs, trachea, larynx, 6 cross sections of the nose, heart, brain, and adrenal glands.

Both the males and females exposed to 90 ppm in the chronic rat study (Reuzel et al., 1987, 1991) exhibited decreased body weight gains compared to concurrent controls, but no treatment-related changes in hematological, biochemical, or urine parameters were observed. Significant concentration-related decreases in male relative kidney weights were reported at 30 and 90 ppm. A decrease in mean absolute brain weight of female rats was reported to occur after 53 and 105 wk of exposure to 90 ppm; there was no change in relative brain weight or in brain histology.

Light microscopic evaluation revealed that the nose, heart, esophagus, and forestomach were the principle targets of bromomethane toxicity (Reuzel et al., 1987, 1991). Degeneration in the olfactory epithelium, accompanied by very slight to moderate hyperplastic changes in the basal cells in the dorso-medial part of the nasal cavity, was observed in all exposed groups of both sexes at 29 mo of exposure. At the lowest concentration, the lesion was described as very slight. Figures 1 and 2 show that these changes were concentration-related in both incidence and severity and were statistically significant at 29 mo. At 29 mo of exposure, the incidence of basal cell lesions in animals exposed to 3 ppm was significantly increased, primarily due to an increase in lesions categorized as very slight. In the 30 and 90 ppm groups exposed for 29 mo, the increase in total basal cell lesions was due to increases in lesions in all severity categories com-

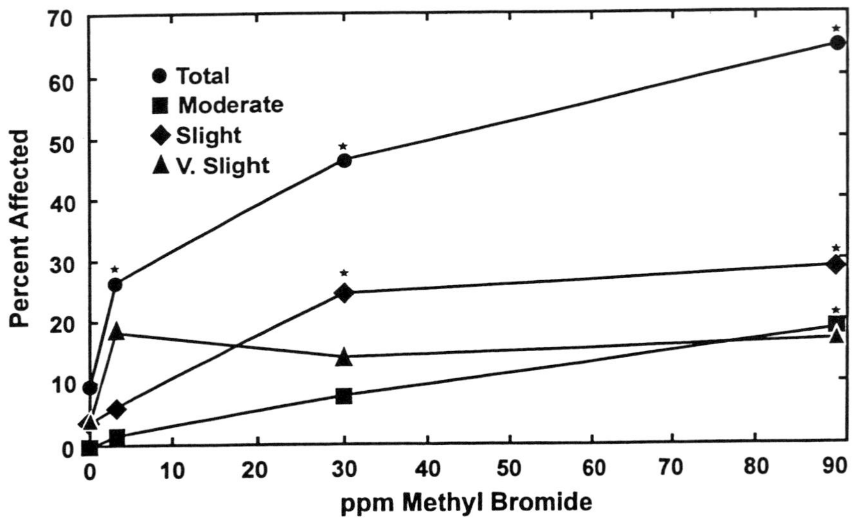

FIGURE 1. Basal-cell hyperplasia of the olfactory epithelium in male rats exposed to methyl bromide (bromomethane) for 29 mo, percent of rats affected; n = 46, 48, 49, and 48 in control, 3, 30, and 90 ppm groups, respectively; $*p < .05$. Data adapted from Reuzel et al. (1991).

pared with concurrent controls. This pattern was the same in males (Figure 1) and females (Figure 2). Lesions in the heart were significantly increased in incidence in the males (cartilaginous metaplasia and thrombus) and the females (myocardial degeneration and thrombus) exposed to 90 ppm. The authors attributed part of the increased mortality at the high concentration

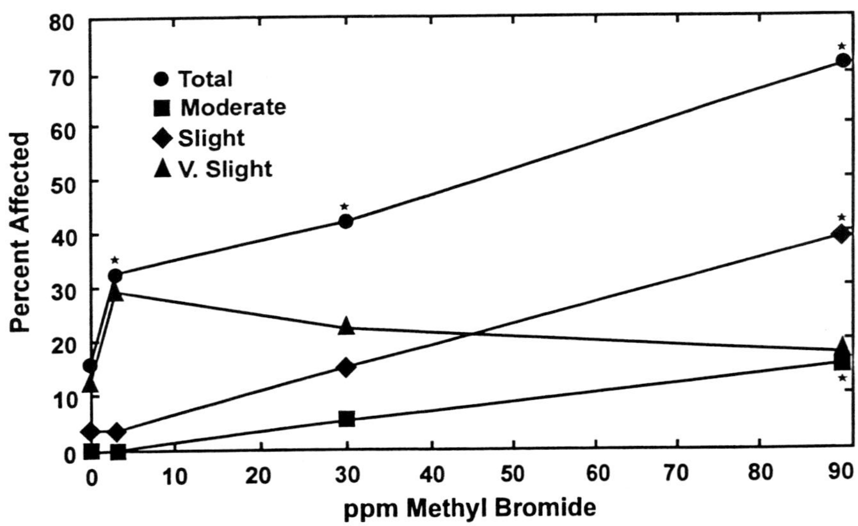

FIGURE 2. Basal-cell hyperplasia of the olfactory epithelium in female rats exposed to methyl bromide (bromomethane) for 29 mo, percent of rats affected; n = 58, 58, 59, and 59 in control, 3, 30, and 90 ppm groups, respectively; $*p < .05$. Data adapted from Reuzel et al. (1991).

to the cardiac lesions. A statistically significant increase in hyperkeratosis of the esophagus was observed in males after 29 mo of exposure to 90 ppm. Slight increases in forestomach lesions at 30 and 90 ppm were not statistically significant. No effects were observed in the tracheobronchial or pulmonary regions of the respiratory tract. No other exposure-related effects were noted. Based on these results, a LOAEL of 3 ppm for effects in the olfactory epithelium of the nasal cavity is established.

In the chronic National Toxicology Program study (NTP, 1992), mice were exposed to 0, 10, or 33, or 100 ppm bromomethane for 6 h/day, 5 days/wk, for either 6, 15, or 24 mo. Exposure to 100 ppm produced 47% mortality in the males and 10% mortality in the females by 20 wk of exposure, so this exposure group was discontinued, at this time, and the surviving animals were held for observation until scheduled sacrifice dates at 15 and 24 mo. The standard complete necropsy procedure was supplemented with neurobehavioral assessments in 16 mice/sex/group and neuropathological examination on 3–8 animals/sex/group at 6, 15, and 24 months. Body weights were significantly depressed in the mice exposed to 100 ppm (33% in the males and 31% in the females), beginning at wk 11 and persisting until exposure termination (20 wk). Significant body weight changes were not observed in the lower exposure groups. Clinical signs of toxicity observed almost exclusively in the 100 ppm mice that persisted throughout the study included tremors, abnormal posture, and limb paralysis. Functional neurobehavioral changes (hypoactivity, a heightened startle response, and higher hindlimb grip scores and hot plate latency) were observed in both sexes exposed to 100 ppm at various times during exposure and were more pronounced in the males.

The target organs of toxicity identified with histologic examination in this study were the brain, heart, and nose, with lesions in these organs occurring more frequently in the males (NTP, 1992). In the brain, there was a statistically significant increase in the incidence of cerebellar degeneration in male and female mice exposed to 100 ppm. Cerebral degeneration was also observed in these animals, but the incidence of this lesion was statistically significant only in males. Because this lesion was observed more frequently in animals that died prior to exposure termination, it may have contributed to the early mortality in this group. Myocardial degeneration and chronic cardiomyopathy were also observed in both males and females exposed to 100 ppm bromomethane and occurred at a higher incidence in those animals dying early. Finally, an increase in the incidence of olfactory epithelial necrosis and metaplasia was seen in the nasal cavities of both the male and female mice exposed to 100 ppm (Figure 3). Necrosis was seen only in the animals dying early, whereas metaplasia was exhibited mainly in those animals surviving until study termination. Mice exposed to lower concentrations did not exhibit significant increases in any of the lesions just described. Based on the results of this study, a NOAEL of 33 ppm and a LOAEL of 100 ppm are established based on toxicity in multiple organs.

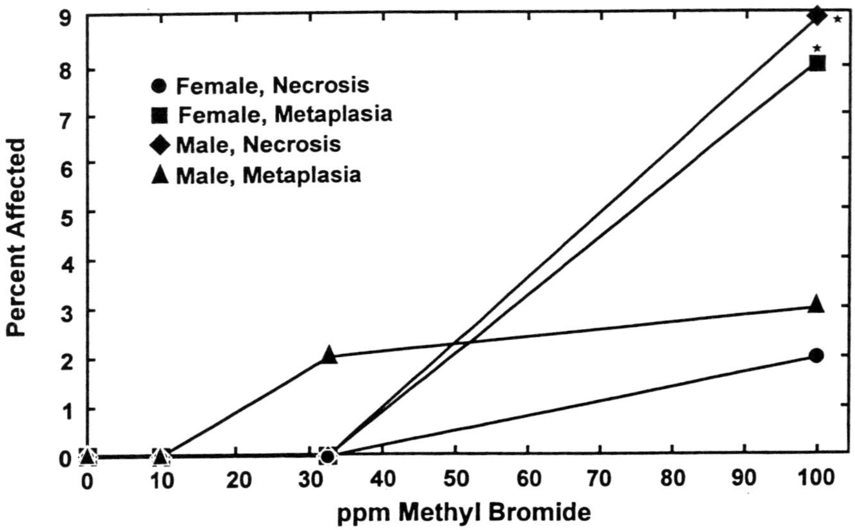

FIGURE 3. Lesions of the olfactory epithelium in mice exposed to methyl bromide (bromomethane) for 2 yr, percent of rats affected; *$p < .05$. Data adapted from NTP (1992).

SUPPORTING STUDIES

The NTP (1992) reported a 13-wk study in which F344 rats (18/sex/group) were exposed to 0, 30, 60, or 120 ppm bromomethane and mice (18–27/sex/group) were exposed to 0, 10, 20, 40, 80, or 120 ppm bromomethane, for 6 h/day, 5 days/wk. Hematological parameters were measured and organ weights were determined. Neurobehavioral testing was conducted on 8 rats and mice per sex per group at wk 0, 6, and 12, and neuromorphological studies were conducted on 4 rats per sex from the control and high-concentration group and on 4 mice per sex for each concentration. Histopathological examination of approximately 40 tissues from control and high concentration exposed animals was carried out, including lungs, bronchi, and nasal turbinates. Exposure-related changes seen in the mice were a significantly lower body weight (88% of control) and a 28% lethality in male mice exposed to 120 ppm bromomethane. Mice exposed to this level exhibited severe curling and crossing of the hindlimbs and twitching of the forelimbs, which were more severe in the males. No exposure-related histopathological effects were seen in mice. In the rats, there was no increase in mortality, but the males and females exposed to 120 ppm bromomethane exhibited significantly lower body weights (87–88% of control). Mild neurobehavioral effects were noted in rats of both sexes exposed to 120 ppm only. Females exposed to 120 ppm were found to have significantly lower hematocrit, hemoglobin, and erythrocytes counts, but the males did not exhibit these changes. The only exposure-related effect noted at histopathological examination was an increase in the incidence of olfactory epithelial dysplasia and cysts in the rats of both sexes exposed to 120 ppm. Based on these results, a NOAEL of 60 ppm for nasal olfactory epithelial changes in rats is established.

Because no significant target organ toxicity was noted in the 14-day or 13-wk studies, a special 6-wk target organ toxicity study at a near lethal concentration was conducted in F344 rats and B6C3F1 mice (Eustis et al., 1988; NTP, 1992). Groups of 5 animals per sex were exposed to 0 or 160 ppm bromomethane for 6 h/day for either 3 consecutive days (rats), 5 days/wk over 2 wk (rats and mice), or 6 wk (rats). Fifteen mice per sex per dose were exposed to 0 or 160 ppm for 6 h/day, 5 days/wk, for 6 wk. End-points studied included clinical observations, mortality, body and organ weights, hematology, clinical chemistry, urinalysis, gross pathology, and histopathology of a standard set of tissues, including the lungs and nasal turbinates. Female rats were the only group to demonstrate >50% survival, with mice being more sensitive than rats (lethality exceeded 50% after 6–8 exposures in both the male and female mice and after 14 exposures in the male rats). Because of the low survival, the male and female mice and male rats were killed after 10, 8, or 14 exposures, respectively. Neurological signs exhibited by both rats and mice, but to a lesser extent in the rats, included lethargy and curling and crossing of hindlimbs, forelimb twitching, and tremors. Decreases in body weight were observed in the exposed mice (74 and 82% of control in males and females, respectively) and rats (68 and 82% of control in males and females, respectively). The mean organ weights of most of the organs measured were significantly reduced in both species.

Target organs affected by exposure to 160 ppm bromomethane were the brain, kidney, nasal cavity, heart, adrenal gland, liver, and testis (Eustis et al., 1988; NTP, 1992). Species differences were noted in the responses of these organs. For example, neuronal necrosis in the cerebral cortex, hippocampus, and thalamus of the brain was seen in the rats, whereas neuronal necrosis was seen predominantly in the internal granular layer of the cerebellum of the mice. Nephrosis, characterized by degeneration, necrosis, and sloughing of the epithelium of the cortical convoluted tubules, was seen in all of the exposed mice and was considered by the authors to be partially responsible for the increase in lethality; these lesions were not observed in the rats. Degeneration and atrophy of the seminiferous tubules were observed in several of the exposed rats and mice, but were less severe in the mice. Olfactory epithelial degeneration was observed in most of the rats of both sexes and male mice, and only one female mouse exhibited this lesion. Myocardial degeneration occurred in rats of both sexes, but to a lesser degree in the male mice. Female mice had atrophy of the inner zone of the adrenal cortex, and rats had cytoplasmic vacuolation of the adrenal cortex.

Comparison of the results of these studies indicates that the rat is more sensitive to nasal effects, although the mouse was more severely affected based on survival and on neurological and renal effects. The NTP studies allow a direct comparison of effects in B6C3F1 mice and F344 rats exposed under the same conditions, and this comparison suggests that the rat is more sensitive to nasal effects. Strain differences between the F344 rats used by the NTP and the Wistar rats used by Reuzel et al. (1987, 1991) cannot be clearly compared due to differences in study protocol.

Studies of the repair of nasal lesions induced by bromomethane and the effect of exposure on olfactory function were examined for corroborative information relevant to the nature of the nasal lesions. Hurtt et al. (1988) investigated the time course of the regeneration of the olfactory epithelium following acute exposure to bromomethane. Male F344 rats were exposed to 0 (n = 5) or 200 ppm 99.9% pure bromomethane for 6 h/day for 1–5 days. Five animals/group were killed after 1, 3, or 5 days of exposure and 1, 2, 3, 5, or 10 wk after cessation of treatment. In a companion study, 6 animals/group were exposed to 0, 90, or 200 ppm bromomethane for 6 h, and olfactory function was studied by determining the effects of bromomethane on the ability of food-deprived animals to locate buried food pellets. Additional animals similarly exposed were killed at various times following a single 6-h exposure to assess the state of morphological regeneration at the time of functional recovery. Only the nasal cavities were examined histopathologically in these studies. Extensive destruction of the olfactory epithelium, characterized by epithelial disruption, fragmentation, and exfoliation, was evident after a single 6-h exposure to 90 or 200 ppm. The most severe effects occurred in the sustentacular and mature sensory cells, and the basal cells remained intact. Regeneration of the olfactory epithelium, characterized at first by replacement with a squamous cell layer that increased in thickness, began by day 3 of exposure and was essentially complete by 10 wk after the last exposure. Olfactory function was impaired in animals exposed to 200 ppm bromomethane, but not 90 ppm. Recovery of this function was evident by 4–6 days after exposure, which preceded morphological regeneration.

Similar results were obtained by Hastings et al. (1991). In this study, rats were exposed to 200 ppm bromomethane for 4 h/day, 4 days/wk for 2 wk. Prior to exposure, rats were food-deprived and trained to find buried food. Morphological, as well as biochemical (carnosine content in the olfactory bulb, which is an indication of the integrity of the olfactory primary sensory neurons), studies were performed to assess the integrity of the olfactory epithelium. Extensive damage to the olfactory epithelium was seen, as evidenced by both morphological analysis and decreased carnosine content after a single 4-h exposure. Olfactory function was also impaired, as evidenced by the inability of the rats to find the buried food at this time. Olfactory function began to improve after the first exposure, and the animals performed as well as their controls by the end of the exposure period. However, regeneration of the olfactory epithelium, as indicated by morphological and biochemical analysis, was not complete until 30–40 days from the start of exposure, consistent with the turnover rate for o factory neurons reported by Boorman et al. (1990).

SELECTION OF CRITICAL STUDY

The studies discussed earlier provide corroborative evidence of functional olfactory deficits and histopathological degeneration of the nasal olfactory epithelial structure, albeit in short-duration studies at higher concentrations, and suggest that these effects are reversible. In general, effects are considered adverse if they are detrimental to the normal function of the target organ, or if they are likely to be precursors to functional impairment. The nasal lesions at higher concentrations are degenerative with functional consequences, and therefore are clearly adverse. The les on increased in incidence and severity with increasing exposure concentration, supporting the conclusion that the effect should be considered adverse. An increase in both severity and incidence of lesions at higher concentrations and the functional changes observed at higher concentrations are strong evidence for considering the mild effect at lower concentrations to be adverse. Derivation of the RfC based on the mild effect is then likely to protect against more severe effects. Likewise, the choice of the most sensitive end point, in this case the nasal epithelial degeneration, as the basis for the RfC derivation should be protective against effects in other target organs. However, the mild severity of the effect and the demonstrated reversibility of the lesion are bases for a reduced uncertainty factor, as discussed later.

Critical components of the RfC derivation for bromomethane include the selection of the critical effect and the principal study to be used as the basis for RfC calculation. Because there are two chronic studies available and bromomethane exhibits toxicity to several organ systems in chronic studies, the selection of the target organ is a critical issue in determining the appropriate LOAEL/NOAEL from the data array. Nasal effects are selected as the critical effect because they are observed at lower concentrations than other effects in chronic/subchronic studies, they are observed in multiple strains/species, and they occur at lower concentrations than reproductive or developmental effects or effects in the heart or central nervous system. The chronic rat study (Reuzel et al., 1991) is selected as the principal study because the effects occur at a lower concentration than in mice, suggesting that rats are a more sensitive species for this effect, and because there is a clear increase in both incidence and severity of olfactory epithelial necrosis with increasing concentration. Although the difference in the LOAEL between the two available chronic studies is large [100 ppm in NTP (1992) compared to 3 ppm in Reuzel et al. (1991)], the results do not show clear difference in the responses of rats and mice, because neither study shows a significant effect in the nose at less than 100 ppm at time points up to 24 mo. However, the lack of effect at 24 mo in the chronic rat study (Reuzel et al., 1987, 1991) may be an artifact of the small numbers of animals and the high control incidence. Increases in incidence of basal-cell hyperplasia in the 30 and 90 ppm groups at 12 and 24 mo were observed, but were not statistically significant due to high control incidences. Thus, the high control incidence and small groups sizes may have precluded observation of a significant effect in the nose at 24 mo. Incidences and severity of basal-cell hyperplasia in the olfactory epithelium were similar in groups exposed to 30 or 90 ppm at 24 and 29 mo, but were not statistically significant at 24 mo because of the high control incidence in groups of 9–10 animals. The effect at 29 mo in groups of 50–60 animals is considerably more robust and is not inconsistent with results at earlier time points in this study.

The data on bromomethane present an unusual picture regarding progression of the nasal lesion. Table 1 shows that inhalation studies with exposures ranging in duration from a 4-h single exposure to subchronic and chronic durations up to 2 yr result in nasal effects at roughly similar concentrations, with NOAELs also at similar concentrations. Hurtt et al. (1987, 1988) reported no effects in rats exposed to 90 ppm after a single or a 5-day exposure. Groups of male Fischer 344 rats were exposed to 0, 90, 175, 250, or 325 ppm bromomethane for 6 h/day for 5 days (Hurtt et al., 1987). A concentration-dependent degeneration of nasal olfactory sensory cells was observed in rats exposed to 175 ppm bromomethane or greater. This degeneration affected 50–80% of the olfactory mucosa and was characterized by complete or partial destruction of the olfactory epithelium at the

TABLE 1. Progression of Nasal Lesion Severity in Rats with Increasing Exposure Duration

Protocol[a]	Sex	LOAEL (ppm)	NOAEL (ppm)	Study
1 day	Male	200	90	Hurtt et al. (1988)
5 days	Male	175	90	Hurtt et al. (1987)
3–14 days	Male	160	None[b]	Eustis et al. (1988)
	Male	160	None[b,,]	Eustis et al. (1988)
2 wk	Male	200	None[b]	Hastings (1991)
13 wk	Male	120	60	NTP (1992)
	Female	120	60	NTP (1992)
12 mo	Male	None	90[c]	Reuzel et al. (1991)
	Female	None	90	Reuzel et al. (1991)
24 mo	Male	None	90[c]	Reuzel et al. (1991)
	Female	None	90[c]	Reuzel et al. (1991)
29 mo	Male	3	None	Reuzel et al. (1991)
	Female	3	None	Reuzel et al. (1991)

[a]Exposures were 6 h/day and 5 days/wk, except Hurtt et al. (1988) (4-h exposure) and Hastings (1991) (4 h/day, 4 days/wk). All studies used F344 rats except Reuzel et al. (1991) (Wistar) and Hastings (1991) (Long-Evans).

[b]Only one concentration, the LOAEL, was tested.

[c]Designation of effect level is not clear; see text.

higher concentrations. No effect in the nasal epithelium was seen in the NTP studies at 60 ppm in males or females after 13 wk of exposure (NTP, 1992), and Reuzel et al. (1991) reported no significant effects in small groups of rats exposed to 90 ppm for 12 or 24 mo. These results suggest that there is limited dependence on exposure duration in this nasal lesion—that is, there is limited progression of the lesion severity with time of exposure. In contrast, the results observed at 29 mo of exposure show a substantial progression, such that the LOAEL for the nasal effect drops from greater than 90 ppm at 24 mo or less, to 3 ppm at 29 mo.

As discussed previously, the assignment of LOAEL and NOAEL levels in the Reuzel et al. (1991) study is complicated by the high control incidences of nasal effects at 12 and 24 mo. The high incidences of the nasal olfactory epithelial effect at 90 ppm suggest a LOAEL at this level in males at 12 mo and in males and females at 24 mo. Similarly, the proportions of animals affected in males and females exposed to 30 ppm for 24 mo are similar to groups at 29 months, but high control values make it difficult to assign a LOAEL. Thus, the difference in the LOAEL between the NTP and the Reuzel studies could be explained strictly by a strain difference, since F344 rats were used in the NTP, Hurtt et al., and Eustis et al. studies. Because this issue could not be resolved with the current database and the chronic studies were of high quality in design, execution, and reporting, the 3 ppm group in the Reuzel et al. (1991) chronic rat study was adopted as the LOAEL for RfC derivation.

RfC DERIVATION

The numerical calculation of the RfC involves three steps. First, the experimental exposure protocol is adjusted to an equivalent continuous exposure. The chronic rat study used exposure for 6 h/day for 5 days/wk. Using the default assumption that an equivalent inhaled dose over a continuous exposure would produce the same effect, the experimental concentration, 3 ppm (11.7 mg/m^3) is adjusted to a continuous exposure to calculate the duration-adjusted LOAEL as follows: 11.7 mg/m^3 × 6 h/24 h × 5 days/7 days = 2.08 mg/m^3. Thus, the duration adjusted LOAEL is 2.08 mg/m^3.

Second, using the default dosimetric adjustments (U.S. EPA, 1990; Jarabek et al., 1990), the human equivalent concentration (HEC) is calculated with the regional gas deposition ratio (RGDR) as follows:

$$\begin{aligned} \text{LOAEL(HEC)} &= 2.08 \text{ mg/m}^3 \times \text{RGDR} \\ &= 2.08 \times [(0.30 \text{ m}^3/\text{day})/11.6 \text{ cm}^2)/(20 \text{ m}^3/\text{day})/177 \text{ cm}^2] \\ &= 2.08 \times 0.23 = 0.48 \text{ mg/m}^3 \end{aligned}$$

The RGDR for this calculation is the ratio of animal ventilation rate over animal surface area of the extrathoracic region of the respiratory tract to human ventilation rate over human surface area of the extrathoracic region.

The HEC from the NTP study is calculated in the same way, and the resultant RGDR for mice is 0.19. The result of the calculation for the NTP study is LOAEL(HEC) = 13 mg/m^3, and NOAEL(HEC) = 4.4 mg/m^3.

The final step in the RfC derivation is the application of uncertainty factors (UF). In this case, an intraspecies factor of 10 to account for sensitive subpopulations is applied because there are no human data suggesting that this UF is not warranted. A UF of 3 is used to adjust from a LOAEL to a NOAEL; it is reduced from the normal default value of 10 due to the mildness of the effect. The interspecies factor of 3 is applied, rather than 10, due to application of default dosimetry. Because the factors of 3 represent operational application of a geometric half of the standard factor of 10, rounded to a single significant figure, multiplication of two factors of 3 results in a composite factor of 10. Thus, $3 \times 3 \times 10 \approx 100$. When the NOAEL(HEC) is divided by the total uncertainty factor of 100 and the result is rounded to one significant figure, the resulting RfC is 5 μg/m^3.

Confidence statements are associated with the principal study, database, and RfC in order to provide a qualitative summary of the robustness of the RfC derivation. The Reuzel et al. (1987, 1991) chronic study was well conducted, used an adequate number of animals and exposure levels, and reported a thorough histopathological examination of the respiratory tract, but it is given a medium confidence rating because it did not identify a NOAEL. The LOAEL identified in this study is supported by the effects seen in rats in the subchronic NTP (1992) study and mice in the chronic NTP (1992) study, as well as in subacute and subchronic studies in rats (Hastings, 1991; Hurtt et al., 1987, 1988). The database is given a high confidence rating because there are chronic inhalation studies in two species, supported by subchronic inhalation studies in several species, and because data are available on the developmental and reproductive effects of bromomethane as well as its pharmacokinetics following inhalation exposure. Based on the confidence in the database and study, high confidence in the RfC follows.

REFERENCES

Alexeef, G. V., and Kilgore, W. W. 1983. Methyl bromide. *Residue Rev.* 88:101–153.

Anger, W. K., Moody, L., Burg, J., Brightwell, W. S., Taylor, B. J., Russo, J. M., Dickerson, N., Setzer, J. V., Johnson, B. L., and Hicks, K. 1986. Neurobehavioral evaluation of soil and structural fumigators using methyl bromide and sulfuryl fluoride. *Neurotoxicology* 7(3):137–156.

Boorman, G. A., Morgan, K. T., and Uriah, L. C. 1990. Nose, larynx, and trachea. In *Pathology of the Fischer rat. Reference and atlas*, eds. G. A. Boorman, S. L. Eustis, M. R. Elwell, C. A. Montgomery, Jr., and W. F. MacKenzie, pp. 315–337. New York: Academic Press.

Eustis, S. L., Haber, S. B., Drew, R. T., and Yang, R. S. H. 1988. Toxicology and pathology of bromomethane in F344 rats and B6C3F1 mice following repeated inhalation exposure. *Fundam. Appl. Toxicol.* 11:594–610.

Hastings, L., Miller, M. L., Minnema, D. J., Evans, J., and Radike, M. 1991. Effects of methyl bromide on the rat olfactory system. *Chem. Senses* 16:43–55.

Herzstein, J., and Cullen, M. R. 1990. Methyl bromide intoxication in four field-workers during removal of soil fumigation sheets. *Am. J. Ind. Med.* 17:321–326.

Hurtt, M. E., Morgan, K. T., and Working, P. K. 1987. Histopathology of acute toxic responses in selected tissues from rats exposed by inhalation to methyl bromide. *Fundam. Appl. Toxicol.* 9:352–365.

Hurtt, M. E., Thomas, D. A., Working, P. K., Monticello, T. M., and Morgan, K. T. 1988. Degeneration and regeneration of the olfactory epithelium following inhalation exposure to methyl bromide: Pathology, cell kinetics, and olfactory function. *Toxicol. Appl. Pharmacol.* 94:311–328.

Jarabek, A. M., Menache, M. G., Overton, J. H., Dourson, M. L., and Miller, F. J. 1990. The U.S. Environmental Protection Agency's Inhalation RfD methodology: Risk assessment for air toxics. *Toxicol. Ind. Health* 6(5):279–301.

National Toxicology Program. 1992. Toxicology and Carcinogenesis Studies of Bromomethane (CAS No. 74-83-9) in B6C3F1 Mice (Inhalation Studies). NTP TR 385, NIH Publication 91-2840.

Reuzel, P. G. J., Kuper, C. F., Dreef-van der Meulen, H. C., and Hollanders, V. M. H. 1987. Chronic (29-Month) Inhalation Toxicity and Carcinogenicity Study of Bromomethane in Rats. Report V86.469/221044, Netherlands Organization for Applied Scientific Research, Division for Nutrition and Food Research, TNO. EPA/OTS Document 86-8700001202.

Reuzel, P. G. J., Dreef-van der Meulen, H. C., Hollanders, V. M. H., Kuper, C. F., Feron, V. J., and van der Heijden, C. A. 1991. Chronic inhalation toxicity and carcinogenicity study of bromomethane in Wistar rats. *Food Chem. Toxicol* 29(1):31–39.

U.S. Environmental Protection Agency. 1986. Health and Environmental Effects Profile for Methyl Bromide. Office of Research and Development, Cincinnati, OH. ECAO-CIN-P182.

U.S. Environmental Protection Agency. 1990. Interim Methods for Development of Inhalation Reference Concentrations (Review Draft). Office of Research And Development, Washington, DC. EPA/600/8-90/066A. August.

EVALUATION OF NASAL TRACT LESIONS IN DERIVATION OF THE INHALATION REFERENCE CONCENTRATION FOR HEXAMETHYLENE DIISOCYANATE

Gary L. Foureman, Mark M. Greenberg

U.S. Environmental Protection Agency, Environmental Criteria and Assessment Office, Hazardous Pollutant Assessment Branch, Research Triangle Park, North Carolina, USA

Ghona K. Sangha, Barry P. Stuart, Ronald N. Shiotsuka, Juergen H. Thyssen

Miles Inc., Agriculture Division, Stilwell, Kansas, USA

Derivation of the inhalation reference concentration (RfC) for HDI, a highly reactive and irritant gas, required consideration of several scientific issues, including toxicological judgment on severity versus incidence of lesions, and occurrence and toxicological significance of adaptive nasal lesions. Analysis of data from a chronic study with hexamethylene diisocyanate (HDI) revealed that the lesion incidence within the nasal cavity was not a clear indicator of toxicologic significance. Although chronic inflammation exhibited a clear concentration-response relationship for incidence, the corresponding severity scores for this lesion were virtually unchanged. Degeneration of the olfactory epithelium did follow a concentration-response relationship for both incidence and severity. Such results indicate that information on both incidence and severity may be necessary to make valid toxicological judgments. A considerable portion of nasal cavity histopathology in the animals exposed to HDI may be considered adaptive. Exposure of nasal tissues to an irritant, such as HDI, may lead to increased production of mucus, morphologic changes in or replacement of sensitive cells, and barriers at the irritant/tissue interface. The histopathological manifestations of these events (mucus hyperplasia, squamous metaplasia, and keratinization of respiratory epithelium) were all observed in this study. The boundary between an adaptive and toxic response is not always clearly delineated, and, in this case, many of these alterations may be considered to be adaptive rather than a functional impairment. Thus, many of these alterations are not clearly adverse toxic responses. On the other hand, degeneration of the olfactory epithelium is clearly adverse. The RfC derived for HDI is 0.01 $\mu g/m^3$ based on a chronic inhalation study in rats with a critical effect of olfactory epithelial degeneration.

The high degree of reactivity of diisocyanates makes this class of chemicals especially useful. Toluene 2,4- and 2,6-diisocyanate (TDI) and diphenylmethane diisocyanate (MDI) are used extensively in the production of polyurethane foams, and hexamethylene 1,6-diisocyanate (HDI) is used as a hardener in automobile paints. The high reactivity is due to the pres-

Address correspondence to Gary L. Foureman, U.S. EPA (MD-52), Research Triangle Park, NC 27711, USA.

ence of the NCO group common to all these chemicals. Unfortunately, this same characteristic is most certainly responsible for the health effects of these chemicals, as isocyanates react with nearly all organic compounds, especially those with hydrogen atoms attached to oxygen, nitrogen, or sulfur (Kennedy & Brown, 1992). Consequently, tissue components and biomolecules may undergo reactions in which secondary or tertiary structure would be altered, causing a decrement or even loss of function. In addition, HDI is a known irritant and a suspect sensitizer of the respiratory tract (ACGIH, 1986). HDI, MDI, and TDI are all on the hazardous air pollutants list in Section 112 of the Clean Air Act Amendments of 1990.

A general discussion of inhalation reference concentration (RfC) derivation is given elsewhere (U.S. EPA, 1990; Shoaf, 1991). RfC derivation, by and large, follows the classical and logical procedures used for many concentration-response assessments: consideration of all data, delimitation of these data based on relevancy and scientific robustness, and application of factors to reflect uncertainty about what is not made known by these data. The specifics of this process for RfC derivation are guided by the stated purpose of the RfC: to provide an estimate of air levels that are protective against adverse health effects occurring from chronic exposure. A principal study, ideally one in which subjects are exposed to low levels for extensive periods of time and in which toxicity is elicited, is selected. A critical effect, the first adverse effect that occurs as concentration increases, is then chosen from this principal study. Analysis of concentrations at which this critical effect is manifest provides effect levels showing its absence (the no-observed-adverse-effect level, NOAEL) and its initial occurrence (the lowest-observed-adverse-effect level, LOAEL). The concentrations at which these effect levels occur are then adjusted based on physiological and anatomical presumptions of what these concentrations would be in humans, resulting in a human equivalent concentration (HEC). Uncertainty factors are then applied to these adjusted effect levels, usually the NOAEL(HEC), to arrive at the actual RfC.

The purpose of this report is to present issues and problems encountered on the main components of the RfC process for HDI—the choice of a principal study and designation of a critical effect. The problems encountered are not those created by the choice of the specific risk assessment process, but are fundamental issues of toxicology. Issues that had to be considered for HDI included the relevance of incidence versus severity of lesions, the occurrence and significance of progression in lesions, and the adaptive/adverse nature of lesions.

RfC DEVELOPMENT: CHOICE OF THE PRINCIPAL STUDY

A comprehensive literature search for toxicological information on inhaled HDI yielded the studies abstracted in Table 1. Basically, two types of studies were found; these were animal studies in which rats were

TABLE 1. Studies Considered in Derivation of the RfC for HDI

	Exposure		
Reference	Duration/frequency	Concentrations	Species and numbers
Miles, Inc. (1989)	6 h/day, 5 days/wk for 2 yr with satellite groups for 1 yr	0, 0.005, 0.025, 0.175 ppm (increased from 0.125 ppm on study day 127)	50 Rats/sex/group for 2-yr study, 10/group for 1-yr study
Miles, Inc. (1984)	Head only, 5 h/day, 5 days/wk for 3 wks	0, 0.005, 0.0175, 0.15, 0.3 ppm	10 Rats/sex/group
Miles, Inc. (1988)	6 h/day, 5 days/wk for 90 days	0, 0.01, 0.04, 0.14 ppm	20 Rats/sex/group
Miles, Inc. (1970)	4 h/day, 5 days/wk for 4 wk	0, 0.2, 2.0 ppm	20 Male rats/group
Karol et al. (1984)	Head only, 2–6 h ppm	0.5, 1.8, or 4.0	Male guinea pigs, 4 to 6/group
Tornling et al. (1990)	8 h/day, 5 days/wk for nearly 13 yr	HDI monomer, 1.5 $\mu g/m^3$ with peaks; HDI trimer, 90 $\mu g/m^3$ with peaks exceeding 2000 $\mu g/m^3$	36 Workers with 115 controls (see Alexandersson et al., 1987)
Grammer et al. (1988)	8 h/day, 5 days/wk ranging from 24 to 51 mo	HDI monomer, from <0.08 to 3.8 $\mu g/m^3$; HDI trimer, from 5.3 to 75 $\mu g/m^3$	150 Workers, no controls
Alexanderrson et al. (1987)	8 h/day, 5 days/wk with an average of 7 yr	HDI monomer, about 1.0 $\mu g/m^3$ with brief peaks, HDI trimer, 15 $\mu g/m^3$ (range 10 to 385 $\mu g/m^3$)	41 Workers with 70 controls matched for age, height, and smoking
Brorson et al. (1990)	7.5 h for 1 day	25 $\mu g/m^3$	5 nonsmoking males
Innocenti et al. (1986)	3 mo, occupational exposure	None given	Case study of a 17-yr- old male worker

exposed to different air concentrations of HDI for varying time periods (from 3 wk up to 2 yr) and occupational studies in which workers exposed to HDI were examined for lung function and sensitization. Karol et al. (1984) examined the potential of HDI to inhibit cholinesterase activity in guinea pig tissues. All of the studies on sensitization were available from published literature. The animal inhalation studies were performed by industry and submitted to the U.S. EPA through the reporting requirements under the Toxic Substances Control Act (TSCA), where they become available through the Toxic Substances Control Act Test Submissions (TSCATS) database. Studies available through TSCATS are usually conducted according to the Good Laboratory Practice Standards and often provide more complete data than studies published in the open literature.

General guidance on choice of a principal study is given in U.S. EPA (1990) and Shoaf (1991). More specific information for selection is considered from what is known of the individual compound and from the stated purpose of the RfC. HDI is a known respiratory tract irritant; therefore, choice of inhalation studies in which effects in this region are given particular attention would be most appropriate. The exposure scenario for the RfC is lifetime continuous exposure; therefore, a study with a long duration of repeated exposures would be most appropriate. Other considerations include use of relevant species, testing a sufficient number of exposure concentrations spaced appropriately to establish a concentration-response relationship, a sufficiently large population of exposed subjects and controls, confirmation and control of chemical purity, controlled exposure conditions, and presentation of sufficient data to support conclusions made.

The longest term animal study available was that of Miles, Inc. (1989). Investigators exposed rats (60/sex/group) to 0.005, 0.025, and 0.175 ppm HDI or filtered air (controls) for 6 h/day, 5 days/wk for 2 yr. An interim sacrifice of 10 animals/sex/group was carried out at 1 yr. Extensive toxicological testing was carried out, including urinalysis, clinical chemistry, and ophthalmology. Although no pulmonary function testing was conducted, the entire respiratory tract was thoroughly scrutinized in all test animals, including those sacrificed at 1 yr. Nasal tract tissues were sectioned at five to six levels for histopathological examination. The levels were designated as follows: I, vestibule; II, posterior to incisor teeth; III, prepapilla; IV, incisive papilla; V, first palatal ridge; and VI, second palatal ridge. These designations are used throughout this report. Complete exposure information and histopathological data to the level of the individual animal is documented in this study.

Available occupational studies examined workers for pulmonary effects and the presence of antibodies against isocyanates. Tornling et al. (1990) reported more wheezing in exposed workers than in nonexposed controls and some excess decrease in pulmonary function among one subgroup. Grammer et al. (1988) demonstrated HDI antibodies in 21% of 150 workers chronically exposed to HDI, although none of those with antibodies had any pulmonary symptoms. Although there is no question that humans would be the most relevant species in which to study effects, several other aspects must be considered. The exposure scenarios in these studies vary widely. Tornling et al. (1990) reported a low average exposure to either HDI or HDI-trimer, but with brief peaks over 1000-fold greater than this average. Thus, it is not clear to which exposure scenario or form (HDI or HDI-trimer) pulmonary effects could be attributed. Some studies suggest that people who have developed pulmonary sensitivity to agents such as isocyanates have usually had one or more episodes of exposure to high concentrations, whereas other workers receiving lesser exposures typically do not develop sensitivity (Karol & Thorne, 1988).

Small cohorts, lack of mechanistic information, and limited information on exposure in the human studies preclude establishment of a concentration-response relationship. Healthy worker effects, such as more sensitive individuals (or those becoming sensitized) leaving exposure areas while acclimatized hardened workers remain, may bias the results. The current information available on occupational asthma from HDI exposure is incomplete; the exposure levels and scenarios eliciting this effect are unclear. What is more clear is that far greater concentrations of an agent are required for sensitization than are required to subsequently elicit a reaction. These limitations and current gaps in knowledge preclude the choice of pulmonary sensitization as the critical effect.

Consideration of the strengths and limitations discussed above led to the selection of the 2-yr chronic study of Miles, Inc. (1989) as the principal study.

RfC DEVELOPMENT: CHOICE OF THE CRITICAL EFFECT

Prolonged exposures to high doses of a toxicant may eventually elicit a myriad of effects, primary and secondary. The critical effect is meant to be that adverse effect occurring earliest as the dose of toxicant increases. What is not indicated in this designation are criteria for what is adverse. The issue of adversity is encountered in several aspects in the principal study. These aspects have to be individually analyzed before a critical effect can be designated. This choice is pivotal for the next step in the RfC derivation process, the designation of a NOAEL and/or LOAEL, terms that both include the descriptor "adverse."

Analysis of the results from the principal study revealed that compound-related effects were limited to histopathology in the nasal passages. The effects observed in the nasal cavity of rats are listed in Table 2. Those indicating tissue degeneration or necrosis are listed first, followed by migration

TABLE 2. Lesions Observed in the Nasal Tract Tissues of Rats After Inhalation of HDI

Necrosis
Olfactory epithelium degeneration
Hyaline droplet degeneration
Chronic-active inflammation
Hyperplasia/metaplasia
Mucus hyperplasia
Epithelial hyperplasia, not otherwise specified (NOS)
Squamous metaplasia
Hyperkeratosis, NOS
Erosion
Ulcer, NOS

of inflammatory cells to the affected site, followed by proliferative and reparative responses of affected cells, followed by the outcomes of this latter process. Depending on the severity of the damage and the continued exposure to the agent, the reparations may be qualitatively dissimilar, ranging from failed reparation (i.e., degeneration), to altered appearance/function, to totally normal regeneration (Boorman et al., 1990). The divergence of the qualitative alterations occurring during this process leads to questions of toxicological relevance. Several of the lesions in Table 2 are examined in detail for relevance in the choice of a critical effect.

Incidence/Severity

Occurrence of lesions as severe in nature as ulceration and hyperkeratosis occurred in every level of the nasal tract examined save the most posterior (level VI). These lesions were largely limited to animals exposed to the highest concentration of HDI, none being present in controls. Average severity scores on a scale of 1 (mild) to 5 (severe) were also recorded for these and the other lesions, with scores for the most severe, such as ulceration, being uniformly high. Data for these severe lesions clearly indicate the highest exposure concentration as an adverse effect level.

Chronic inflammation was noted in the nasal tract of animals from all exposure groups and controls. Incidence data for this lesion are presented for male animals in Figure 1. The incidence of this lesion in animals

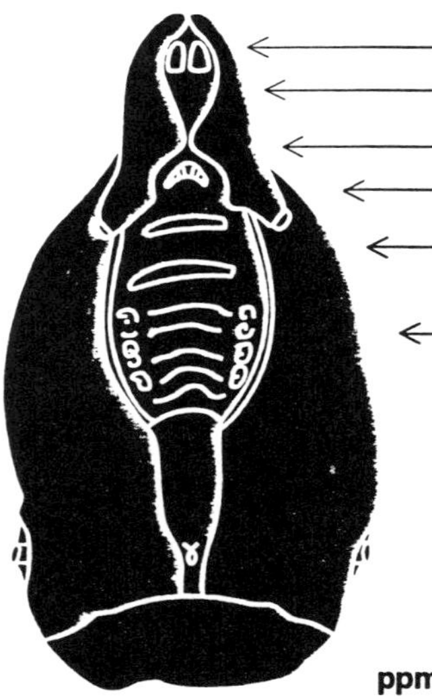

Level	Incidence (%)			
I	27	42	43	52
II	35	35	75	98
III	10	13	23	65
IV	20	18	27	55
V	18	20	32	63
VI	0	10	0	0
ppm HDI	0	0.005	0.025	0.175

FIGURE 1. Incidence of chronic inflammation in the nasal tract tissues of male rats exposed to HDI. Arrowheads point to levels of sectioning, I–VI. See text for further description of levels sectioned. Abstracted from Miles, Inc. (1989).

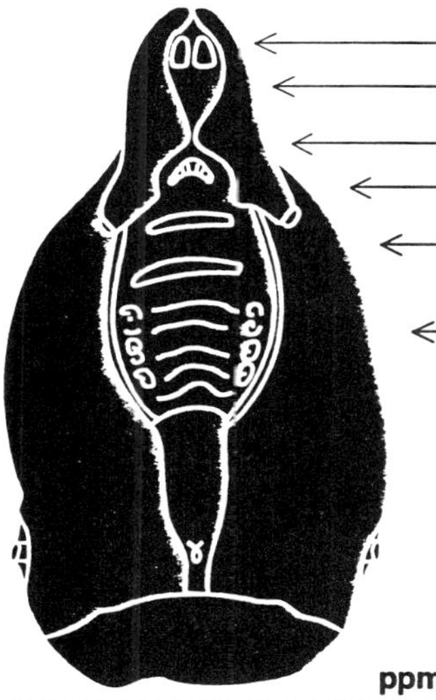

Level	Mean Severity Score			
I	1.1	1.1	1.1	1.2
II	1.6	1.7	1.4	2.7
III	1.2	1.4	1.3	1.4
IV	1.8	1.4	1.3	1.4
V	1.3	1.3	1.4	1.4
VI	–	1.3	–	–
ppm HDI	0	0.005	0.025	0.175

FIGURE 2. Mean severity scores of chronic inflammation in the nasal tract of male rats exposed to HDI. Arrowheads point to levels of sectioning, I–VI. See text for further description of levels sectioned. Abstracted from Miles, Inc. (1989).

exposed to the highest concentration shows a clear increase over control values in all levels save VI. The incidence in animals exposed to the mid concentration shows doubling over controls at levels II and III. At level I the incidence in animals exposed to the lowest concentration is about 1.5 times the controls. Thus, it would seem that the lowest concentration tested could be considered a LOAEL. The severity scores for this lesion make this conclusion less clear. Inspection of these scores, presented in Figure 2, shows that with a single exception, they are not different from the control values. Only posterior to the incisors (level II) and only at the highest concentration is there a clear increase in the severity score, and it would seem that the highest concentration would be a LOAEL. This nonconcordance in incidence and severity confounds the choice of a NOAEL or LOAEL. A similar disparity between incidence and severity occurred for other lesions in the nasal tract, including mucus hyperplasia, epithelial hyperplasia, hyaline droplet degeneration, and squamous metaplasia.

Olfactory epithelial degeneration, on the other hand, demonstrated concordancy between incidence and severity. This lesion was noted at levels IV, V, and VI in the nasal tract of exposed animals only. Figure 3 is representative of data for this lesion, showing its absence at the lowest concentration with parallel increases in both incidence and severity at the two highest concentrations. The concentration-related increase in both parameters lends further biological credence in considering this lesion as a

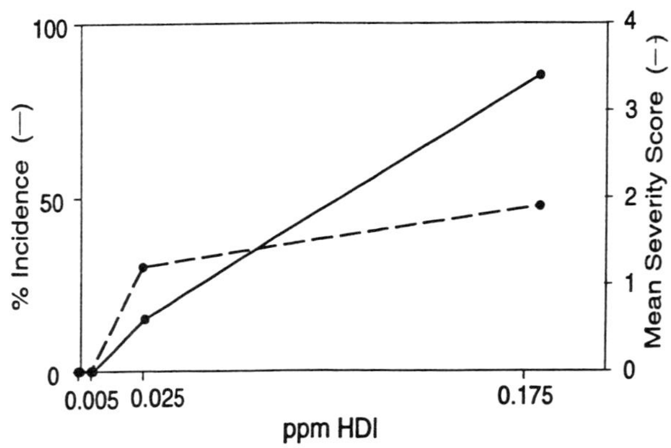

FIGURE 3. Mean severity scores and incidence of olfactory epithelial degeneration (at incisive papilla, level IV) in male rats exposed to HDI. Abstracted from Miles, Inc. (1989).

critical effect, and one from which effect levels could be chosen unambiguously.

Stipulating concordancy between the lesion parameters of incidence and severity has several consequences. This practice could be viewed as implying that occurrence of a clear increase of lesion incidence is not toxicologically significant. A more correct interpretation, however, would be that the toxicological significance of an increase in one parameter is uncertain and/or unclear, whereas an increase in both of these parameters is less uncertain and more clear. The consequences of making decisions without complete information on the parameters of incidence and severity are readily apparent from this example, as the difference in effect levels chosen would be 35-fold (0.005 vs. 0.175) if incidence were the sole parameter available.

Progression

Progression of a lesion has several connotations. Reference may be to anatomical progression where, with time, the same lesion may spread from one to other sites in an exposed subject or to the same site in a greater proportion of exposed animals (i.e., incidence). Alternatively, reference may be to alterations of the lesion itself, where either the extent of involvement of proximal tissues may be expanded or the character of the individual lesion exacerbated (i.e., severity). The interim sacrifice performed at 1 yr in the principal study provided information for analysis of progression.

Analysis of temporal progression of some nasal tract lesions is shown in Table 3. Inspection of both incidence and severity data indicates that mucus hyperplasia progressed minimally, as neither incidence or severity was altered at any exposure level from the end of the first year to the end of the second year. A similar analysis of squamous metaplasia in males (level II) revealed the same condition. Data for degeneration of the olfactory epithelium, however, clearly showed progression in incidence and severity from one year to the next. The lack of progression in mucus hyperplasia and squamous metaplasia may indicate that the lesions induced were either static, at equilibrium with the reparative capacity of the tissue, or that functional alteration had possibly occurred.

The foregoing interpretation implies that occurrence of a lesion that progresses only slightly with time is not toxicologically relevant. A more correct statement, however, would be that the toxicological significance of this lack of progression is unclear, especially when compared with a lesion that does progress with time in more than one parameter.

Adaptive/Adverse

Toxic insult to healthy tissues stimulates their reparative and defensive capabilities. The divergence of the qualitative alterations occurring during this process engenders questions as to their classification as either adaptive or adverse. This difficulty in classification becomes even greater as mechanistic investigations have traced lesion etiology to biochemical and genetic events, prompting the views from Ruben and Rousseaux (1991) that "The boundary between an adaptive and a toxic response is not well delineated" and "the borderline between adaptive physiology (adaptive function) and toxicity (functional impairment) is not always clear."

Lesion classification also has to be considered in terms of individual organs; metaplasia in the pancreas versus the tracheal epithelium has markedly different consequences and potential progressional outcomes. This dilemma of categorization, as applied to respiratory tract tissues, has been recognized (Burger et al., 1989; Monticello et al., 1990). Several nasal tract lesions observed in the principal study, hyaline droplet degeneration, mucus hyperplasia, and squamous hyperplasia, have had differing interpretations on the issue of adverse versus adaptive.

Hyaline droplet degeneration was observed in both control and exposed male animals at levels III–VI. The incidence was markedly increased over

TABLE 3. Temporal Progression of Nasal Tract Lesions in Male Rats Exposed to HDI

		ppm HDI			
Lesion Level	Year	0	0.005	0.025	0.175
Mucus hyperplasia	1	0	30 (1.0)	30 (1.0)	60 (1.0)
Prepapilla (III)	2	13 (1.5)	27 (1.2)	47 (1.4)	63 (1.4)
Squamous metaplasia	1	0	0	0	90 (3.4)
Postincisor (II)	2	7 (2.8)	15 (2.0)	10 (2.2)	93 (3.5)
Olfactory epithelium degeneration	1	0	0	10 (1.0)	10 (1.0)
Palatal ridge (V)	2	0	2 (1.0)	12 (1.6)	92 (2.3)

Note. Data are percent incidence with lesion severity scores in parentheses.

controls even at the lowest exposure concentration. Mean severity scores were not increased over controls in any exposure group. Thus, limited justification exists for designating the lowest exposure concentration as an effect-level. To further confound issues about this lesion the following commentary is provided by Monticello et al. (1990): "Eosinophilic globules (also termed intracytoplasmic hyaline droplets) are normally found in low numbers in control rats and mice and increase with age. They increase in quantity and severity following chronic exposure to certain toxicants . . . perhaps constituting an adaptive response." Lewis et al. (1994) present evidence that these droplets contain xenobiotic-metabolizing enzymes. Thus, considerable uncertainty exists about the classification of this lesion, both for this specific instance and in the scientific community at large. The strength of evidence in this case indicates that this lesion has more of an adaptive rather than adverse character.

Mucus hyperplasia was also observed in both control and exposed male animals, predominantly at levels II–V. Again, incidence was markedly increased over controls even at the lowest concentration. The response was clearly concentration-related in only one section (III), as incidences at the highest concentration were either static (levels IV and V) or decreased (level II). Mean severity scores were not increased over control values, and thus limited justification exists for designating the lowest exposure concentration as an effect level. In addition, Monticello et al. (1990) provided the following views on this lesion: "Secretory cell and respiratory epithelial cell hypertrophy/hyperplasia are common responses to certain gaseous irritants. . . . "Secretory cell hyperplasia and hypertrophy may provide increased amounts of mucus to the nasal airways and may thus represent an adaptive response." Thus, considerable uncertainty also exists about the classification of this lesion, as the consequences of increased mucus production are ambivalent. Increased production would provide greater protection to the sensitive underlying tissues, whereas superfluous production may result in the occurrence of mucostasis and the pathophysiological consequences of this condition. The strength of evidence in this specific instance indicates this lesion to be more adaptive rather than adverse.

The occurrence of squamous metaplasia was limited to levels II and V in both sexes. Incidence of this lesion was increased over control values (7% in males, 15% in females) only in those animals exposed to the highest concentration (93% in males, 70% in females). As mentioned earlier, this situation also existed at the 1-yr interim sacrifice (Table 3). Within the levels of the nasal tract, the mean severity scores were not different between control and exposed animals. Whereas no doubt exists that this lesion was compound related, its adverse nature is called into question not only by the nonconcordance of incidence and severity, but also by several comments on metaplasia in the nasal tract:

"Adaptive squamous metaplasia . . . is a common response observed in the rodent nasal passages following chronic exposure to cytotoxic irritants" (Monticello et al., 1990).

"The change [metaplasia] described in the respiratory epithelium is often reversible and adaptive" (Ruben & Rousseaux, 1991).

These statements clearly indicate the lesion as adaptive and are relevant, as HDI is a known upper respiratory tract irritant.

"During the reparative process, . . . lesions may progress to squamous metaplasia with chronic persistent injury" (Boorman et al.,1990).

"Metaplasia may be considered an adaptive response, but the predictive value for lesion progression and secondary effects on other organs is not always clear" (Ruben & Rousseaux, 1991).

These two statements reflect an ambivalence about the nature of metaplasia. The first associates the presence of squamous metaplasia with repair of injury, with its presence implying the occurrence of earlier adverse effects. The second refers to outcomes specific to the site of the metaplasia (e.g., metaplasia in the pancreas vs. the respiratory tract as mentioned earlier) and indicates that the full history of the specific metaplasia, including consequences to other related systems, should be known before classification is attempted.

The in-depth analysis by Burger et al. (1989) of squamous metaplasia thoroughly describes the dual character of this lesion as it may occur in the respiratory tract. These authors elicited what could be termed adaptive squamous metaplasia in the ventral larynx in rats exposed to aerosolized glycerol or to smoke and from cigarettes that heat, but do not burn, tobacco. This metaplastic response was observed to (1) regress significantly after 6 wk of recovery, (2) be well differentiated, and (3) lack areas of atypia and/or dysplasia. These descriptors are in contrast to those of Boorman et al. (1990), who note that respiratory tract squamous metaplasia with disorganization of cell layers, cellular pleomorphism, and atypia may be preneoplastic. Given the possible consequences of progression (i.e., potentially to squamous-cell carcinoma) and the results of the principal study showing this eventuality not to occur, the strength of evidence indicates this lesion should be considered more adaptive than adverse in nature, the effect level being the highest concentration.

The olfactory epithelium is sensitive to toxicants; the olfactory neurons are extremely so toward noxious gases (Boorman et al., 1990). Recent work (Hastings et al., 1994) has demonstrated a functional olfactory deficit corroborative of histopathological damage to the olfactory epithelium. Therefore, the nature of degeneration of the olfactory epithelium, noted at levels IV, V, and VI in the nasal tract of exposed animals only (Figure 3), is clearly adverse. As noted earlier, this situation gives an unambiguous indication of effect levels. The clearly adverse nature of this lesion lends further biological credence to consider it as the critical effect.

Designation of the Critical Effect and Effect Levels

From consideration of these myriad effects, the uncertainties about adversity, the availability of exposure information, and the relevance of the lesions to humans and to the purpose of the RfC, degeneration of the olfactory epithelium is designated the critical effect, with a NOAEL of 0.005 ppm (0.035 mg/m^3) and a LOAEL of 0.025 ppm (0.17 mg/m^3).

These effect levels are adjusted to reflect the stated purpose of the RfC (i.e., continuous exposure). Currently, this procedure is accomplished by factoring the concentration both by the fraction of a day and of a week that the exposure duration represents. The resultant concentrations are termed "adjusted" and are designated as either NOAEL(ADJ) or LOAEL(ADJ). For HDI, this gives NOAEL(ADJ) = 0.006 mg/m^3 and LOAEL(ADJ) = 0.030 mg/m^3.

These levels are further adjusted to derive a human equivalent concentration (HEC) based on physiological and anatomical presumptions of what these concentrations would be in humans (U.S. EPA, 1990; Shoaf, 1991). For a gas such as HDI, these presumptions are codified in a factor termed the regional gas dose ratio (RGDR). The region affected was the nasal tract, which is encompassed by the term "extrathoracic." Using values of 11.6 cm^2 and 177 cm^2 for the surface areas (*S*) of this region and volumes per day (MV) of 0.24 m^3 and 20 m^3 in rats and humans, respectively, the RGDR was calculated as follows:

$$\text{RGDR} = (\text{MV}_a/S_a)/(\text{MV}_h/S_h)$$

$$\text{RGDR} = (0.24\ \text{m}^3)/11.6\ \text{cm}^2/(20\ \text{m}^3/177\ \text{cm}^2) = 0.183$$

The HEC is then calculated as follows:

$$\text{NOAEL(HEC)} = \text{NOAEL(ADJ)} \times \text{RGDR}$$

$$= 0.006\ \text{mg/m}^3 \times 0.183 = 0.001\ \text{mg/m}^3$$

$$\text{LOAEL(HEC)} = 0.030\ \text{mg/m}^3 \times 0.183 = 0.005\ \text{mg/m}^3$$

UNCERTAINTY FACTORS

General information on uncertainty factors and their application is given in U.S. EPA (1990) and in Shoaf (1991). Of five possible areas of uncertainty, three are applied to the current RfC for HDI.

No information is available on potential sensitive human subpopulations leading to application of a full or normal factor of 10 for this area.

Uncertainty inherent in the process of extrapolation of animals to humans has been partly addressed as exposure concentrations were adjusted to the HEC. Therefore, a partial factor is evoked. As the assumption is made that these parameters are log-normally distributed, the partial factor evoked is the half-log of the normal factor, $10^{0.5}$.

One intent of the RfC process is to consider the total toxicity data currently available for the chemical under consideration. Ideally, reproductive, developmental, and any special studies (such as those examining a specific target organ or in which special tests are performed) would be available in addition to information from long-term inhalation exposures. These data would at least ameliorate the uncertainty about these toxicities. Aside from long-term inhalation studies, no other studies are available for HDI. This situation would ordinarily indicate maximum uncertainty about the potential of these toxicities. However, the high reactivity of the isocyanates argues against circulation of intact compound beyond the site of contact (i.e., the respiratory tract). Assuming the metabolites of isocyanates not to be toxic, little basis exists for suspecting toxicity would be found in developmental or reproductive studies. This situation also ameliorates uncertainty in toxicity information, and, as a consequence, another partial factor of $10^{0.5}$ is applied.

The other areas of uncertainty that do not apply to the HDI RfC are extrapolation from a subchronic to chronic study (as a chronic study was chosen as the principal study) and use of a LOAEL (as a NOAEL was available for the principal study).

Current practice in the RfC process is to multiply the uncertainty factors together to obtain the total uncertainty, which in this case would be $10^1 \times 10^{0.5} \times 10^{0.5} = 100$.

This factor is then applied to the NOAEL(HEC) to arrive at the RfC:

$$\text{RfC} = \text{NOAEL(HEC)} / \text{Uncertainty factor}$$

$$(0.001 \text{ mg/m}^3 / 100 = 10^{-5} \text{ mg/m}^3 \text{ or } 0.01\ \mu\text{g/m}^3$$

CONFIDENCE IN THE RfC

Confidence statements are meant and constructed to reflect the likelihood that the RfC would change based on any new information that may become available. They also serve as a repository of caveats and limitations about the principal study and the database. Confidence levels are assigned both to the principal study and to the database. In concordance with the intent of the RfC process to consider the totality of toxicity data available, the overall confidence rating of the RfC is no higher than that determined for the database.

The rating given to the principal study was medium, the accompanying statement pointing out the excellent quality of the study in examining the target tissues. The rating given to the database was also medium, primarily due to the unresolved issue of pulmonary sensitization. The overall rating for the RfC is therefore medium.

GENERAL CONCLUSIONS

The current RfC for HDI is 0.01 μg/m^3 based on a 2-yr inhalation study in rats (Miles, Inc., 1989) with a critical effect of olfactory epithelium degeneration. Specific information on lesions described to the maximum level of detail can be utilized not only in the process of deriving an RfC, but in other similar risk assessment processes. Omission of certain parameters in reporting of lesions may result in a specious designation of effect levels not only in the RfC derivation but also in other similar risk assessment processes. In response to an irritant, the character of lesions in the nasal tract such as squamous metaplasia, mucus hyperplasia, and hyaline droplet formation appears to be more adaptive than adverse. More consistent and definitive guidance from the toxicologic pathology community on the nature of these lesions is desirable.

REFERENCES

Alexandersson, R., Hedenstierna, G., Plato, N., and Kolmodin-Hedman, B. 1987. Exposure, lung function, and symptoms in car painters exposed to hexamethylendiisocyanate and biuret modified hexamethylendiisocyanate. *Arch. Environ. Health* 42(6):367–373.

American Conference of Governmental Industrial Hygienists. 1986. Documentation of the Threshold Limit Values (TLVs). Washington, DC: ACGIH.

Boorman, G. A., Morgan, K. T., and Uriah, L. C. 1990. Nose, larynx, and trachea. In *Pathology of the Fischer rat*, pp. 315–337. San Diego: Academic Press.

Brorson, T., Skarping, G., and Nielsen, J. 1990. Biological monitoring of isocyanates and related amines. II. Test chamber exposure of humans to 1,6-hexamethylene diisocyanate (HDI). *Int. Arch. Occup. Environ. Health* 62(5):385–389.

Burger, G. T., Renne, R. A., Sagartz, J. W., Ayres, P. H., Coggins, C. R. E., Mosberg, A. T., and Hayes, A. W. 1989. Histologic changes in the respiratory tract induced by inhalation of xenobiotics: Physiologic adaptation or toxicity? *Toxicol. Appl. Pharmacol.* 101:521–542.

Grammer, L., Eggum, P., Silverstein, M., Shaughnessy, M., Liotta, J., and Patterson, R. 1988. Prospective immunologic and clinical study of a population exposed to hexamethylene diisocyanate. *J. Allergy Clin. Immunol.* 82(4):627–633.

Hastings, L., Andringa, A., and Miller, M. L. 1994. Exposure of the olfactory system to toxic compounds. *Inhal. Toxicol.* 6(suppl):437–440.

Innocenti, A., Mariano, A., and Valiani, M. 1986. Occupational hexamethylene diisocyanate (HDI) asthma. *Med. Lav.* 77(2):191–194.

Karol, M. H., and Thorne, P. A. 1988. Pulmonary hypersensitivity and hyperreactivity: implications for assessing allergic responses. In *Toxicology of the lung*, eds. D. E. Gardner, J. D. Crapo, and E. J. Massero, pp. 427–448. New York: Raven Press.

Karol, M., Hansen, G., and Brown, W. 1984. Effects of inhaled hexamethylene diisocyanate (HDI) on guinea pig cholinesterases. *Fundam. Appl. Toxicol.* 4(2 Part 1):284–287.

Kennedy, A. L., and Brown, W. E. 1992. Isocyanates and lung disease: Experimental approaches to molecular mechanisms. *Occup. Med. State of the Art Rev.* 2:301–329.

Lewis, J. L., Nikula, K. J., and Sachetti, L. A. 1994. Induced xenobiotic-metabolizing enzymes localized to eosinophilic globules in olfactory epithelium of toxicant-exposed F344 rats. *Inhal. Toxicol.* 6(suppl):422–425

Miles, Inc. 1970. Toxicological Experiments, with cover letter dated 07/29/87. TSCATS/308406. EPA/OTS Document 86-870001255.

Miles, Inc. 1984. 21-day Inhalation Toxicity Study with Hexamethylene Diisocyanate (HDI) in Rats, with cover letter dated 07/29/87. TSCATS/308356. EPA/OTS Document 86-870001230.

Miles, Inc. 1988. 90-Day Inhalation Toxicity Study with 1,6-Hexamethylene Diisocyanate in Rats (with attached appendices and cover letter dated 01/18/89). TSCATS/401508. EPA/OTS Document 86-890000080.

Miles, Inc. 1989. Chronic Inhalation Toxicity and Oncogenicity Study with 1,6-Hexamethylene Diisocyanate (HDI) in Rats (Final Report), with attached appendices and cover letter dated 12/20/89. TSCATS/405187. EPA/OTS Document 86-900000055.

Monticello, T. M., Morgan, K. T., and Uriah, L. 1990. Nonneoplastic lesions in rats and mice. *Environ. Health Perspect.* 85:249–274.

Ruben, Z., and Rousseaux, C. G. 1991. The limitations of toxicologic pathology. In *Handbook of toxicologic pathology*, pp. 131–142. San Diego: Academic Press.

Shoaf, C. R. 1991. Current assessment practices for noncancer end points. *Environ. Health Perspect.* 95:111–119.

Tornling, G., Alexandersson, R., Hedenstierna, G., and Plato, N. 1990. Decreased lung function and exposure to diisocyanates (HDI and HDI-BT) in car repair painters: Observations on re-examination 6 years after initial study. *Am. J. Ind. Med.* 17(3):299–310.

U.S. Environmental Protection Agency. 1990. Interim Methods for Development of Inhalation Reference Concentrations (Review Draft). Office of Research and Development, Washington, DC. EPA/600/8-90/066A. August.

ADVANCES IN COMPARATIVE NASAL ENZYMOLOGY BETWEEN F344 RATS AND HUMANS

Alan R. Dahl, Devon J. Zastrow, Jon A. Hotchkiss

Inhalation Toxicology Research Institute, Albuquerque, New Mexico, USA

Species differences in metabolism of xenobiotics contribute to dissimilarities in toxic response. Rats are a standard species for toxicity testing, including agents that cause nasal lesions; however, there is minimal evidence to confirm that rat and human nasal tissues handle xenobiotics identically. One primary impediment to establishing a comparative relationship between rat and human nasal tissues is that nasal tissues are much smaller than the tissues of most organs, and traditional techniques involving enzyme isolation and characterization are impractical—especially for human nasal enzymes. Therefore, we are pursuing a two-pronged approach to resolve this issue.

To isolate and characterize xenobiotic-metabolizing enzymes, we have produced tissue-specific cDNA libraries (Hotchkiss et al., 1994) from F344 rat nasal transitional and olfactory epithelium and human nasal respiratory epithelium. Using antibodies or cDNA probes, we have obtained cDNAs that code for xenobiotic enzymes from this library. The cDNA libraries were constructed by Clontech Laboratories, Inc. (Palo Alto, CA) in the Lambda ZAP II (Stratagene; La Jolla, CA) bacteriophage cloning vector using unidirectional priming (oligo dT-Xho I primer) to increase the number of clones that could be detected by antibody screening. Lambda ZAP II is an insertion vector that can accept up to 10 kB of foreign DNA at six unique cloning sites located in the N-terminal region of the *lacZ* gene. The recombinant phages are recognized with blue/white color identification when placed on lac^- hosts in the presence of IPTG and X-gal. In addition, DNA sequences cloned into the multiple cloning site may be expressed as fusion proteins under the control of the lacZ promoter. The characteristics of the three cDNA libraries are summarized in Table 1. As an alternative to using the cDNA libraries, we at times have directly isolated the cDNAs by reverse transcription and polymerase chain reaction amplification using enzyme-specific primers. The cDNA from either source can be inserted into a eukaryotic expression vector and expressed in mammalian COS-1 cells. The metabolic activity of the COS-1 cells can then be examined to characterize the enzyme.

The second prong of the approach is to localize the enzymes of interest in the nasal tissues using in situ hybridization and immunohistochemistry. This aspect of the work will be enhanced by use of confocal microscopy.

Devon J. Zastrow was a D.O.E./A.W.U. summer student research participant.

Research supported by PHS grant ES 5910 from NIEHS and the Office of Health and Environmental Research of the U.S. Department of Energy under contract DE-AC04-76EV01013 in facilities fully accredited by the American Association for Laboratory Animal Care.

TABLE 1. Summary of the Characteristics of the F344 Rat and Human Nasal cDNA Libraries

Library name	Number of independent clones	Insert size range	Average insert size
Rat respiratory/transitional	1.6×10^6	0.8–5 K	1.6 K
Rat olfactory	1.5×10^6	0.6–5 K	1.6 K
Human respiratory	1.7×10^6	1.0–5 K	1.5 K

In progress toward comparing nasal enzyme activities, we have expressed the rat form of olfactory-specific P-450 2G1 and are examining its metabolic capacity toward testosterone (Figure 1). We also have screened the human nasal respiratory mucosa library for P-450 family 4 using a probe constructed for this purpose (Figure 2), but we were unable to find evidence for this P-450 family. The family 4 enzymes are involved in the regulation of inflammation (4a forms) and in the activation of aromatic amines (4b form). The 4b form is known to occur in rabbit nasal respiratory and olfactory mucosa (Dahl & Hadley, 1991). Finding no evidence for this family in the library, we went back to the original RNA (from 10 male and 10 female patients) and looked for message by Northern blotting. Once again, we found no evidence for this isozyme in human nasal respiratory tissue.

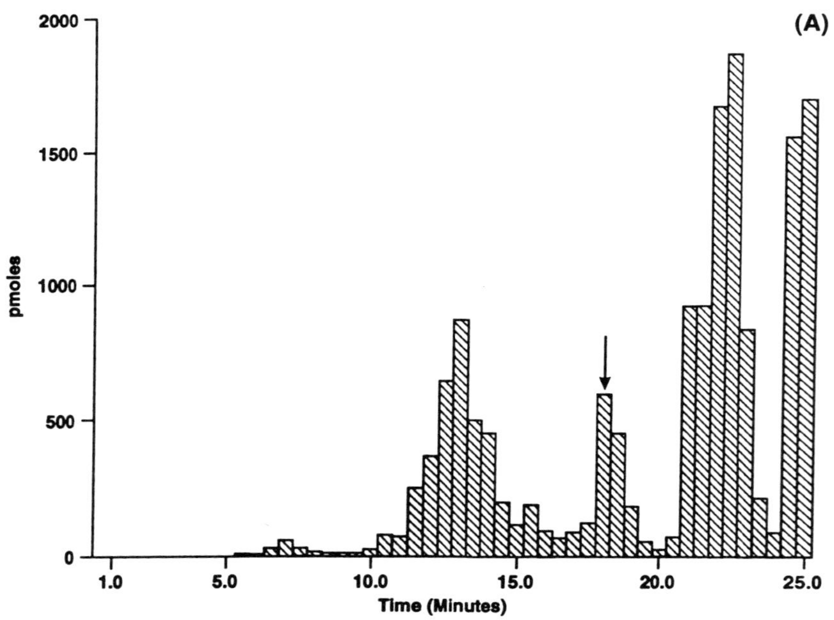

FIGURE 1. Metabolism of testosterone by COS cells. (A) High-performance liquid chromatography (HPLC) profile of ^{14}C-labeled metabolites from COS cells transfected with a pSVL/2G1 construct.

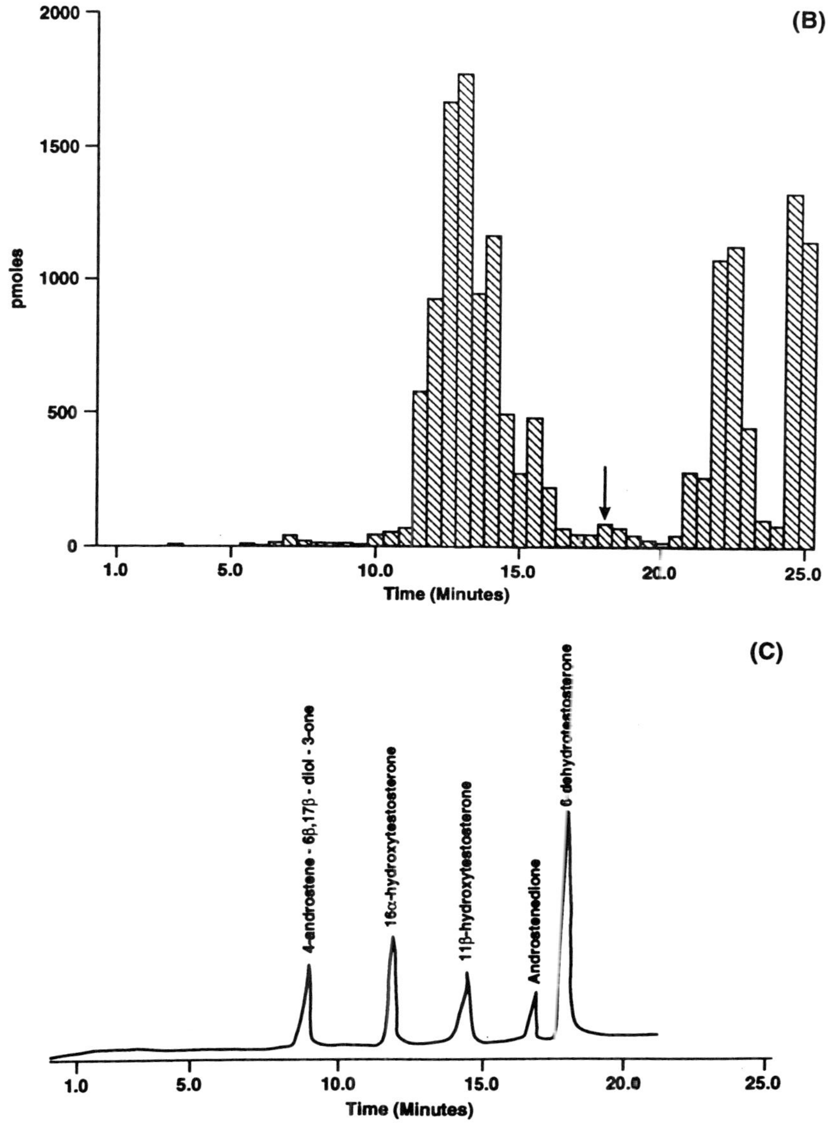

FIGURE 1. Metabolism of testosterone by COS cells (*Continued*). (B) CCS cells transfected with the pSVL plasmid. (C) HPLC profile of testosterone standards.

We have also developed a probe (Figure 2B) for screening the libraries for members of P450 family 3, which are involved in steroid metabolism.

With regard to the second prong of our approach, that of comparative enzyme localization, we have recently published studies for rhodanese (Lewis et al., 1991, 1992) and carboxylesterase (Lewis et al., 1994), and we are continuing these efforts for members of P-450 families 2, 3, and 4.

(A)

Peptide	***Phe Met Phe Glu Gly His Asp Thr***
mRNA	**5′-UUC AUG UUC GAA GGA CAC GAC ACA-3′**
Oligo	**3′-AAR TAC AAR CTY CCX GTR CTR TG-5′**

R=Purines Y=Pyrimidines X=A,G,C,T

(B)

Peptide	***Lys Asn Lys Asp Asn Ile Asp Pro***
mRNA	**5′-AAA AAC AAA GAC AAC AUA GAC CCA-3′**
Oligo	**3′-TTY TTR TTY CTR TTR TAX CTR GG-5′**

R=Purines Y=Pyrimidines X=A,G,T

FIGURE 2. Oligo probes for screening libraries. (A) P-450 family 4 probe based on a 16-amino-acid sequence present in family 4 members but with <50% identity with other families. (B) P-450 family 3 probe based on a 17-amino-acid sequence present in family 3 members but with <50% identity with other families.

REFERENCES

Dahl, A. R., and Hadley, W. M. 1991. Nasal cavity enzymes involved in xenobiotic metabolism: Effects on the toxicity of inhalants. *CRC Crit. Rev. Toxicol.* 21:345–372.

Hotchkiss, J. A., Lowell, S. H., and Dahl, A. R. 1994. Construction of F344 rat and human cDNA libraries and isolation of species-specific cytochrome P450 family 4 cDNAs. In *Inhalation Toxicology Research Institute Annual Report 1992–1993*, eds. K. J. Nikula, S. A. Belinsky, and P. L. Bradley, pp. 113–115. Springfield, VA: National Technical Information Service.

Lewis, J. L., Rhoades, C. E., Gervasi, P.-G., Griffith, W. C., and Dahl, A. R. 1991. The cyanide-metabolizing enzyme rhodanese in human nasal respiratory mucosa. *Toxicol. Appl. Pharmacol.* 108:114–120.

Lewis, J. L., Rhoades, C. E., Bice, D. E., Harkema, J. R., Hotchkiss, J. A., Sylvester, D. M., and Dahl, A. R. 1992. Interspecies comparison of cellular localization of the cyanide metabolizing enzyme rhodanese within olfactory mucosa. *Anat. Rec.* 232:620–627.

Lewis, J. L., Nikula, K. J., Novak, R., and Dahl, A. R. 1994. Comparative localization of carboxylesterase in F344 rat, beagle dog, and human nasal tissue. *Anat. Rec.* 239:55–64.

COMPARATIVE ANATOMY AND PHYSIOLOGY OF RAT, MONKEY, AND HUMAN NASAL PASSAGES

John M. DeSesso

MITRE Corporation, Center for Environment, Resources and Space, McLean, Virginia, USA

Supported by MITRE Sponsored Research Project 9587V and EPA contract 68-DO-0020. The views expressed herein are those of the author and should not be construed as those of EPA.

Rats are the usual animal models used in inhalation bioassays to study the toxicity or carcinogenicity of chemicals. The anatomy and physiology of the nasal passages of rats, however, differ significantly from those of humans. To investigate the dosimetry of inhaled chemicals and to determine the causes of nasal toxicity, an appropriate animal model for the human upper respiratory regions is required. The anatomy and physiology of the nasal passages and upper airways of humans, rats, and monkeys have been described, compared, and assessed for the purpose of determining a relevant animal model in which to investigate nasal toxicity and dosimetry. For a more detailed analysis, see DeSesso (1993).

GEOMETRY OF THE UPPER RESPIRATORY REGIONS

The upper respiratory airways serve as a conduit for ambient air to enter the lungs. In humans and monkeys, the arrangement of the upper airways is L-shaped, exhibiting a 90-degree turn in the pharynx. The larynx is inferior to the oral cavity, and the epiglottis does not appose the soft palate. Both the oral cavity and nasal passages open into the pharynx, allowing breathing to occur through both the mouth and nose (oronasal breathers). In contrast, the layout of the upper respiratory structures of the rat is tandem, with the larynx residing close to the posterior edge of the oral and nasal cavities. The epiglottis abuts the soft palate, isolating the oral cavity from the respiratory airways and making the rat an obligatory nose breather.

ANATOMY OF THE NASAL PASSAGES

The nasal passages are the site for olfactory stimulation and conditioning of air. Entry occurs through a pair of openings (nares) leading into the vestibule. In humans, admission to the nasal cavity proper occurs through the nasal valve. While the rat does not possess a nasal valve, it does exhibit small cartilaginous projections into the vestibule just inside the nares, the atrioturbinates. The vestibule of the monkey is considerably less complicated than the rat, although monkeys possess an anteriorly located fold of tissue, the vestibular wing, which creates a dorsal lateral meatus within the vestibule.

In all three species, the lateral walls of the nasal cavity proper exhibit three bony processes, termed turbinate bones or conchae, that project into the passageway. The most striking anatomical differences occur in the rat turbinate bones, which possess scroll-like dorsal and ventral lamellae. This results in a greatly increased nasal passage surface area compared to monkeys and humans. The relative nasal surface area ratios (calculated by dividing the nasal surface area by the nasal cavity volume) of humans and monkeys are nearly equal (6.4 and 7.75, respectively), whereas the rat relative nasal surface area (51.7) is about eight times larger than either of the primate species.

In rats, the inferior portion of the nasal septum contains swell bodies composed of vascular spaces beneath the epithelium and capable of rapid engorgement with blood. Engorged swell bodies obstruct the lower portion of the airway and deflect air toward the olfactory region. Swell bodies do not exist in monkeys or humans.

MUCOSA

The nasal passages are lined by a well-vascularized mucosa that is composed of an epithelial sheet underlain with a thin layer of loose connective tissue containing blood and lymphatic vessels. The nasal passages proper are lined by three major types of epithelia: squamous, respiratory, and olfactory. As one proceeds deeper within the nasal cavity proper, the epithelial covering becomes progressively more delicate, in terms of its ability to respond to stress, and more specialized, in terms of its functions of conditioning air and olfaction. The zone of transition between adjacent areas of squamous and respiratory epithelia is not abrupt, and the transitional epithelia in these zones share the characteristics (in graded fashion) of the adjacent epithelial types. The percentage of nasal epithelium that is olfactory is high in rats (50%) but much smaller in humans (8%) and monkeys.

NASAL SECRETIONS

Nasal secretions are produced by respiratory mucosa. In both monkeys and humans, the predominant direction of mucociliary transport is toward the nasopharynx, where nasal secretions are swallowed. In humans, the secretions produced in the posterior two-thirds of the nasal passages are replaced every 10 min, whereas those from the anterior one-third are replaced once an hour. The rat does not show a great disparity in anterior versus posterior mucociliary transport rates; consequently a much greater percentage of nasal secretion is directed toward the nares.

INSPIRATORY AIRFLOW

In humans, air entering the nasal vestibule travels cephalically, then turns approximately 60 degrees as it enters the nasal cavity proper. Most air traverses a corridor between the medial edges of the middle and inferior turbinates and the nasal septum. In monkeys, air enters the nasal passage nearly horizontally, but the vestibular wing causes a significant vortex that directs some of the air into the dorsal lateral meatus. Within the monkey nasal passages, air that is not deflected by the vestibular wing traverses corridors that are similar to the human. The situation is more complex in the rat due to the atrioturbinates and the swell bodies. Air enters the nose in a nearly horizontal orientation, but the atrioturbinates and swell bodies deflect the air laterally and superiorly so that the air traverses the olfactory region of the nasal cavity, including the lateral meatus and the upper lamellae of the superior and middle turbinates.

CONCLUSIONS

Based on the mode of breathing, overall geometry of the nasal passages and turbinate bones, relative nasal surface areas, proportions of nasal surfaces lined by various epithelia, mucociliary clearance patterns, and inspiratory airflow routes, the rat differs greatly from humans and therefore is a poor model. In contrast, the monkey exhibits many similarities to humans, including oronasal mode of breathing, nasal passages with turbinates that resemble the human, relative nasal surface areas and epithelial surface areas that approach those of humans, and mucociliary clearance routes and inspiratory airflow routes that parallel those of the human. The monkey does differ from humans in that its vestibular anatomy creates a vortex during inspiration. There are other physiological differences, including a more rapid respiratory rate and smaller minute and tidal volumes compared to humans. Nevertheless, the monkey is a more appropriate model than the rat for studying the toxic effects of inhaled substances on the nasal passages and extrapolating the findings to humans.

REFERENCES

DeSesso, J. M. 1993. The relevance to humans of animal models for inhalation studies of cancer in the nose and upper airways. *Qual. Assur.* 2:213–231.

COMPARISON OF FISCHER 344 RAT NASAL AIRWAY COMPUTATIONAL RECONSTRUCTION WITH MORPHOMETRIC DATA

Matthew N. Godo, Elizabeth A. Gross, Regina B. Richardson, Donald R. Joyner, Kevin T. Morgan, Julia S. Kimbell

Chemical Industry Institute of Toxicology, Research Triangle Park, North Carolina, USA

The effect of nasal airflow patterns on dosimetry was previously investigated using computational fluid dynamic simulations of airflow and inhaled gas transport in the anterior section of the Fischer 344 rat nasal passages (Kimbell et al., 1993). The Fischer 344 rat is a species for which a great deal of toxicological data regarding inhaled xenobiotics is already available. Simulations of inhaled formaldehyde gas predicted that regions of high uptake correlate with the location of major nasal lesions. To examine the role of ethmoid turbinate geometry in airflow-driven dosimetry, an improved method was used to reconstruct the nasal airway of a Fischer 344 rat from serial step cross-sectional data (Godo et al., 1993). In general, these data may be obtained from histological sections, magnetic resonance imaging (MRI), or x-ray computer-assisted tomography (CAT) scans. Such computer reconstructions may subsequently be used to address chemical and particle dosimetry issues through computational fluid dynamic simulations of momentum and mass transfer and particle transport. In this work, morphometric measurements (surface area and volume) from the computer reconstruction were compared with published values to demonstrate that accurate three-dimensional (3-D) computer reconstruction of the entire upper respiratory tract geometry is possible. A long-term objective is the refinement of the Fischer 344 computer reconstruction to permit 3-D airflow and gas transport simulations. Parallel extensions of this numerical method can then be used to make risk assessment predictions for humans.

APPROACH

An 18-wk-old male Fischer 344 rat weighing 315 g at the time of sacrifice was used in this study. From a set of 596 50-μm-thick step sections of the nasal airway, a subset of 77, depicting significant structural changes, was selected. Airway data were obtained by tracing the perimeter of the selected sections. Each section was converted from a set of XY coordinates to a two-dimensional (2-D) finite-element mesh (FEM) using a commercial computational fluid dynamics software package (FIDAP, FDI, Evanston, IL). An intermediate computer program called GRID2D (developed at CIIT) was used to build the input files to the commercial program. The 2-D FEMs were then recombined to produce a 3-D FEM. This was again accomplished using FIDAP through an additional intermediate computer program called GRID3D (developed at CIIT). Manual refinement of the 3-D FEM was performed to ensure that alignment between the 2-D cross sections matched the original airway in profile. Also, the finite element structure was checked to ensure that it was free of distorted elements, a necessary prerequisite to fluid dynamic simulations.

RESULTS AND DISCUSSION

A qualitative comparison between the numerical reconstruction and a truncated metal nasal airway cast (Morgan et al., 1989) is illustrated in Figure 1. The overall shape of the computer reconstruction compared well with the metal cast. Reproduction of the six ethmoid meatuses (labeled b through g in Figure 1), visible from the lateral aspect of the airway, demonstrated the ability of the reconstruction algorithms to handle complex 3-D geometrical domains. Other features, including the maxillary sinus (h), incisive duct (i), ventral meatuses (k and l), and the lateral recess (m and n), were also accurately reproduced.

A quantitative comparison with morphometric measurements was also performed. Only one side of the nasal passage was reconstructed. The surface area was calculated by doubling the value obtained by subtracting the area of the nostril surface and the exit and the midplane bisection of the nasopharynx from the total reconstructed surface area. The reported volume was obtained by simply doubling the reconstruction volume. The surface area and volume of 1742 mm^2 and 314 mm^3, respectively, were slightly larger than the published results of Gross et al. (1982) (1344 mm^2 and 257 mm^3) for 16-wk-old Fischer 344 rats. These differences are probably related to the smaller size (288 g) of the rats in the latter study. The cross-sectional area, plotted as a function of the axial airway distance from the nostril in Figure 2, was compared to morphometric data for two male Fischer 344 rats, aged 50 and 414 days, reported by Patra et al. (1987). A peak in cross-sectional area for Patra et al. occurred approximately 8 mm proximal to the point at which a similar peak occurred in the computational construct. Further comparison with hollow acrylic casts of the rat nasal airway of Morgan et al. (1989) indicated that the beginning of the cross-sectional area peak in the computational construct was coincident with the beginning of the ethmoid turbinates. The difference between these

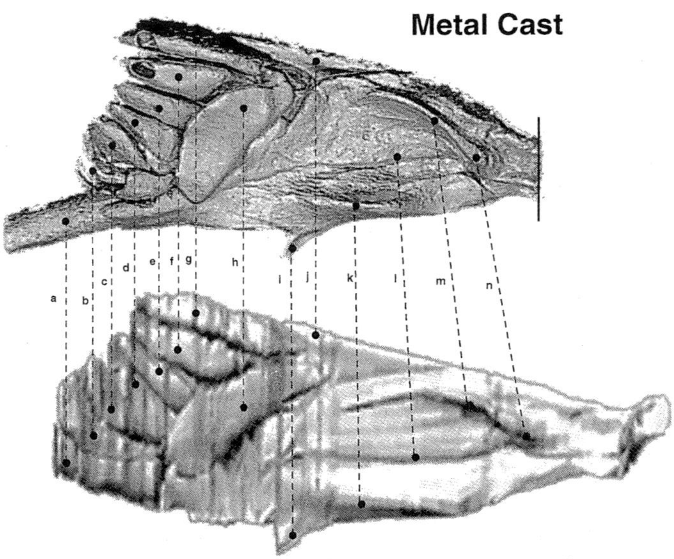

FIGURE 1. Comparison between metal casts and numerical reconstructions of the Fischer 344 rat upper respiratory tract: a, nasopharyngeal meatus; b through g, lateral ethmoid meatuses; h, maxillary sinus; i, incisive duct; j, dorsal medial meatus; k, ventral meatus; l, ventral lateral meatus; m, lateral recess of lateral meatus; and n, anterior lateral recess.

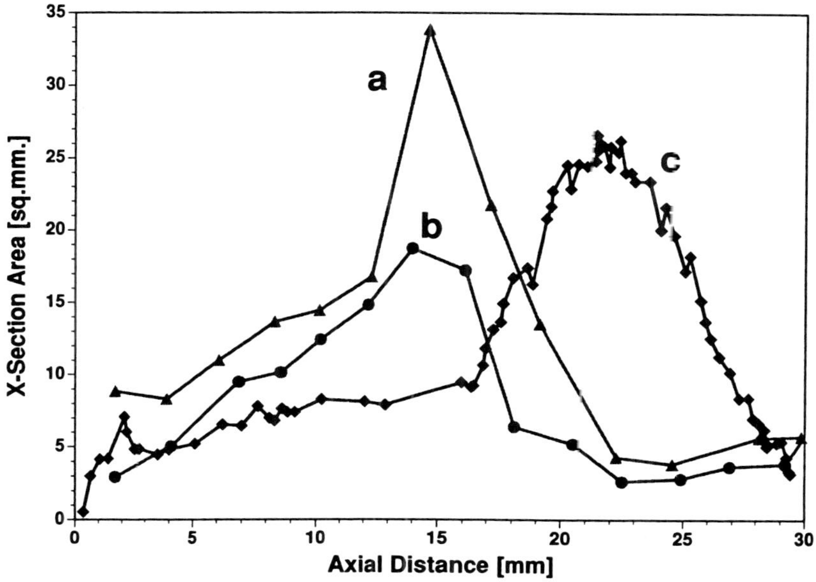

FIGURE 2. Cross-sectional area versus axial distance for the upper respiratory tract of the Fischer 344 rat: (a) data of Patra et al. (1987), 414-day-old rat; (b) data of Patra et al. (1987), 50-day-old rat; (c) computer reconstruction.

data sets was attributed to sectioning technique. In the work of Patra et al. (1987), measurements were obtained for cross sections that were not parallel. The present study emphasizes the need to consider axial correspondence in the acquisition of cross-sectional data that will be used for three-dimensional reconstruction.

The present studies, which develop tools needed to predict airflow-driven regional nasal dosimetry, will permit more meaningful interspecies comparisons and predictions for toxic responses that are a consequence of high local dose resulting from species-specific nasal airflow characteristics. This work provides the framework for simulation of 3-D airflow in the nose, which will then be used to determine specific regional uptake patterns of inhaled gases and particles. It is important to recognize, however, that the mathematical prediction of airflow-driven regional gas deposition patterns in the nasal passages of rats, other laboratory animals, and humans is dependent upon the generation of accurate geometric data. In this regard, correlation of computer-based reconstructions with other sources of anatomical information, such as nasal casts and morphometric data bases, will continue to be essential for detection of reconstruction artifacts. Development of increasingly efficient and cost-effective methods for the computational reconstruction process will play a critical role in future applications of this research to the field of inhalation toxicology.

REFERENCES

Godo, M. N., Morgan, K. T., Gross, E. A., and Kimbell, J. S. 1993. Airflow-driven regional dosimetry in the upper respiratory tract: Algorithms for reconstruction of the nasal airway geometry. *Toxicologist* 13(1): 259.

Gross, E. A., Swenberg, J. A ., Fields, S., and Popp, J. A. 1982. Comparative morphometry of the nasal cavity of rats and mice. *J. Anat.* 135(1):83–88.

Kimbell, J. S., Gross, E. A., Joyner, D. J., Godo, M. N., and Morgan, K. T. 1993. Applications of computational fluid dynamics to regional dosimetry of inhaled chemicals in the upper respiratory tract of the rat. *Toxicol. Appl. Pharmacol.* 121:253–263.

Morgan, K. T., Monticello, T. M., Fleishman, A., and Patra, A. L. 1989. Preparation of rat nasal airway casts and their application to studies of nasal airflow. In *Extrapolation of dosimetric relationships for inhaled particles and gases*, eds. J. D. Crapo, E. D. Smolko, F. J. Miller, J. A. Graham, and A. W. Hayes, pp. 45–48. San Diego, CA: Academic Press.

Patra, A. L., Ménache, M. G., Shaka, N. B., and Gooya, A. 1987. A morphometric study of nasal-pharyngeal growth for particle deposition in the rat. *Am. Ind. Hyg. Assoc. J.* 48(6):556–562.

SPECIES DIFFERENCES IN THE TOXICITY AND METABOLISM OF 3-TRIFLUOROMETHYLPYRIDINE (3-FMP) IN OLFACTORY TISSUE

Paul M. Hext, Barbara A. Gaskell, Edward A. Lock, Geoffrey H. Pigott

Zeneca Central Toxicology Laboratory, Alderley Park, Macclesfield, Cheshire, UK

3-Trifluoromethylpyridine (3-FMP) is a semivolatile by-product from the synthesis of a process intermediate. Routine toxicological evaluation in the rat showed that in inhalation studies of up to 90 days duration, the olfactory epithelium was a primary target tissue with an additional response in the liver (Gaskell et al., 1988). Oral administration produced similar lesions, and 3-FMP has been shown to be metabolized by rat olfactory and hepatic tissues (Hext & Lock, 1992). In order to establish the relevance of these findings in rats to humans and to investigate further the biochemical mechanisms responsible for this selective toxicity, a series of toxicity, tissue accumulation, and metabolism studies was conducted in a range of species.

METHODS

Rats, mice, and rabbits were exposed by inhalation for 6 h/day for up to 10 days to 0.5, 5.0, or 50 ppm 3-FMP. Cynomolgus monkeys were exposed to 50 ppm for a single 6-h exposure or to 0.1, 5.0, or 50 ppm 3-FMP for 6 h/day for 10 days. Marmosets were administered 100 mg/kg 3-FMP in corn oil per os. At appropriate times after exposure or dosing, animals were killed and nasal and hepatic tissues processed and stained by standard techniques for histopathological evaluation.

Rats and marmosets were also administered 100 mg/kg [methyl-^{14}C]-3-FMP (^{14}C-FMP; specific activity 0.29 mCi/mmol) in corn oil per os. Tissue distribution of radioactivity at selected times after dosing was determined using whole-body autoradiography.

Isolated olfactory tissue from the rat and cynomolgus monkey and nasal tissue from the marmoset (total nasal tissue was used due to difficulties in dissecting out olfactory tissue) were incubated at 37°C for up to 1 h in the presence of 10 μM ^{14}C-FMP (specific activity 33.4 mCi/mmol) in oxygen-saturated Krebs-Ringer phosphate buffer, pH 7.4. The tissue was then removed for determination of accumulated radioactivity, and the medium was analyzed for known metabolites of 3-FMP by HPLC.

RESULTS

Mice exposed to 50 ppm died during or following the first 6-h exposure. Rats, rabbits, and the remaining groups of mice all survived the 10-day exposure but developed histological lesions specific to the olfactory epithelium. The rat was the most sensitive species examined. The sites of tissue injury in the olfactory epithelium were the dorsal meatus, dorsal upper septum, tips of the ethmoturbinates, and lateral walls. The lesions consisted of a variety of changes—focal loss of apical cytoplasm, a reduction in the number of cell layers, and disorganization of the normal cellular arrangement and areas where only basal cells remained. Periodic acid/Schiff/alcian blue staining of the Bowmans glands was lost or reduced in all three species, but no changes were seen in axon bundles or in the olfactory bulb. There was little evidence for a dose-response relationship, with 0.5 ppm being the lowest effect concentration in all species. No lesions were seen in any region of the nasal passages of cynomolgus monkeys or marmosets.

Hepatotoxicity was seen in rats, mice, and rabbits exposed to 5 ppm or higher, cynomolgus monkeys exposed to 50 ppm, and marmosets dosed orally.

Whole-body autoradiograms of rats and marmosets dosed with ^{14}C-FMP showed specific accumulation of radioactivity into the ethmoturbinates and liver. Incubation in vitro of isolated olfactory tissue from rats and nasal tissue from marmosets showed a time-dependent accumulation of tissue-bound radioactivity. Accumulation rates ($\bar{x} \pm$ SD, $n \geq 3$) in vitro were 0.44 ± 0.05 nmol ^{14}C-FMP equivalents/mg protein/h and 0.1 ± 0.02 nmol ^{14}C-FMP equivalents/mg protein/h for rat olfactory and marmoset nasal tissues, respectively. Nasal olfactory and respiratory tissues from the cynomolgus monkey showed no accumulation.

The major metabolite identified in the medium in which rat olfactory or marmoset nasal tissue had been incubated was 3-FMP *N*-oxide (FMPNO). Rates of synthesis were 1.14 ± 0.16 and 0.61 ± 0.16 nmol/mg protein/h for rat and marmoset tissues, respectively. Traces of 3-FMP 4-pyridone were also detected in incubations from rat olfactory tissue. FMPNO was not detected in the medium in which cynomolgus monkey tissue had been incubated, the main metabolite being identified as 3-FMP 4-pyridone. The rate of synthesis for this metabolite was relatively low, being approximately 0.01 nmol/mg protein/h. Microsomes prepared from rat and monkey olfactory tissues produced metabolite profiles similar to those described for intact tissues. Limited studies on the metabolism of ^{14}C-FMP by microsomes prepared from olfactory tissue from the pig-tailed macaque (*Macaca nemistrina*) or baboon (*Papio hamadryas*) showed similar metabolite profiles to those described for the cynomolgus monkey.

Preparations of microsomes from rat, cynomolgus monkey, pig-tailed macaque, and marmoset liver all metabolized ^{14}C-FMP to FMPNO.

DISCUSSION

Species differences in the toxicity of 3-FMP to olfactory tissue were evident and were consistent with the ability of olfactory tissue to accumulate and metabolize 3-FMP. Thus, rat olfactory tissue developed lesions, readily accumulated radioactivity derived from ^{14}C-FMP, and metabolized this predominantly to the *N*-oxide derivative, which has been shown by Green, Nash, and Gaskell (unpublished data) to produce olfactory toxicity identical to that produced by 3-FMP. In contrast, cynomolgus monkey tissue was unaffected by exposure to 3-FMP, lacked any capacity to accumulate radioactivity, and metabolized 3-FMP to the 4-pyridone derivative, which is not toxic to olfactory tissue (Green, Nash, and Gaskell, unpublished data). It would appear, therefore, that the ability of a tissue to accumulate and metabolize 3-FMP to 3-FMPNO is a prerequisite for the expression of toxicity. This is supported by and provides an explanation for the observed hepatotoxicity of 3-FMP, all species developing liver lesions and all metabolizing of 3-FMP to 3-FMPNO in the liver.

Results from the studies described here, including the limited information from the microsomal metabolism of 3-FMP, point to a fundamental difference between the metabolism of 3-FMP by olfactory tissue of the rat (and probably the mouse and rabbit) and many primate species. Interestingly, marmoset nasal tissue both accumulated radioactivity from ^{14}C-FMP and synthesized appreciable quantities of FMPNO in vitro but did not develop nasal lesions. Since metabolism to 3-FMPNO is considered to be the key step in the activation of 3-FMP to a toxic metabolite, other factors must also influence the expression of toxicity by 3-FMP. These probably include the conversion rate in vivo, the probable need for further metabolism of the *N*-oxide derivative, and the role of detoxification pathways.

In conclusion, the general lack of toxicity in primate olfactory tissue compared with the rat suggests that humans are unlikely to develop an olfactory lesion if exposed to 3-FMP.

REFERENCES

Gaskell, B. A., Hext, P. M., Pigott, G. H., Hodge, M. C. E., and Tinston, D. J. 1988. Olfactory and hepatic changes following inhalation of 3-trifluoromethylpyridine. *Toxicology* 50:57–68.

Hext, P. M., and Lock, E. A. 1992. The accumulation and metabolism of 3-trifluoromethylpyridine by rat olfactory and hepatic tissues. *Toxicology* 72:61–75.

ANTIOXIDANTS AND BIOTRANSFORMATION POTENTIAL IN HUMAN NASAL LAVAGE

David G. Housley

Department of Biochemistry, University of Wales, Cardiff, Wales, UK

Ronald Eccles

Common Cold and Nasal Research Centre, University of Wales, Cardiff, Wales, UK

Roy J. Richards

Department of Biochemistry, University of Wales, Cardiff, Wales, UK

The most common methods used to lavage the human nasal cavity are limited by the fact that the lavage fluid does not remain in contact with the mucosa for long periods and most methods require the subject to tilt their heads back, which contaminates the nasal lavage with nasopharyngeal secretions. We wished to develop a method that would allow a prolonged lavage and avoid contamination from the nasopharynx. Three components, protein, uric acid, and lysozyme, all previously shown to be present in high levels in human nasal lavages (Peden et al., 1990; Raphael et al., 1989), were assayed to establish what effect this prolonged lavage had upon the composition of the material recovered. Major biotransformation enzymes have been described in bronchoalveolar lavage from mice (Clouter & Richards, 1992), and therefore attempts were made to establish their presence in concentrated human nasal secretions. Concentrated human nasal lavage was also examined for reduced glutathione (GSH), which can act as an antioxidant and is capable of nonenzymic biotransformation.

STUDY DESIGN

Ethical approval was obtained from South Glamorgan Area Health Authority, Wales, UK. Six consenting healthy male subjects between the ages of 19 and 25 yr were screened by an ear, nose, and throat (ENT) physician and were excluded from the study if they had gross anatomic nasal deformity; a history of allergy or asthma; an upper respiratory tract infection in the previous 6 wk; suffered from cardiovascular, respiratory, endocrine, or other conditions considered by the physician as reason for exclusion; were currently taking medication; were smokers; or passive smokers or worked in an environment where occupational exposure to dust or fumes may occur. Each subject was lavaged on five separate occasions, the lavage time increasing on each occasion. Lavage times were 0 (saline instilled and then removed immediately), 1, 2, 5, and 10 min. Three subjects were lavaged on days 1, 2, 3, 4, and 5, 2 subjects were lavaged on days 1, 2, 3, 4, and 7 and, one subject was lavaged on days 1, 2, 4, 7, and 8.

NASAL LAVAGE/ASSAYS

Subjects were asked to gently blow their nose and then report which nasal cavity (if any) had the greatest resistance to airflow. These cavities were then recorded as open (least resistance) or closed (greatest resistance). The balloon of a modified Foley catheter was inflated inside one of the nostrils to occlude it. Subjects were asked to lean forward at 45° and 5.0 ml of 0.9% sterile saline at room temperature was passed into the cavity and left for periods of up to 10 min. The procedure was then repeated for the other cavity, the two samples being kept separate. The lavage fluid was filtered through a 150-μm mesh and centrifuged at 300 × g for 10 min at 10°C and the supernatant fraction analyzed. Protein (PR) (Bradford, 1976) and lysozyme (LY) (Metcalf et al., 1986) were measured as previously described and uric acid (UA) by means of Sigma Diagnostics procedure 685. Lavage supernatant was concentrated by lyophilization, and assays were carried out for cytochrome P-450 activity (ethoxycoumarin *o*-deethylase) (Ullrich & Weber, 1972), NADPH-dependent cytochrome c reductase (Williams & Kamin, 1962), and gamma-glutamyl transpeptidase (Sigma Diagnostics procedure 419). Lyophilized samples were resuspended in water, the <3 kD fraction was collected by 5 serial centrifugations through a centricon concentrator, and GSH was assayed (Reed et al., 1980).

RESULTS/DISCUSSION

Biotransformation enzyme activity was absent in all samples. GSH was present in some of the samples at concentrations of 2 μM. The method of detection of GSH

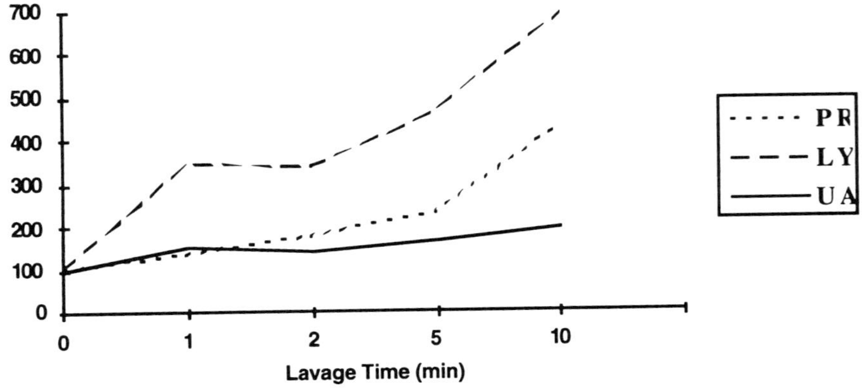

FIGURE 1. Changes in the concentration of lavage constituents with increasing duration of lavage time. The results are expressed as percentage increase from the zero time lavage, which consisted of a single "in and out" wash with saline. PR = protein, LY = lysozyme, UA = uric acid.

also showed two unidentified compounds to be present at levels much higher than GSH. While these compounds contained amino groups, subsequent analysis suggested that they were not one of the common amino acids.

At time zero, recovery of fluid was 4.1 ±.14 ml and at 10 min was 3.2 ± 0.23 ml. The PR, LY, and UA all increased with prolonged lavage time (Figure 1), with little difference between the cavities being detected. Because both cavities in a subject were found to behave the same, they were assumed to be separate samples for calculations of concentration. Thus, at each time point, data are the mean ± SEM of both cavities from six subjects. Time zero PR concentrations (considered 100%, Figure 1) were 49.6 ± 17.4 μg/ml and rose to 430% or 214 ± 66.5 μg/ml at 10 min. The LY rose from 12.36 ± 4.5 μg/ml at 0 to 85.8 ± 15.7 μg/ml at 10 min, a 590% increase. The UA rose from 10.31 ± 1.82 *μM* at 0 to 20.37 ± 3.21 *μM* at 10 min, a 100% increase.

CONCLUSIONS

The method described allows each nasal cavity to be lavaged separately for periods of up to 10 min, avoiding contamination with nasopharyngeal secretions. Irrespective of lavage time, major biotransformation enzymes were not present, suggesting enzymatically catalyzed biotransformation of inhaled xenobiotics cannot occur in the secretions lining the human nasal cavity. Although the effect of wash time on collection of GSH is not known, the results indicate that nonenzymic conjugation of GSH with inhaled compounds may be able to occur. The GSH present may also be able to participate in antioxidant defense, but the results suggest that UA is the major antioxidant in human nasal secretions. The observation that PR, LY, and UA increase with lavage time could be a result of the prolonged lavage removing more material from the mucosal surface or as a result of secretion. The rate of accumulation of these three components in the lavage fluid differs for each, suggesting that if secretion was the source of the material, the mechanisms of release or the pharmacological control of release were different for lysozyme and uric acid.

REFERENCES

Bradford, N. M. 1976. A rapid and sensitive method for the quantitation of microgram quantities of protein utilising the principle of protein-dye binding. *Anal. Biochem.* 72:248–254.

Clouter, A., and Richards, R. J. 1992. Change in detoxication status in airway hypophase following administration of 4-ipomeanol to mice. *Respir. Med.* 86(1):80.

Metcalf, J. A., Gallin, J. I., Nauseef, W. M., and Root, R. K. 1986. *Laboratory manual of neutrophil function*. New York: Raven Press.

Peden, D. B., Hohman, R., Brown, M. E., Mason, R. T., Berkebile, C., Fales, H. M., and Kaliner, M. A. 1990. Uric acid is a major antioxidant in human nasal airway secretions. *Proc. Natl. Acad. Sci. USA* 87:7638–7642.

Raphael, G. D., Jeney, E. N., Baranjuk, K. I., Meredith, S. D., and Kaliner, M. A. 1989. Pathophysiology of rhinitis-lactoferrin and lysozyme in nasal secretions. *J. Clin. Invest.* 84:1528–1535.

Reed, D. J., Babsom, J. R., Beatty, P. W., Brodie, A. E., Ellis, W. W., and Potter, D. W. 1980. High performance liquid chromatography of nanomole levels of glutathione, glutathione disulphide and related thiols and disulphides. *Anal. Biochem.* 106:55–62.

Ullrich, V., and Weber, P. 1972. The O-dealkylation of 7-ethoxycoumarin by liver microsomes. *Physiol. Chem.* 353s:1171–1177.

Williams, C. H., and Kamin, H. 1962. Microsomal triphosphopyridine nucleotide-cytochrome C reductase of liver. *J. Biol. Chem.* 237:587–595.

INHALATION TERATOLOGY STUDIES OF PHARMACEUTICAL PREPARATIONS IN RABBITS: A SNOUT-ONLY EXPOSURE TECHNIQUE

David J. Alexander, Gillian D. Dines, Shaun R. Maguire, Mark A. Crowther

Glaxo Group Research Ltd., Ware, Hertfordshire, England

Rabbits are routinely used for teratology studies but are rarely used if the primary route of administration is by inhalation, since dosing rabbits by that route has, in the past, presented significant problems. As part of a development program for a novel metered-dose pharmaceutical aerosol formulation, a method of snout-only inhalation exposure of rabbits has been developed and validated. The technique is described, and clinical and reproductive data from a study performed using this novel technique are presented. No fetal abnormality data are presented.

MATERIALS AND METHODS

New Zealand White rabbits at a minimum of 3.5 mo were used. The test material was a metered-dose aerosol formulation delivering 100 μg drug/metered dose in a propellant.

The apparatus was from ADG Developments (Codicote, Hitchin, Hertfordshire, UK) a snout-only exposure apparatus fitted with an automatic aerosol pack discharge unit (Greenacre/ADG Developments, Stanstead Abbotts, Hertfordshire UK) mounted on the top of the chamber. The apparatus was operated under conditions of dynamic airflow. The only modifications to the chamber, as supplied, were to invert the planet supporting rings used to hold the restraint tubes, which lowered the level of the planet ring to 6 cm below the exposure ports. Additionally, a blank chamber section was required between each pair of exposure rings due to the size of the restraint tubes.

Specially designed polycarbonate restraint tubes, fitted with a tapered nose cone section, were developed for the rabbits, which were held in place by a nylon disc attached to an aluminium rod. The rod passed through a Plastazote closed-cell-structure foam tube closure.

Twenty-five female rabbits were allocated, 5 to each of 5 groups, and were exposed to 0 (air controls), 0 (placebo controls), 6.8 (Group 3), 13.2 (Group 4) or 36.4, (Group 5) μg/L drug for 1 h daily between days 8 and 20 of pregnancy. Day of mating was day 1 of pregnancy. Rabbits were acclimatized to the restraint procedures on days 5, 6, and 7.

Routine aerosol characterization, body weight, clinical signs, food consumption, plasma drug assays, and macroscopic postmortem examinations were performed.

RESULTS

The mass median diameters (and geometric standard deviations) of the active drug for Groups 3, 4, and 5 were 1.76 μm (2.13), 1.72 μm (1.99), and 1.93 μm (1.98), respectively, indicating that they were in the range considered respirable for rabbits and consistent between groups.

One female from Group 3 was found dead on day 6 of pregnancy after returning to the cage following tube acclimatization.

All animals were generally unaffected by treatment. Between days 8 and 10 of the study there was an initial treatment-related body weight loss in Group 5. After day 10, body weight gains of Group 5 were slightly less than that of the control group. No other drug-exposed groups were affected.

Plasma samples taken on days 8 and 20 of the study indicated systemic exposure 5 min and 23 h following exposure in Groups 3, 4, and 5, with plasma drug levels up to a mean value of 86 ng/ml. Due to the limited group size, there was no clear evidence of a dose-related increase; however, the mean value for Group 5 was greater than those of Groups 3 and 4.

In the caesarean data, the only possible effect of treatment was an increase in the postimplantation loss in Group 5 (33.3%) when compared to Group 1 (7.1%).

DISCUSSION

This study clearly demonstrated that, using an especially designed restraint tube, acclimatizing the rabbits to the procedures prior to exposure and exposing the rabbits for 1-h daily using these techniques is a satisfactory method for inhalation dosing of rabbits.

One death occurred in the acclimatization period, which may have indicated an individual animal's intolerance to the procedure; however, no other clinical signs or struggling was observed during the exposure procedure. There were no abortions and no effects on the incidence of resorptions, numbers of live fetuses, or mean fetal weight, confirming the general acceptability of the procedure.

Clear signs of a drug-related response were seen in the effects on body weight gain and postimplantation losses in Group 5.

Drug absorption studies indicated evidence of systemic exposure to the drug, and a comparison with oral teratology studies conducted with the same drug substance indicated that an inhaled dose of approximately 0.49 mg/kg gave plasma concentrations of the same order as an oral dose of 50 mg/kg. This indicates that the drug used in this study was some 100 times more systemically bioavailable through the inhaled route compared to oral administration.

CONCLUSION

Exposure of rabbits using this snout-only exposure technique is an acceptable method for conducting inhalation teratology studies to meet regulatory requirements for safety evaluation of novel inhaled pharmaceutical products.

UTILITY OF NASAL LAVAGE IN DETECTING AN INFLAMMATORY RESPONSE AFTER OCCUPATIONAL EXPOSURE TO FUEL-OIL ASH

Russ Hauser, P. Barry Ryan, Karl T. Kelsey, David C. Christiani

Occupational Health Program, Harvard School of Public Health, Boston, Massachusetts, USA

Nasal lavage (NL) has been utilized in investigations on the inflammatory response of the upper airway to ambient contaminants. Previous studies conducted on subjects exposed to ozone, grain sorghum dust, or cotton dust demonstrated changes in NL inflammatory cell counts, specifically polymorphonuclear leukocytes (PMNs) (Koren et al., 1990; von Eissen et al., 1993; Harder et al., 1993). We investigated the utility of NL in detecting an inflammatory response after occupational exposure to fuel-oil ash in electric utility workers. Fuel-oil ash contains, among other substances, vanadium oxides, a known respiratory irritant (Levy et al., 1984). Exposure to ambient vanadium pentoxide dust is known to produce irritation of the eyes, nose, and throat (Zenz & Borg, 1967; Sjoberg, 1950). Therefore, we hypothesized that workers occupationally exposed to fuel-oil ash would demonstrate an inflammatory response in the upper airways that could be detected with NL.

METHODS

Thirty-seven male workers, aged 26–60 years old, volunteered to participate. A baseline lavage was performed on the morning of the first day back to work after at least 36 h away from work. A follow-up lavage was performed in the morning 3 days after the baseline lavage. Occupational exposures between baseline and follow-up lavage were assessed with a personal sampling device for particulate matter with a 10 μm (PM10) cut-point. The length of time each subject wore the personal

Supported by NIH grants ES05947, ES07069, and ES0002.

sampling device varied from 3 study days to none. To estimate exposure levels for all subjects, we had the subjects complete a detailed work history on each day between the baseline and follow-up lavage. If a subject reported cold or flu symptoms at the time of the NL or in the previous 2 wk, their NL results were excluded from analyses.

NASAL LAVAGE PROCEDURE

The NL procedure used in the study was adapted from that used by Koren et al (1990). While seated, each subject tilted his or her head backward (at a 45° angle) Using a needleless syringe, 5 ml of warm (37°C), sterile, phosphate-buffered saline (PBS) without Ca^{+2} and Mg^{+2} was instilled into one nostril. The subject held his breath and performed a partial swallow that closed off the nasopharyngeal cavity improving the ability to hold the fluid in the nasal cavity. The saline was retained for 10 s, and then passively allowed to drain for 30 s into a sterile specimen cup This was repeated in the other nostril.

The volume of lavage fluid recovered was recorded and used to volume adjust the cell counts, which were expressed as cells per milliliter recovered fluid. After obtaining each sample, 0.5 ml of *N*-acetylcysteine (Mucomyst-10, Bristol Lab), a mucous digestant, was added. The sample was then vortexed for 30 s and allowed to stand at room temperature for 30 min. The sample was then centrifuged at 1000 × g for 10 min and the supernatant was separated. The cells were resuspended in 0.5 m of phosphate-buffered saline and 0.1 ml was counted on a hemocytometer.

From the remaining resuspended cells, 0.15 ml was used to prepare slides for differential cell counts. The slides were prepared by cytospin (Cytospin 2, Shandor Corp.) and were stained with WrightGiemsa prior to counting. Under oil immersion, 200 cells were examined. The differential cell count was used to calculate PMNs per milliliter lavage fluid.

ANALYSIS

The change in PMNs per milliliter was calculated by subtracting the baseline lavage PMN count from the follow-up lavage PMN count. The change in PMNs per milliliter was adjusted for the average of the baseline and follow-up lavage PMN counts. Using both the personal sampling data (PM10 in $\mu g/m^3$) and work history diaries, we estimated occupational exposures (time-weighted average, TWA) for each subject during each of the 3 days between lavages. The adjusted change in PMNs per milliliter was regressed, after controlling for whether the subject smoked cigarettes in the morning of the lavage, on mean and peak (TWA) exposure between baseline and follow-up lavage, as well as on exposure during day 1, 2, or 3 after baseline lavage.

RESULTS

The exposure levels measured ranged from 60 to 4500 $\mu g/m^3$. The mean, median, and standard deviation of the five exposure indices (day 1, day 2, day 3, peak, and mean) are presented in Table 1. The baseline lavage PMN/ml counts for the 37 subjects ranged from 0 to 33,700, with a mean and standard deviation of 2600 ± 5900 and a median of 400. The follow-up lavage counts ranged from 0 to 24,600, with a mean and standard deviation of 2700 ± 4800 and a median of 600. The unadjusted changes in PMNs per milliliter had a mean and standard deviation of

TABLE 1. Mean, Median, and Standard Deviation (SD) of the Five Exposure Indices ($\mu g/m^3$)

	Day 1	Day 2	Day 3	Peak	Mean
Mean	1880	1980	2020	2280	1980
Median	1460	1520	1550	2130	1850
SD	1660	1980	1830	1880	1660

Note. Mean refers to mean exposure during the 3 days between baseline and follow-up lavage. Peak refers to peak exposure (TWA) during the 3 days between baseline and follow-up lavage. Day 1 refers to exposure on the day of the morning baseline lavage. Day 2 refers to exposure on day 2 after the baseline lavage. Day 3 refers to exposure on day 3 after the baseline lavage.

120 ± 6350 PMNs/ml, ranging from –26,300 to +24,400. Using multiple regression models, none of the PM10 exposure indices significantly predicted the adjusted or unadjusted change in PMN counts (all p values ≥ .22).

CONCLUSION

In this investigation we did not demonstrate a relationship between exposure to fuel-oil ash PM10 levels and adjusted or unadjusted change in PMNs per milliliter. A possible explanation for these results is that PM10 measures particles less than 10 μm, while the nasal cavity will have particles deposited with larger aerodynamic diameters. If the PM10 exposure levels do not accurately reflect the dose to the nasal cavity, they will not be predictive of the inflammatory cell response as measured by NL.

A second possible explanation is that sampling did not specifically measure particle subtypes, such as vanadium or nickel, that have been shown to be respiratory irritants and may better relate to change in upper airway inflammatory cell changes. The personal air samples will be analyzed for total vanadium and nickel, allowing for further analyses relating change in PMNs per milliliter to total vanadium and nickel.

A final explanation is that occupational exposure to fuel-oil ash either may not induce an inflammatory response in the nasal cavity or induces an inflammatory response that is not detectable with the present NL technique Further study is required on the utility of NL in occupational and environmental studies.

REFERENCES

Harder, S., Boehlecke, B., Jacobs, R. R., Becker, S., and Devlin, R. B. 1993. Spirometric and inflammatory response of humans exposed to cotton dust containing different amounts of LPS. *Am. Rev. Respir. Dis.* 147(4):A529.

Koren, H. S., Hatch, G. E., and Graham, D. E. 1990. Nasal lavage as a tool in assessing acute inflammation in response to inhaled pollutants. *Toxicology* 60:1525.

Levy, B. S., Hoffman, L , and Gottsegen, S. 1984. Boilermakers' bronchitis: Respiratory tract irritation associated with vanadium pentoxide exposure during oil-to-coal conversion of a power plant. *J. Occup. Med.* 26(8) 567–570.

Sjoberg, S. G. 1950. Vanadium pentoxide dust: A clinical and experimental investigation of its effects after inhalation. *Acta Med. Scand.* (Suppl 238) 138:1–188.

Von Eissen, S., O'Neill, D., and Robbins, R. 1993. Grain sorghum dust inhalation at harvest causes nasal inflammation and peripheral blood neutrophilia. *Am. Rev. Respir. Dis.* 147(4):A528.

Zenz, C., and Borg, B. A. 1967. Human responses to controlled vanadium pentoxide exposure. *Arch. Environ. Health* 14:709–712.

NASAL LAVAGE ANTIOXIDANTS IN GUINEA PIGS, RATS, AND MICE

Joel Norwood, Jr., Kay Crissman, Gary E. Hatch

Pulmonary Toxicology Branch, Environmental Toxicology Division, Health Effects Research Laboratory, U.S. Environmental Protection Agency, Research Triangle Park, North Carolina, USA

Nasal lavage (NL) may be a valuable technique for assessing the toxic effects of air pollutants on nasal epithelial tissues. In particular, measurements of antioxidants in NL may provide a useful tool for assessing oxidative effects. We used a noninvasive and rapid method for NL in small animals that employs a stainless steel plug that is inserted into the oropharynx, through which saline is injected into the posterior pharynx and out of the nares. We have reported preliminary measurements of antioxidant substances (ascorbate, glutathione, and urate) in NL (Hatch, 1992) and here report a comparison between guinea pigs, rats, and mice in the washout pattern and concentration of these substances in NL. For purposes of establishing a denominator for antioxidants, protein and lysozyme were also measured.

METHODS

Animals

Specific pathogen-free Hartley strain guinea pigs (males, 520–745 g; females, 518–629 g), SpragueDawley rats (males, 264–307 g; females, 211–285 g), and CD1 mice (males, 31–49 g; females, 29–35 g), were purchased from Charles River Breeding Labs (Raleigh, NC). All animals were housed in plastic cages with heat-treated wood chip bedding and kept in temperature- and humidity-controlled rooms with a 12-h light/dark cycle. They were given food (Continental Grain Co., Chicago, or Ralston Purina Co., St. Louis, MO) and water ad libitum or until 16 h prior to lavage.

Nasal Lavage Procedure

Animal nasal lavage was performed by using a specially designed stainless steel "plug" that was inserted into the posterior pharynx. The plug sealed off the oropharynx, forming a cup over the epiglottis while providing a passage for saline to be injected through the posterior pharynx and out of the nares. Saline was collected as it dripped from the nares into a tube. Saline solution was warmed to 37.5°C, injected through the nose via the plug and collected from the nares. The nasal lavage was kept briefly on ice and centrifuged (400 × g, 15 min) to assure that no cells were present in the supernatant. There was never a visible cell pellet in any of the samples.

The authors gratefully acknowledge the technical assistance of Mrs. Judy Richards of ManTech Environmental Technologies, Inc.

The research described in this article has been reviewed by the Health Effects Research Laboratory, U.S. Environmental Protection Agency, and approved for publication. Approval does not signify that the contents necessarily reflect the view and policies of the agency, nor does mention of trade names or commercial products constitute endorsement or recommendation for use.

Nasal Lavage Biochemistry

The supernatant of the lavage was analyzed for protein (Coomassie blue binding method of Bradford, 1976, using bovine serum albumin as a standard) and lysozyme (micrococcus lysis method of Konstan et al., 1982). Perchloric acid (60%) was added to the supernatant to a final concentration of 3% and protein was pelleted by centrifugation (27,000 × g, 20 min). The supernatant was stored at –70°C until analysis for ascorbic and uric acids (Kutnink et al., 1985) by high-pressure liquid chromatography with electrochemical detection. Glutathione was assayed by the glutathione disulfide reductase recycling method of Anderson (1985).

Experimental Design

Four experiments were performed in this study. In the first, equal numbers of male and female guinea pigs (n = 12 total) and rats (n = 12 total) were lavaged with 5 sequential 1.0-ml aliquots of saline. In the second, a 16-h fast preceded the lavage, due to our finding food particles in the guinea pig NL that could possibly affect the NL constituents. Male rats (n = 8), and equal numbers of male and female mice (n = 14 total) and guinea pigs (n = 16 total) were lavaged. The third experiment was done in male mice (n = 10) with a smaller volume per wash (0.5 ml). This was done because the protein levels in later washes of Experiment 2 mice were too low to be measured. In the fourth, all three species (male guinea pigs and rats, n = 6 per group, and mice, n = 7), were lavaged similarly to Experiment 3 and each successive nasal wash was analyzed for water soluble antioxidants.

RESULTS

Figures 1–4 show the washout pattern of protein and antioxidants measured in NL. Table 1 shows the total amount of lavagable protein and antioxidant (the sum of the amounts in five washes) expressed per kilogram of body weight. Comparisons between males and females showed no significant differences ($p > .05$, t-test); therefore, data were pooled across sex. Similarly, fasting did not significantly affect the total amount of protein recovered (Table 1).

The NL protein exhibited a washout curve in all three species (Figure 1). Mice had about two times more lavagable protein per kilogram than guinea pigs, and guinea pigs had about twice the amount as rats (Table 1).

Lysozyme was usually below the minimal detectable limits of the assay (3 μg/ml). Only about 13–25% of the first wash contained detectable lysozyme activity (data not shown), and there were no obvious species differences noted.

The washout curve for NL ascorbic acid appeared to approach completion after five washes (Figure 2). Table 1 indicates that mice had about two times more lavagable ascorbic acid (per kilogram body weight) than guinea pigs, which had about two times more than rats. The lowest detectable concentration of ascorbic acid was 0.03 μg/ml.

The washout curve for glutathione was quite variable, and concentrations tended to increase in later washes in all three species, suggesting an influx of glutathione during lavage (Figure 3). Table 1 indicates that mice had 11–13 times more lavagable glutathione (per kilogram body weight) than guinea pigs and rats. The lowest detectable concentration of glutathione was about 0.10 μg/ml.

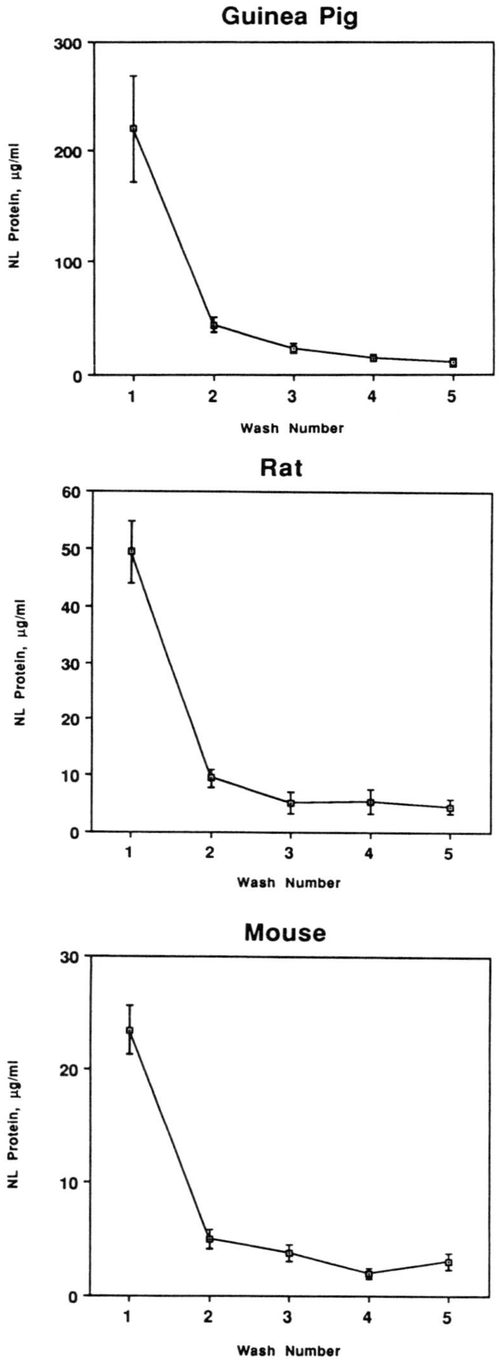

FIGURE 1. Concentrations of protein in sequential nasal lavages of guinea pigs, rats, and mice. Each aliquot of lavage was 1.0 ml in guinea pigs and rats and 0.5 ml in mice.

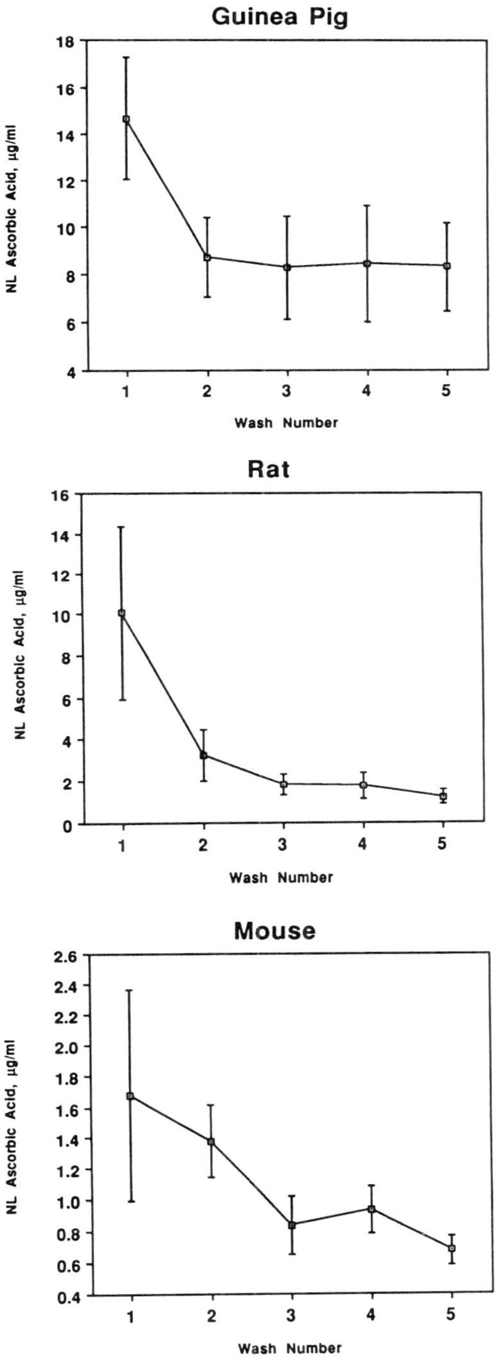

FIGURE 2. Concentrations of ascorbate in sequential nasal lavages of guinea pigs, rats, and mice. Each aliquot of lavage was 1.0 ml in guinea pigs and rats and 0.5 ml in mice.

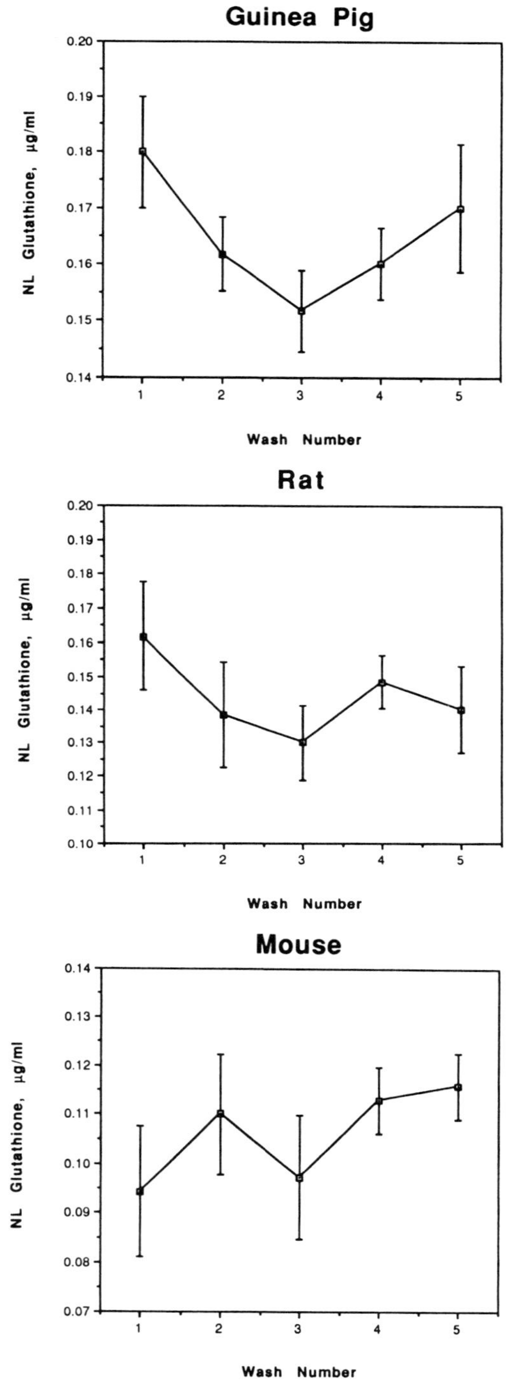

FIGURE 3. Concentrations of glutathione in sequential nasal lavages of guinea pigs, rats, and mice. Each aliquot of lavage was 1.0 ml in guinea pigs and rats and 0.5 ml in mice.

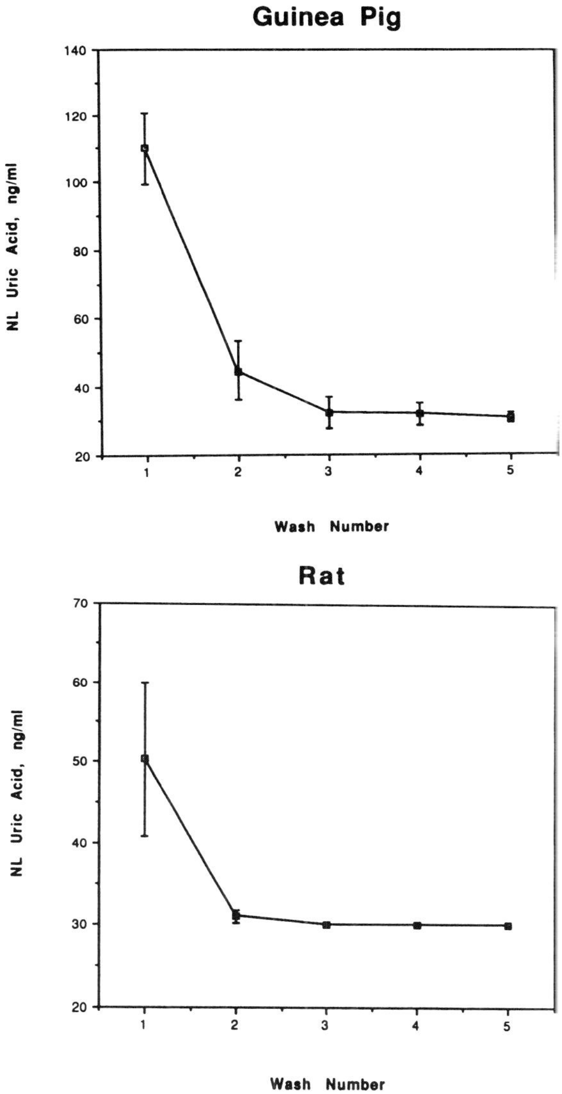

FIGURE 4. Concentrations of urate in sequential nasal lavages of guinea pigs and rats. Urate was undetectable in mouse nasal lavage. Each aliquot of lavage was 1.0 ml in guinea pigs and rats and 0.5 ml in mice.

The washout curve for uric acid fell to below the minimal detectable limits of 30 ng/ml in all species (Figure 4). Guinea pigs appeared to have the highest levels of uric acid (110 ng/ml) in the first wash, while rats had lower levels (50 ng/ml). None of the washes from mice had detectable uric acid (not shown). Rats and guinea pigs had approximately equal amounts of uric acid per kilogram body weight (Table 1).

TABLE 1. Species Comparison of Total Lavagable Constituents

Measurement	Guinea pigs	Rats	Mice
Experiment 1 (nonfasting) protein, μg/kg	472 ± 50	254 ± 36	—
Experiment 2 (Fasting) protein, μg/kg	499 ± 99	272 ± 26	[a]
Experiment 3 (nonfasting) protein, μg/kg	—	—	924 ± 74
Experiment 4 (nonfasting) ascorbate, μg/kg	49.8 ± 9.5	27 ± 11	117 ± 21
Glutathione, μg/kg	0.86 ± 0.05	1.04 ± 0.12	11.3 ± 0.9
Uric Acid, μg/kg	0.26 ± 0.02	0.24 ± 0.02	[a]

Note. Data reported as mean ± standard error. —, Analysis not performed on these animals.
[a]Most washes were below the detectable limits of assay.

DISCUSSION

This study demonstrates the feasibility of measuring antioxidants in NL with the eventual goal of using antioxidant measurements to assess toxic effects of pollutants on the nose. Antioxidants located in the lining layer of the nasal cavity may be important indicators both of sensitivity of the airways to inhaled pollutants, and of toxicant reactions occurring in the lining fluids. Protein, ascorbate, and uric acid were almost always measurable in NL, and a washout curve was usually observed. The complete washout suggests that the constituents examined are probably extracellular. Glutathione was the only constituent that appeared to leak into the NL during later washes. There appeared to be some differences in the lavagable amounts of protein and antioxidants when expressed per kilogram of body weight. Mice tended to have the highest levels, followed by guinea pigs, then rats. Protein, as well as the antioxidants, seemed to follow a similar species pattern, with mice having the highest levels per body weight. This may be due to differences in the nasal surface area per kilogram or to differences in concentration within the lining fluid. The major antioxidant in NL in this study was ascorbate, followed by glutathione, then urate (molar concentrations would approximate the microgram per millilter values because molecular weights are somewhat similar: ascorbate, 176, glutathione, 307; and urate, 168). Ascorbate might contribute more total reducing equivalents because of its potential for two-electron oxidation that is not shared by glutathione and urate. The same concentration pattern of extracellular antioxidant substances was observed in bronchoalveolar lavage of rats and guinea pigs (Slade et al., 1993). Humans appear to have much higher levels of urate in NL than laboratory animals (Hatch, 1992). Thus, large species differences exist in NL fluid antioxidants that could affect the relative susceptibility to oxidant pollutants.

REFERENCES

Anderson, M. E. 1985. Determination of glutathione and glutathione disulfide in biological samples. *Methods Enzymol.* 113:48–55.

Bradford, M. M. 1976. A rapid and sensitive method for the quantitation of microgram quantities of protein utilizing the principle of proteindye binding. *Anal. Biochem.* 72:248–254.

Hatch, G. E. 1992. Comparative biology of the normal lung. In *Treatise on pulmonary toxicology*, ed. R. A. Parent, Vol. I, pp. 617–632. Boca Raton: CRC Press.

Konstan, M. W., Cheng, P. W., and Boat, T. F. 1982. A comparative study of lysozyme and its secretion by tracheal epithelium. *Exp. Lung Res.* 3:175–181.

Kutnink, M. A., Skala, J. H., Sauberlich, H. E., and Omaye, S. T. 1985. Simultaneous determination of ascorbic acid, isoascorbic acid (erythorbic acid) and uric acid in human plasma by high-performance liquid chromatography with amperometric detection. *J. Liquid Chromatogr.* 8:31–46.
Slade, R., Crissman, K., Norwood, J., and Hatch, G. 1993. Comparison of antioxidants in bronchoalveolar lavage fluid from human, guinea pigs and rats. *Exp. Lung Res.* 19:469–484.

EFFECT OF TYPE OF FIXATIVE ON LABELING RATE OF RAT NASAL EPITHELIUM WITH A MONOCLONAL ANTIBODY TO PROLIFERATING CELL NUCLEAR ANTIGEN

Roger A. Renne, Katherine M. Gideon, Kim E. Saunders, Ray L. Buschbom

Pacific Northwest Laboratory, Richland, Washington, USA

There is a large volume of data to indicate a correlation between carcinogenicity induced by xenobiotics and prolonged cell proliferation induced in the target organ by the same chemical or physical insult (Butterworth et al., 1991). These observations have led to the development of assays to quantify cell turnover in target tissues. Proliferating cell nuclear antigen (PCNA) is an endogenous nuclear protein present in proliferating cells. Development of monoclonal antibodies to PCNA have led to its use as a marker of cell proliferation.

Studies comparing cell replication data using PCNA labeling with exogenous markers such as bromodeoxyuridine or tritiated thymidine have found very good correlation using alcohol-based fixatives, but less consistent correlation with exogenous markers when formalin was used as the fixative. This difference is thought to be due in part to the existence of two different forms of PCNA: an alcohol-insoluble fraction associated with the site of DNA synthesis and thus present in high concentrations during G1, S, G2, and early M phases of the cell cycle but absent or in very low concentrations during other phases or in quiescent cells, and an alcohol-soluble fraction present throughout the cell cycle (Dietrich, 1993). The choice of monoclonal antibody used, the duration of fixation, and the type of tissue being examined are also important variables affecting results of PCNA assays.

Our experiment compared the PCNA labeling in nasal tissues of Fischer rats exposed by inhalation for 18 mo to a nasal irritant chemical and fixed in an alcohol-based fixative (Carnoy's fixative; 60% ethanol, 30% chloroform, and 10% acetic acid) or an aldehyde-based fixative (10% neutral buffered formalin, NBF).

STUDY DESIGN

Groups of 12–14 Fischer 344 rats were necropsied following inhalation exposure to 1 of 3 concentrations of an irritant chemical for 18 mo. Respiratory tract tissues from half the rats in each exposed or control group were fixed in one of the two fixatives. Tissues were fixed for 3 days, washed in running water overnight, then decalcified in 5% formic acid/ion exchange resin (Rexyn, Fisher Scientific,

Pittsburgh, PA). Tissues fixed in Carnoy's fixative required 5 days of decalcification; tissues fixed in NBF required 7 days.

Nasal tissues were trimmed using the method of Young (1981). Sections from levels 1, 2, and 3 were placed on slides coated with 3-aminopropyltriethoxysilane (APTS; Sigma, St. Louis, MO).

For immunohistochemical labeling, we utilized a monoclonal anti-rat PCNA antibody of mouse origin (PC10; DAKO Corp., Carpinteria, CA). Different antibody dilutions and lengths of incubation of tissues in primary antibody were tried to maximize the staining intensity of target cells while minimizing the amount of non-specific staining of adjacent tissue and the amount of primary antibody required. A 1 : 400 dilution at 4°C for 18 h provided optimal staining. Adjacent sections were stained with hematoxylin and eosin (H&E) for histopathologic examination.

Sections were deparaffinized in xylene, hydrated, rinsed in 1× Automation Buffer (Biomedia Corp., Foster City, CA) for 5 min, and placed in 3% hydrogen peroxide for 5 min to quench endogenous peroxidase. After rinsing in distilled water for 5 min, slides were placed in undiluted Antigen Retrieval Decal solution for 30 min at room temperature, then rinsed for 5 min each in 2 changes of 100% methanol and 1 change of 70% methanol. Following two 5-min rinses in 1× Automation Buffer, slides were incubated in a moist chamber for 18 h at 4°C in anti-PCNA antibody (PC10) diluted 1 : 400. Following two 5-min rinses in 1× Automation Buffer, undiluted link antibody (Biogenex) was applied for 30 min in a moist incubation chamber at room temperature. After another two 5-min rinses in 1× Automation Buffer, the slides were incubated for 30 min at room temperature with label antibody (Biogenex). Slides were then rinsed twice in 1× Automation Buffer and once in distilled water, and incubated for 6 min in the dark at room temperature in a moist chamber with diaminobenzidine (DAB) chromogen. After rinsing in running tap water for 5 min, slides were counterstained with Richard Allen Hematoxylin-1 for 15 s, washed in tap water for 5 min, rinsed in 1× Automation Buffer until sections turned blue, then rinsed in tap water, dehydrated, cleared, and mounted.

Numbers of labeled epithelial nuclei were counted in areas of the nasal cavity in which proliferative lesions were most severe, as determined by examination of H&E sections. These areas were the respiratory epithelium lining the nasoturbinates and median septum in the level 1 section, and the olfactory epithelium lining the dorsal meatus in the level 2 and level 3 sections. We utilized the unit length labeling index method of Monticello et al. (1990), measuring basement membrane length of areas counted with a digital camera (Javelin Electronics, Torrance, CA) and an automated image analysis program (Optimas Bioscan, Edmonds, WA). Results were analyzed statistically using analysis of variance; significant differences among group means were delineated by Tukey's studentized range test.

RESULTS

Statistical analyses indicated no significant differences ($p > .05$) in labeling between tissues fixed in either of the two fixatives, for any site or any group. Labeling rates were significantly increased compared to controls in level 1 nasal turbinates of rats exposed to the high or middle concentration and in level 1 median septa of rats exposed to the high concentration (Figure 1). Olfactory epithelial labeling rates in the dorsal meatus of level 2 nasal sections were significantly increased in rats exposed to the high concentration. Labeling rates in the dorsal

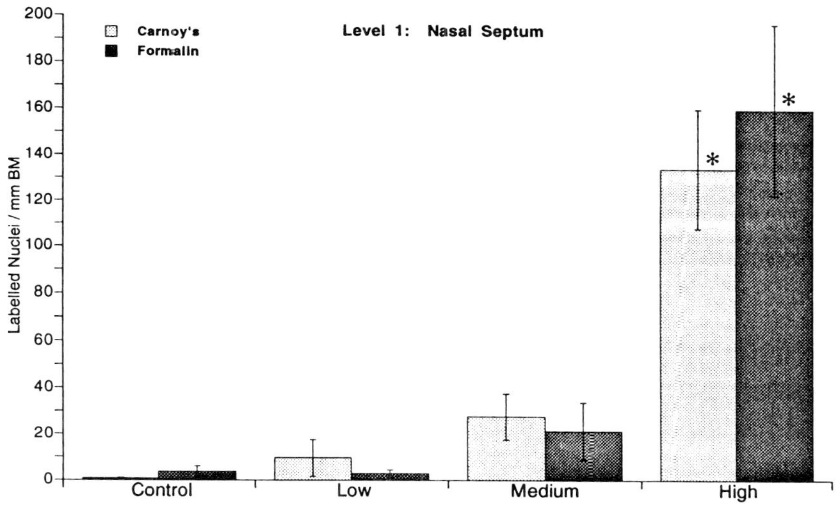

FIGURE 1. PCNA labeling data from median septa at level 1, fixed in NBF or Carnoy's fixative. Asterisk indicates statistically significant ($p < .05$) increase compared to control group data from the same site.

meatus and in the ethmoturbinates in level 3 nasal sections were not significantly increased in exposed groups.

A concentration-related hyperplasia and squamous metaplasia of respiratory epithelium were evident in H&E-stained sections of nose from exposed rats. These lesions were most severe in the epithelium lining the nasal turbinates and the adjacent median septum in the level 1 nasal sections. The epithelium lining the nasal turbinates was hyperplastic in all exposed groups, but only the tips of the turbinates were affected in rats exposed to the lowest concentration. The median septum was affected in only 6 of 12 rats exposed to the middle concentration and 2 of 12 rats exposed to the low concentration. Degeneration of olfactory epithelium lining the dorsal meatus at level 2 was present in rats exposed to the high or middle concentration. The olfactory epithelium lining the dorsal meatus at level 3 was affected only at the high concentration. Basal-cell hyperplasia and respiratory metaplasia were present in some areas of degeneration. In rats exposed to the lowest concentration, the olfactory epithelium was not affected.

DISCUSSION/CONCLUSIONS

Fixation of rat nasal tissues in NBF under the conditions of this experiment provided PCNA labeling data that were comparable with data from tissues fixed in Carnoy's fixative. Information in the literature indicates that it is important to minimize the time tissues to be assayed for PCNA remain in fixative (Dietrich, 1993). Our data indicate that fixation in NBF or Carnoy's fixative for 3 days provides adequate samples for immunohistochemical labeling with PC10 monoclonal antibodies. These results suggest that the PCNA assay can be a valuable tool to assess cell proliferation rates in NBF-fixed respiratory tract tissues from rodents on inhalation toxicity studies.

Cell proliferation rates in affected nasal epithelium as measured with PCNA labeling correlated in general with epithelial hyperplasia observed on hematoxylin and eosin (H&E) sections, with the following exceptions:

- Minimal to mild hyperplasia was evident in H&E sections on the tips of the nasal turbinates in rats exposed to the lowest concentration and on the median septa of some rats exposed to the middle concentration. Although the PCNA labeling rates at these sites were increased compared to controls, these increases were not statistically significant.
- Although basal-cell hyperplasia was evident in H&E-stained sections in areas of olfactory epithelial degeneration in the level 2 section of rats exposed to the middle concentration and in the level 3 section of some rats exposed to the high concentration, there was not a statistically significant increase in PCNA labeling in these tissues.

REFERENCES

Butterworth, B. E., Slaga, T. J., Farland, W., and McClain, M., eds. 1991. *Chemically induced cell proliferation: Implications for risk assessment.* New York: Wiley-Liss.

Dietrich, D. R. 1993. Toxicological and pathological applications of proliferating cell nuclear antigen (PCNA), a novel endogenous marker for cell proliferation. *Crit. Rev. Toxicol.* 23:77–109.

Monticello, T. M., Morgan, K. T., and Hurtt, M. E. 1990. Unit length as the denominator for quantitation of cell proliferation in nasal epithelia. *Toxicol. Pathol.* 18:24–31.

Young, J. T. 1981. Histopathologic examination of the rat nasal cavity. *Fundam. Appl. Toxicol.* 1:309–312.

REGIONAL HUMAN LUNG DOSIMETRY OF INHALED OZONE: THE SCRUBBING EFFICIENCY OF THE NOSE

Abdellaziz Ben-Jebria, Joseph R. Kabel, James S. Ultman

Department of Chemical Engineering, Pennsylvania State University, University Park, Pennsylvania, USA

Ozone (O_3) resulting from the photochemical reaction of automobile engine exhaust is a major component of urban air pollution that can have adverse effects on human lungs. Both the nature and severity of these health effects depend upon the distribution of O_3 to various regions of the respiratory tract. Although people normally breathe through their nose while at rest, an increase in physical activity or the presence of a nasal obstruction can cause a switch to oral breathing (Niinimaa et al., 1980). Therefore, an important issue in O_3 dosimetry is the relative efficiency

This research was supported in part by a Health Effects Institute (HEI) research agreement 90-12 and a National Institute of Environmental Health Sciences grant ES06075. HEI is an organization jointly funded by the U.S. Environmental Protection Agency (EPA) (Assistance Agreement X-812059) and automotive manufacturers. The contents of this article do not necessarily reflect the views of the HEI, nor do they necessarily reflect the policies of EPA or automotive manufacturers.

Dr. Ben-Jebria is also affiliated with INSERM, Laboratoire de Physiologie, Université de Bordeaux-II, Bordeaux, France.

of the nasopharyngeal and oropharyngeal pathways in scrubbing O_3 from inhaled air before it reaches the respiratory airways. The current research was designed to address this issue by comparing the spatial distribution of O_3 uptake in the lungs during nasal and oral breathing at a constant quiet respiratory airflow.

METHODS

The longitudinal distribution of O_3 absorption was measured in intact lungs by employing a bolus-response technique we previously developed (Hu et al., 1992). The apparatus consisted of a fast-responding chemiluminescent O_3 analyzer of our design (Ben-Jebria et al., 1990), a small-scale O_3 bolus generator, and a breathing assembly that terminates in interchangeable mouthpiece and nasal cannula fixtures. The breathing assembly incorporates an injection port for introducing the O_3 bolus, a Fleisch pneumotachograph for monitoring respiratory flow, and a sampling port for monitoring the O_3 concentration curves of the inhaled bolus and the exhaled response.

In each single-breath test, a 10-ml bolus of a 2 ppm O_3-air mixture was rapidly introduced into an inhaled airstream, and O_3 concentration was continually recorded as a function of respired volume throughout the remainder of the breath (Figure 1). The amount of absorbed O_3 expressed as a fraction of the inhaled O_3 (Λ) was computed from the integrals under the bolus and response curves (i.e., $\Lambda = 1 - M_R/M_B$). The penetration, V_P, was computed as the mean volume of the bolus curve relative to the end of inspiration. This parameter characterizes the airway volume that the bolus reaches during inspiration, and by performing a series of test breaths with earlier and earlier bolus injection times, the variation of Λ with V_P was determined.

The experiments were carried out on 9 healthy male nonsmokers with ages ranging from 21 to 29 yr. For each subject, the oral and nasal experiments were completed during one 2 to 4-h session. Half the subjects performed the oral breathing experiments before the nasal measurements, whereas the order of experiments was reversed in the remaining subjects. In all experiments, the subjects viewed a real-time respired volume monitor as an aid in controlling their breathing at a constant flow of 250 ml/s.

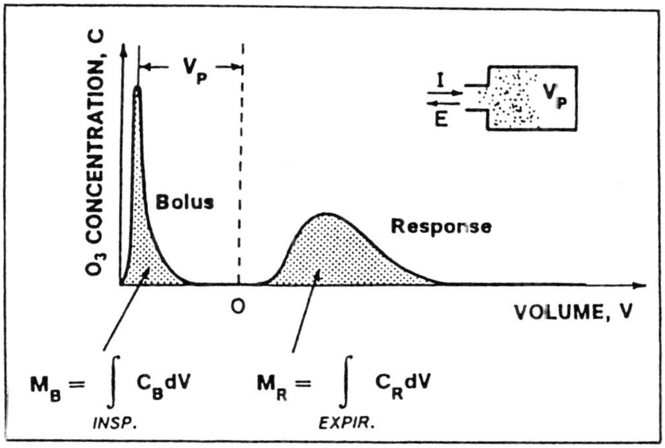

FIGURE 1. Schematic representation of inhaled O3 bolus and exhaled concentration response curves.

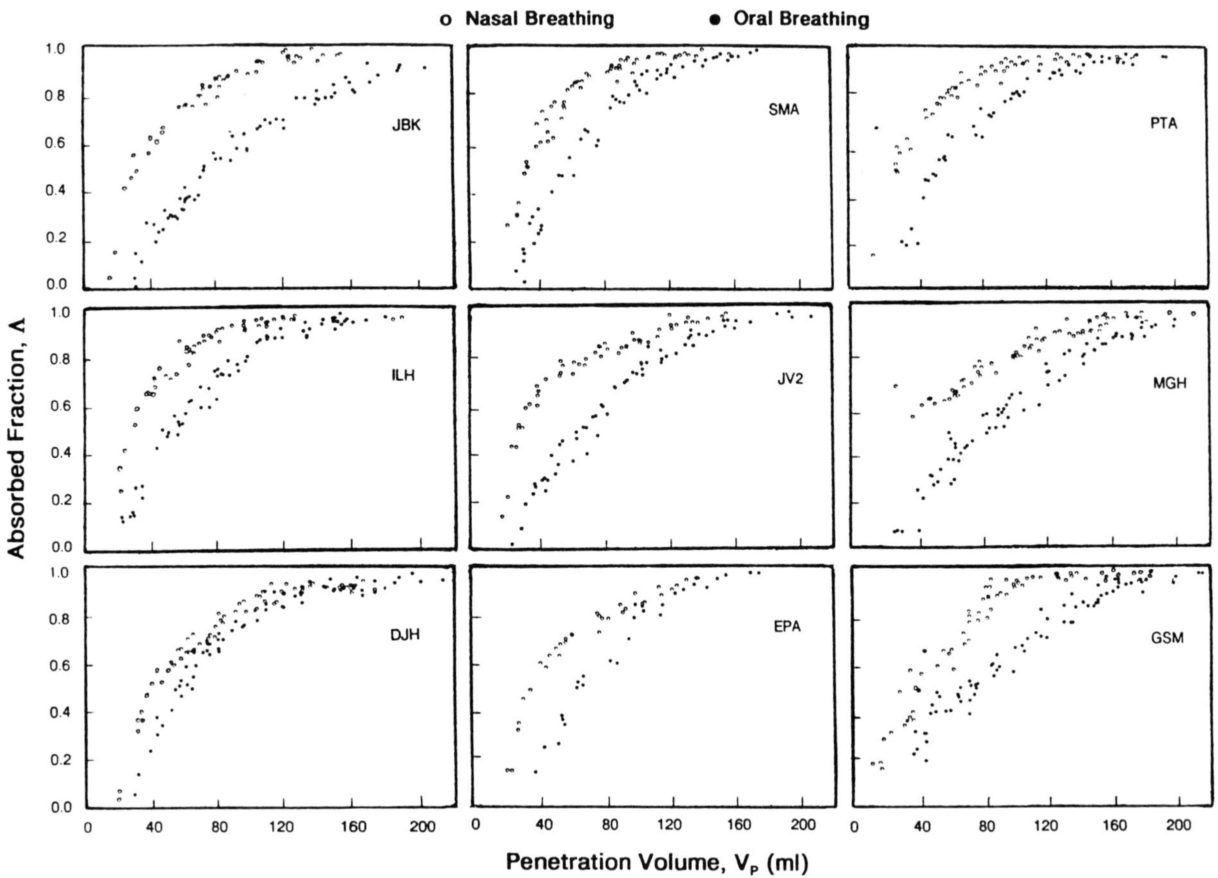

FIGURE 2. Longitudinal distribution of O_3 absorption in the lungs, comparison between oral and nasal breathing in nine healthy men.

RESULTS AND DISCUSSION

For each of the nine subjects, the Λ-V_P distribution for nasal breathing rose more steeply than for oral breathing (Figure 2). Whereas Λ during oral breathing reached its asymptotic value of 100% at a penetration volume around 170 ml, Λ for nasal breathing reached 100% at a V_P around 130 ml. Thus, it is clear that O_3 uptake during nasal breathing is greater than during oral breathing up to a V_P of at least 150 ml. The results from the 9 subjects were pooled by sorting the Λ data into 10-ml increments of V_P and then averaging them to obtain overall sample means within each increment. These averaged data indicate that during oral breathing, 50% of inhaled O_3 was absorbed by the upper airways (i.e., $70 > V_P > 20$ ml) and 100% by the distal end of the conducting airways (i.e., $V_P = 180$ ml). During nasal breathing, however, Λ increased to 80% in the upper airways and the remaining O_3 was absorbed in the central conducting airways.

Our results corroborate the data obtained by Yokoyama and Frank (1972) in the isolated upper airways of dogs. These investigators demonstrated that the nose is significantly more efficient in scrubbing O_3 than the mouth. However, Gerrity et al. (1988) found that O_3 absorption in human lungs was slightly higher during oral than during nasal breathing. The use of a slow-responding O_3 analyzer, which probably underestimated the gas concentration, and a pharyngeal O_3 sampling tube, which can be obstructed by water condensation and mucus accumulation, may explain the unexpected result of Gerrity and co-investigators.

One can argue that the proximal shift in the Λ-V_P distribution by the nasal breathing is due to the difference in the dead-space volume, V_D, between the nose and the mouth. If this was true, then V_D of the nasal flow path would have to be 40 ml smaller than V_D of the oral flow path. Values of V_D determined by applying the Fowler method (Fowler, 1948) to CO_2 expirograms obtained during nasal breathing at the entrance of the nose compared to those obtained during oral breathing at the entrance of the mouth did not exhibit a systematic difference from subject to subject. It is therefore unlikely that the nose has a significantly smaller volume than the mouth in the subjects that we tested.

We conclude that the fractions of inhaled O_3 absorbed by the nose and the mouth are 0.8 and 0.5, respectively, so that nasal breathing is about 60% more efficient than oral breathing. It follows that the nose is more effective in protecting the lower airways from O_3 than is the mouth, and a switch from nasal to oral breathing increases the risk of O_3 damage to lower airway tissue.

REFERENCES

Ben-Jebria, A., Hu, S.-C., and Ultman, J. S. 1990. Improvements in a chemiluminescent analyzer for respiratory applications. *Rev. Sci. Instrum.* 61:3435–3439.

Fowler, W. S. 1948. Lung function studies: II. The respiratory dead space. *Am. J. Physiol.* 154:405–416.

Gerrity, T. R., Weaver, R. A., Berntsen, J., House, D. E., and O'Neil, J. J. 1988. Extrathoracic and intrathoracic removal of ozone in tidal breathing humans. *J. Appl. Physiol.* 65:393–400.

Hu, S.-C., Ben-Jebria, A., and Ultman, J. S. 1992. Longitudinal distribution of ozone absorption in the lung: quiet respiration in healthy subjects. *J. Appl. Physiol.* 73:1655–1661.

Niinimaa, V., Cole, P., Mintz, S., and Sheppard, R. J. 1980. The switching point from nasal to oronasal breathing. *Respir. Physiol.* 42:61–71.

Yokoyama, E., and Frank, R. 1972. Respiratory uptake of ozone in dogs. *Arch. Environ. Health* 25:132–133.

REGIONAL DIFFERENCES IN OZONE-INDUCED NASAL EPITHELIAL CELL PROLIFERATION IN F344 RATS: COMPARISON WITH COMPUTATIONAL MASS FLUX PREDICTIONS OF OZONE DOSIMETRY

Jon A. Hotchkiss, Lois K. Herrera, Jack R. Harkema

Inhalation Toxicology Research Institute, Albuquerque, New Mexico, USA

Julia S. Kimbell, Kevin T. Morgan

Chemical Industry Institute of Toxicology, Research Triangle Park, North Carolina, USA

Gary E. Hatch

U.S. Environmental Protection Agency, Research Triangle Park, North Carolina, USA

Acute inhalation exposure to ozone induces site-specific epithelial DNA synthesis (Johnson et al., 1990), hyperplasia, and mucous-cell metaplasia (Harkema et al., 1989) in the nasal cavity of F344 rats. This response is restricted to the nasal transitional epithelium (NTE) lining the maxilloturbinates, lateral wall, and lateral aspect of the nasoturbinates. Other epithelia (squamous, olfactory, and respiratory) present at the same anatomic level are minimally affected. The observed differences in response to ozone could be due to inherent differences in the susceptibility of specific types of epithelium to ozone-induced injury or could reflect regional differences in the absorbed dose of ozone. Using a three-dimensional computational model of airflow-driven regional dosimetry in the F344 nasal cavity developed by Kimbell et al. (1993), we examined the contribution of regional dosimetry and tissue-specific sensitivity to ozone-induced increases in epithelial DNA synthesis.

A total of 12 male F344/N rats were exposed to 0 ($n = 6$) or 0.66 ppm ozone ($n = 6$) for 6 h in whole-body inhalation chambers. Rats were injected ip 18 h after exposure with bromodeoxyuridine in saline (BrdU; 50 mg/kg body weight) to identify epithelial cells in the S-phase of the cell cycle. Two hours after BrdU injection, the rats were deeply anesthetized with 4% halothane in oxygen and euthanized by exsanguination. The nasal cavities were fixed, decalcified, and embedded in paraffin. Five-micrometer tissue sections taken from immediately posterior to the incisor teeth (T_1) and at the level of the incisive papilla (T_2) were immunohistochemically stained to detect BrdU (Johnson et al., 1990). Ozone-induced nasal epithelial cell proliferation was estimated by quantitating the number of BrdU-labeled epithelial nuclei/mm basal lamina (unit length labeling index, ULLI; Monticello et al., 1990).

A three-dimensional model of the anterior nasal airways of an F344 rat was constructed from digitized images of transverse sections taken at 50-μm intervals from the external nares to the nasopharynx (Kimbell et al.,1993). At each anatomic

Research sponsored by the Office of Health and Environmental Research, U.S. Department of Energy, under contract DE-AC04-76EV01013.

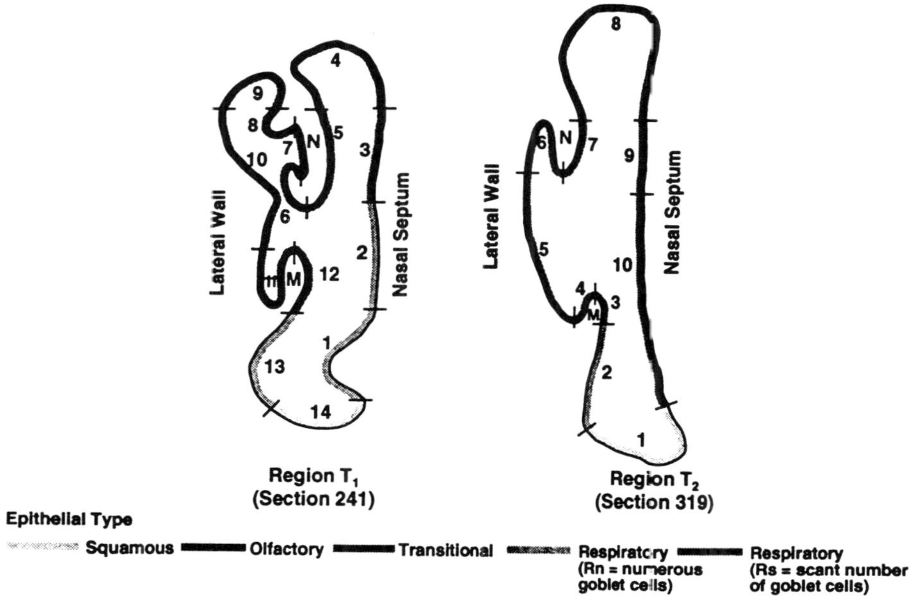

FIGURE 1. Sites analyzed for ozone-induced DNA synthesis and predicted ozone uptake in the anterior nasal cavity of F344 rats. The right nasal airways from regions T_1 and T_2 are shown with the types of surface epithelia present and the sites analyzed.

site, the surface epithelium lining the nasal lumen was subdivided into smaller regions (Figure 1; T_1, 14 regions from 1.2 to 5.4 mm in linear length; T_2, 10 regions from 0.4 to 13 mm in linear length), and the mass flux and ULLI were determined.

Region T_1 was lined by squamous (S), olfactory (O), and nasal transitional

TABLE 1. Comparison of Local Ozone Uptake and Epithelial DNA Synthesis (Region T_1)

ULLI (times greater than control)	O_3 Uptake (ng/mm^2)	Region number	Epithelial type
43.4	0.16	12	NTE
12.4	0.11	9	NTE
11.8	0.06	11	NTE
10.7	0.07	8	NTE
9.5	0.12	10	NTE
8.9	0.10	13	NTE
8.3	0.10	7	NTE
6.1	0.08	6	NTE
4.6	0.08	1	Rn
1.4	0.07	14	S
0	0.11	2	Rn
0	0.07	3	Rn
0	0.08	5	Rs
0	0.10	4	O

TABLE 2. Comparison of Local Ozone Uptake and Epithelial DNA Synthesis (Region T_2)

ULLI (times greater than control)	O_3 Uptake (ng/mm^2)	Region number	Epithelial type
11.0	0.08	5	Rs
8.0	0.0	4	Rs
7.0	0.17	3	Rs
6.2	0.08	2	Rn
6.0	0.06	6	Rs
3.75	0.09	10	Rs
0	0.09	8	O
0	0.08	1	S
0	0.08	9	Rs
0	0.06	7	Rs

epithelium (NTE), and ciliated respiratory epithelium with numerous (Rn) or scant (Rs) numbers of mucous goblet cells. Region T_2 was lined by S, O, Rn, and Rs. There was a proximal-to-distal gradient in the predicted ozone dose and epithelial response (e.g., DNA synthesis). In region T_1, there were significant increases in ULLI in all regions lined by NTE (6- to 43-fold increase over air-exposed controls), in S (1.4-fold increase), and in one of three regions on the nasal septum lined by Rn (4.6-fold increase), but not in regions lined by O or Rs (Table 1). In region T_2, there were significant increases in ULLI in respiratory epithelium with both few and numerous goblet cells, but not regions lined by S or O (Table 2). In region T_1 the two sites with the highest predicted ozone uptake had the greatest increase in DNA synthesis. However, a ranking check of the data (Spearman rank order test) indicated that there was no overall correlation between the presence or magnitude of ozone-induced cell proliferation and the predicted ozone uptake for that location.

This study directly compared a local cellular response to ozone-induced epithelial injury (i.e., DNA synthesis) with predicted ozone dosimetry based on airflow-driven mass flux patterns. The results suggest that while dosimetry may play some role in the observed pattern of ozone-induced epithelial injury, much of the observed response may be due to inherent differences in epithelial susceptibility to ozone-induced injury.

REFERENCES

Harkema, J. R., Hotchkiss, J. A., and Henderson, R. F. 1989. Effects of 0.12 and 0.8 ppm ozone on rat nasal and nasopharyngeal epithelial mucosubstances: Quantitative histochemistry. *Toxicol. Pathol.* 17:525–535.

Johnson, N. F., Hotchkiss, J. A., Harkema, J., and Henderson, R. F. 1990. Proliferative responses of rat nasal epithelia to ozone. *Toxicol. Appl. Pharmacol.* 103:143–155.

Kimbell, J. S., Gross, E. A., Joyner, D. R., Godo, M. N., and Morgan, K. T. 1993. Application of computational fluid dynamics to regional dosimetry of inhaled chemicals in the upper respiratory tract of the rat. *Toxicol. Appl. Pharmacol.* 121:253–263.

Monticello, T. M., Morgan, K. T., and Hurrt, M. E. 1990. Unit length as the denominator for quantitation of cell proliferation in nasal epithelial. *Toxicol. Pathol.* 18:24–31.

LOCAL MASS TRANSFER COEFFICIENTS OF THE HUMAN NASAL PASSAGE FROM MEASUREMENTS OF AEROSOL DEPOSITION IN A NASAL CAST

K.-H. Cheng, D. L. Swift

Division of Environmental Health Engineering, Johns Hopkins University School of Hygiene and Public Health, Baltimore, Maryland, USA

Y.-S. Cheng, Y.-F. Su, H.-C. Yeh

Inhalation Toxicology Research Institute, Albuquerque, New Mexico, USA

The human nasal airway acts as the first protective compartment of the entire respiratory tract by conditioning inhaled air and by removing noxious particles and gases before they can penetrate deeply into the lung. Because airborne toxic substances deposited or absorbed in the nasal airway may produce local biological effects, there is a need to determine the local dosimetric distribution of inhaled gases, vapors, and particles. Calculation of the local mass transfer coefficients provides a quantitative approach to predicting the regional uptake and local dose due to inhaled vapors or ultrafine particles whose transport is dominated by the mechanism of diffusion. The determination of local and regional dose distribution from which the dose-response relationship can be established is fundamentally important for assessing potential health risks.

In this study, we have calculated the local mass transfer coefficient using deposition measurements of ultrafine particles in a replicate nasal passage model. The cast was constructed from 22 coronal, 5-mm-thick magnetic resonance images (MRIs) of the nasal passage of a healthy, nonsmoking, adult male. The model consists of 22 "sandwich elements" of 5-mm-thick clear, rigid plastic plates. It extends from the nostril tip to the junction of the nasopharynx with the pharynx. The passages were hand-sculpted from successive MRIs drawn on the two sides of each plate. The cross-sectional area and perimeter of the nasal passage images were described by Swift et al. (1992).

The experimental setup for generation, size characterization, and deposition measurements of ultrafine particles was described by Cheng et al. (1993). The aerosol used was unattached thoron progeny (^{212}Pb), a gamma emitter with a half life of 10.6 h. A monodisperse ^{212}Pb aerosol (1.7 nm mean diameter) was produced from the decay of a 0.3-mCi dry ^{228}Th source. Particle size distribution was determined using a five-stage graded diffusion battery (Cheng et al., 1992); this particle's size corresponded to a diffusion coefficient of 0.022 cm^2/s. The aerosol was passed through the cast at a constant flow rate of 20 L/min. The overall deposition efficiency (ξ) of the cast was calculated from the concentration on the filter sampled directly from the exposure box (C_{in}) and the concentration on the filter placed after the cast (C_{out}): $\xi \doteq 1 - (C_{out}/C_{in})$. The regional aerosol deposition was measured by the radioactivity on each 5-mm slice of the cast using an NaI detector and a multichannel analyzer.

Current address for Y.-F. Su is Pacific Northwest Laboratory, P.O. Box 999, Richland, WA 99352.

The convective mass flux, N (units of mass/cm^2/s), of a flowing substance to the surrounding surface is proportional to the concentration difference between the concentration in the free stream and the concentration at the airway surface, ΔC, with the proportionality constant defined as the convective mass transfer coefficient, K_c:

$$N = K_c \, \xi C \tag{1}$$

The mass flux, N, is also equal to the mass flow rate divided by the surface area upon which the substances are deposited or absorbed. The mass flow rate, M, is defined as the product of the volumetric flow rate, Q, and the concentration difference, $dC_B(x)$, between the element inlet and outlet. Combining the above two relationships, N can be expressed as

$$N = -\frac{Q \, dC_B(x)}{P(x) \, dx} \tag{2}$$

where $P(x)$ is the perimeter of the airways as a function of distance, x. Based on Eqs. (1) and (2), K_c is related to $C_B(x)$ in the following expression:

$$K_c \, [C_B(x) - C_W] = -\frac{Q \, dC_B(x)}{P(x) \, dx} \tag{3}$$

where C_w is the wall concentration (zero for diffusing particles). The term K_c is obtained by integrating Eq. (3) over the element thickness, ΔL. The resulting equation is thus:

$$K_c = -\frac{Q}{P \Delta L} \ln \left(\frac{C_{B,o}}{C_{B,i}} \right) \tag{4}$$

where $C_{B,o}$ and $C_{B,i}$ are the outlet and inlet particle concentrations of the element and P is the average perimeter of the element. The local Sherwood (Sh) number, characteristic of convective diffusion, can be calculated from the definition:

$$\mathrm{Sh} = \frac{K_c d}{D} \tag{5}$$

in which d (in cm) is the hydraulic diameter of the element (four times the hydraulic radius) and D is the particle diffusion coefficient.

The results of the calculations are shown in Table 1. The maximum transport coefficients in the main nasal cavity were found in the region between 1.0 and 3.0 cm from the nasal entrance, corresponding to the proximal turbinates. Distal to the proximal turbinates, the Sherwood number decreased abruptly and remained relatively low throughout the rest of the nasal turbinates, between 3.0 to 8.0 cm from the nostrils. After a distance of 9.0 cm into the nasal airway, the transport rate started to rise again and reached even higher values than those obtained in the proximal turbinates. The average Sherwood numbers for nasal vestibule and proximal turbinates (0–3.0 cm), main turbinates (3.0–8.0 cm), and nasopharynx (8.0–10.5 cm) are 89.6, 20.0, and 199.7, respectively.

REFERENCES

Cheng, Y.-S., Su, Y.-F., Newton, G. J., and Yeh, H.-C. 1992. Use of a graded diffusion battery in measuring the activity size distributions of thoron progeny. *J. Aerosol Sci.* 23(4):361–372.

TABLE 1. Calculated Nasal Passage Mass Transfer Characteristics

Distance from nose tip (cm)	Local deposition fraction (%)	Mass transfer coefficient (cm/sec)	Sherwood number	Reynolds number
0–0.5	0.20	0.128	3.1	1337.3
0.5–1.0	3.73	2.488	73.3	1255.6
1.0–1.5	6.15	3.449	137.3	1020.7
1.5–2.0	7.94	3.372	103.5	733.0
2.0–2.5	11.92	4.453	114.8	595.3
2.5–3.0	12.12	4.687	105.4	554.6
3.0–3.5	4.50	1.686	28.7	496.6
3.5–4.0	5.90	1.985	26.9	422.7
4.0–4.5	6.70	1.945	22.7	340.4
4.5–5.0	8.20	2.083	21.3	273.1
5.0–5.5	4.41	1.289	12.2	289.4
5.5–6.0	3.70	1.290	15.0	326.4
6.0–6.5	3.46	1.197	16.3	307.3
6.5–7.0	4.68	1.573	22.8	280.6
7.0–7.5	3.39	1.124	15.3	259.1
7.5–8.0	2.90	1.206	19.2	307.5
8.5–8.5	2.53	1.795	92.8	498.9
8.5–9.0	1.50	1.839	198.5	829.2
9.0–9.5	2.02	2.793	292.3	899.3
9.5–10.0	1.51	2.167	222.3	902.0
10.0–10.5	2.51	4.058	192.6	973.5

Cheng, Y.-S., Su, Y.-F., Yeh, H.-C., and Swift, D. L. 1993. Deposition of thoron progeny in human head airways. *Aerosol Sci. Technol.* 18:359–375.

Swift, D. L., Cheng, Y.-S., Su, Y.-F., and Yeh, H.-C. 1992. Design, characterization and use of replicate human upper airways for radon dosimetry studies. In *Indoor radon and lung cancer: Realty or myth? Part 1, Twenty-ninth Hanford Symposium on Health and the Environment,* ed. F. T. Cross, pp. 213–226. Richland, WA: Batelle Press.

DEPOSITION OF 0.005-12 μm MONODISPERSE PARTICLES IN A COMPUTER-MILLED, MRI-BASED NASAL AIRWAY REPLICA

R. A. Guilmette, Y. S. Cheng, H. C. Yeh

Inhalation Toxicology Research Institute, Albuquerque, New Mexico, USA

D. L. Swift

Johns Hopkins University, Baltimore, Maryland, USA

The nasal airways (NA) are the first tissue sites available for deposition of inhaled aerosols in normally nose-breathing people. These airways are also known

Ms. Margo Allen was responsible for constructing the nasal airway model and for performing the deposition measurements using the large-sized particles. Ms. Shawna Smith performed the deposition measurements with the ultrafine particles. Their expert technical assistance is gratefully acknowledged. Work was supported by the U.S. Department of Energy under contract DE-AC04-76EV01013.

sites for the development of cancer from inhaled particles and vapors in humans and experimental animals. The complex anatomy of the nasal airways has made the prediction of deposition difficult. However, sparse data indicate that the deposition patterns will be heterogeneous within the airways and will likely depend strongly on particle size, flow rate, and anatomy.

To better understand the patterns of particle deposition in nasal airways, we have developed a method for constructing anatomically accurate replica models that use in vivo morphometric data obtained from magnetic resonance (MR) imaging. The replica model used in this study was derived from MR images of a nonsmoking Caucasian male, 53 years of age, 73 kg body mass, and 173 cm height, with no obvious nasal airway deformity or disease. The NA perimeters were digitized from contiguous, coronal MR images; lofted surfaces were constructed in three dimensions from adjacent perimeter traces using personal-computer-based CAD (computer-assisted design) software (SmartCAM, Point Control Co., Eugene, OR); and the replica model was milled from 1.5 mm-thick acrylic plastic sheets using a computer-controlled micromilling machine (CAMM 3; Roland DG, Los Angeles, CA).

Two methods were used to generate monodisperse particles. Ultrafine aerosols (0.005–0.20 μm) of NaCl or Ag were produced by a previously described vaporization/condensation method, in which the condensed, singly charged aerosol particles were segregated into monodisperse size fractions by an electrostatic classifier (Cheng et al., 1988). The aerosol was then drawn in both inspiratory and expiratory directions through the replica model under constant-flow conditions (4, 10, or 20 L min^{-1}), and aerosol concentrations at the inlet and outlet of the model were measured using a continuous-flow condensation nucleus counter.

The large-size aerosol particles (0.6–12 mm MMD) consisted of monodisperse liquid droplets of uranine-labeled di(2-ethylhexyl) sebacate (DES), and were produced by a vibrating orifice monodisperse aerosol generator (TSI model 3050, St. Paul, MN). After passing through an ^{85}Kr discharger, the aerosols were drawn through a split-flow system in which half of the aerosol was collected on a membrane filter and the other half was passed through the NA model. The aerosol that did not deposit in the NA replica was then collected on a filter at the outlet. Both inspiratory and expiratory depositions were measured at constant flow rates (10, 20, 30, or 40 L min^{-1}) by measuring the amounts of fluorescent DES deposited on the filters using standard fluorimetric techniques. Deposition efficiencies (E) were calculated according to the relationship $E = 1 - (C_{out}/C_{in})$, where C represents either concentration for the ultrafine particles or amount of fluorescent DES on the filter for the large-sized particles.

The data sets for the ultrafine and large particles were fitted separately using a Marquardt nonlinear regression procedure (NLIN, SAS Statistical Software, Cary, NC). A single model formulation was used to fit all the data simultaneously. This model, based on the earlier work of Cheng et al., (1991), was

$$E = 1 - e^{\Sigma(-a_i D^{0.5} Q^{-0.125})} \quad (1)$$

where E is the deposition efficiency, D is the particle diffusion coefficient, Q is the flow rate, and a_i are the flow-rate-specific fit parameters. In terms of the large-particle data, all four flow-rate-specific data sets were fit simultaneously using the following formulation:

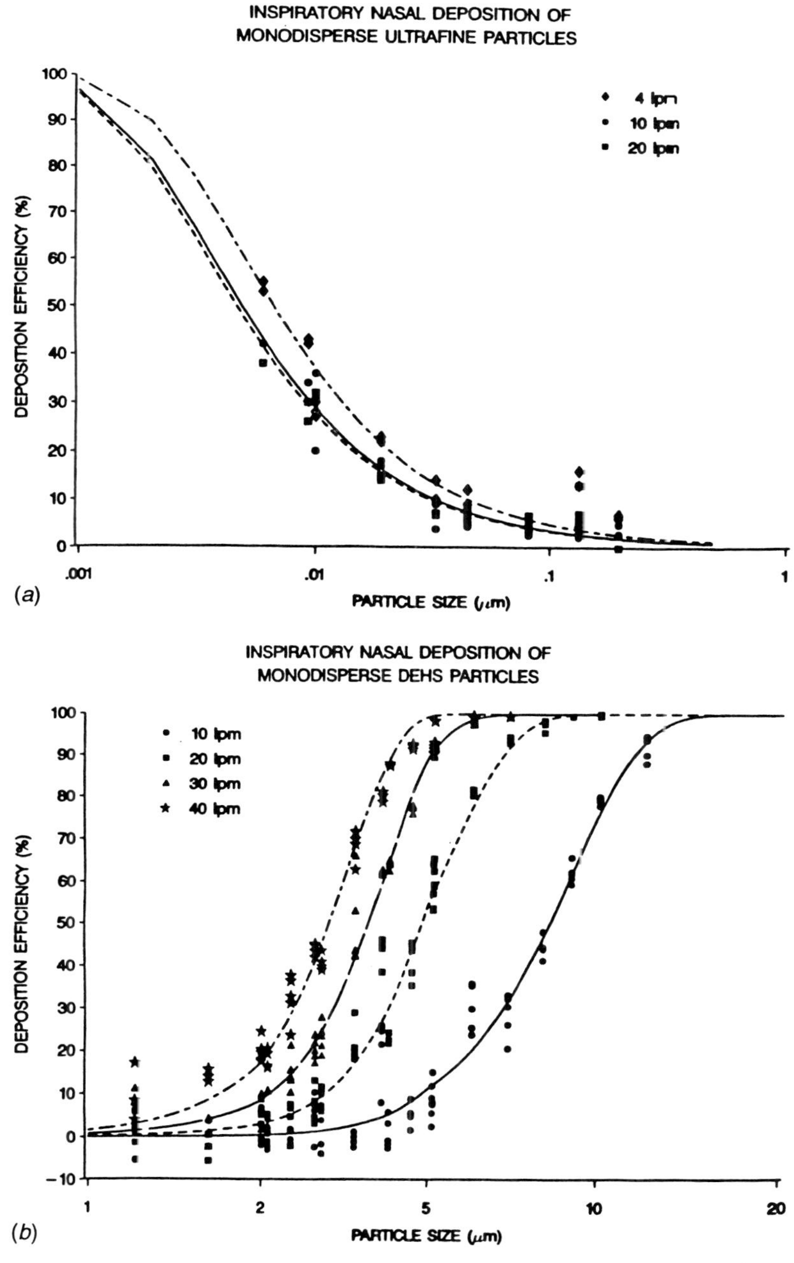

FIGURE 1. Deposition efficiencies in the MR-imaging-based nasal airway mode for (a) ultrafine and (b) large-sized monodisperse particles. Data are stratified by flow rate. Curves represent the fits obtained using Eqs. (1) and (2) respectively.

$$E = 1 - e^{\Sigma(-a_i d^b Q^{-c})} \quad (2)$$

where a_i are the flow-rate-specific fit coefficients, d is the aerodynamic diameter (μm), b is the fitted exponent to the particle size, and c is the exponent of the flow rate. Several approaches to fitting Eq. (2) were used, such as fitting a_i, b, and c simultaneously, fixing $b = 2$ and $c = 1$, or fixing $c = 1$. The optimal fits, based on the magnitude of the residual sums of squares and the random pattern of the residuals, were obtained using Eq. (2) in which the exponent c was fixed to 1.0.

The results of the deposition measurements are illustrated in Figure 1 for inspiratory deposition, and the fitted parameter values are summarized in Table 1 for the ultrafine and large-sized particles and for inspiratory and expiratory depositions. In terms of the deposition of ultrafine particles, the relationship of deposition efficiency and particle size was similar to that observed by other researchers using cadaver-based models, that is, deposition increased with decreasing particle size. Deposition efficiencies were not different at 10 and 20 L min^{-1} for either inspiratory or expiratory flows, but were higher at 4 L min-1. There were no significant differences seen between inspiratory and expiratory depositions at the same flow rates. The range of fit parameter values obtained in this study (17.5–21.4) was somewhat greater than the value obtained previously by Cheng et al. (1991), that is, 12.8. This difference may have been due to the extended range of particle sizes included in the latter study, which included thoron progeny aerosols at sizes less than 5 nm, or it may have been due to the larger nasal airway dimensions present in the cadaver-based model. More data with the MR-imaging-based model are needed to resolve this issue.

The results of the deposition study using the larger sized particles were similar to those obtained in other in vitro and in vivo deposition studies. The sigmoidal shapes of the deposition curves are a common feature of the relationship between particle size and deposition fraction. However, in this study, the exponential dependence of deposition on particle size was significantly greater (3.56 for inspiratory, 3.00 for expiratory) than the quadratic relationship (d^2Q) commonly assumed for inertial deposition mechanisms. Part of this difference may be attributed to constraining the magnitude of the exponent c to 1.0. However, plotting the deposition efficiency

TABLE 1. Fitted Parameters (± Asymptotic Standard Errors) for Particle Deposition in MR-Imaging-Based Nasal Airway Model

Inspiratory deposition			Expiratory deposition		
Q (L min^{-1})	a_i	b	Q (L min^{-1})	a_i	b
		Ultrafine Particles			
4	21.2 ± 0.8		4	21.4 ± 0.9	
10	17.5 ± 0.8		10	19.8 ± 0.9	
20	18.1 ± 0.8		20	19.7 ± 1.0	
		Large Particles			
10	3.9 (±0.5) × 10^{-5}	3.56 ± 0.06	10	5.5 (±0.9) × 10^{-5}	3.00 ± 0.07
20	1.3 (±0.1) × 10^{-4}		20	1.7 (±0.2) × 10^{-4}	
30	2.5 (±0.2) × 10^{-4}		30	3.0 (±0.3) × 10^{-4}	
40	4.0 (±0.3) × 10^{-4}		40	3.8 (±0.4) × 10^{-4}	

versus the impaction parameter d^2Q did not result in convergence of the flow-rate-specific data sets, indicating that there was still a marked dependence of deposition efficiency on flow rate for both inspiratory and expiratory air flows. Our deposition data show flow-specific patterns similar to those observed in vivo by Heyder and Rudolf (1977), and together suggest that (1) the mathematical formulation used in fitting the data may be inappropriate, (2) the method of constraining the fits may be inappropriate, or (3) other deposition mechanisms for large-sized particles, such as turbulence dependence, may need to be considered.

REFERENCES

Cheng, Y. S.,Yamada, Y., Yeh, H. C., and Swift, D. L. 1988. Diffusion deposition of ultrafine aerosols in a human nasal cast. *J. Aerosol Sci.* 19:741–752.

Cheng, Y. S., Yeh, H. C., and Swift, D. L. 1991. Aerosol deposition in human nasal airway for particles 1 nm to 20 mm: a model study. *Radiat. Protect. Dosim.* 38:41–47.

Heyder, J., and Rudolf, G. 1977. Deposition of aerosol particles in the human nose. In *Inhaled Particles*, IV, ed. W. H. Walton, p. 107. Oxford: Pergamon Press.

LOCAL DEPOSITION OF INERTIAL SIZE PARTICLES IN THE ADULT HUMAN NASAL PASSAGE MEASURED IN REPLICATE NASAL PASSAGE CASTS

David L. Swift, Mark A. Bickert

Division of Environmental Health Engineering, Johns Hopkins University School of Hygiene and Public Health, Baltimore, Maryland, USA

Toxic aerosols in the size range 1–10 μm are subject to inertial deposition in the human nasal passage during breathing. Such particles may produce acute or chronic effects in the nasal passage, be transported via the nasal epithelium to the systemic circulation, or be cleared to the gastrointestinal (GI) tract by mucociliary clearance. Their fate depends on both their physiochemical properties and their site of deposition in the nasal passage.

The nasal passage has three types of epithelium with respect to clearance and epithelial transport: (1) the anterior nares, whose surface is cutaneous, (2) the main nasal passage and anterior nasopharynx with ciliated and mucus secreting cells, and (3) the posterior nasopharynx with secretory cells but no ciliated cells. Particles that deposit on each surface type undergo a different kind of clearance and/or transepithelial transport.

Few investigators have systematically studied local nasal deposition of aerosols to measure the percent of particles that deposit on each surface type. In this study, we have employed a nasal replicate model, whose construction and properties were described by Swift et al. (1992), to measure local aerosol deposition with a resolution of 3 mm.

The model was attached to a chamber from which monodisperse aerosols of di-2-ethyl hexyl sebacate (DEHS) are drawn at a steady flow rate of 30 L min^{-1}. The monodisperse aerosols were produced by a spinning top aerosol generator; the aerosols contained ^{99m}Tc, a tracer for particle number concentration. Aerosols were drawn through the model for periods up to 5 min. Following the aerosol exposure, the model elements were disassembled and counted for radioactivity in a scintillation detector.

Three monodisperse aerosol diameters were employed: 2.0, 4.0, and 8.4 μm. In the range of inertial particles, these represent small, medium, and large particles, the largest diameter having a deposition percent during normal inspiratory breathing that approaches 90% in most human subjects.

For all three particle sizes, there was a dominant zone of deposition in the anterior region, just at and beyond the nasal valve (Figure 1). For the 4.0- and 8.4-μm particles, there was a single dominant peak of deposition with smaller amounts of deposition occurring throughout the remainder of the nasal passage. The 2.0–μm diameter particles showed a double peak of deposition, at the nasal valve and somewhat further downstream of the valve. The explanation for this observation is not evident, but it is interesting to note that similar observations were made by Itoh et al. (1985); in this case the three sizes studied were 4.0, 2.0, and 0.4 μm diameter. The 2.0-μm diameter aerosol demonstrated similar double peaks of deposition, while the 4.0-μm diameter particles showed a single dominant deposition peak.

Particles that deposit in the anterior nares are, for the most part, not subject to mucociliary clearance. Such particles are assumed to be cleared physically by nasal wiping, sneezing, blowing, and so forth. Although there is no experimental evidence for the mean duration of such particles, mathematical models of nasal clearance in the anterior region have assumed a clearance time of 12–24 h.

The fate of particles deposited in the anterior nares with respect to transepithelial transport is not known; the portion of the surface that is cutaneous would be expected to have low to negligible rates of transport. Particles that deposit in the posterior nasal epithelium would be expected to be subject to rapid mucociliary clearance and epithelial transport if their molecular structure were favorable for such transport.

Anterior deposition of polydisperse aerosols was determined by Rasmussen et al (1990) by measuring the fraction of the nasal deposited aerosol that could be readily removed by anterior nasal wiping to a depth of approximately 1.5 cm. They

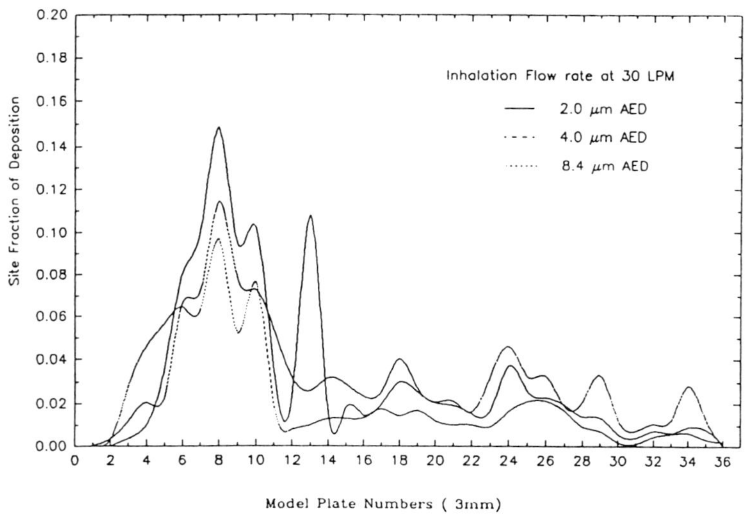

FIGURE 1. Local deposition fraction of aerosols in nasal replicate model. Anterior nares, 0–5; nasal valve, 6; anterior turbinate zone, 7–12; main nasal passage, 13–26; nasopharynx, 26–36.

found that the fraction of the deposited aerosol that could be recovered by this means averaged 70% but varied among 10 normal subjects from 31 to 92%. There may be particles deposited in the anterior, nonciliated region that are not accessible to such wiping maneuvers, so these figures are not necessarily representative of the local deposition characteristics. Nevertheless, they are in general agreement with the present study findings.

It may be concluded from the study presented here that a majority of particles in the 1–10 μm diameter range are deposited in the anterior nasal region, with a smaller fraction depositing in the main nasal passage and anterior nasopharynx. A single peak of deposition is observed for 4.0 μm and 8.1 μm diameter particles, whereas for 2.0-μm diameter particles, a double peak is observed.

REFERENCES

Itoh, H., Smaldone, G. C., Swift, D., and Wagner, H. N. 1985. Mechanisms of aerosol deposition in a nasal model. *J. Aerosol Sci.* 16:367–371.

Rasmussen, T. R., Swift, D. L., Hilberg, O., and Pedersen, O. F. 1990. Influence of nasal passage geometry on aerosol particle deposition in the nose. *J. Aerosol Med.* 3:15–25.

Swift, D. L., Cheng, Y.-S., Su, Y.-F., and Yeh, H.-C. 1992. Design, characterization and use of replicate human upper airways for radon dosimetry studies. In *Indoor radon and lung cancer: Reality or myth?, Part 1, 29th Hanford Symposium on Heath and the Environment*, ed. F. T. Cross, pp. 213–226. Richland, WA: Batelle Press.

CONSIDERATIONS FOR DEVELOPING INHALABILITY FUNCTIONS IN F344 RATS

M. G. Ménache

Duke University Medical Center, Durham, North Carolina, USA

F. J. Miller

Chemical Industry Institute of Toxicology, Research Triangle Park, North Carolina, USA

A. M. Jarabek

U.S. Environmental Protection Agency, Research Triangle Park, North Carolina, USA

O. G. Raabe

University of California, Davis, California, USA

The authors would like to thank attendees at the Nasal Toxicity and Dosimetry Symposium for discussions on appropriate shapes for inhalability curves. The authors thank Mrs. Linda Frye for technical assistance in the preparation of this manuscript. This study was supported, in part, by funds provided by the U.S. Environmental Protection Agency under cooperative agreement CR819093 with Duke University. The research described in this article has been reviewed by the Health Effects Research Laboratory, U.S. Environmental Protection Agency and approved for publication. Approval does not signify that the contents necessarily reflect the views and policies of the agency, nor does mention of trade names or commercial products constitute endorsement or recommendation for use.

The definition of inhalability or inspirability is that fraction of particles in an aerosol that can enter the nose or mouth upon inhalation. This is in contrast to the definition of respirability, that is, those particles that, after inhalation, can penetrate past the terminal bronchioles and thereby reach the gas exchange region (Köhler and Oberdörster, 1991).

The importance of inhalability functions has long been recognized in industrial and occupational settings. The American Conference of Governmental Industrial Hygienists (ACGIH), as well as other organizations, has developed inhalability curves as a function of particle size. Inhalability is important to health risk assessment for two reasons. First, for "large" particles, doses to the upper respiratory tract (URT) will be overestimated if an adjustment for inhalability is not made. If the toxic effect is in the URT, then the true dose delivered to these tissues must be determined. Second, inhalability functions can differ among species due to differences in size and orientation of the nasal airway entrance and in airflow characteristics. In extrapolating from laboratory animals to humans, therefore, it is important to understand how inhalability differs among species.

No published studies have been designed specifically to develop inhalability functions for animals, although it has been shown that pulmonary deposition of large particles (>4–5 μm) via inhalation is minimal in laboratory animals (Raabe et al., 1988). Using data for F344 rats from this study, measured internal deposition was compared to estimated inhaled concentration in order to consider how inhalability functions might be derived for rodents.

ACGIH DATA AND CURVE

Based on experimental data from three published studies (Ogden & Birkett, 1977; Armbruster & Breuer, 1982; Vincent & Mark, 1982), the ACGIH (1985) adopted a simple mathematical function to describe the inhalable fraction of an aerosol:

$$SI(d_{ae}) = 0.5(1 + e^{-0.06d_{ae}}) \quad (1)$$

where $SI(d_{ae})$ is the sampling efficiency of an "ideal inspirable sampler" and d_{ae} is the aerodynamic diameter of a monodisperse aerosol. Details of these studies are summarized in Table 1. Further discussion may be found in the ACGIH (1985) report and the original publications. Two important points concerning these experiments are that the exposures were performed using mannequins, not human subjects, and that the exposures took place in wind tunnels. The wind speeds used are appropriate to industrial settings but might not be reflective of the exposure scenarios of interest for health risk assessment in the general population. An interesting finding of these experiments was that inhalability did not continue to decline toward zero as particle sizes became large but remained at about 50% for particles as large as 100 μm (Figure 1). Behavior of larger particles was not examined.

BREYSSE AND SWIFT DATA AND CURVE

Breysse and Swift (1990) examined inhalability in human subjects breathing normally in still air (Table 1). The authors fit the experimental observations to a function used to describe sampling efficiency of an inverted sampling funnel in still air:

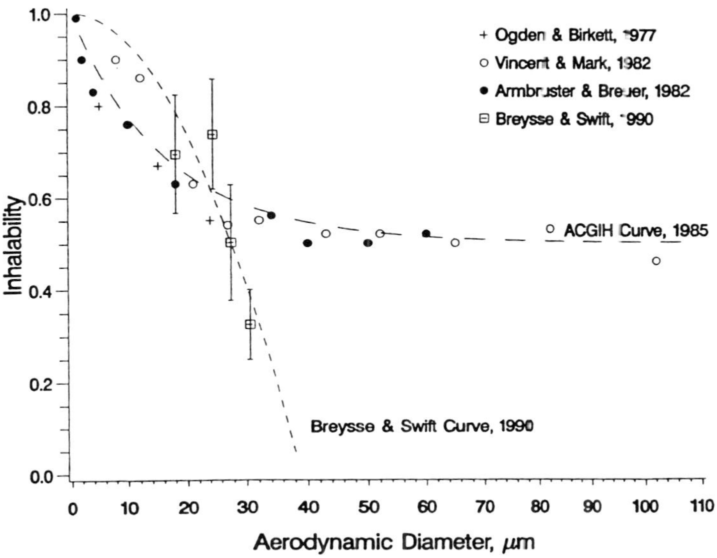

FIGURE 1. Data from three studies using model human heads in wind tunnels have been used to estimate inhalability in an occupational setting (ACGIH curve). A single study using four human subjects breathing in still air has been used to provide an alternative estimate of inhalability (Breysse & Swift, 1990, curve).

$$E = 1 - 0.00066d_p^2 \quad (2)$$

where E is inhalation efficiency and d_p is particle diameter in μm. Predicted inhalability in humans is very different using this equation than it is when using the ACGIH equation (Figure 1). As Breysse and Swift (1990) discussed, one of the advantages of this function is that it "behaves correctly at small diameters." Their curve, however, would predict that no particles larger than 39 μm would be inhaled. There is certainly anecdotal evidence that such particles are not only inspirable but also respirable. Michel and colleagues (1977) found grains of pollens as large as 70 μm in the peripheral areas of the parenchyma in human lung specimens.

RAT DATA

Raabe et al. (1988) exposed five species of laboratory animals to monodisperse particles of up to 10 μm d_{ae}. Although this study was designed to determine deposition patterns in animals, the data may be used to examine inhalability. To illustrate the approach, experimental deposition data for the rat will be used. The ratio of internally deposited radioactivity to the estimated inhaled concentration was calculated. If all particles of a given size are inhalable, then this ratio will approach an upper asymptote of 1. If all particles are not inhalable, this ratio will first approach 1, but then decrease as particle size continues to increase. The ratio is assumed to be an estimate of inhalability for "large" particles, that is, for those particles for

TABLE 1. Summary of "Human" Inhalability Data

Study	Subject	Wind speed (m/s)	Head orientation (°)	Particle size	Breathing pattern
Ogden and Birkett (1977)	Tailor's dummy head on a box	0.75 and 2.75	0, 45, 90, 135, 180	Monodisperse 5–30 μm d_{ae} bis(2-ethylhexyl) sebacate	Steady: 84, 335, 670, 1340 ml/s, inhalation or exhalation
Armbruster and Breuer (1982)	Full-scale model of human head (no torso)	1, 2, 4, 6, 8	0, 90, 180	Polydisperse coal dust <60 μm	Nose/mouth sinusoidal pattern: resting, 16 bpm, 10.8 l/min; normal, 20 bpm, 20.0 l/min; hard, 25 bpm, 37.5 l/min
Vincent and Mark (1982)	Full-size head and torso of a tailor's dummy	1, 2, 4	Slowly rotating during entire exposure	"Narrowly graded" dry aloxite <100 μm	Cyclic mouth 22 bpm, 20 l/min
Breysse and Swift (1990)	Three male and one female human subjects	Still air	—	Monodisperse: 18, 27.5, 30.5 μm Polydisperse: 24.5 μm	Nasal inhalation/ oral exhalation 15 bpm, average minute volume 15.5 l/min

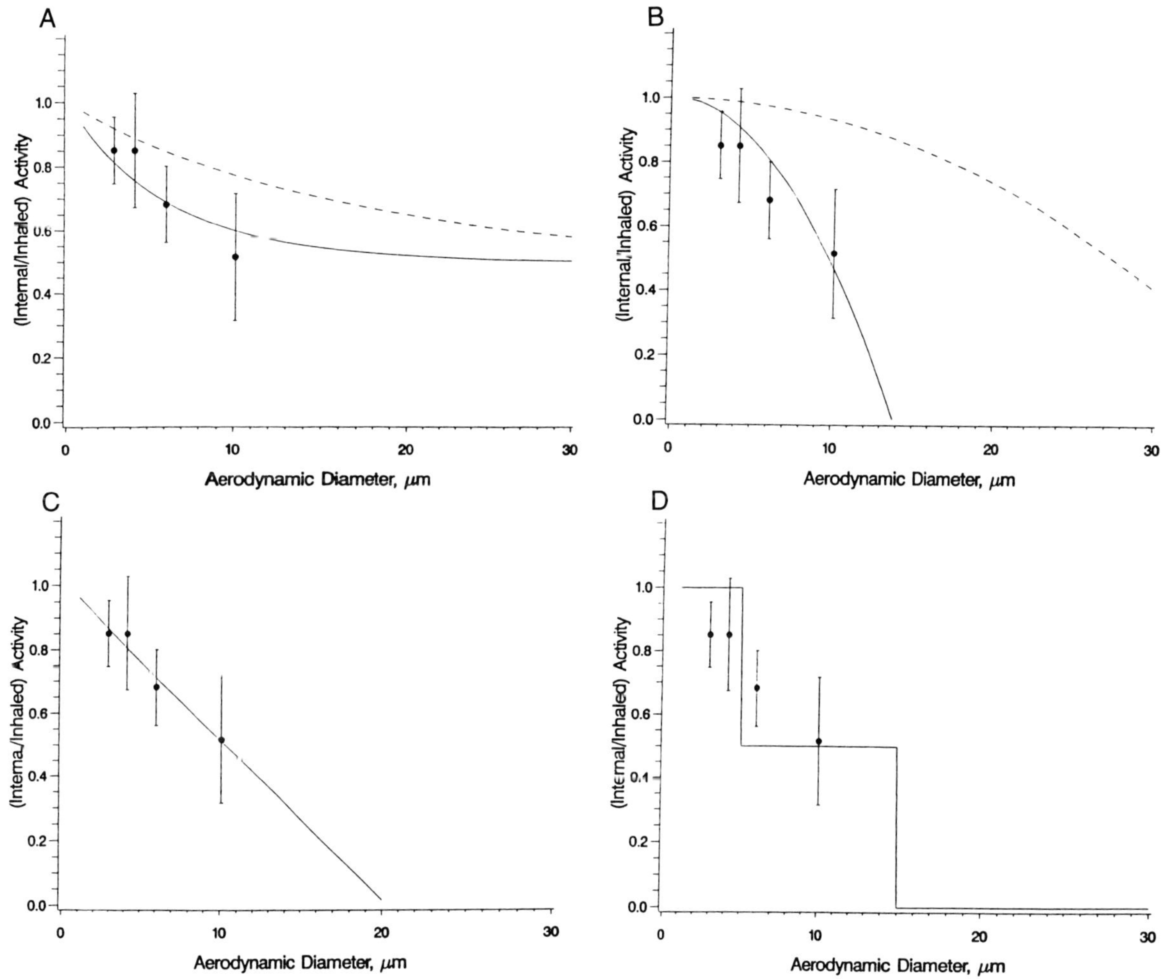

FIGURE 2. Ratio of measured internal radioactivity relative to estimated inhaled radioactivity in the rat. The points represent the mean ± standard error (N = 10) of the data of Raabe et al. (1988). (A) Data are compared to the ACGIH curve (dashed) and fit to that function (solid). (B) Data are compared to the Breysse and Swift (1990) curve (dashed) and fit to that function (solid). (C) Data are fit to a linear function. (D) Data are plotted against a step function.

which the upper asymptote of 1 has been reached and for which, with increasing particle size, the ratio declines. As shown in Figure 2, the ratio is near 1 for 3–4 μm d_{ae} particles but declines for particles with larger d_{ae}. The ratio is therefore assumed to represent an estimate of inhalability in rats for particles larger than 3 μm. Clearly, experiments designed specifically to examine inhalability in laboratory animals would provide more precise data. Until such data are available, these data provide an approximation of inspirability in small animals.

Selecting an Inhalability Curve for the Rat

The panels of Figure 2 illustrate four possible shapes for inhalability functions that could be fit to the particle deposition data for the rat. Panels A and B reproduce the ACGIH and Breysse and Swift curves. As can be seen from the dashed lines, inhalability in humans is greater than estimated inhalability (expressed as the ratio of internal to inhaled activity) in rats at all particle sizes. The solid lines indicate that the four rat data points could be fit to either a convex or concave to the origin curve. Linear (panel C) and step (panel D) functions are also shown. As part of the poster session presentation, symposium participants were asked to "vote" for the curve they felt was most appropriate for these rat data. Important points that came out of this interaction were:

- With only four data points, it is very difficult to defend or not defend any curve on the basis of an empirical model fit.
- A linear function fits the data well and describes the general pattern of decreasing inhalability as particle size increases.
- A concave-to-the-origin curve also reflects the general expected shape of declining inhalability but allows "small" particles to be nearly completely inhalable.
- A convex-to-the-origin curve appeared inappropriate for the exposure scenarios of interest to toxicologists and for health risk assessment.
- While it was felt there was not really enough data for model fitting, there seemed to be enough information to "do better" than a step function.
- An alternate shape, such as the logistic function, that would allow both for concavity with small particles and convexity for large particles was suggested.

REFERENCES

American Conference of Governmental Industrial Hygienists. 1985. Particle size-selective sampling in the workplace. Report of the ACGIH Technical Committee on Air Sampling Procedures. ACGIH, Cincinnati, Ohio.

Armbruster, L., and Breuer, H. 1982. Investigations into defining inhalable dust. *Ann. Occup. Hyg.* 26:21–32.

Breysse, P. N., and Swift, D. L. 1990. Inhalability of large particles into the human nasal passage: in vivo studies in still air. *Aerosol Sci. Technol.* 13:459–464.

Köhler, D., and Oberdörster, G. 1991. Definition of different therapeutic, diagnostic, and environmental aerosols. Expert meeting III: Summary statement of co-chairs. *J. Aerosol Med.* 4(3):269–272.

Michel, F. B., Marty, J. P., Quet, L., and Cour, P. 1977. Penetration of inhaled pollen into the respiratory tract. *Am. Rev. Respir. Dis.* 115:609–616.

Ogden, T. L., and Birkett, J. L. 1977. The human head as a dust sampler. In *Inhaled particles IV*, ed. W. H. Walton, pp. 93–105. Oxford: Pergamon Press.

Raabe, O. G., Al-Bayati, M. A., Teague, S. V., and Rasolt, A. 1988. Regional deposition of inhaled monodisperse coarse and fine aerosol particles in small laboratory animals. Ann. Occup. Hyg. 32(S1):53–63.

Vincent, J. H., and Mark, D. 1982. Applications of blunt sampler theory to the definition and measurement of inhalable dust. *Ann. Occup. Hyg.* 26:3–19.

PHYSIOLOGICALLY BASED PHARMACOKINETIC MODEL FOR NASAL UPTAKE AND METABOLISM OF NONREACTIVE VAPORS

J. B. Morris, D. N. Hassett and K. T. Blanchard

Toxicology Program, University of Connecticut, Storrs, Connecticut, USA

A physiologically based pharmacokinetic (PBPK) model has been developed for nasal vapor uptake in the F344 rat that incorporates nasal enzyme distribution, as well as nasal airflow patterns (Morris et al., 1993). As discussed by Kety (1951), diffusion of vapor molecules through tissues can be approximated by a compartmental approach, provided the compartments are 10 μm or less in depth. Therefore, for the current model, nasal tissue is modeled as stacks of 10-μm compartments with respiratory mucosa containing 1 mucus, 2 epithelial and 2 submucosal compartments and olfactory mucosa containing 1 mucus, 4 epithelial, and 2 submucosal compartments. It is assumed that airborne vapor is in Henry's law equilibration with mucus. Vapor transfer between compartments is based on the molecular diffusivity. Blood is allowed to perfuse the submucosal compartments only and is assumed to be equilibrated with the compartment through which it passes. Metabolic activities are distributed among the compartments in accordance with published histochemical data, and metabolism rate is assumed to follow Michaelis–Menten kinetics. Two airflow paths are assumed: a lateral/ventral path that passes over only respiratory mucosa, and a dorsal medial path that passes over both respiratory and olfactory mucosa (Kimbell et al., 1993).

This model was fit to F344 rat nasal uptake data obtained for five vapors: acetone (an unmetabolized vapor), isoamyl alcohol (alcohol dehydrogenase substrate), ethyl acetate (carboxylesterase substrate), and *o*-xylene and bromobenzene (mixed-function oxidase substrates). Uptake data for the metabolized vapors were obtained in normal as well as metabolic-inhibitor pretreated animals. These vapors cover a wide span of partition coefficients (25–1300) and deposition efficiencies (1–80%). The model accurately predicted the fractional uptake of these vapors in both normal and metabolically inhibited animals, the maximal deviation between actual and predicted deposition efficiency being 4% (Figure 1). The best fit for the data was obtained assuming: (1) an asymmetric blood flow pattern (with the majority of the blood flow being apportioned to the respiratory mucosa) and (2) only a small fraction of the total airflow (8%) following the dorsal medial pathway to the olfactory mucosa. This amount of airflow over the olfactory mucosa is nearly identical to that predicted by Kimbell et al. (1993).

In this model, tissue diffusion is estimated on the basis of transfer between 10-μm compartments. This simplifying assumption allows treatment of nasal tissue by standard PBPK approaches. Although decreasing the depth of each compartment would be anticipated to increase the degree of precision of the model, doing so greatly increases the model complexity, as more compartments would be needed to model the same total tissue depth. For comparative purposes, nasal uptake was also modeled assuming 5-μm deep compartments. The 5-μm compartment-based model predictions for total nasal uptake differed from the 10-μm compartment-based model on the average by 1.003-fold (SD 0.027). Thus, as suggested by Kety

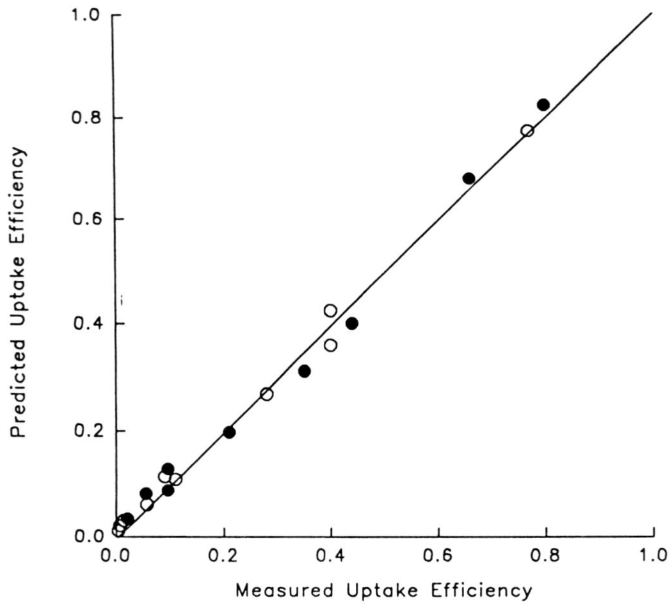

FIGURE 1. Predicted nasal uptake efficiency is plotted against that measured experimentally for isoamyl alcohol, acetone, ethyl acetate, bromobenzene, and *o*-xylene. Uptake was measured in normal (closed circles) and metabolically inhibited (open circles) animals under unidirectional inspiratory flow conditions at 50, 100, or 300 ml/min. The line of perfect fit (slope of unity, *y*-intercept of zero) is also shown.

(1951), 10-μm deep compartments appear to be sufficiently narrow to allow for reasonable estimations. Certainly, doubling the number of compartments does not alter model predictions significantly.

Model simulations were performed under cyclic flow conditions to examine potential effects of vapor desorption during exhalation on uptake during the inspiratory phase. Model simulations (10-μm compartments) revealed that desorption of significant amounts of vapor occurred during exhalation and that this process served to greatly enhance uptake during the inspiratory phase, a phenomenon described by Dahl et al. (1991). For example, predicted acetone uptake under unidirectional flow (200 ml/min), where no desorption can occur, is 16%, compared to 53% under cyclic flow (1-ml tidal volume, 100 breaths per minute, mean inspiratory flow 200 ml/min). Similar results were obtained for the metabolized vapors. For all vapors, uptake during inspiration was strongly influenced by vapor concentration entering the nose from the trachea during exhalation. Nasal inspiratory uptake efficiencies of 42–53% are predicted for acetone, for example, if lung uptake is high (70–100%) and desorption is maximal, compared to nasal uptake efficiencies of only 24% if lung uptake is low (20% or less) and minimal desorption is predicted to occur. These modeling predictions emphasize not only the importance of desorption in controlling uptake during inspiration, but also the importance of pulmonary uptake and exhaled air vapor concentration in influencing nasal uptake of inspired vapors.

REFERENCES

Dahl, A. R., Snipes, M. B., and Gerde, P. 1991. Sites for uptake of inhaled vapors in beagle dogs. *Toxicol. Appl. Pharmacol.* 109:263–275.

Kety, S. 1951. The theory and applications of the exchange of inert gas at the lungs and tissues. *Pharmacol. Rev.* 3:1–42.

Kimbell, J. S., Gross, E. A., Joyner, D. R. Godo, M. N., and Morgan, K. T. 1993. Application of computational fluid dynamics to regional dosimetry of inhaled chemicals in the upper respiratory tract of the rat. *Toxicol. Appl. Pharmacol.* 121:253–263.

Morris, J. B., Hassett, D. N., and Blanchard, K. T. 1993. A physiologically based pharmacokinetic model for nasal uptake and metabolism of non-reactive vapors. *Toxicol. Appl. Pharmacol.* 123: 120–129.

DOSIMETRY OF VINYL ACETATE IN THE NASAL PASSAGES OF RATS: COMPARISON OF IN VITRO AND IN VIVO METHODS OF QUANTIFYING METABOLISM

Matthew S. Bogdanffy, David R. Plowchalk, Anne E. Harman, Marie L. Taylor, Douglas A. Keller

Haskell Laboratory for Toxicology and Industrial Medicine,
E. I. du Pont de Nemours and Company, Newark, Delaware, USA

Vinyl acetate (VA) is used as a polyvinyl acetate emulsion in latex paints, adhesives, and in paper and paperboard coatings and also in the manufacture of polyvinyl alcohol. Chronic exposure to VA by inhalation caused nasal tumors in rats and nonneoplastic nasal lesions in both rats and mice. The tumors included squamous-cell carcinoma, carcinoma in situ, and benign endophytic and exophytic papilloma (Bogdanffy et al., 1994). Our broad objective is to develop a dosimetry model that allows extrapolation to humans of the nasal cancer risks observed in rats.

Vinyl acetate has mutagenic potential, although the activity appears to be related to the hydrolytic release of the nasal carcinogen acetaldehyde, rather than to VA itself (see ECETOC, 1991, for review). Our working hypothesis of the mechanism of action is that VA deposited on the nasal epithelium is hydrolyzed via nasal carboxylesterases to yield acetaldehyde, a genotoxicant, and acetic acid, a cytotoxicant (Kuykendall et al., 1993). Thus, carboxylesterase-mediated hydrolysis is a critical biochemical step in the activation of VA and must be evaluated in rat and human tissue for valid interspecies extrapolation. Since human tissue can only be studied using in vitro methods, these methods must be validated against in vivo measures. The present experiments compare carboxylesterase kinetic constants obtained using

This work was supported in part by the Society of the Plastics Industry, Inc.

two different in vitro techniques with constants generated using an established in vivo technique.

METHODS

Kinetics of VA hydrolysis were measured in vitro using (1) homogenates of nasal mucosa and (2) individual turbinates. Homogenates from regions of the nasal cavity lined with respiratory and olfactory mucosa were prepared and assayed for carboxylesterase activity, as described by Bogdanffy and Taylor (1993). Vinyl acetate metabolism was measured in whole turbinates by placing a single maxilloturbinate (respiratory mucosa), or endoturbinate-1 (olfactory mucosa) in a sealed scintillation vial. Approximately 1 ml of vial headspace was replaced with VA vapor to achieve nominal concentrations ranging from 300 to 30,000 ppm. The vial was then incubated in a rotary shaker at 37°C for up to 30 min, during which headspace samples were removed and the VA concentration was quantified by gas chromatography (GC). Loss of VA from and the appearance of acetaldehyde into the headspace were modeled with SimuSolv using a one-compartment description containing removal terms for VA (carboxylesterase) and AA (aldehyde dehydrogenase, not presented here). Kinetic constants describing the removal of VA were scaled to whole-nose-equivalents by normalizing the terms for wet tissue content.

In vivo kinetic constants for metabolism of VA were generated using a unidirectional flow vapor deposition technique described by Morris (1990). Nasal airflow was maintained at the physiological rate of 100 ml/min. Concentrations of VA in the chamber and in the air passing through the nasal cavity, or in gas bags containing VA vapor standards, were quantified by trapping the VA in midget impingers containing propylene glycol monomethyl ether acetate and analyzing aliquots of the trapping solution by GC. Data were collected for 20 min during steady-state uptake. From these data, absolute deposition (D_{abs}, μg/min) was calculated. As defined by Morris (1990), D_{abs} is an additive combination of factors that contribute to vapor removal from the airstream, such as reaction with mucus and other macromolecules, removal via the bloodstream, and removal via metabolism. Since VA is nonreactive, direct reaction with mucus and other macromolecules is assumed to be negligible, leaving removal via the bloodstream and metabolism as the only significant contributors to overall deposition. Complete inhibition of metabolism enables estimation of blood flow removal and, by subtraction, the contribution of metabolism. Thus, differences in D_{abs} between naive rats and rats pretreated with *bis*(*p*-nitrophenyl) phosphate (BNPP) enable an estimation to be made of the contribution of blood flow and carboxylesterase to total VA deposition. Carboxylesterase activity was inhibited by 5% BNPP injections, as described previously (Kuykendall et al., 1993), deposition studies were carried out, and residual carboxylesterase activity in nasal mucosa was measured in respiratory and olfactory tissue homogenates as cited earlier.

RESULTS AND DISCUSSION

Kinetic constants for the hydrolysis of vinyl acetate that were generated for nasal respiratory and olfactory mucosa using the two in vitro techniques are summarized in Table 1.

In vivo estimates of *V*/*K* were derived over two concentration ranges in naive

TABLE 1. Comparison of Carboxylesterase Kinetic Constants Obtained using Two In Vitro Techniques

	Respiratory tissue		Olfactory tissue	
	Homogenates[a]	Gas uptake	Homogenates[a]	Gas uptake
Vmax (mg/hr)	35.66	5.40	133.08	4.62
Km (mg/mL)	0.037	0.008	0.028	0.014
V/K (mL/hr)	963	684	4684	325

[a]Data adapted from Bogdanffy and Taylor (1993).

(uninhibited) rats. The first evaluation included VA concentrations ranging from 257 to 1936 μg/L. At VA concentrations above approximately 880 μg/L, acetaldehyde concentrations in the expired air appeared to deviate from linearity and approach a plateau, suggesting metabolic saturation. Similar results were observed by Simon et al. (1985) at approximately 700 μg/L. Thus, for the second analysis, the four lowest exposure concentrations, ranging from 257–880 μg/L, were evaluated. The results, presented in Table 2, show that analysis of the deposition data collected at the low concentrations yields *V/K* values that are in reasonable agreement with in vitro *V/K* values presented in Table 1, with the exception of the in vitro value derived from olfactory homogenates.

Deposition of nonreactive vapors in the upper airways is largely dependent on blood flow and metabolic removal from the tissues. Morris (1990) showed that the contribution of each of these factors could be assessed by use of a unidirectional airflow model in which vapor deposition data is obtained from naive and metabolically inhibited rats under conditions of steady-state removal. Thus, metabolic removal of ethyl acetate, described as *V/K*, was determined to be 29 ml/h. Deposition data collected for VA yield distinctly different *V/K* values when evaluated below inspired concentrations of 1936 or 880 μg/L. Although the higher concentration range yields *V/K* values comparable to ethyl acetate (67 ml/h for vinyl acetate vs. 29 ml/h for ethyl acetate), the lower concentration range yielded a *V/K* value

TABLE 2. Derivation of in Vivo *V/K* from Nasal Deposition Studies

Exposure range (μg/L)	C_{ex}[a] (μg/L)	Slope of dabs vs. C_{ex} (L/hr) Naive	BNPP	Naive-BNPP[b]	Percent inhibition[c]	Metabolic removal	Circulatory removal	*V/K*[d]
257–1936	17.3–1114	2.24	0.678	1.56	80.6	1.94 L/h	0.300 L/h	67 ml/h
257–880	17.3–165	18.84	0.678	18.16	80.6	22.5 L/h	0 L/h	776 mL/h

[a]Concentration of VA in the air leaving the nasal cavity. Refer to Morris (1990) for details.

[b]Difference of columns 3 and 4.

[c]BNPP inhibited carboxylesterase 90.5%, and 24.7% in respiratory and olfactory tissues, respectively; 80.6% inhibition is a weighted average inhibition of the two mucosae based on 85% and 15% of the inspired airstream passing over regions lined with respiratory and olfactory mucosa, respectively (Kimbell et al., 1993).

[d] *V/K* is calculated by dividing the metabolic removal term by the blood–air partition coefficient. The blood:air partition coefficient was determined experimentally to be 29 using the method of Sato and Nakajima (1979). Blood esterases were inhibited by treatment with paraoxon.

that is agreeable with *V/K* measured by two different in vitro techniques (776 ml/h in vivo vs. in vitro values of 963 or 675 ml/h for respiratory tissue). Support for the use of deposition data in the low concentration range comes from (1) the y-intercept of the linear regression on D_{abs} versus C_{ex} over the low concentration range was not statistically different from zero (which was not true over the higher concentration range) ($p < .01$ was used for determining statistical difference from zero), and (2) acetaldehyde concentrations in expired air were linear below VA concentrations of 880 μg/L (880 μg/L exposures yielded C_{ex} values of 165 μg/L, i.e., approximately 50-fold below the lowest K_m measured in vitro).

The *V/K* values obtained from olfactory tissue homogenates were approximately 14 times higher than those obtained when using the gas uptake technique, the difference being largely due to differences in V_{max}. The discrepancy can be attributed to the release of esterases in Bowman's glands into the homogenate preparation. Histochemical studies have shown the carboxylesterase content of Bowman's glands to be much greater than respiratory epithelial cells or sustentacular cells (Bogdanffy et al., 1987). In contrast to the results from homogenates, comparison of *V/K* values for respiratory and olfactory tissue generated using the gas uptake technique shows that respiratory tissue is more efficient than olfactory tissue at hydrolyzing vinyl acetate. Thus, esterases in sustentacular cells probably account for most, if not all, of the metabolism of VA in olfactory tissue, with very little, if any, contribution from the Bowman's glands.

In conclusion, these results show good agreement of *V/K* values obtained using a whole turbinate gas uptake assay with those obtained using an established in vivo technique. Further, the results demonstrate that *V/K* obtained from the gas uptake technique can be expected to adequately predict an in vivo measure of *V/K*. Thus, the gas uptake technique should be useful for in vitro measures of human nasal carboxylesterase.

REFERENCES

Bogdanffy, M. S., and Taylor, M. L. 1993. Kinetics of nasal carboxylesterase-mediated metabolism of vinyl acetate. *Drug Metab. Dispos.* 21:1107–1111.

Bogdanffy, M. S., Randall, H. W., and Morgan, K. T. 1987. Biochemical quantitation and histochemical localization of carboxylesterase in the nasal passages of the Fisher-344 rat and B6C3F1 mouse. *Toxicol. Appl. Pharmacol.* 88:183–194.

Bogdanffy, M. S., Dreef-van der Meulen, H. C., Beems, R. B., Feron, V. J., Cascieri, T. C., Tyler, T. R., Vinegar, M. B., and Rickard, R. W. 1994. Chronic toxicity and oncogenicity inhalation study with vinyl acetate in the rat and mouse. *Fundam. Appl. Toxicol.* 23:215–229.

ECETOC. 1991. Joint Assessment of Commodity Chemicals. No. 18. Vinyl Acetate. European Chemical Industry Ecology and Toxicology Centre, Brussels, Belgium.

Kimbell, J. S., Gross, E. A., Joyner, D. R., Godo, M. N., and Morgan, K. T. 1993. Application of computational fluid dynamics to regional dosimetry of inhaled chemicals in the upper respiratory tract of the rat. *Toxicol. Appl. Pharmacol.* 121:253–263.

Kuykendall, J. R., Taylor, M. L., and Bogdanffy, M. S. 1993. Cytotoxicity and DNA-protein crosslink formation in rat nasal tissues exposed to vinyl acetate are carboxylesterase-mediated. *Toxicol. Appl. Pharmacol.* 123:283–292.

Morris, J. B. 1990. First-pass metabolism of inspired ethyl acetate in the upper respiratory tracts of the Fisher 344 rat and Syrian hamster. *Toxicol. Appl. Pharmacol.* 102:331–345.

Sato, A., and Nakajima, T. 1979. A vial equilibration method to evaluate the drug-metabolizing enzyme activity for volatile hydrocarbons. *Toxicol. Appl. Pharmacol.* 47:41–46.

Simon, P., Filser, J. G., and Bolt, H. M. 1985. Metabolism and pharmacokinetics of vinyl acetate. *Arch. Toxicol.* 57:191–195.

GASEOUS TRANSPORT AND UPTAKE IN A BULLFROG NASAL CAVITY MODEL

Kee H. Pyon, Peter W. Scherer

Department of Bioengineering, University of Pennsylvania, Philadelphia, Pennsylvania, and Monell Chemical Senses Center, Philadelphia, Pennsylvania, USA

Maxwell M. Mozell

Health Science Center at Syracuse, Syracuse, New York, and Monell Chemical Senses Center, Philadelphia, Pennsylvania, USA

George Preti

Department of Dermatology, University of Pennsylvania, Philadelphia, Pennsylvania, and Monell Chemical Senses Center, Philadelphia, Pennsylvania, USA

A scaled-up (52-fold) anatomically correct model of a bullfrog (*Rana catesbeiana*) left principal nasal cavity has been constructed (see Figure 1). The patterns of gaseous odorant distribution in the model were studied for three odorants representing high, medium, and low solubility in the bullfrog olfactory mucosa as determined by Mozell & Jagodowicz (1973). High and medium sniff flow rates were simulated in the model by kinematically matching flow parameters in the real and model nasal cavity. The specific aim was to analyze odorant transport and uptake at the mucosal level at different flow rates to determine the mucosal activity patterns of soluble and insoluble odorants.

Construction and operation of the bullfrog nasal cavity model were similar to those of the human nasal cavity model described in Hahn et al. (1993).

Guaran gel was used as the mucus simulant applied on the walls. It has absorbent qualities similar to mucus, adheres evenly to the wall of the model, and displays transport rates similar to frog palate mucus (King et al., 1974). Guaran gel has also been used as a mucus simulant in studies of cough simulations (Zahm et al., 1991) and aerosol deposition (Kim & Eldridge, 1988).

Table 1 shows the odorants used in the experiments. These were dissolved in light mineral oil to yield 0.1% (w/v) solutions.

The experimental procedure was as follows:

1. Mucus simulant was applied on the walls of the model.
2. As the odorant flowed into the model, gas samples were withdrawn for a duration of 9 min 30 s [which corresponds to $(52)^2$ times the real bullfrog high sniff, as determined by kinematically matching nondimensional flow parameters] from the model (via a Variac vacuum pump) into glass tubes filled with Tenax, an organic polymer for collecting organic compounds (Novotny et al., 1974). The fan was set at 0.404 ft^3/min for high flow and 0.235 ft^3/min for medium flow.

This work was supported by NIH grants DC 00220 and DC 00072.

Address correspondence to Peter W. Scherer, Department of Bioengineering, University of Pennsylvania, 240 S. 33rd St., Philadelphia, PA 19104, USA.

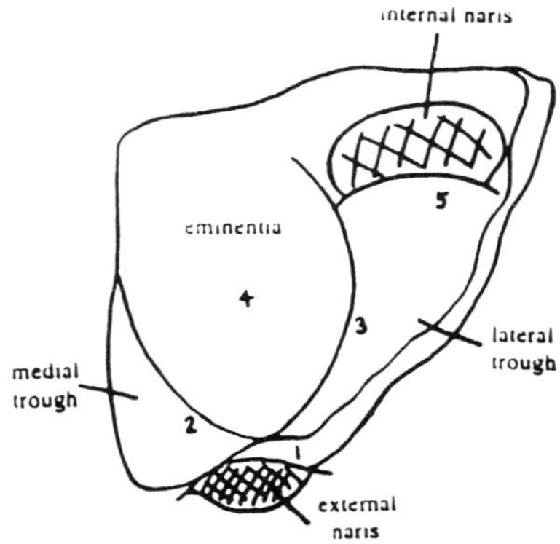

FIGURE 1. Left principal bullfrog nasal cavity. The numbers represent the regions where the odorants were collected.

3. Gas samples were analyzed using a Finnigan 4510 gas chromatograph/mass spectrometer/data system utilizing Incos software for data acquisition.
4. The Incos software package located and quantitated individual compounds based on the molecular ion obtained in the mass spectrum by employing retention times (scan numbers) and the intensities of key ions (see Table 1) designated by the user.

RESULTS AND DISCUSSION

Effect of Mucus Simulant on Air-Phase Concentration of Odorant

The odorants carvone, geraniol, and *d*-limonene are highly, moderately, and weakly soluble, respectively, in the real bullfrog olfactory mucosa (Mozell & Jagodowicz, 1973). In Figure 2, the results show the air-phase concentration of odorants along the lateral trough at high flow rate with and without the application of mucus simulant. Calculating average K (ratio of the air-phase odorant concentration without mucus simulant to air-phase concentration collected with mucus simulant at the collection regions), high K corresponds to high solubility in the mucus simulant: $K_{\text{carvone-avg}} = 4.5$, $K_{\text{geraniol-avg}} = 2.1$, and $K_{d\text{-limonene-avg}} = 2.3$.

From the order of K values and the air-phase concentrations measured, we see that carvone is more soluble than either geraniol or *d*-limonene in the mucus simulant. Although the K value for *d*-limonene is slightly higher than that of geraniol, the actual air-phase concentration of *d*-limonene is much higher (see Figure

TABLE 1. Odorants Utilized

Odorant	Concentration (mg/ml)	β partition coefficient[a]	Ions
Carvone	0.88	0.00013 (high solubility)	82, 108[b]
Geraniol	0.77	0.001 (med solubility)	41,[b] 69, 93
d-Limonene	0.68	0.006 (low solubility)	68, 136[b]

[a]Where β is the air/bullfrog olfactory mucosa partition coefficient determined by using radioactive isotopes (Hornung et al., 1980).

[b]Key ions; see text.

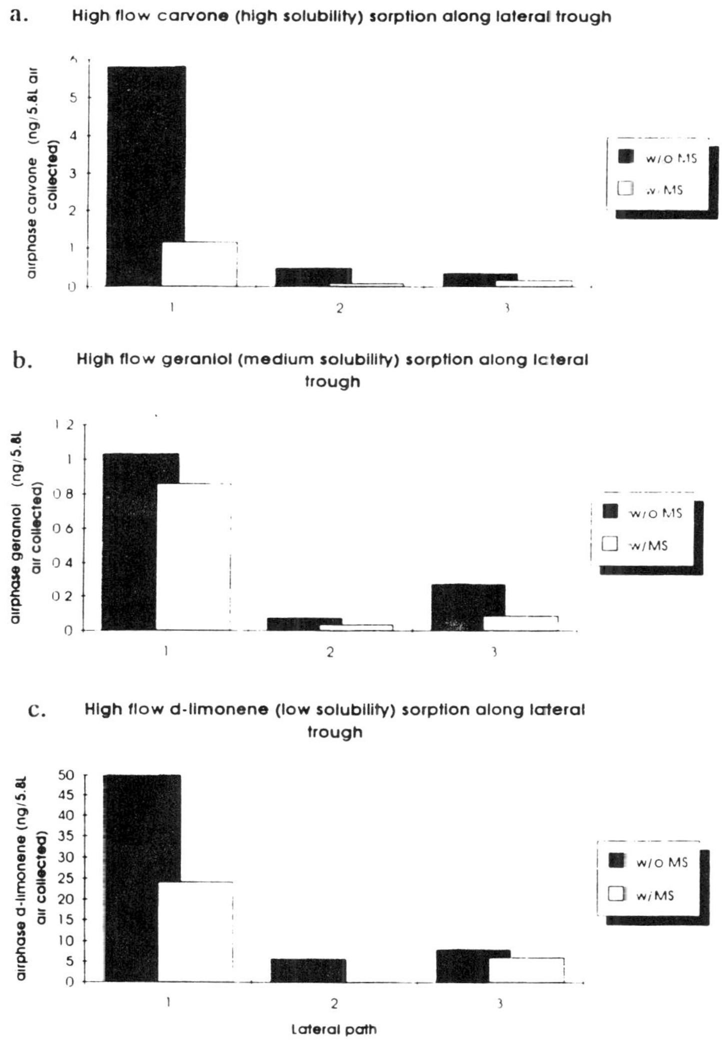

FIGURE 2. Effect of mucus simulant on air-phase concentration of odorants along the lateral trough. MS, mucus simulant.

2c), suggesting that it may be less soluble in mucus simulant than geraniol and/or more volatile than geraniol, thereby presenting more molecules to the vapor phase.

Effect of Flow Rate on Air-Phase Odorant Distribution with Mucus Simulant

Carvone High air-phase concentration is found at the earlier regions of the model for both medium and high flow rates. With the exception of region 2, there is more air-phase odorant measured at medium flow rate. The decreasing concentration after region 2 shows that the carvone is being sorbed into the mucus simulant at both flow rates. The carvone distribution at high flow rate could be due to the highly soluble molecules being pulled quickly to the farther end so that fewer

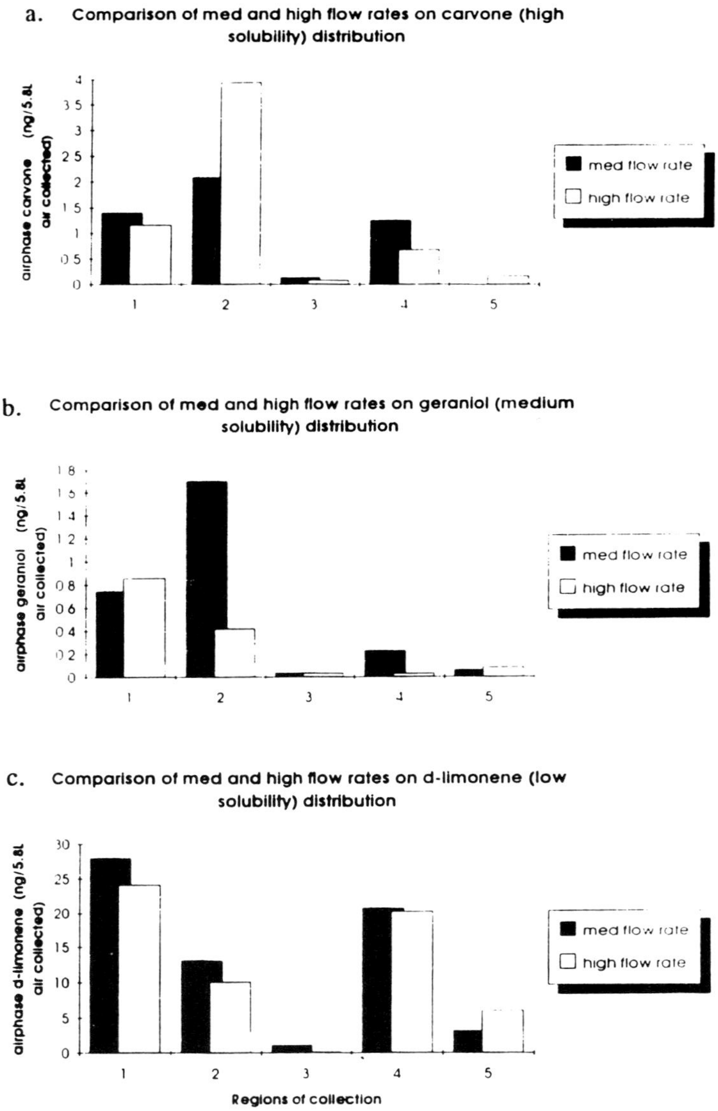

FIGURE 3. Effect of flow rate on air-phase odorant distribution at all collection regions.

molecules are sorbed early in the flow path. But for medium flow rate, many of the molecules are sorbed early in the flow path (region 2).

Geraniol Again, high air-phase concentration is found at the earlier regions of the model for both medium and high flow rates. Although these results are similar to that of carvone, the distribution difference at region 2 is due to geraniol being less soluble than carvone in the mucus simulant, resulting in higher air-phase concentration at medium flow rate as opposed to the concentration at high flow rate.

***d*-Limonene** There does not seem to be much change in air-phase concentration due to flow rate. Overall, *d*-limonene is not very soluble in the mucus simulant except along the lateral path (regions 3 and 5). This distribution could be due to the fact that the weakly soluble *d*-limonene molecules must be pulled farther along the flow path before some of it is sorbed. The higher concentration of *d*-limonene at region 4 may also reflect the volatility of *d*-limonene and not just less solubility in the mucus simulant.

CONCLUSION

The odorant collections through the model show that there is a relationship between the effect of flow rate and the sorption strength of the odorant, going from a negative effect for the weakly sorbed odorant to a positive effect for the strongly sorbed odorant, at least in the front part of the model. This distribution corresponds to experimental data from recordings of olfactory nerve activity in the bullfrog (Mozell et al., 1991) and the theoretical olfaction model of Hahn et al. (1994). This odorant collection procedure could be implemented to study pollutant transport in the bullfrog nasal cavity (also in the rat and monkey noses) and the large-scale human nasal cavity model already available in our laboratory. Numerical simulations of gaseous uptake processes will be done using FIDAP (Fluid Dynamics Analysis Package) to study more detailed odorant and pollutant transport activities and compare the numerical results with the experimental data.

REFERENCES

Hahn, I., Scherer, P. W., and Mozell, M. M. 1993. Velocity profiles measured for airflow through a large-scale model of the human nasal cavity. *J. Appl. Physiol.* 75(5):2273–2287.

Hahn, I., Scherer, P. W., and Mozell, M. M. 1994. A mass transport model of olfaction. *J. Theor. Biol.* 167:115–128.

Hornung, D. E., Mozell, M. M., and Serio, J. A. 1980. Olfactory mucosa/air partitioning of odorant. In *Olfaction and taste VII*, ed. H. van der Starre, pp. 167–170. London: IRL Press.

Kim, C. S., and Eldridge, M. A. 1988. Airway responsiveness to inhaled and intravenous carbachol in sheep: effect of airway mucus. *J. Appl. Physiol.* 65:2744–2751.

King, M., Gilboa, A., Meyer, F. A., and Silberberg, A. 1974. On the transport of mucus and its rheologic simulants in ciliated systems. *Am. Rev. Respir. Dis.* 110:740–745.

Mozell, M. M., and Jugodowicz, M. 1973. Chromatographic separation of odorants by the nose: Retention times measured across in vivo olfactory mucosa. *Science* 181:1247–1249.

Mozell, M. M., Kent, P. F., and Murphy, S. J. 1991. The effect of flow rate upon the magnitude of the olfactory response differs for different odorants. *Chemical Senses* 16:631–649.

Novotny, M., Lee, M. L., and Bartle, K. D. 1974. Some aspects of the chromatographic headspace concentration method using a porous polymer. *Chromatographia* 7:333.

Zahm, J. M., King, M., Duvivier, C., Pierro, D., Girod, S., and Puchelle, E. 1991. Role of simulated repetitive coughing in mucus clearance. *Eur. Respir. J.* 4:311–315.

SIMULATION OF TRANSPORT PROCESSES WITHIN THE HUMAN NASAL AIRWAY

Z. Zhang, G. Yu, L. Chu, R. Lessmann

Department of Mechanical Engineering, University of Rhode Island, Kingston, Rhode Island, USA

Experimental studies of particle deposition and airflow patterns within the human nasal cavity have been published by a great number of researchers. At the same time, there are few numerical investigations of these topics, due primarily to the complicated nature of nasal geometry and limitations of computer capacity. Elad et al. (1992) reported three-dimensional (3-D) numerical simulations of human nasal flow patterns generated by using finite-element computer software (FIDAP). This study was based on a noselike model that resembled the complex structure of the nasal cavity. The convective terms in the momentum equation were neglected, restricting the results to low Reynolds numbers. Kimbell et al. (1993) used the same software to compute the airflow and gas transport patterns within a rat's nasal airway. The simulated velocity patterns matched with experimental results. A similar numerical study is currently being conducted by Scherer (1993).

In this article, we present a 3-D finite-difference simulation using a different commercial software package (PHOENICS). Both finite-difference and finite-element schemes have been developed to simulate viscous flows, but finite differences are somewhat more popular in the field of computational fluid dynamics. This is due to the fact that finite-element simulations, though more adaptable to handling irregular geometries, are slower to converge and require more computer storage. Finite-difference methods, especially the control volume scheme, converge more easily and are numerically stabile for higher Reynolds number flows.

NASAL MODEL AND NUMERICAL METHOD

To provide a realistic numerical simulation, we used a model in which a nasal-like structure is formed from a combination of curved plates to resemble the complex nasal turbinate. This nasal structure is assumed to be symmetrical, and only the left nasal airway flow was simulated. The nasal turbinates are winglike bones emerging from the lateral wall of the nasal cavity. They are located in the middle section of the nasal cavity. The inferior turbinate, occupying most of the lower part of the nasal cavity, is about 4.5 cm long and emerges from the lateral wall at a height of 2.5 cm from the floor. The superior turbinate is about 3.5 cm long and is located above the inferior. The coronal cross-sectional geometries (in the direction of air flow) of our nasal model are constructed section by section based on the tomographic pictures published by Montgomery et al. (1979). A primary trapezoidal shape with a base of 1 cm at the nasal floor and a width of 0.2–0.4 cm at the top is assumed for all nasal passage cross sections. The spatial distribution of sagittal cross-sectional shapes is shown in Figure 1. A body-fitted 3-D grid system is employed in order to match the computational domain with the outline of a realistic sagittal cross-section nasal geometry. The Navier–Stokes (N-S) equations and the concentration equation are simulated by using PHOENICS. Airway flow is assumed to be steady, incompressible, and Newtonian. No slip and zero concentration conditions are applied at the wall.

RESULTS

Respiratory flow rates of 15, 30, and 60 L/min, which correspond to various activity conditions of the human adult have been simulated. Local velocity, concentration, and pressure distribution patterns are obtained. These results show that there is a significant discrepancy between the predictions of direct simulation and those using the *k-e* turbulence model, where *k* and e are turbulent energy and dissipation, respectively. Pressure drop values using turbulent flow models are unrealistically low.

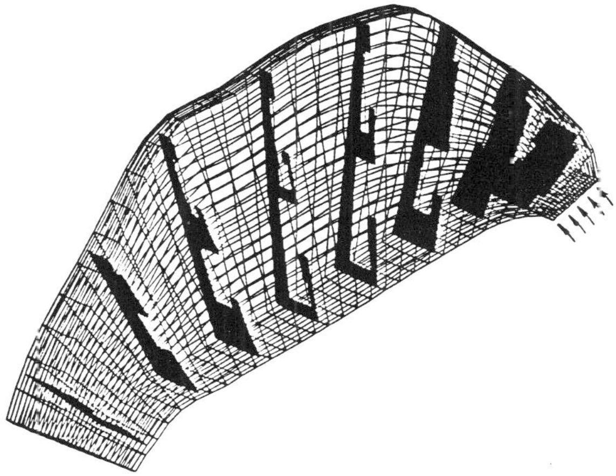

FIGURE 1. Spatial distribution of sagittal cross-sectional shapes.

At a flow rate of 15 L/min, inhaled air smoothly passes the turbinate structure region, showing a clearly laminar flow pattern. When the flow rate increases, a flow recirculation zone is formed immediately distal to the inferior turbinate. At a flow rate of 60 L/min, fluid particles in the recirculation zone are isolated from the main stream flow, and mass transfer is dominated by diffusion in this region. Since the bulk of the fluid within the recirculation zone is nearly stagnant, trapped pollutants such as ozone have sufficient time to diffuse away, and the concentration value here is very low.

The total ozone mass transfer efficiency, which is defined as the ratio of local to inlet concentration with a zero concentration boundary condition, is related inversely to activity levels. For flow rates of 15, 30, and 60 L/min, the mass transfer efficiencies are found to be 0.52, 0.66, and 0.76, respectively.

SUMMARY

The 3-D airflow and mass transfer patterns within the human nasal cavity have been simulated. The new features of this work include: (1) use of a finite difference scheme in conjunction with a body-fitted grid system and selected grid block-off, (2) both direct computation of the N-S equation and computation with the *k*-e turbulence model are performed and compared, and (3) the Reynolds numbers in our direct simulation range from 500 to 2000, which covers most adult human respiratory conditions. It may be concluded that the general k-e model within the software is not suitable for nasal airway simulations.

REFERENCES

Kimbell, J. S., Gross, E. A., Joyner, D. R., Godo, M. N., and Morgan, K. T. 1993. Application of computational fluid dynamics to regional dosimetry of inhaled chemicals in the upper respiratory tract of the rat. *Toxicol. Appl. Pharmacol.* 121:253–263.

Montgomery, W. M., Vig, P. S., Staab, E. V., and Matteson, S. R. 1979. Computed tomography: A three-dimensional study of the nasal airway. *Am. J. Orthodontics* 76:363–375.

Scherer, P. 1993. Dosimetry modeling for humans. *Symp. Nasal Toxicity and Dosimetry of Inhaled Xenobiotics: Implication for Human Health*, September, Durham, NC.

STRAIN-RELATED DIFFERENCES IN OZONE-INDUCED SECRETORY METAPLASIA IN THE NASAL EPITHELIUM OF RATS

Jack R. Harkema, Carol M. Wierenga, Lois K. Herrera, Jon A. Hotchkiss, W. Alan Evans, David G. Burt, Charles H. Hobbs

Inhalation Toxicology Research Institute, Albuquerque, New Mexico, USA

Ozone is a common oxidant pollutant in urban centers. Nasal epithelium undergoes secretory metaplasia in rats repeatedly exposed to ambient concentrations of ozone (Harkema et al., 1989). This is a response to oxidant injury within the nose that is restricted to the nasal transitional epithelium (NTE) lining the lateral wall and naso- and maxilloturbinates in the proximal aspect of the rat nasal airway.

Recently it has been reported that ozone-induced inflammatory and epithelial responses in the pulmonary airways of rats and mice vary among different strains (Kleeberger et al., 1990; Pino et al., 1991). In the rat, ozone exposure causes more severe lung airway injury in Wistar rats than in Sprague-Dawley rats (Pino et al., 1991). The effect of strain on the susceptibility of nasal epithelium to ozone-induced injury has not been examined. The purpose of the present experiment was to determine if ozone-induced secretory metaplasia is different among inbred strains/outbred stocks of rats.

Fifty-six male rats, 8–12 mo of age, were used in this study. The rats came from two different outbred stocks (Sprague-Dawley rats from Taconic, Germantown, NY; Wistar rats from Charles River Laboratories, Wilmington, MA) and two inbred substrains of F344 rats (F344/N rats from the ITRI breeding colony; F344/NTac rats [Taconic]). The rats were divided into two groups ($n = 7$ per strain/stock) and were exposed to 0 or 0.8 ppm ozone in whole-body chambers for 6 h/d for 7 days. The rats were anesthetized using 5% halothane and killed by exsanguination via the abdominal aorta or renal arteries, 24 h after the end of exposure. Immediately after death, the head of each rat was removed from the carcass and the nasal airways were fixed in 10% neutral buffered formalin.

After fixation, the heads were decalcified using a 13% aqueous formic acid solution. After decalcification, the nasal cavity was transversely sectioned immediately posterior to the upper incisor teeth and through the incisive papilla of the hard palate, as previously described by Young (1981). The selected tissue block from the proximal aspect of the nose was embedded in paraffin, sectioned at 6 μm, and stained with alcian blue (pH 2.4)/periodic acid Schiff (AB/PAS) for identification of acidic and neutral intraepithelial mucosubstances (IM). The volume density of AB/PAS-stained mucosubstance in the NTE was determined by image analysis, as previously described by Harkema and Hotchkiss (1991). The data were expressed as the mean volume density/surface area (Vs; nl/mm^2 basal lamina) of AB/PAS-positive mucosubstances within the NTE ± the standard error of the mean (Figure 1).

Carol M. Wierenga was a Department of Energy/Associated Western Universities summer student research participant, W. Alan Evans was a postdoctoral fellow.

Research sponsored by the Office of Health and Environmental Research, U.S. Department of Energy, under Contract DE-AC04-76EV01013.

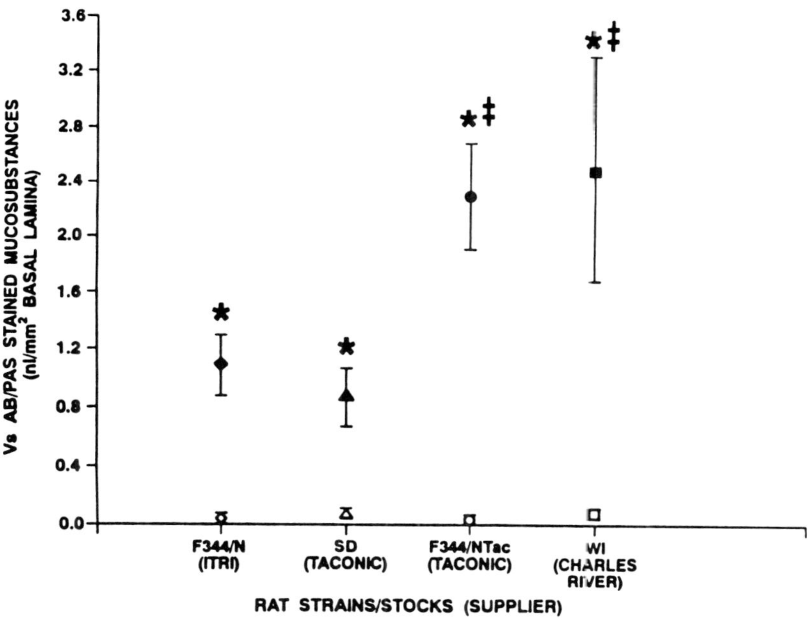

FIGURE 1. Amounts of intraepithelial mucosubstances in the maxilloturbinates of 4 different strains/stocks of rats after exposure to 0.8 ppm ozone (closed data points and bars = group means ± standard error of the means) or 0 ppm ozone (open data points and bars = group means ± standard error of the means); V_s = volume density of intraepithelial mucosubstances/surface area; asterisk indicates significantly different from air-exposed (0 ppm) controls, $p < .05$; double dagger indicates significantly different from group means of ozone-exposed F344/N rats from ITRI colony and Sprague-Dawley rats exposed to ozone (SD), $p < .05$; WI, Wistar rats.

No measurable amounts of IM were found in control rats of any strain or stock. Ozone-induced secretory metaplasia with measurable amounts of IM was present in all strains/stocks of rats. The severity of the secretory metaplasia, however, varied among the ozone-exposed rat strains/stocks. Ozone-exposed F344/NTac rats and ozone-exposed Wistar rats had significantly more IM than the ozone-exposed F344/N rats from the ITRI colony and the ozone-exposed Sprague-Dawley rats (Figure 1). The amounts of IM after exposure to 0.8 ppm ozone ranged from 0.87 ± 0.20 nl/mm^2 (mean ± SEM) in Sprague-Dawley rats to 2.49 ± 0.38 nl/mm^2 in Wistar rats.

The reason for the strain/stock-related variability in the severity of ozone-induced secretory metaplasia in the NTE is unknown. The results of this study indicate that the Wistar rats and the F344/NTac rats are more sensitive to ozone exposure compared to the F344/N rats from the ITRI colony and the Sprague-Dawley rats. It is interesting that Pino et al. (1991) also found that Wistar rats were more sensitive to ozone-induced injury compared to Sprague-Dawley rats, when they examined the epithelium lining the lung centriacinus. In addition, the results of the present study demonstrated that there can be substrain-related differences in response to ozone between F344/N rats from two different colonies. Ozone-induced secretory metaplasia was more pronounced (i.e., twice as much IM in the NTE) in F344/NTac rats than in F344/N rats from the ITRI breeding colony. These results raise questions about the most appropriate strain (or substrain) or stock of rat to be used in future animal toxicology studies designed to help establish a new NAAQS for ozone.

REFERENCES

Harkema, J. R., and Hotchkiss, J. A. 1991. In vivo effects of endotoxin on nasal epithelial mucosubstances: Quantitative histochemistry. *Exp. Lung Res.* 17:743–761.

Harkema, J. R., Hotchkiss, J. A., and Henderson, R. F. 1989. Effects of 0.12 and 0.80 ppm ozone on rat nasal and nasopharyngeal epithelial mucosubstances: Quantitative histochemistry. *Toxicol. Pathol.* 17:525–535.

Kleeberger, S. R., Bassett, D. J. P., Jakab, G. J., and Levitt, R. C. 1990. A genetic model for evaluation of the susceptibility of ozone-induced inflammation. *Am. J. Physiol.* 258:L313.

Pino, M. V., Hyde, D. M. and Stovall, M. Y. 1991. Strain differences in the response of the rat lung to acute ozone exposure. *Am. Rev. Respir. Dis.* 143:A698.

Young J. T. 1981. Histopathologic examination of the rat nasal cavity. *Fundam. Appl. Toxicol.* 1:309–312.

INDUCED XENOBIOTIC-METABOLIZING ENZYMES LOCALIZED TO EOSINOPHILIC GLOBULES IN OLFACTORY EPITHELIUM OF TOXICANT-EXPOSED F344 RATS

Johnnye L. Lewis, Kristen J. Nikula, Laurel A. Sachetti

Inhalation Toxicology Research Institute, Albuquerque, New Mexico, USA

Eosinophilic globules develop in secretory cells of respiratory and olfactory mucosae in rodents and primates following chronic inhalation of a variety of toxicants, including dimethylamine (rats, mice; Buckley et al., 1985), cigarette smoke (rats, unpublished results), and ozone (monkeys; Harkema et al., 1987). The number and size of the globules increase with increased dose and/or exposure duration. These globules have been described by Monticello et al. (1990) as homogeneous,

FIGURE 1. All photomicrographs 300x magnification. (A) H&E: Hematoxylin- and eosin-stained section of an ethmoturbinate from a 32-wk smoke-exposed rat showing eosinophilic globules, which appear as homogeneous, light gray areas of cytoplasm (arrows), within sustentacular cells. (B) Four weeks, CE immunohistochemistry (IHC): Section of the same turbinate from a 4-wk smoke-exposed animal. The CE immunoreactivity occurs in the periphery of globules as well as in Bowman's glands (arrowheads), where it is normally observed. Hematoxylin counterstain of nuclei. (C) Thirty-two weeks, CE IHC: Same staining technique as (B). Following 32-wk exposure, immunoreactivity for CE more densely fills globules, which have increased in number and size. (D) Control: Same staining technique as (B) and (C), except normal goat serum substituted for CE antibody. No reaction product is seen in this serum-control section from a smoke-exposed rat. (E) Rhodanase IHC: Serial section to (C) incubated with RH antibody. There is no immunoreactivity in the globules, but the normal staining of the apices of sustentacular cells is present (arrows).

The CE antibody was provided by Dr. Raymond Novak, Wayne State University; P-450 NMa and NMb antibodies were provided by Xinxin Ding, University of Michigan. Research supported by NIH grant RO1-DC01714 from the National Institute on Deafness & Other Communication Disorders and by the U.S. Department of Energy, Office of Health Effects Research, under Contract DE-AC04-76EV01013.

Laurel A. Sachetti is a D.O.E./A.W.U. teacher research associate program participant.

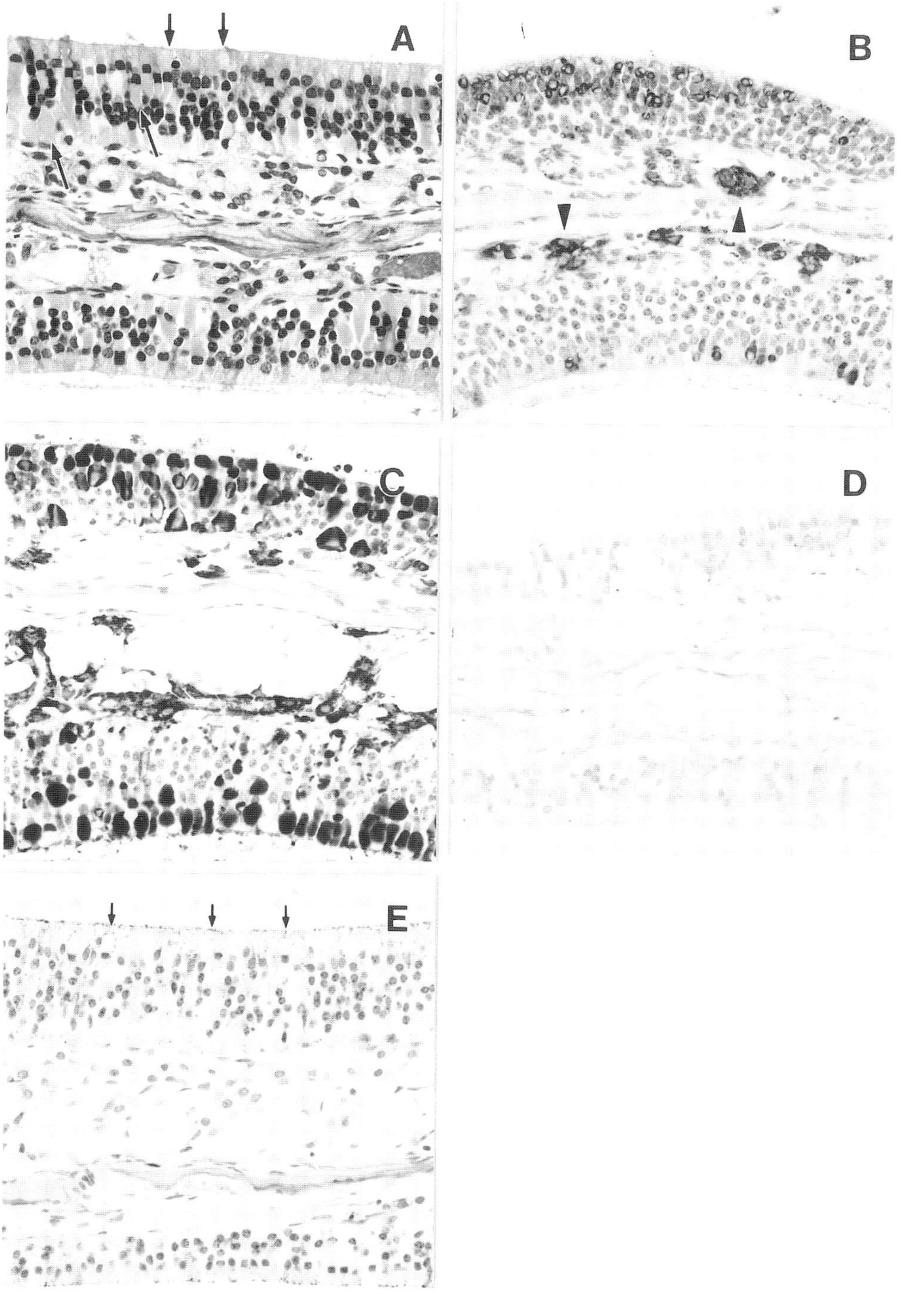
A
B
C
D
E

electron-dense material within dilated rough endoplasmic reticulum. These authors suggested that the globules may consist of proteinaceous, secretory material and represent an adaptive response. However, the precise nature of the material is unknown. Nasal mucosae have a high capacity for xenobiotic metabolism. Although most nasal enzymes have been refractory to classic enzyme inducers, at least one form of carboxylesterase (CE) in the nasal mucosae can be induced by inhalation of toxicants (Lewis et al., 1993). In addition, nasal CEs may be secreted (Bogdanffy et al., 1987; Lewis et al., 1994). Thus, the proteinaceous material in the eosinophilic globules may consist of induced CE and possibly other secretory enzymes.

Initial studies to define the nature of the proteinaceous material examined eosinophilic globules in tissues from CDF(F344)/CrlBR rats exposed to cigarette smoke for the presence of CE and rhodanese (RH), a mitochondrial enzyme that probably is not secreted. Serial sections (5 μm) from air- and cigarette-smoke-exposed rats were stained with hematoxylin and eosin (H&E), or immunohistochemically stained with antibody to a purified 62-kD rabbit CE or antibody to purified rat RH. Olfactory mucosae from identically treated animals were processed for electron microscopy or homogenized. Immunoreactivity with P-450 NMb (rabbit form of 2G1), NMa (another rabbit family 2 isozyme with activity against nicotine), and CE antibodies in homogenates were compared on Western blots.

A high density of CE immunoreactivity was seen within the globules (Figure 1, A–D). The density of the CE staining, the proportion of the globule stained with CE, and the number of globules increased in animals exposed for a longer duration (Figure 1, B and C). Anti-RH immunoreactivity was localized to the apical portion of the sustentacular cell cytoplasm as expected, but no staining occurred within the globules (Figure 1E). In contrast to other reports (see Monticello et al., 1990), transmission electron micrographs of tissues with eosinophilic globules showed dilation of smooth endoplasmic reticulum. Western blots reacted with CE antisera showed a single immunoreactive band of the same molecular mass as the purified CE (62 kD). The intensity of this band was greater in homogenates from animals in groups with many large globules. Isozymes of P-450 (NMa, NMb) that have been identified in high concentrations in nasal tissue were decreased in the tissues rich in eosinophilic globules. Sodium dodecyl sulfate polyacrylamide gel electrphoresis (SDS-PAGE) protein separations indicate that the greatest induction occurs in a 36-kD protein band not yet identified.

The presence of CE within the eosinophilic globules associated with toxicant exposure suggests that the globule response may represent induction of xenobiotic-metabolizing enzymes. Treatment with phenobarbital induces a broad spectrum of xenobiotic-metabolizing enzymes in liver and produces a similar dilation of the endoplasmic reticulum. Localization of CE and other xenobiotic-metabolizing enzymes in tissues containing globules induced by other toxicant exposures is underway to determine the consistency, pattern, and specificity of the induction response.

REFERENCES

Bogdanffy, M. S., Randall, H. W., and Morgan, K. T. 1987. Biochemical quantitation and histochemical localization of carboxylesterase in the nasal passages of the Fischer-344 rat and B6C3F1 mouse. *Toxicol. Appl. Pharmacol.* 88:183–194.

Buckley, L. A., Morgan, K. T., Swenberg, J. A., James, R. A., Hamm, T. E., and Barrow, C. S. 1985.

The toxicity of dimethylamine in F-344 rats and B6C3F1 mice following a 1-year inhalation exposure. *Fundam. Appl. Toxicol.* 5:341–352.
Harkema, J. R., Plopper, C. G., Hyde, D. M., St. George, J. A., Wilson, D. W., and Dungworth, D. L. 1987. Response of the macaque nasal epithelium to ambient levels of ozone: a morphologic and morphometric study of the transitional and respiratory epithelium. *Am. J. Pathol.* 128:29–44.
Lewis, J. L., Nikula, K. J., Novak, R., Chang, I. Y., and Dahl, A. R. 1993. Induction of nasal carboxylesterase following inhalation exposure to pyridine. *Toxicologist* 13:262.
Lewis, J. L., Nikula, K. J., Novak, R., and Dahl, A. R. 1994. Comparative localization of carboxylesterase in F344 rat, beagle dog, and human nasal tissue. *Anat. Rec.* 239:55–64.
Monticello, T. M., Morgan, K. T., and Uraih, L. 1990. Nonneoplastic nasal lesions in rats and mice. *Environ. Health Perspect.* 85:249–274.

CHARACTERIZATION OF NASAL EPITHELIAL DAMAGE AND REPAIR IN RATS FOLLOWING NEONATAL INOCULATION WITH A RAT-ADAPTED INFLUENZA VIRUS

Janice A. Dye

Center for Environmental Medicine and Lung Biology, University of North Carolina, Chapel Hill, North Carolina, USA

Daniel L. Costa, Donna L. Neldon, Gary R. Burleson

U.S. Environmental Protection Agency, Research Triangle Park, North Carolina, USA

Jeffery S. Tepper

ManTech Environmental, Research Triangle Park, North Carolina, USA

Kevin T. Morgan

Chemical Industry Institute of Toxicology, Research Triangle Park, North Carolina, USA

Respiratory tract infections in early childhood have been associated with the development of airway inflammation and obstruction and increases in airway responsiveness. These changes may persist for years beyond the initial infection. Epithelial damage induced by viral infections in the respiratory tract may significantly influence regional uptake of inhaled gases due to local disruption of normal barrier functions, and could thus influence disease processes associated with exposure to air pollutants. During studies designed to develop a rodent model of persistently increased airway responsiveness (AR), using a rat-adapted human influenza virus (RAIV-10), a range of nasal lesions was detected in neonatally infected rats.

Six-day-old F344 rats were intranasally (IN) inoculated with 25 μl of either RAIV-10 in Eagle's minimal essential medium (MEM), or Eagle's MEM alone. At 1, 3, 6, and 13 days postinoculation (DPI), nasal and lower respiratory tract (LRT) tissues were fixed in situ with formalin. Serum samples were also collected to assess

This abstract does not reflect U.S. Environmental Protection Agency (EPA) policy. Supported by EPA CR 817643 through the CEM and Lung Biology.

virus specific antibody production. A second group of F344 rats received two inoculations (at 6 and 30 days of age). Nasal and lower respiratory tract (LRT) histologic changes were assessed at 6 and 13 DPI in half of these subjects, while the remaining rats underwent bronchoprovocation testing at 8–10 wk of age to detect persistent changes in airway responsiveness.

There was minimal morbidity and no mortality noted in these rat pups. Although IN inoculation was used, viral replication in the trachea/lungs was detected via a viral plaque assay. Ultimately, however, the twice RAIV-inoculated animals did not exhibit airway responsiveness increases over those of the sham-inoculated subjects, when challenged with either intravenous acetylcholine or inhaled 5-hydroxytryptamine or substance P.

Following the initial virus inoculation, histologic changes were most severe in the nasal mucosa involving both respiratory and olfactory epithelial regions. Lesions were present as early as 1 DPI, and showed characteristic patterns of progression and resolution with some interanimal variation in severity. Respiratory epithelial damage, which was most severe in the anterior dorsal medial meatus and the adjacent dorsal atrio- and nasoturbinates, was associated with epithelial cell exfoliation and necrosis, mild erosions, suppurative and nonsuppurative inflammation, intraepithelial neutrophilic accumulations, and an overlying intraluminal exudate. Olfactory epithelial damage was multifocal or locally diffuse, with degeneration of sensory cells, epithelial disruption, and minimal inflammation. These lesions were beginning to resolve by 6 DPI, and except for the presence of prominent nasal-associated lymphoid tissue (NALT) regions, lesions were largely repaired by 13 DPI.

No histologic changes were noted in the group receiving a second RAIV inoculation; thus our attempt to induce more significant respiratory disease by infecting these rats more than once during their development (as occurs throughout childhood in many humans) was unsuccessful. Effects were presumably thwarted by specific anti-RAIV immunologic responses in these pups, despite their relatively immature immune status at the time. This assumption is supported by the serology data suggesting an increase in anti-RAIV immunoglobulin production, primarily immunoglobulin A (IgA), at 11 DPI. Although this increase was relatively small, it should be noted that in a previous study of adult F344 rats, relatively large increases in locally produced IgA (i.e., nasal and lung lavage fluid IgA) were associated with only small increases in concurrent serum IgA enzyme-linked immunosorbent assay (ELISA) determinations.

In conclusion, RAIV is a potentially valuable tool for studies of the influence of virally induced nasal epithelial damage on local dosimetry of inhaled air pollutants and subsequent toxic responses.

REGIONAL CYTOTOXICITY AND NONPROTEIN SULFHYDRYL DEPLETION IN THE RAT NASAL CAVITY FOLLOWING ACUTE INHALATION EXPOSURE TO ETHYL ACRYLATE

Clay B. Frederick, David W. Brown, Larry G. Lomax

Rohm and Haas Company, Spring House, Pennsylvania, USA

Methyl, ethyl, and butyl acrylate vapors have been evaluated in chronic inhalation studies (Miller et al., 1985; Reininghaus et al., 1991). No treatment-related increases in tumors were observed in these studies. However, cytotoxicity was observed in the olfactory epithelium of the nasal cavity of both rats and mice at high inhalation exposure concentrations. Toxicity was not observed in the respiratory epithelium.

The current study was designed to explore the association of regional nonprotein sulfhydryl (NPSH) depletion to cytotoxicity in the rat nasal cavity following acute inhalation exposure to ethyl acetate (EA). These data are anticipated to be useful for comparison to the output of a comprehensive physiologically based pharmacokinetic model to describe inhalation exposure to acrylic monomers.

MATERIALS AND METHODS

Groups of male Fischer 344 rats were administered a single nose-only inhalation exposure of 0, 5, 25, or 75 ppm ethyl acrylate (EA) for 1, 3, or 6 h. The animals were randomly divided into groups of 5 animals for either histopathological analysis (immediate or after a 6-wk recovery period) or NPSH analysis. For histopathology, the skull was fixed in formalin and decalcified by standard procedures. The nasal cavity was transversely sectioned according to the method of Young (1981), with an additional section taken rostral to the incisor teeth (Section I in this study). Five sections were prepared for each animal with Levels II–V of this study corresponding to Levels I–IV of Young (1981). The tissue sections were routinely processed and evaluated by light microscopy. Nasal lesions were carefully mapped using a set of representative cross-sectional diagrams provided by Drs. K. T. Morgan, J. S. Kimbell, M. N. Godo, and S. Méry (Chemical Industry Institute of Toxicology, Research Triangle Park, NC). For NPSH analysis, the rat skull was carefully opened along the midline, and nasal epithelial tissue was collected from five regions. The tissues were weighed in tared containers and immediately frozen for subsequent Ellman's assay.

RESULTS

The histopathologic lesions observed in the nasal cavity were confined to the olfactory epithelium (Table 1). Histologic lesions were not observed in rats exposed to EA for 1 h or in rats exposed to 0 or 5 ppm EA for 3 or 6 h. Sections I and II did not show evidence of microscopic changes in any of the animals involved in this study. Rats exposed to 25 ppm or 75 ppm EA for 6 h had sustentacular cell necrosis, olfactory neuron degeneration and desquamation, and epithelial folding in sections III, IV, and V. After a 6-wk recovery period, the olfactory epithelium was normal except for the sporadic occurrence of neuroepithelial rosettelike structures in animals exposed to 25 or 75 ppm EA for 3 or 6 h. These structures were interpreted as evidence for ongoing regeneration of olfactory neuroepithelium.

The concentration response for the depletion of NFSH concentration in various regions of the rat nasal cavity following EA inhalation exposure indicates that significant concentration-related NPSH depletion is observed throughout the nasal cavity (Figure 1). The olfactory epithelium of the dorsal meatus region of the nasal cavity appears to be a preferential site for NPSH depletion based on percentage depletion relative to control values.

TABLE 1. Summary by Treatment Group of the Number of Animals Exhibiting an Olfactory Epithelial Lesion Following Acute Inhalation Exposure to Ethyl Acrylate

Vapor concentration (ppm)	Duration of Exposure				
	1 hr	3 hr	6 hr	3 hr Recovery	6 hr Recovery
0	0/5	0/5	0/5	0/5	0/5
5	0/5	0/5	0/5	0/5	0/5
25	0/5	2/5	3/5	0/5	0/5
75	0/5	5/5	5/5	0/5	0/5

DISCUSSION

The acid released by the carboxylesterases in olfactory sustentacular cells (Bogdanffy et al., 1987) may contribute to the toxic effects induced by organic esters following inhalation exposure. However, comparison of the dosimetry for toxicity of the esters relative to the corresponding acids would suggest that the ester moiety contributes to the toxic response (Frederick et al., 1994). Extrapolation of these results to humans will require evaluation of the sensitivity of human (or primate) olfactory tissue to acids and esters.

The localization of the olfactory toxic effects of EA to the dorsal meatus region of the nasal cavity appears to be primarily based on dosimetry considerations. This conclusion is supported by the more widespread distribution of olfactory toxicity observed at high inhalation concentrations of EA (Miller et al., 1985). Therefore, an understanding of regional air flow in the rat nasal cavity and its relationship to dosimetry is critical to understanding the olfactory toxicity. Only with this level of understanding can a reasonable comparison be made in the dosimetry in rodent

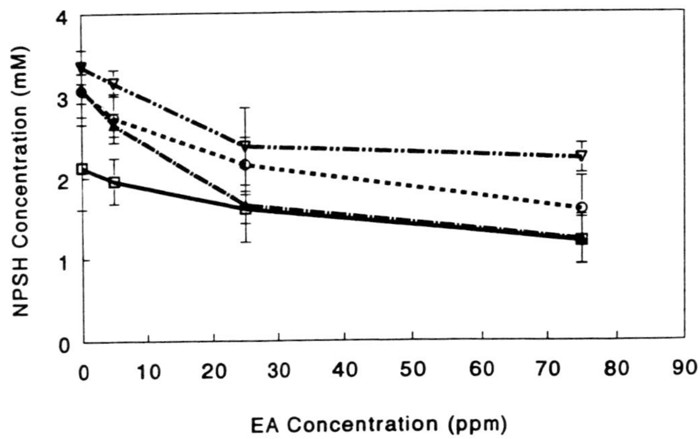

FIGURE 1. Concentration-response for the regional depletion of NPSH in the rat nasal cavity following inhalation exposure to EA for 3 h. The regions of the nasal cavity that were evaluated included respiratory epithelium on the septum (open square), respiratory epithelium from the lateral walls (open circle), olfactory epithelium from the dorsal meatus region (solid triangle), and olfactory epithelium from the septum (open triangle). The NPSH depletion was most severe in the dorsal meatus olfactory epithelium, which also exhibited cytotoxicity.

and human olfactory epithelium under various inhalation exposure scenarios, since the available data suggest that there is a very different air flow pattern in the human nose compared to the rodent nose (Kimbell et al., 1993; Hahn et al., 1993).

REFERENCES

Bogdanffy, M. S, Randall, H. W., and Morgan, K. T. 1987. Biochemical quantitation and histochemical localization of carboxylesterase in the nasal passages of the Fischer-344 rat and the B6C3F1 mouse. *Toxicol. Appl. Pharmacol.* 88:183–194.

Frederick, C. B., Udinsky, J. R., and Finch, L. 1994. The regional hydrolysis of ethyl acrylate to acrylic acid in the rat nasal cavity. *Toxicol. Lett.* 70:49–56.

Hahn, I., Scherer, P. W., and Mozell, M. M. 1993. Velocity profiles measured for air flow through a large-scale model of the human nasal cavity. *J. Appl. Physiol.* 75:2273–2287.

Kimbell, J. S., Gross, E. A., Joyner, D. R., Godo, M. N., and Morgan, K. T. 1993. Application of computational fluid dynamics to regional dosimetry of inhaled chemicals in the upper respiratory tract of the rat. *Toxicol. Appl. Pharmacol.* 121:253–263.

Miller, R. R., Young, J. T., Kociba, R. J., Keyes, D. G., Bodner, K. M., Calhoun, L. L., and Ayres, J. A. 1985. Chronic toxicity and oncogenicity bioassay of inhaled ethyl acrylate in Fischer 344 rats and B6C3F1 mice. *Drug Chem. Toxicol.* 8:1–42. Reininghaus, W., Koestner, A., and Klimisch, H.-K. 1991. Chronic toxicity and oncogenicity bioassay of inhaled methyl acrylate and n-butyl acrylate in Sprague-Dawley rats. *Food Chem. Toxicol.* 29:329–339.

Young, J. T. 1981. Histopathological examination of the rat nasal cavity. *Fundam. Appl. Toxicol.* 1:309–312.

OLFACTORY MUCOSAL DAMAGE FOLLOWING DERMAL EXPOSURE TO DICHLOBENIL: PROPOSED MECHANISM AND FACTORS TO CONSIDER WHEN PREDICTING POTENTIAL HUMAN RISK

Mary Beth Genter, Nora J. Deamer

Department of Toxicology, NC State University, Raleigh, North Carolina, USA

2,6-Dichlorobenzonitrile (dichlobenil) is a herbicide active ingredient shown by other investigators to bind avidly to and cause degeneration of the olfactory mucosa in the C57Bl mouse at doses of >12 mg/kg ip (Brandt et al., 1990). Studies to elucidate the mechanism of olfactory toxicity of this compound and to determine whether humans exposed to dichlobenil are at risk of olfactory mucosal damage are ongoing in this laboratory.

EXPERIMENTAL DESIGN AND RESULTS

To eliminate one confounding factor in the extrapolation of potential health risks of this chemical to humans, we performed a dermal study in the same strain of mice, thus mimicking a more relevant route of exposure for agricultural chemicals. Dermal administration of dichlobenil resulted in olfactory epithelial damage at a dose of 50 mg/kg and greater (no-observed-adverse-effect level for olfactory mucosal damage following dermal exposure is 25 mg/kg; application rates in agricultural settings range from 1.4 to 20 lb active ingredient/acre [U.S. EPA, 1987]).

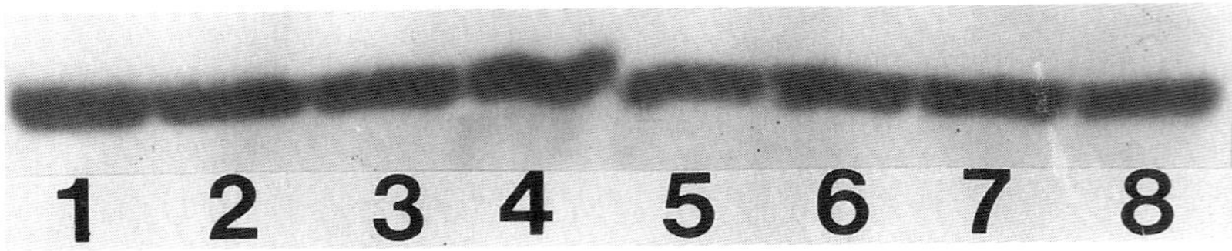

FIGURE 1. Western blot showing relative amounts of microsomal epoxide hydrolase (mEH) in rat olfactory mucosal (lanes 1–4) and liver (lanes 5–8) microsomes. Note that on a per milligram protein basis (8 μg microsomal protein/lane), olfactory mucosa and liver contain comparable amounts of mEH. Anti-mEH antibody was the kind gift of Dr. Franz Oesch.

Another important issue in extrapolating risk across species is understanding the mechanism of toxic action so that the likelihood of similar behavior of the compound in humans can be predicted. Dichlobenil is metabolized to a number of different metabolites (Griffiths et al., 1966), and the major hydroxylated metabolite is 3-hydroxy (3-OH) dichlobenil, a species that itself is not an olfactory toxicant (unpublished observation from this laboratory). Incubation of either olfactory mucosal or liver microsomes with [ring-^{14}C]dichlobenil in the presence of an NADPH generating system, followed by silica-gel thin-layer chromatography of a chloroform-methanol-acetic acid extract, revealed that both liver and olfactory mucosal microsomes produce the same metabolites, the predominant of which are the 3- and 4-hydroxy dichlobenil metabolites. Preliminary data suggest that olfactory mucosal microsomes may produce a greater amount of the 3-OH metabolite than liver. Clearly, identification of the cytochrome P-450 isozyme(s) responsible for dichlobenil metabolism in the rodent and characterization of the human olfactory mucosal P-450 profile are necessary in order to complete this objective.

As the bioactivation of dichlobenil is clearly P-450 dependent (Brandt et al., 1990; Eriksson & Brittebo, 1991), we decided to test the hypothesis that olfactory mucosal P-450 enzymes activate dichlobenil to a toxic metabolite and that some crucial detoxication enzyme was absent from the rodent olfactory mucosa. Because previous metabolic studies were strongly suggestive of an epoxide intermediate in the metabolism of dichlobenil, microsomal epoxide hydrolase (mEH) was the enzyme that we decided to investigate. Although both liver and olfactory mucosa contain comparable levels of mEH on a per milligram of tissue basis, as shown in Figure 1, we recently demonstrated that the region of the olfactory mucosa (i.e., that lining the dorsal medial meatus) affected by dichlobenil, as well as numerous other olfactory toxicants (e.g., bromobenzene, 4-ipomeanol), is immunohistochemically void of mEH (Figure 2); epoxides are likely metabolites in the case of bromobenzene and 4-ipomeanol as well.

DISCUSSION

The results of these studies, coupled with the work of previous investigators, have demonstrated the importance of determining the presence, distribution, and relative ratios of both mEH and the P-450 isozyme responsible for the bioactivation of these compounds in humans in order to predict potential human risk due to systemic exposure to dichlobenil and other similar olfactory toxicants. We have pro-

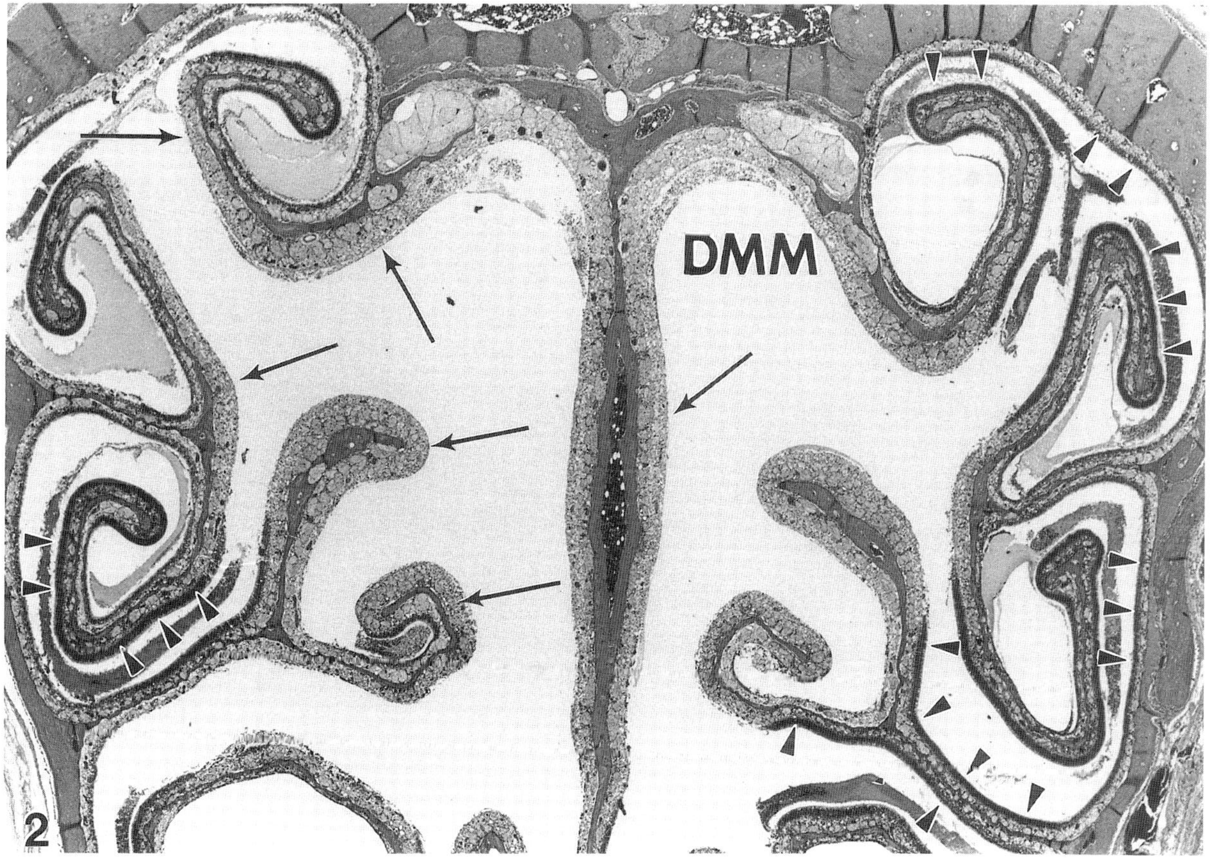

FIGURE 2. Hematoxylin and eosin (H&E) stained transverse section revealing the distribution of olfactory mucosal degeneration in the Long-Evans rat following ip administration of 100 mg/kg dichlobenil or 200 mg/kg IDPN. Note that the epithelium lining the dorsal medial meatus (DMM) is most severely affected (note absence of epithelium in regions denoted by arrows), while the epithelium lining the more ventral and lateral airways is largely spared. The distribution of undamaged mucosa in this micrograph corresponds well with the immunohistochemical distribution of microsomal epoxide hydrolase (denoted by arrowheads in the micrograph).

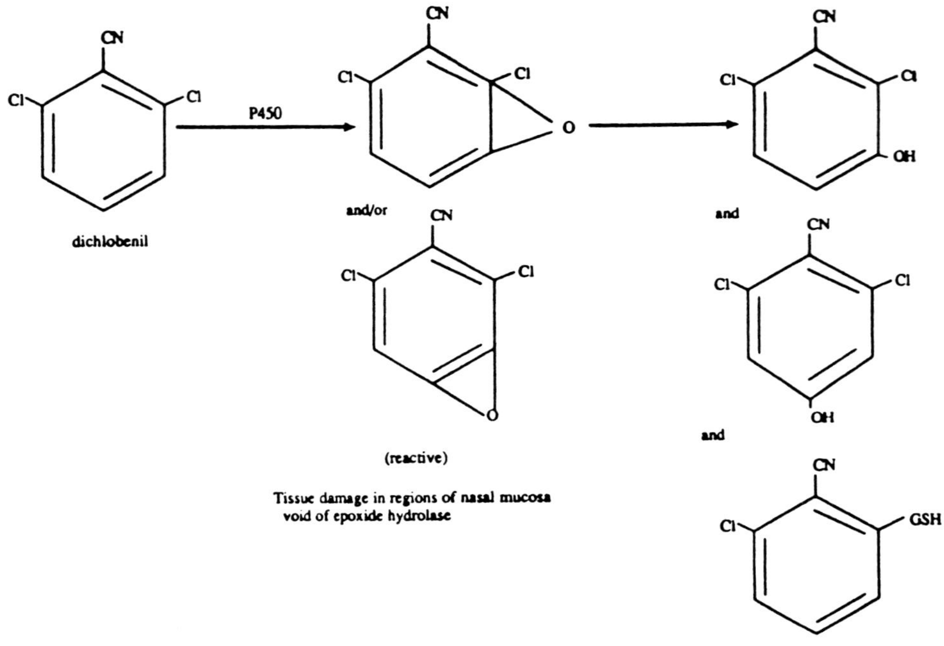

FIGURE 3. Proposed mechanism by which dichlobenil produces olfactory mucosal damage.

posed a scheme (outlined in Figure 3), involving activation of dichlobenil by cytochrome P-450 enzymes, production of a reactive epoxide, and incomplete detoxication of the epoxide in the affected regions of the nasal cavity, as the mechanism underlying the olfactory toxicity of dichlobenil. In this scheme, the liver and the ventral and lateral regions of the nasal cavity would be spared tissue damage by dichlobenil because of the presence of sufficient mEH to detoxify the epoxides produced. However, in the mucosa lining the dorsal medial airway, insufficient (if any) mEH is present to detoxify the epoxides produced by cytochrome P-450 activity, allowing reaction between the epoxides and tissue macromolecules, ultimately leading to olfactory mucosal destruction.

REFERENCES

Brandt, I., Brittebo, E. B., Feil, V. J., and Bakke, J. E. 1990. Irreversible binding and toxicity of the herbicide dichlobenil (2,6-dichlorobenzonitrile) in the olfactory mucosa of mice. *Toxicol. Appl. Pharmacol.* 103:491–501.

Eriksson, C., and Brittebo, E. B. 1991. Metabolic activation of the herbicide dichlobenil in the olfactory mucosa of mice and rats. *Chem. Biol. Interact.* 79:165–177.

Griffiths, M. H., Moss, J. A., Rose, J. A., and Hathaway, D. E. 1966. The comparative metabolism of 2,6-dichlorothiobenzamide (Prefix) and 2,6-dichlorobenzonitrile in the dog and the rat. *Biochem. J.* 98:770–781.

U. S. Environmental Protection Agency. 1987. Pesticide Fact Sheet (Dichlobenil). Washington, DC: Office of Pesticide Programs.

OLFACTORY TOXICITY OF METHIMAZOLE: RESULT OF INTRAPERITONEAL AND ORAL DOSING STUDIES

Mary Beth Genter, Theodor A. Zainal, Nora J. Deamer

Department of Toxicology, NC State University, Raleigh, North Carolina, USA

Methimazole (Figure 1) is a compound used in the treatment of hyperthyroid conditions in humans at an oral dose of approximately 1–2 mg/kg/day. The *S*-oxidation of methimazole is carried out by the flavin-containing monooxygenase (FMO) system, with no apparent contribution from the cytochrome P-450 enzymes. Because the secondary amine nitrogen of β,β'-iminodipropionitrile (IDPN; Figure 1) is purportedly an FMO substrate (Morandi et al., 1987), we used methimazole (300 mg/kg.ip, single dose; rats were sacrificed 36 h later) to try to competitively inhibit the FMO-mediated metabolism of IDPN and thereby decrease the olfactory toxicity of IDPN. Instead of achieving this goal, we discovered that methimazole itself is an olfactory mucosal toxicant, causing complete destruction of the olfactory epithelium in the male Long-Evans rat nasal cavity at a dose of 300 mg/kg.

EXPERIMENTAL DESIGN AND RESULTS

Dose-response analyses were carried out for methimazole administered by ip injection and by gavage (to mimic the human route of exposure). The no-observed-adverse-effect level (NOAEL) for olfactory mucosal damage is 25 mg/kg by ip administration (Table 1). Interestingly, the severity of the lesion (i.e., the area of the olfactory mucosa damaged) diminished as dose decreased (Figure 2), with the lesion present only in the epithelium lining the dorsal medial meatus of the nasal cavity in rats receiving lower doses of methimazole (25 or 50 mg/kg ip). Results of the gavage study revealed that methimazole is also an olfactory toxicant by oral

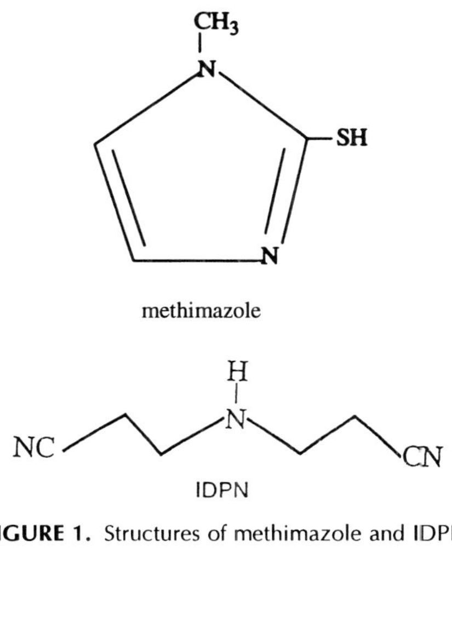

FIGURE 1. Structures of methimazole and IDPN.

TABLE 1. Methimazole-Induced Olfactory Mucosal Destruction: Dose-Response Studies

ip Exposure		Oral (gavage) exposure	
Dose (mg/kg)	Response	Dose (mg/kg)	Response
0	—[a]	0	—
2	— NOAEL	2	—
		5	—
		10	—
25	DMM[b] damage	25	— NOAEL[c]
50	DMM damage	50	DMM damage
100	DMM + peripheral damage		
150	DMM + peripheral damage		
300	100% mucosal destruction		

[a]No damage.
[b]Epithelium lining the dorsal medial meatus.
[c]No observed adverse effect level for the respective route of exposure.

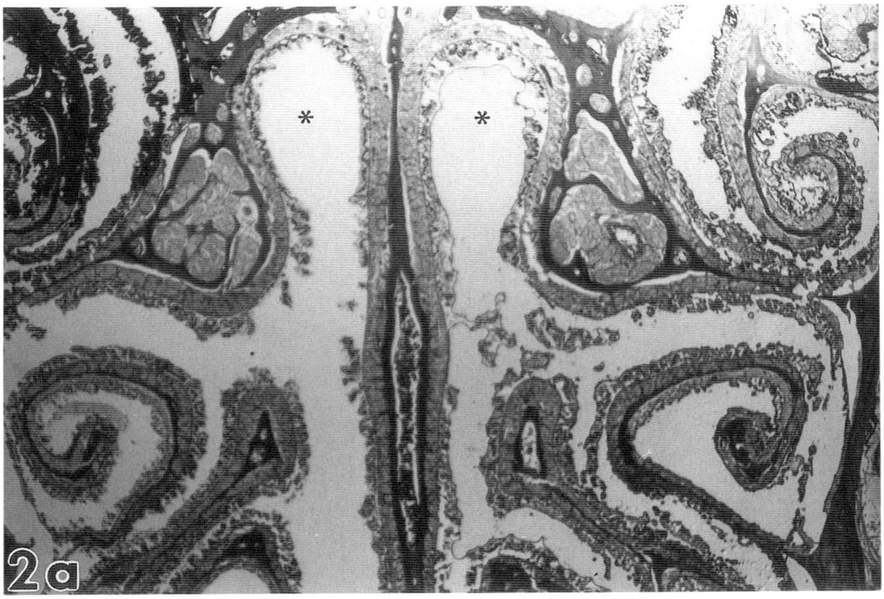

FIGURE 2. Hematoxylin and eosin (H&E) stained transverse sections from male Long-Evans rats receiving a single ip dose of methimazole: (a) 300 mg/kg. Note that the severity of the lesion, in terms of the proportion of the olfactory mucosa affected, decreases as the dose of methimazole is reduced. Although 100% of the olfactory mucosa is destroyed in rats receiving 300 mg/kg, a dose of 25 mg/kg results in destruction of only the mucosa lining the dorsal medial airways (asterisk).

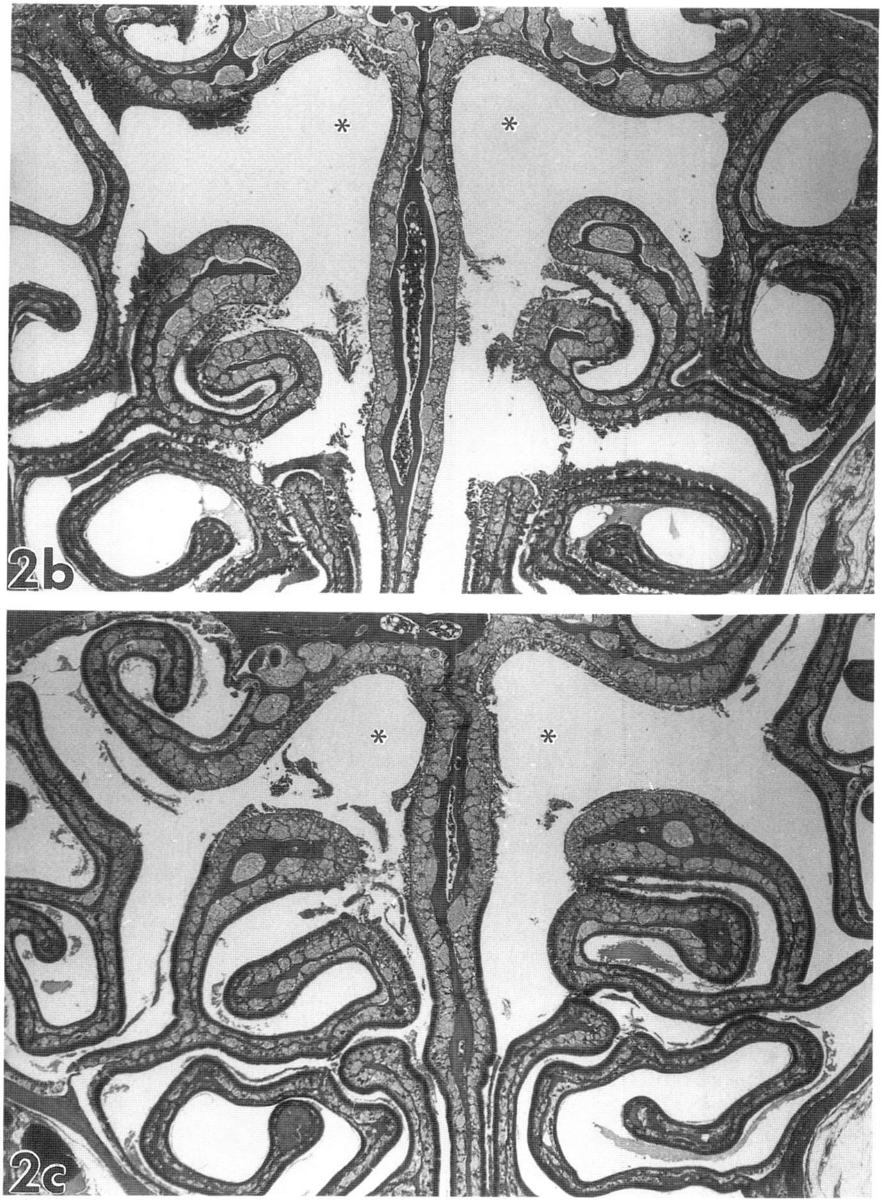

FIGURE 2. (*Continued*) Hematoxylin and eosin (H&E) stained transverse sections from male Long-Evans rats receiving a single ip dose of methimazole: (b) 100 mg/kg and (c) 25 mg/kg.

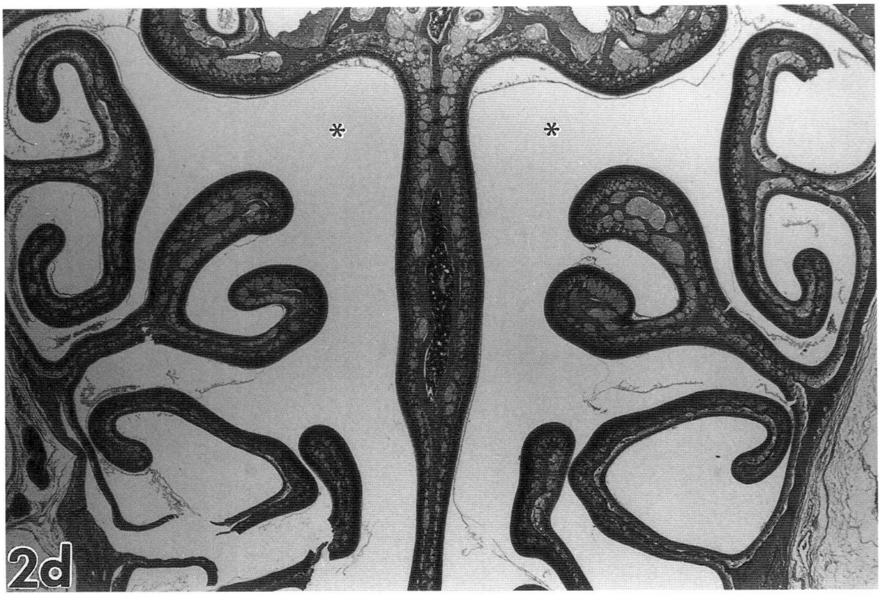

FIGURE 2. (*Continued*) Hematoxylin and eosin (H&E) stained transverse sections from male Long-Evans rats receiving a single ip dose of methimazole: (d) 0 mg/kg (control).

exposure. As shown in Table 1, methimazole (single dose in corn oil; rats were sacrificed 36 h later) caused olfactory mucosal damage at an oral dose of 50 mg/kg (highest dose tested by this route), with a NOAEL of 25 mg/kg.

Immunohistochemical studies revealed that both the mouse lung and mouse liver forms of FMO are detectable in the rat olfactory mucosa, with the lung form distributed equally throughout the mucosa and localized in the Bowman's glands. The liver form was likewise present in the Bowman's glands, but was found primarily in the Bowman's glands of the mucosa lining the dorsal medial meatus.

DISCUSSION

Previous work with methimazole *S*-oxidation reactions in vitro revealed that an unknown reactive intermediate is produced; this species is eliminated with the addition of glutathione to the reaction mixture (Sabourin & Hodgson, 1984). We hypothesize that this as yet unidentified reactive species is also the olfactory toxic metabolite. Work with compounds structurally related to methimazole is currently underway in this laboratory in order to test this hypothesis.

There are a number of factors to consider when determining potential risk of olfactory mucosal damage to the human olfactory mucosa by methimazole. Is the olfactory toxicity of methimazole dependent upon FMO-mediated metabolism? Are FMO enzymes present in the human olfactory mucosa? As the studies detailed here are acute (i.e., single administration), it would be interesting to know whether the chronic toxicity of methimazole differs from the acute toxicity (i.e., are the effects cumulative)? If the reactive metabolite is indeed formed through FMO mediated oxidation and detoxified by conjugation with glutathione, it would also be important to know whether the human olfactory mucosa contains sufficient glutathione to detoxify the putative reactive toxic metabolite.

REFERENCES

Morandi, A., Gambetti, P., Arora, P. K., and Sayre, L. M. 1987. Mechanism of neurotoxic action of β,β'-iminodipropionitrile (IDPN): *N*-hydroxylation enhances neurotoxic potency. *Brain Res.* 437:69–76.

Sabourin, P. J., and Hodgson, E. 1984. Characterization of the purified microsomal FAD-containing monooxygenase from mouse and pig liver. *Chem. Biol. Interact.* 51:125–139

EXPOSURE OF THE OLFACTORY SYSTEM TO TOXIC COMPOUNDS: STRUCTURAL AND FUNCTIONAL CONSEQUENCES

Lloyd Hastings, Anastasia Andringa, Marian L. Miller

Department of Environmental Health, University of Cincinnati, Cincinnati, Ohio, USA

Although research on the effects of toxic agents on the olfactory tissue has increased in recent years, there is still little information available concerning the effects of these compounds on the function of the system, that is, olfaction. Numerous studies have examined nasal pathology resulting from exposure to olfactory toxicants; many others have investigated the mechanisms underlying the toxicity, most notably metabolism of xenobiotics by the nasal mucosa (Barrow, 1986). These studies, though important, fail to relate the degree of morphological or biochemical insult to olfactory deficits. The primary objective of this research was to investigate the effects of a variety of toxic compounds on olfactory function.

The effects of these toxic compounds—from metals to solvents to pesticides—on the olfactory system were studied using a multidisciplinary approach assessing behavioral, morphological, and biochemical parameters. Carnosine content of the olfactory bulb was used as a biochemical marker for the integrity of the olfactory epithelium. [Carnosine is produced and transported to the olfactory bulb only by mature, functional primary sensory neurons (Harding et al., 1978). Loss of primary sensory neurons is mirrored by a corresponding decrease of carnosine in the olfactory bulb.] By examining all three aspects—morphology, behavior (olfaction), and biochemistry (carnosine)—we have been able to obtain a more complete understanding of the impact of toxic compounds on the olfactory system.

The same general approach was used for all compounds tested. Rats were trained on a task to allow assessment of olfactory function. These rats, along with other rats for histology and carnosine assay, were then exposed and evaluated at selected time points for up to 4 mo postexposure.

Three techniques for evaluating olfactory function were employed. These were (1) a buried-food pellet task (used to assess the effects of methyl bromide, MeBr); (2) an absolute threshold measurement based on a conditioned suppression para-

This research was supported by NIEHS grants ES 04099 and P30 ES06096.

digm (cadmium); and (3) an absolute threshold measurement and an odor discrimination task based on a discrimination paradigm [MeBr, 3-methylindole (3-MI), dichlobenil (DCB), and nickel]. Odor stimuli for the last two paradigms were generated by a precision flow-dilution olfactometer.

For morphological assessment, fixative was instilled into the nares of the animals after euthanasia, but before dissection. Tissues were placed in the same isoosmolar fixative (phosphate-buffered, 2.0% paraformaldehyde/2.5% glutaraldehyde, with calcium salts and dextrose) for at least 24 h; tissues were not decalcified. After postfixation in 1% osmium tetroxide for 2 h and dehydration in an ascending series of ethanols and propylene oxide, blocks were embedded in Spurr and oriented to provide frontal sections (perpendicular to the basement membrane). One micrometer-thick sections were cut with glass knives on an MT-7000 Ultra (RMC) ultramicrotome, placed on glass slides and stained with toluidine blue.

Tissue carnosine concentration was analyzed using a high-performance liquid chromatography (HPLC) system that rapidly removed the amino acids of no concern. Following dissection, the olfactory bulbs were placed on ice, weighed, and stored at –70°C. The frozen tissues were homogenized and the homogenates centrifuged for 5 min in a refrigerated (2°C) centrifuge. A 50-μl aliquot of *ortho*-ophthaldialdehyde/2-mercaptoethanol (OPT) reagent was added to an aliquot of each supernatant, the mixture rapidly vortexed, and incubated for 1.0 min. Following incubation, an aliquot of the mixture was injected onto a Beckman 4.5 cm Ultrasil ODS (10 μm C_{18}) precolumn followed by a Supelco 15 cm Supelcosil LC-18 (5 μm C_{18}) analytical column. The OPT-derivatized amino acids, specifically L-carnosine, were detected using a chromatograph fluorescence detector with an excitation wavelength of 330 nm and an emission wavelength window of 400–500 nm.

The following exposure protocols were used for the various compounds:

MeBr: 200 ppm, 4 h/day for either 1 or 8 days; inhalation
Cadmium: 500 μg/m3, 6 h/day, 5 days/wk for 20 wk; inhalation
Nickel: 625 μg/m3, 6 h/day for 16 days; inhalation
Chromium: 200 μg/m3, 6 h/day for 40 days; inhalation
3-MI: 400 mg/kg, single ip injection
DCB: 50 or 200 mg/kg, single/multiple ip injection

RESULTS

Methyl Bromide

On the first day following MeBr exposure, extensive damage to the olfactory epithelium as well as greatly impaired olfactory function were observed. However, even with continuous exposure, olfactory function quickly recovered. This recovery occurred despite persistent thinning and disorganization of the olfactory epithelium and long-term depression of carnosine levels in the olfactory bulb. When MeBr-exposed rats were tested using more sophisticated techniques, no changes were seen in odor discrimination and only minimal changes were observed in absolute thresholds.

Cadmium

Despite being exposed to cadmium oxide aerosol at 10 times the threshold limit value (TLV), rats displayed no alteration in olfactory function. No morphologcal changes were observed in the olfactory epithelium, nor was there a significant decrease in carnosine content in the olfactory bulb. While there was no effect on olfactory function, olfactory bulbs from cadmium-exposed rats showed a significant increase in cadmium content. Cadmium exposure was also sufficient to produce lesions in both the heart and lungs.

Nickel

Exposure to nickel sulfate aerosol at six times the TLV resulted in alterations in nasal morphology (severe reduction in sustentacular cells), a significant reduction in carnosine in the olfactory bulb, but no effect on olfactory function. The olfactory epithelium regained its normal appearance soon after cessation of exposure.

Chromium

Forty days of exposure to sodium dichromate at four times the TLV produced no effects on nasal morphology. Olfactory function was not tested, nor was olfactory bulb carnosine content measured.

3-Methylindole

Rats exposed to a single injection of 3-MI (400 mg/kg) suffered extensive and prolonged damage to the olfactory epithelium. This was accompanied by a long-lasting anosmia as measured by an odor discrimination task; a similar discrimination task based on visual cues was not affected. Olfactory function eventually recovered, but not until almost 3 mo after the initial exposure. Carnosine content in the olfactory bulb at 30 days postexposure showed a severe depression. The most striking observation was the profound depletion of Bowman's glands in the lamina propria after 3-MI exposure.

Dichlobenil

Rats exposed to DCB (50 mg/kg, ip) experienced severe morphological damage, but no effect was observed on olfactory function. Subsequent injections of 50 or 100 mg/kg were not cumulative and produced no additional effect. A single injection of 200 mg/kg resulted in morphological damage and a transient anosmia. An injection of 300 mg/kg produced morbidity, but again only a transient anosmia was noted. Bowman's glands were also affected but not as severely as with 3-MI. During recovery there was a pronounced increase in globose basal cells, the presumptive progenitor cells of the olfactory receptors.

These results are summarized in Table 1.

SUMMARY AND CONCLUSIONS

Although a wide variety of compounds produces severe toxic responses in the rat olfactory system, olfactory function usually recovers quite quickly. Recovery of function also usually occurs long before there is substantial regeneration of the olfactory epithelium. One factor that appears to be a major determinant concerning how quickly the system recovers is the extent of damage to the Bowman's glands and/or damage to structures beneath the basement membrane. Rats exposed to MeBr, which denudes the olfactory epithelium above the basement membrane,

TABLE 1. Various Effects of Toxic Compounds on the Olfactory System

Toxicant	Target cells	Bowman's glands	Olfactory epithelium	Lamina propria	Carnosine (% control)	Functional recovery
MeBr, 200 ppm/ 4 h	Sustentacular cells	Present	Absent	Inflammation	5–10%	1–4 d
3-MI, 400 mg/kg, ip	Bowman's glands	Decreased/ absent	Absent	Inflammation	<5%	45–60 d
DCB, 200 mg/kg, ip	Bowman's glands	Decreased	Increased globose	Inflammation	Not measured	4 d
Cadmium, 500 μg/m^3 20 wk	None	Present	No loss	Minor inflammation	>90%	No loss
Nickel 625 μg/m^3 16 d cells	Sustentacular	Present	No loss	None	>80%	No loss
Chromium 200μg/m^3 40 d	None	Present	No loss	None	Not measured	Not measured

quickly regain their sense of olfaction. On the other hand, rats exposed to 3-MI, which destroys the Bowman's glands along with the olfactory epithelium, take months to recover their sense of smell. These results suggest that rodents, with their well-developed olfactory system, possess an exceedingly resilient and redundant system. This poses somewhat unique problems for investigations of olfactory toxicology. This is especially true concerning study of the functional capacity of the system after toxic insult.

REFERENCES

Barrow, C. S. 1986. Toxicology of the nasal passages. New York: Hemisphere Publishing.

Harding, J. W., Getchell, T. V., and Margolis, F. 1978. Denervation of the primary olfactory pathway in mice. V. Long-term effects of intranasal $ZnSO_4$ irrigation on behavior, biochemistry, and morphology. *Brain Res.* 140:271–285.

REGIONAL DIFFERENCES IN THE EFFECTS OF CIGARETTE SMOKE ON STORED MUCOSUBSTANCES IN RESPIRATORY EPITHELIUM OF F344 RAT NASAL SEPTUM

Jon A. Hotchkiss, W. Alun Evans, Kirk R. Maples, Bean T. Chen, Gregory L. Finch, Jack R. Harkema

Inhalation Toxicology Research Institute, Albuquerque, New Mexico, USA

Inhalation of mainstream cigarette smoke (MCS) results in significant morphologic alterations to surface epithelia lining the upper and lower respiratory tract in laboratory rodents. Acute exposure of rats to MCS results in the appearance of increased numbers of mucous goblet cells in bronchi and intrapulmonary airways (Jones et al., 1978). This study was designed to assess the effects of repeated MCS exposure on the quantity of stored intraepithelial mucosubstances (IM) within the respiratory epithelium lining the anterior nasal septum of F344 rats.

Twelve male F344/N rats, 12–13 wk of age, were used in this study. Eight rats were exposed to MCS in Hazelton 2000 exposure chambers (Hazelton Systems, Inc., Aberdeen, MD) over a 2-wk period. Smoke exposure concentrations were determined by gravimetric analysis of filter samples taken from the exposure chamber. During the first week, rats were exposed to 150 mg/m3 (total particulate matter) MCS, 6 h/day, for 2 days, followed by 250 mg/m3 MCS, 6 h/day, for 3 days. The second week, rats were exposed to 250 mg/m3 MCS, 6 h/day, for 4 days. At all other times, the rats were maintained in filtered air. Four control rats were maintained in an identical whole-body exposure chamber supplied continuously with filtered air during the 2-wk MCS exposure.

All rats were sacrificed 18 h after the end of the last exposure. The nasal cavity of each rat was fixed, decalcified, and processed for light microscopy. Nasal tissue sections taken from immediately posterior to the incisor teeth were histochemically stained using alcian blue (pH 2.5)/periodic acid Schiff's sequence (AB/PAS) to identify acidic and neutral IM.

Qualitative assessment of AB/PAS-stained sections suggested that there were regional differences in the response of the respiratory epithelium lining the nasal septum to MCS exposure. Therefore, the dorsal, mid, and ventral septum were morphometrically analyzed as separate intranasal regions (Figure 1). The volume density of IM within the nasal respiratory epithelium lining these septal regions was estimated using the method of Harkema et al. (1987). The data were expressed as the mean volume density (V_s; nl/mm^2 basal lamina) of AB/PAS-positive IM ± standard error of the mean (SEM).

No significant MCS-induced change in the quantity of IM was observed when the entire septum was analyzed as a single anatomic structure (Figure 2). Rats exposed to MCS and sacrificed 1 d after the last exposure had 270% more IM in the dorsal septum, 58% less IM in the mid septum, and the same amount of IM in the ventral septum, compared to air-exposed controls. The loss of IM in the mid septum was associated with MCS-induced squamous metaplasia of the surface epithelium.

Repeated exposures of F344 rats to MCS resulted in significant alterations in stored IM and focal squamous metaplasia in several regions of the surface epithelium lining the anterior nasal septum. The dorso-ventral variation in response to MCS exposure may be due to inherent differences in the susceptibility of the surface epithelium in each region to MCS-induced injury, due to regional differences in the cell populations present, or may be due to differences in the dose of MCS-toxicants delivered to the epithelium in those regions.

Kimbell et al. (1993) have reported a threefold difference in the velocity of air flowing over the surface epithelia lining the three regions analyzed in the present study (mid septum > ventral septum > dorsal septum). Considering the regional differences in nasal airflow and respiratory epithelial morphology, our data can be interpreted as follows. The dorsal septal region experiences the lowest airflow and

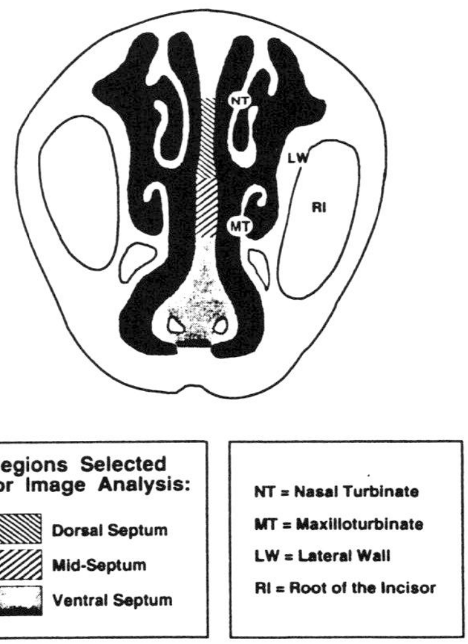

FIGURE 1. Coronal section of the proximal airway of the F344 rat showing the regions of the nasal septum analyzed for stored intraepithelial mucosubstances.

thus receives the smallest total dose of MCS. MCS-induced injury/irritation resulted in locally increased expression of airway mucins. The ventral septal epithelium received a greater dose of MCS due to higher regional airflow patterns, but also had significantly more IM prior to MCS exposure. The dose of MCS to this region was insufficient to alter the mucociliary phenotype of the epithelium. The mid septum

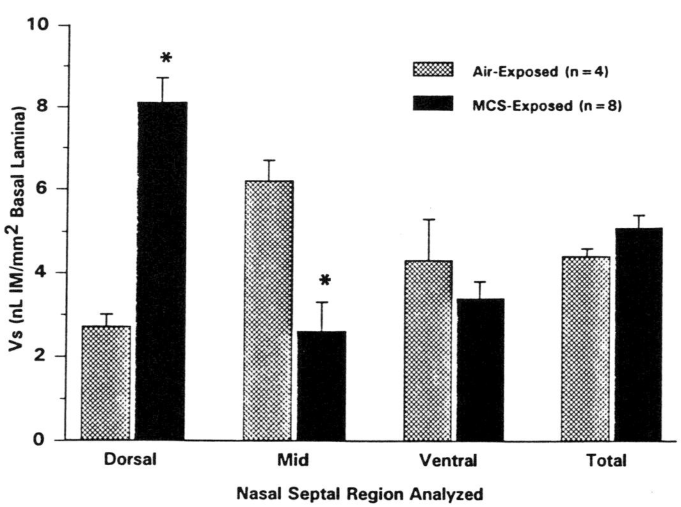

FIGURE 2. Effect of repeated mainstream cigarette smoke (MCS) exposure on the amount of stored intraepithelial mucosubstances in the anterior nasal septum of F344 rats. Rats were exposed to filtered air or MCS for 9 days and sacrificed 1 day after the last exposure. The values shown represent the mean ± standard error of the mean. Asterisk indicates significantly different than air-exposed control (p ± 0.05).

was subjected to the highest airflow and the highest dose of MCS. We hypothesize that MCS-induced toxicity in this region resulted in significant respiratory epithelial cell injury and subsequent squamous metaplasia.

In summary, the results of this study indicate that the mucus-producing apparatus in the rat anterior nasal cavity is markedly altered by repeated exposure to MCS, but the type and magnitude of the alteration depend on the intranasal location of the airway epithelium examined.

REFERENCES

Harkema, J. R., Plopper, C. G., Hyde, D. M., St. George, J. A., Wilson, D. W., and Dungworth, D. L. 1987. Response of the macaque nasal epithelium to ambient levels of ozone: A morphologic and morphometric study of the transitional and respiratory epithelium. *Am. J. Pathol.* 128:29–44.

Jones, R., Phil, M., and Reid, L. 1978. Secretory hyperplasia and modification of intracellular glycoprotein in rat airways induced by short periods of exposure to tobacco smoke and the effect of anti-inflammatory agent phenylmethyloxidiazole. *Lab. Invest.* 39:41–49.

Kimbell, J. S., Gross, E. A., Joyner, D. R., Godo, M. N., and Morgan, K. T. 1993. Application of computational fluid dynamics to regional dosimetry of inhaled chemicals in the upper respiratory tract of the rat. *Toxicol. Appl. Pharmacol.* 121:253–263.

STUDIES OF THE TOXICITY OF VINYL ACETATE AND ACETALDEHYDE AND INDUCTION OF DNA-PROTEIN CROSS-LINK FORMATION IN RAT NASAL EPITHELIAL TISSUES

Jim R. Kuykendall, Marie L. Taylor, Matthew S. Bogdanffy

Haskell Laboratory for Toxicology and Industrial Medicine, E. I. du Pont de Nemours and Company, Newark, Delaware, USA

Vinyl acetate monomer is used to make polymers for the paint, adhesive, and paper-board industries. Vinyl acetate is a nasal carcinogen in rats exposed by inhalation for 2 yr to 200 or 600 ppm, but not 50 ppm (Bogdanffy et al., 1993). Vinyl acetate is hydrolyzed to acetic acid and acetaldehyde, a rearrangement product of vinyl alcohol (Simon et al., 1985) in a reaction catalyzed by carboxylesterase, an enzyme known to be present at high quantities in specific cell types of the nasal mucosa (Bogdanffy et al., 1987). Both acetic acid (Stott & McKenna, 1985) and acetaldehyde (Appelman et al., 1982) are nasal toxicants and acetaldehyde was shown to cause DNA-protein cross-links (DPXs) (Lam et al., 1986) and tumors (Woutersen et al., 1986) in the rat nasal mucosa. Tumors arose from both respiratory and olfactory regions of the nose and included squamous-cell carcinoma, carcinoma in situ, and benign endophytic and exophytic papilloma.

METHODS

The formation and stability of DPXs formed by incubation of pUC13 plasmid DNA and calf thymus histones with acetaldehyde and vinyl acetate were studied (Kuykendall & Bogdanffy, 1992a, 1992b) using a modification of a filter binding

This work was supported in part by the Society of the Plastics Industry, Inc.

assay (Trask et al., 1984). In assays containing 1–100 m*M* acetaldehyde, DPXs were formed at a rate of 127 DPXs/plasmid molecule/mmol acetaldehyde in a reaction containing 1 μg histones and 0.33 μg DNA incubated at 37°C for 1 h.

RESULTS AND DISCUSSION

Acetaldehyde-induced DPXs were unstable at 37°C, with loss of up to 75% by 8 h. Stimulation of 100 m*M* acetaldehyde-induced DPX formation was observed at lower pH, with the accumulation of three- and twofold higher levels of DPXs at pH 5 and 6, respectively, than at pH 7.5 (100 m*M* phosphate buffer). Induction of DPX formation by 1–100 m*M* vinyl acetate in the presence of rat liver microsomes (at 37°C over 3 hrs) followed S-phase enzymatic kinetics, with a rate of formation of 1.1 DPXs/plasmid molecule/mmol vinyl acetate/μg microsomal protein/μg DNA.

Vinyl acetate did not cause formation of DPXs in the absence of microsomes. A carboxylesterase inhibitor, bis(*p*-nitrophenyl) phosphate (BNPP), blocked DPX formation by vinyl acetate and microsomes. These preliminary studies suggested that rat liver microsome-activated vinyl acetate induces plasmid DNA-histone cross-links, in vitro, through esterase-mediated metabolism. Since nasal tissues contain high levels of carboxylesterase, tumorigenesis may be related to in situ production of the hydrolysis products acetaldehyde and acetic acid. This possibility was studied by incubating rat nasal turbinate explants or isolated rat nasal epithelial cells with acetaldehyde and vinyl acetate in vitro (Kuykendall et al., 1993). Vinyl acetate was cytotoxic to both respiratory and olfactory tissues (turbinates) in vitro at 50–200 m*M*, but not 25 m*M*, after 2 h of exposure. Pretreatment of rats with BNPP attenuated the cytotoxic effects and metabolism of vinyl acetate in both tissue types. Semicarbazide, an aldehyde scavenger, was unable to protect the tissues from vinyl acetate-induced cytotoxicity. When the metabolites were tested, acetic acid, but not acetaldehyde, was cytotoxic to both tissues.

The induction of DPX formation by acetaldehyde and vinyl acetate in isolated rat nasal epithelial cells was detected using a sodium dodecyl sulfate (SDS)/KCl precipitation technique (Zhitkovich & Costa, 1992). Endogenous cross-link levels ranged from 0.5 to 2.0% of total DNA and were considered background. Epithelial cells isolated from both olfactory and respiratory turbinates exhibited dose- and time-dependent increases in DPX formation when exposed to 10–150 m*M* acetaldehyde for 1–2 h at 37°C. Similarly, respiratory and olfactory epithelial cells exposed to 5–75 m*M* vinyl acetate for 1–2 h accumulated up to 15-fold higher cross-link levels than untreated cells. Pretreatment of nasal cells with 1 m*M* BNPP reduced 25 m*M* vinyl acetate-induced DPX formation by over 75% in both tissues. In assays designed to study DPX formation in isolated epithelial cells, as well as plasmid-histone DPX assays, vinyl acetate appears to induce much higher levels of DPXs at equimolar doses than acetaldehyde. This is thought to be related to stimulation of acetaldehyde-induced DPX formation by the pH-lowering effect of acetic acid production (via vinyl acetate hydrolysis). In plasmid-histone cross-link assays containing 100 m*M* vinyl acetate and rat liver microsomes, the pH was found to decrease from 7.5 to levels near 6.0 over a 3-h incubation. Addition of acetic acid to plasmid-histone assays containing 100 m*M* acetaldehyde resulted in a dose-dependent increase in DPX formation.

These data support a hypothesis that carboxylesterase-mediated hydrolysis of vinyl acetate is necessary to generate the active intracellular cross-linking agent, acetaldehyde, and the cytotoxic metabolite, acetic acid. In addition, the possibility that acetic acid production may stimulate acetaldehyde-induced DPX formation may also serve to indirectly implicate acetic acid in the genotoxic process. The results provide a basis for a mechanistic model of vinyl acetate-induced nasal toxicity.

REFERENCES

Appelman, L. M., Woutersen, R. A., and Feron, V. J. 1982. Inhalation toxicity of acetaldehyde in rats. I. Acute and subacute studies. *Toxicology* 23:293–307.

Bogdanffy, M. S., Randall, H. W., and Morgan, K. T. 1987. Biochemical quantitation and histochemical localization of carboxylesterase in the nasal passages of the Fischer-344 rat and B6C3F1 mouse. *Toxicol. Appl. Pharmacol.* 88:183–194.

Bogdanffy, M. S., Dreef-van der Meulen, H. C., Beems, R. B., Feron, V. J., Cascieri, T. C., Tyler, T. R., Vinegar, M. B., and Rickard, R. W. 1994. Chronic toxicity and oncogenicity inhalation study with vinyl acetate in the rat and mouse. *Fundam. Appl. Toxicol.* 23:215–229.

Kuykendall, J. R., and Bogdanffy, M. S. 1992a. Reaction kinetics of DNA-histone crosslinking by vinyl acetate and acetaldehyde. *Carcinogenesis* 13:2095–2100.

Kuykendall, J. R., and Bogdanffy, M. S. 1992b. Efficiency of DNA-histone crosslinking induced by saturated and unsaturated aldehydes in vitro. *Mutation Res.* 283:131–136.

Kuykendall, J. R., Taylor, M. L. and Bogdanffy, M. S. 1993. Cytotoxicity and DNA-protein crosslink formation in rat nasal tissues exposed to vinyl acetate are carboxylesterase-mediated. *Toxicol. Appl. Pharmacol.* 23:283–292.

Lam, C.-W., Casanova, M., and Heck, H. d'A. 1986. Decreased extractability of DNA from proteins in the rat nasal mucosa after acetaldehyde exposure. *Fundam. Appl. Toxicol.* 6:541–550.

Simon, P., Filser, J. G., and Bolt, H. M. 1985. Metabolism and pharmacokinetics of vinyl acetate. *Arch. Toxicol.* 57:191–195.

Stott, W. T., and McKenna, M. J. 1985. Hydrolysis of several glycol ether acetates and acrylate esters by nasal mucosal carboxylesterase in vitro. *Fundam. Appl. Toxicol.* 5:399–404.

Trask, D. K., DiDonato, J. A., and Muller, M. T. 1984. Rapid detection and isolation of covalent DNA/protein complexes: application to topoisomerase I and II. *EMBO J.* 3:671–767.

Woutersen, R. A., Appelman, L. M., VanGarderen-Hoetmer, A., and Feron, V. J. 1986. Inhalation toxicity of acetaldehyde in rats. III. Carcinogenicity study. *Toxicology* 41:213–231.

Zhitkovich, A., and Costa, M. 1992. A simple, sensitive assay to detect DNA-protein crosslinks in intact cells and in vivo. *Carcinogenesis* 13:1485–1489.

REGIONAL HISTOPATHOLOGY OF THE MOUSE NASAL CAVITY FOLLOWING TWO WEEKS OF EXPOSURE TO ACRYLIC ACID FOR EITHER 6 OR 22 HOURS PER DAY

Larry G. Lomax, David W. Brown, Clay B. Frederick

Rohm and Haas Co., Spring House, Pennsylvania, USA

Chronic inhalation studies have been conducted with rodents for several of the esters of acrylic acid (AA) (Miller et al., 1985; Reininghaus et al., 1991) and a subchronic inhalation study has been conducted with AA (Miller et al., 1981). No treatment-related increases in tumors were observed in these studies. However, cytotoxicity was observed in the olfactory epithelium in the dorsal meatus region of

This study was supported by the Basic Acrylic Monomer Manufacturers (BAMM).

the nasal cavity of both rats and mice at high exposure levels of the monomers. No toxic effects were observed in either the nasal respiratory epithelium or other organ systems and tissues.

This study was designed to address the application of concentration times time ("$C \times T$ product" or Haber's rule) extrapolation relative to nasal irritation and to evaluate the reversibility of nasal lesions.

MATERIALS AND METHODS

Groups of female B6C3F1 mice were administered whole-body inhalation exposure to acrylic acid for 2 wk at concentrations of 0, 5, or 25 ppm for 6 or 22 h/day. An additional group of mice was exposed for 2 wk to 25 ppm for 4.4 h/day. Animals from each dose group were scheduled for necropsy upon the termination of exposure or after a 6-wk recovery period. For histopathology, the skull was fixed in formalin and decalcified by standard procedures. The nasal cavity was transversely sectioned according to the method of Young (1981) with an additional section taken rostral to the incisor teeth (Section I in this study). Five sections were prepared for each animal with Levels II–V of this study corresponding to Levels I–IV of Young (1981). The tissue sections were routinely processed and evaluated by light microscopy. Nasal lesions were carefully mapped using a set of representative nasal cavity cross-sectional diagrams (modified for mouse nasal cavity anatomy) provided by Drs. K. T. Morgan, J. S. Kimbell, M. N. Godo, and S. Méry (Chemical Industry Institute of Toxicology, Research Triangle Park, NC).

RESULTS AND DISCUSSION

The olfactory epithelium in the dorsal meatus region was the only target tissue in the nasal cavity of mice inhaling 5 ppm AA for 22 h or 25 ppm AA for 4.4, 6, or 22 h (Figure 1). The histopathologic lesions observed were disorganization and atrophy of the olfactory epithelium, basal-cell hypertrophy, necrosis and desquamation of olfactory epithelium, and Bowman's gland degeneration and attenuation. Histologic lesions were not observed in control mice or in mice exposed to 5 ppm AA for 6 h/day. Section I did not show evidence of microscopic changes in any of the animals involved in this study. After a 6-wk recovery period, the olfactory epithelium was normal for all animals in the study except those exposed to 25 ppm for 22 h/day. These animals exhibited regions of respiratory metaplasia (replacement of sensitive olfactory epithelium with resistant respiratory-like epithelium) that have previously been reported for rodent subchronic and chronic studies of nasal irritants that induce olfactory toxicity.

Three of the treatment groups (Table 1) had a very similar $C \times T$ product (5 ppm × 22 h = 110 ppm h/day; 25 ppm × 4.4 h = 110 ppm h/day; 25 ppm × 6 h = 150 ppm h/day), and these groups all had a very similar incidence and severity of lesions in the nasal cavity. The toxic effects observed for these treatment groups were fully reversible in the recovery animals. It can therefore be concluded that in this $C \times T$ range, an approximately linear extrapolation of C or T would give a proportionate change in a toxic response that is completely reversible based on histopathological analysis.

However, the other two treatment groups indicated that there were limitations to the proportionality of the $C \times T$ relationship to nasal cavity lesions. The lowest

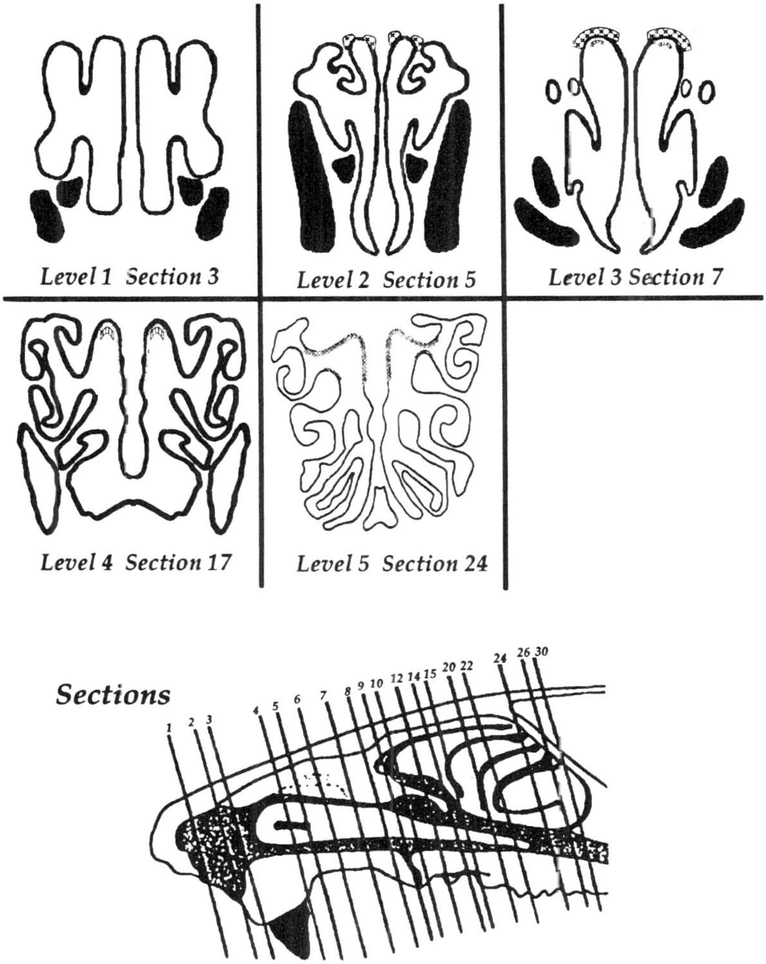

FIGURE 1. Composite map of lesions in the mouse nasal cavity following inhalation exposure to 25 ppm AA for 2 wk at 22 h/day. The lesions observed were limited to the dorsal meatus region and included disorganization and atrophy of the olfactory epithelium (shaded), Bowman's gland degeneration and atrophy (cross-hatched), and basal-cell hyperplasia (vertically hatched). Following a 6-wk recovery period, some replacement of olfactory epithelium with respiratory epithelium was observed in the dorsal meatus region ("respiratory metaplasia").

dose group exhibited no histopathological indications of toxicity at a $C \times T$ product (5 ppm × 6 h = 30 ppm h/day) that was only fourfold less than the dose groups exhibiting histopathology. Conversely, the highest $C \times T$ product group (25 ppm × 22 h = 550 ppm h/day) exhibited a high incidence of severe olfactory damage characterized by involvement of the olfactory epithelium in the dorsal meatus region extending onto the ethmoid turbinates. The recovery animals for this group exhibited replacement of the damaged epithelium by resistant respiratory-like epithelium. The increased incidence, severity, involvement, and lack of complete

TABLE 1. Summary of Histopathology Findings by Nasal Cavity Region, Exposure Time, and Dose for Animals Examined Immediatley Following 2 Weeks of Inhalation Exposure to Acrylic Acid

Exposure concentration	0 ppm AA		5 ppm AA		25 ppm AA		
Daily duration of exposure	6 h	22 h	6 h	22 h	4.4 h	6 h	22 h
Section 1							
Olfactory atrophy and disorganization	0/10	0/10	0/10	0/10	0/10	0/10	0/10
Necrosis and desquamation of olfactory epithelium	0/10	0/10	0/10	0/10	0/10	0/10	0/10
Bowman's gland degeneration and attenuation	0/10	0/10	0/10	0/10	0/10	0/10	0/10
Basal-cell hyperplasia with squamous differentiation	0/10	0/10	0/10	0/10	0/10	0/10	0/10
Section 2							
Olfactory atrophy and disorganization	0/10	0/10	0/10	7/10	9/10	8/10	9/10
Necrosis and desquamation of olfactory epithelium	0/10	0/10	0/10	1/10	7/10	1/10	0/10
Bowman's gland degeneration and attenuation	0/10	0/10	0/10	0/10	2/10	0/10	9/10
Basal-cell hyperplasia with squamous differentiation	0/10	0/10	0/10	0/10	1/10	0/10	9/10
Section 3							
Olfactory atrophy and disorganization	0/10	0/10	0/10	10/10	10/10	10/10	10/10
Necrosis and desquamation of olfactory epithelium	0/10	0/10	0/10	1/10	6/10	1/10	1/10
Bowman's gland degeneration and attenuation	0/10	0/10	0/10	4/10	3/10	1/10	10/10
Basal-cell hyperplasia with squamous differentiation	0/10	0/10	0/10	1/10	0/10	0/10	5/10
Section 4							
Olfactory atrophy and disorganization	0/10	0/10	0/10	10/10	9/10	10/10	10/10
Necrosis and desquamation of olfactory epithelium	0/10	0/10	0/10	0/10	7/10	1/10	2/10
Bowman's gland degeneration and attenuation	0/10	0/10	0/10	0/10	1/10	0/10	8/10
Basal-cell hyperplasia with squamous differentiation	0/10	0/10	0/10	0/10	1/10	0/10	1/10
Section 5							
Olfactory atrophy and disorganization	0/10	0/10	0/10	8/10	6/10	7/10	8/10
Necrosis and desquamation of olfactory epithelium	0/10	0/10	0/10	0/10	4/10	2/10	3/10
Bowman's gland degeneration and attenuation	0/10	0/10	0/10	0/10	0/10	0/10	3/10
Basal-cell hyperplasia with squamous differentiation	0/10	0/10	0/10	0/10	1/10	0/10	0/10

recovery indicate that exposure to a $C \times T$ product in the 550 ppm h/day region is associated with a more severe response than would be predicted by the linear extrapolation of effects in the 100–150 ppm h/day region.

REFERENCES

Miller, R. R., Ayres, J. A., Jersey, G. C., and McKenna, M. J. 1981. Inhalation toxicity of acrylic acid. *Fundam. Appl. Toxicol.* 1:271–277.

Miller, R. R., Young, J. T., Kociba, R. J., Keyes, D. G., Bodner, K. M., Calhoun, L. L., and Ayres, J. A. 1985. Chronic toxicity and oncogenicity bioassay of inhaled ethyl acrylate in Fischer 344 rats and B6C3F1 mice. *Drug Chem. Toxicol.* 8:1–42.

Reininghaus, W., Koestner, A., and Klimisch, H.-K. 1991. Chronic toxicity and oncogenicity bioassay of inhaled methyl acrylate and n-butyl acrylate in Sprague-Dawley rats. *Food Chem. Toxicol.* 29:329–339.

Young, J. T. 1981. Histopathological examination of the rat nasal cavity. *Fundam. Appl. Toxicol.* 1:309–312.

COMPARISON AND REVIEW OF CHEMICALS CAUSING NASAL TOXICITY

Robert H. Ross, Po-Yung Lu

Biomedical and Environmental Information Analysis Section,
Oak Ridge National Laboratory, Oak Ridge, Tennessee, USA

As one of the body's front-line defense mechanisms against xenobiotics, the nasal tissues come into contact with a variety of chemicals. In most instances this encounter goes unnoticed, but when certain chemicals are inhaled at critical concentrations and for sufficient duration, the result is damage to nasal tissues. Historically, most consideration has been given to pulmonary effects, but current interest and investigations are also focusing on nasal tissues. This article analyses the results of studies in which 26 chemicals caused noncarcinogenic nasal toxicity and 13 chemicals caused nasal tumors.

BACKGROUND

The information for this article was taken primarily from the U.S. Environmental Protection Agency Integrated Risk Information System (IRIS) and the Health Effects Assessment Summary Tables (HEAST). These data sources do not contain every chemical known to be a nasal toxicant, but do have data for a sufficient number of chemicals to provide an adequate presentation of relevant information regarding nasal toxicity.

RESULTS

The results are presented in Table 1 (noncarcinogenic effects) and Table 2 (carcinogenic effects). As shown in Tables 1 and 2, the U.S. EPA source for most of the information came from IRIS, which is an on-line computerized data file available

TABLE 1. Noncarcinogenic Nasal Toxicity of Selected Chemical

Chemical (CAS No.)	Species	RfC (mg/m^3)	Nasal effect
Acrolein (107-02-8)	Rat	0.00002	Metaplasia, neutrophilic infiltration of nasal epithelium
Acrylic acid (79-10-7)	Mouse	0.0003	Lesions of mucosa
Ammonia (7664-41-7)	Human	0.1	Rhinitis
1,3-Dichloropropene (542-75-6)	Mouse	0.02	Hypertrophy and hyperplasia of mucosa
Epichlorhydrin (106-89-8)	Rat, mouse	0.001	Lesions in epithelium
Hexamethylene 1,6-diiso-cynate (822-06-0)	Rat	0.00001	Hyperplasia and metaplasia
Hydrogen chloride (7647-01-0)	Rat	0.007	Hyperplasia
Hydrogen sulfide (7783-06-4)	Mouse	0.0009	Inflammation of mucosa
4,4-Methylene diphenyl)-isocyanate (101-68-8)	Rat	0.00002	Lesions
Propylene (115-07-1)	Rat	None	Metaplasia and hyperplasia
Propylene oxide (75-56-9)	Rat	0.03	Nest like folds
Sarin (107-44-8)	Dog	None	Rhinorrhea
Triethylamine (121-44-8)	Rat	0.007	Inflammation
Vinyl acetate (108-05-4)	Rat	0.2	Hyperplasia, metaplasia, atrophy

Note. RfC, reference concentration; IRIS, Integrated Risk Information System; HEAST, Health Effects Assessment Summary Tables; RQ, Reportable Quantity.

TABLE 2. Oncogenic Nasal Toxicity of Selected Chemicals

Chemical (CAS No.)	Species	Inhalation unit risk ($(\mu g/m^3)^{-1}$)	Nasal effect
Bis(chloromethyl) ether (542-88-1)	Rat	0.062	Carcinoma of turbinates and epithelium
1,2-Dibromoethane (106-9304)	Rat	0.00022	Carcinoma of nasal cavity
Dimethyl sulfate (77-78-1)	Hamster, rat, mouse	None	Tumors
Epichlorhydrin (106-89-8)	Rat	0.0000012	Tumors of nasal cavity
Formaldehyde(50-00-0)	Rat	0.000013	Tumors
Hydrazine (302-01-2)	Rat	0.0049	Adenoma/adenocarcinoma
Nickel (refinery dust) (7440-2-0)	Human	0.00024	Tumors
Nickel subsulfide (12035-72-2)	Human	0.00048	Tumors
Propylene oxide (75-56-9)	Mouse	0.0000037	Nasal cavity hemangioma or hemangiosarcoma

Note. Information obtained from U.S. Environmental Protection Agency's Integrated Risk Information System (IRIS). NA, not available.

Other effects	Effective concentration (mg/m³))	Protocol	EPA source
None	0.92	13 wk, 6 h/days, 5 days/wk	IRIS
None	14.9	13 wk, 6 h/days, 5 days/wk	IRIS
Lesions in lungs, pneumonia	6.4	Intermittent, 12 yr	IRIS
Hyperplasia in forestomach and urinary bladder	83.5	2 yr, 6 h/days,5 days/wk	IRIS
None	95	61–62 exposures 6 h/day, 5 days/wk	IRIS
Eye irritation	0.17	Chronic	IRIS
Larynx and trachea hyperplasia	15	Lifetime, 6 h/day, 4.7 days/wk	IRIS
Kidney, liver, and spleen wt. changes	110	13 wk, 6 h/day, 5 days/wk	IRIS
None	0.2 (NOAEL)	24 mo, intermittent	HEAST
Fibrosis of heart, thyroid and lymph node hyperplasia	8605	103 wk, 6 h/day, 5 days/wk	RQ
None	71	123–124 wk. 6 h/day 5 days/wk	IRIS
Numerous	0.04	6 mo, 6 h/day, 5 or 7 days/wk	
None	4139	28 wk, 6 h/day, 5 days/wk	IRIS
None	704	104 wk, 6 h/day, 5 days/wk	IRIS

Other effects	Effective concentration (mg/m³))	Protocol	EPA source group
Carcinoma of lung	0.47	Intermittent, up to 100 6-h exposures	A
Numerous	307	Intermittent, 88–103 wk	B2
Tumors of rat lung and thorax	2.6	Intermittent, 15 mo	B2
None	114	30 days 6 h/day	B2
None	6.9	24 mo, 6 h/days, 5 days/wk	B1
Thyroid adenocarcinoma	1.3	1 yr, 6 h/day, 5 days/wk	B2
Lung tumors	NA	At least 13 yr	A
Lung tumors	NA	Intermittent, occupational	A
Mammary gland tumors in females	950	2 yr, intermittent	B2

through the National Library of Medicine. One entry came from HEAST, which is published three times a year by the Office of Research and Development of the U.S. EPA. Both IRIS and HEAST contain toxicity values, such as Reference Concentrations and slope factors, and supporting information. The data from one other entry were from a Reportable Quantity Document. Reportable Quantity documents are prepared by the Office of Solid Waste and Energency Response of the U.S. EPA. They are intended to provide health-related limits for emergency actions (environmental releases) under Section 101 of the Comprehensive Emergency Response, Compensation and Liability Act.

DISCUSSION AND CONCLUSIONS

The following are the more significant observations/conclusions from this analysis.

- Very few chemicals have sufficient human data to quantify exposures.
- The rat was the most commonly used test species.
- Both organic and inorganic chemicals are capable of damaging nasal tissues.
- In addition to nasal effects, damage to other tissues was recorded for 16/26 chemicals causing noncancer effects.
- The chemical with the lowest Reference Concentration (RfC) was hexamethylene 1,6-diisocyanate, 1×10^{-5} mg/m^3, followed by acrolein and 4,4-methylene diphenyl isocyanate, 2×10^{-5} mg/m^3 (see Table 1).
- The chemical with the lowest effective concentration was sarin, 0.04 mg/m^3 (no RfC available), followed by hexamethylene 1,6-diisocyanate, 0.17 mg/m^3 (see Table 1).
- The chemical with the highest RfC was vinyl acetate, 0.2 mg/m^3 (see Table 1).
- The chemical with the highest effective concentration was propylene, 8605 mg/m^3 (no RfC available).
- Both an RfC and unit risk were available for epichlorhydrin and propylene oxide.
- The chemical of most concern based on unit risk is *bis*(chloromethyl)ether, 6.2×10^{-2} $(\mu g/m^3)^{-1}$.
- The chemical of least concern based on unit risk is epichlorhydrin, 1.2×10^{-6} $(\mu g/m^3)^{-1}$, followed closely by propylene oxide, 3.7×10^{-6} $(\mu g/m^3)^{-1}$.
- The EPA cancer group distribution was as follows: three chemicals in Group A, one in Group B1, and nine in Group B2.
- Tumors were present both in nasal tissues and other tissues for 9/13 chemicals.

INTERIM EPA POLICY FOR PARTICLE SIZE AND LIMIT CONCENTRATION ISSUES IN INHALATION TOXICITY STUDIES

John E. Whalan, John C. Redden

Health Effects Division (7509C), U.S. Environmental Protection Agency, Washington, DC, USA

U.S. Environmental Protection Agency (EPA) recommendations for particle size (25% of the particles <1 μm) and limit concentration (5 mg/L) have been difficult to achieve and have complicated the performance and interpretation of inhalation toxicity studies. The Pesticide Reregistration Rejection Rate Analysis–Toxicology (U.S. EPA, 1993) revealed that 17.4% of the inhalation studies reviewed between 1985 and 1991 were rejected. Of these, 55% failed to meet the particle size criteria, and 29% failed to define an LC50 and/or achieve a limit concentration.

The Society of Toxicology Inhalation Specialty Section Position Paper (1992) considered the particle size requirement to be unrealistic because physical laws limit the minimum size of particles at a 5 mg/L concentration. Since deposition in the respiratory tract is influenced by particle size and chemical characteristics, it is impossible to predict the most harmful particle size. Particles should deposit throughout the rodent respiratory tract. The SOT recommended limit testing at the highest achievable concentration (≤5 mg/L), with particles between 1 and 4 μm mass median aerodynamic diameter (MMAD).

Raabe and co-workers studied the correlation between aerodynamic particle sizes and deposition sites in anesthetized Long-Evans rats (Raabe et al., 1977) and unanesthetized Fischer 344 rats (Raabe et al., 1988) exposed nose-only to radiolabeled monodisperse spherical aluminosilicate particles. These data are portrayed in Figures 1 and 2.

Pulmonary deposition in rats remains fairly constant for 1–4 μm particles, but nasopharyngeal deposition sharply increases for particles >2 μm. For particles >1 μm, the number of particles reaching the pulmonary region decreases, but these larger particles have a greater mass. For example, a 3-μm particle has 27 times the mass of a 1 μm particle, and thus 27 times the toxic potential if toxicity is mediated by mass.

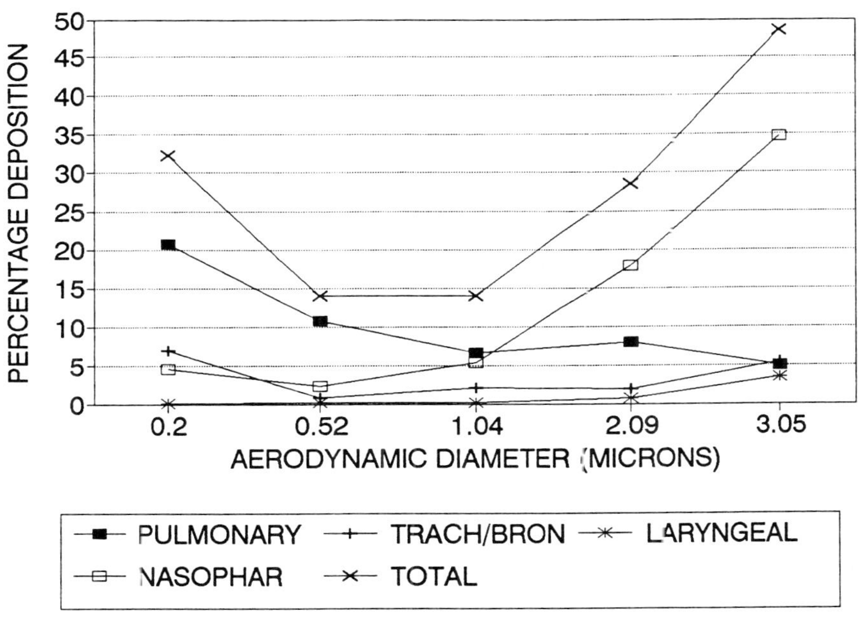

FIGURE 1. Correlation between particle size and deposition site in Long-Evans hooded rats (from Raabe et al., 1977, with permission).

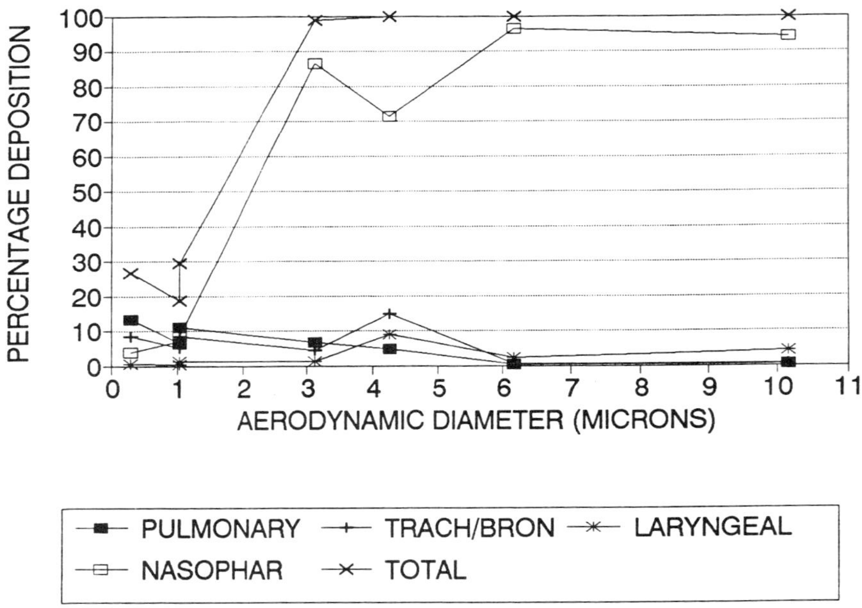

FIGURE 2. Correlation between particle size and deposition site in Fischer 344 rats (from Raabe et al., 1988, with permission).

In order to maximize deep lung deposition for pesticides, an MMAD of 2–3 μm is recommended (personal communication, Dr. H. C. Yeh, ITRI, Albuquerque, NM). Lewis et. al. (1989) recommend a test aerosol with an MMAD $\leq$3 μm (geometric standard deviation $\leq$3), acknowledging that 1 or 2 μm MMAD is preferable. The nature of the material may make this impossible to attain. For hygroscopic aerosols, 1 μm MMAD is preferred.

Before the release of revised inhalation guidelines in 1995, the Health Effects Division (HED) drafted the following interim policy for particle size and limit concentration. If these recommendations cannot be achieved, the study report should detail the reasons why.

1. Aerodynamic particle sizes are acceptable if MMADs are 1–4 μm in an acute study and 1–3 μm in a subchronic or chronic study. Hygroscopic materials should be small enough to allow for pulmonary deposition once the particles grow in the respiratory tract. The report should include particle size distribution data, MMAD values, and a description of the generation methods and equipment.
2. The analytical limit concentration for aerosols, gases, and vapors in acute studies is 2 mg/L. If a concentration of 2 mg/L cannot be achieved, the maximum attainable concentration should be used. Toxicity Categories III and IV will be revised as follows: Category III, >0.5 to 2 mg/L; and Category IV, >2 mg/L.
3. It is recommended, but not required, that nose-only or head-only exposure be used for aerosol studies to minimize oral exposure. Individual housing must be used during whole-body exposure.

4. Whenever the test article is a formulation, the analytical concentration must be reported for the total formulation, and not just for the active ingredient (a.i.). If there is some difficulty in measuring chamber analytical concentration due to precipitation, nonhomogeneous mixtures, volatile components, or other factors, additional analyses of inert components may be necessary.

REFERENCES

Lewis, T. R., Morrow, P. E., McClellan, R. O., Raabe, O. G., Kennedy, G. L., Schwetz, B. A., Goehl, T. J., Roycroft, J. H., and Chhabra, R. S. 1989. Establishing aerosol exposure concentrations for inhalation toxicity studies. *Toxicol. Appl. Pharmacol.* 99:377–383.

Raabe, O. G. Yeh, H. C., Newton, G. J., Phalen, R. F., and Velasquez, D. J. 1977. Deposition of inhaled monodisperse aerosols in small rodents. In *Inhaled Particles IV*, ed. W. H. Walton, pp. 3–21. New York: Pergamon Press.

Raabe, O. G., Mohamed, A. A., Teague, S. V., and Rasolt, A. 1988. Regional deposition of inhaled monodisperse coarse and fine aerosol particles in small laboratory animals. *Ann. Occup. Hyg.* 32(1):53–63.

Society of Toxicology, Technical Committee of the Inhalation Specialty Section. 1992. Recommendations for the conduct of acute inhalation limit tests. *Fundam. Appl. Toxicol.* 18:321–327.

U.S. EPA. 1993. Pesticide Reregistration Rejection Rate Analysis—Toxicology. July, EPA 738-R-93-004.

U.S. EPA REFERENCE CONCENTRATION FOR ACRYLONITRILE

Jeffrey S. Gift, Annie M. Jarabek

U.S. EPA, Office of Health and Environmental Assessment,
Environmental Criteria and Assessment Office, Research Triangle Park,
North Carolina, USA

Acrylonitrile (ACN) is one of the 189 chemicals listed as a hazardous air pollutant under the 1990 Clean Air Act Amendments (CAAA). Efforts are currently underway within EPA to complete health assessments, including inhalation reference concentrations (RfCs), for all CAAA designated pollutants. The RfC is an estimate of a daily inhalation exposure to the human population (including sensitive subgroups) that is likely to be without an appreciable risk of deleterious noncancer effects during a lifetime.

The methods used to derive inhalation reference concentrations (RfCs) are described in detail elsewhere (Jarabek et al., 1990). Briefly, an RfC derivation begins with determination of a no-observed-adverse-effect level (NOAEL) and a lowest-observed-adverse-effect level (LOAEL) from the exposure levels of a given individual study. The RfC methods then require conversion by dosimetric adjustment of the NOAELs and LOAELs observed in the laboratory animal experiments or in human epidemiological or occupational studies to human equivalent concentrations (HECs) for ambient exposure conditions. These ambient conditions are currently assumed to be 24 h/day for a lifetime of 70 years. The dosimetric conversion to an HEC is necessary before the different adverse effects in the data array can be evaluated. Different default dosimetric adjustments are made based on whether the observed toxicity is in the respiratory tract or at remote sites. The optimal approach for dosimetric adjustment is considered to be a physiologically based pharmacokinetic (PBPK) model, but models under development for ACN (Gargas et al., 1990) have not yet adequately described respiratory tract dosimetry for use in extrapolating portal-of-entry effects (such as nasal toxicity) of inhalation exposures.

The various NOAELs and LOAELs in the data array, defined as the toxicity profile of adverse effects observed in all the quantitative data available, are then evaluated to determine the critical toxic effect. The critical effect is generally characterized by the lowest NOAEL(HEC) that is also representative of the threshold region (the region where toxicity is apparent from the available data) for the data array. The RfC is then derived from the NOAEL(HEC) by application of uncertainty factors (UFs). Confidence levels of high, medium, or low are assigned to the study used in the derivation, to the overall database, and to the RfC itself.

STUDIES EVALUATED

Several epidemiology studies have examined the morbidity and mortality patterns of workers exposed to ACN, but no significant excess mortality has been noted in these studies for any noncarcinogenic end point, including respiratory disease (U.S. EPA, 1991). Although human data are preferred as the basis for an RfC, the noncancer health effect and exposure data presented in these studies are not adequate for use in the derivation of an RfC. Of the remaining toxicologic literature, the three studies discussed next are considered key to the RfC derivation.

Sakurai et al. (1978) performed a cross-sectional investigation of workers in Japanese acrylic fiber factories. A slight, but not statistically significant, increase (3/10 controls; 9/18 exposed) in reddening of conjunctiva and pharynx was detected in workers at a factory where exposure levels (as determined via personal sampling devices) averaged 4.2 mg/m^3 (HEC = 3 mg/m^3). The study results are not considered conclusive given the small increase, small sample size, potential for examiner bias (examiners were not blind to the exposure status of the subjects), and healthy worker effect. However, the study results do provide qualitative support for the critical effect used in the derivation of the RfC.

Quast et al. (1980) exposed Sprague-Dawley rats (100/sex/group) 6 h/day, 5 days/wk for 2 yr to concentrations of 0, 43, or 172 mg/m^3. Significant early mortality occurred within the first year for male and female rats exposed to 172 mg/m^3, indicating that this exposure level exceeded the maximum tolerated dose for ACN. The early mortality that was observed in female rats exposed to 43 mg/m^3 during the last few weeks of the study is not considered relevant to a human risk assessment because it was determined to have resulted from an increased incidence and size of mammary gland tumors that occur at a high spontaneous rate in this rat strain (U.S. EPA, 1991).

Based on gross and histopathologic evaluation of tissues from over 40 different organs evaluated by Quast et al. (1980), only two tissues exhibited a treatment-related adverse effect from ACN exposure: (1) the nasal respiratory epithelium (four transverse sections of the nasal cavity were cut and examined) and (2) the brain (nine sections through the central nervous system were cut and examined). Proliferative changes in glial cells, both tumors (astrocytomas) and early proliferation suggestive of tumors, were significantly increased in females exposed to 43 mg/m^3. This end point was not used in the derivation of an RfC because it was considered a preneoplastic lesion, and because the corresponding LOAEL(HEC) for this remote effect (7.7 mg/m^3) was above the 1.9 mg/m^3 LOAEL(HEC) derived for the respiratory effect.

Inflammatory and degenerative changes of the upper respiratory tract were characterized by suppurative rhinitis, hyperplasia, focal erosions, and squamous metaplasia of the respiratory epithelium, with hyperplasia of the mucous secreting cells. These changes were confined to the nasal region extending from the external nares into the region lined by respiratory epithelium. The lesion did not extend into the olfactory epithelium. For male rats, increased incidence of respiratory epithelium hyperplasia in the nasal cavity was statistically significant at 172 mg/m^3 and an increase in hyperplasia of the mucous secreting cells was statistically significant at 43 mg/m^3. For female rats, a slight, but not statistically significant, increase in focal inflammation in the nasal cavity and a statistically significant increase in flattening of the respiratory epithelium were observed at 43 mg/m^3. A LOAEL of 43 mg/m^3 was identified, and the corresponding LOAEL(HEC) was 1.9 mg/m^3.

Murray et al. (1978) conducted an inhalation developmental toxicity study in rats that determined that exposure of dams to ACN can cause fetal skeletal malformations. However, the 87 mg/m^3 NOAEL(HEC) was well above the 1.9 mg/m^3 LOAEL(HEC) determined from Quast et al. (1980) for respiratory effects, suggesting that portal-of-entry effects are more sensitive.

RfC DERIVATION/UNCERTAINTY FACTOR

The RfC for ACN (U.S. EPA, 1991) was based on the NOAEL of 43 mg/m^3 for nasal toxicity using Quast et al. (1980) as the principal study. The uncertainty factor (UF) of 1000 reflects a factor of 10 to protect unusually sensitive individuals, and a reduced UF of 3 was used to adjust from a minimally adverse LOAEL to a NOAEL. A UF of 3 for interspecies variability is used because the use of the dosimetric adjustments accounts for part of this area of uncertainty. An additional factor of 10 was applied to account for an incomplete database, or more specifically, the lack of an inhalation bioassay in a second species and the lack of reproductive data by the inhalation route. The LOAEL(HEC) of 1.9 mg/m^3 was divided by this 1000-fold UF to obtain the RfC of 0.002 mg/m^3.

DISCUSSION

The critical study was a well-conducted chronic study in an appropriate number of animals (100/sex/group). However, medium confidence is given to this study because it was performed on only one species, did not identify a NOAEL, was confounded by the early sacrifice of rats with large mammary gland tumors, and the target organ (nasal epithelium) was only examined at the end of the study in relatively few animals (10–12/sex/group). The database is given medium confidence due to a lack of chronic or subchronic inhalation data in a second species, and the lack of reproductive data by the inhalation route coupled with the concern for this end point, since reproductive toxicity has been observed in an oral study (Beliles et al., 1980). Medium confidence in the RfC reflects the other two ratings. Future rat and mouse studies (e.g., subchronic bioassays that identify NOAELs; inhalation reproductive toxicity studies) could increase the confidence and reduce the overall UF associated with this RfC.

REFERENCES

Beliles, R. P., Paulin, H. J., Makris, N. G., and Weir, R. J. 1980. Three-Generation Reproduction Study of Rats Receiving Acrylonitrile in Drinking Water. Chemical Manufacturers Association, Washington, DC, LBI Project 2660, February.

Gargas, M. L., Kedderis, G. L., Fennell, T. R., and Andersen, M. E. 1990. A physiologically-based pharmacokinetic model (PBPK) model for acrylonitrile (ACN) in the rat. *Toxicologist* 10(1):216 abstr. 863.

Jarabek, A. M., Menache, M. G., Overton, J. H., Dourson, M. L., and Miller, F. J. 1990. The U.S Environmental Protection Agency's inhalation RfD methodology: Risk assessment for air toxics. Toxicol. Ind. Health 6(5):279–301.

Murray, F. J., Schwetz, B. A., Nitschke, K. D., John, J. A., Norris, J. M., and Gehring P. J. 1978. Teratogenicity of acrylonitrile given to rats by gavage or by inhalation. *Food Cosmet. Toxicol.* 16(6):547–552.

Quast, J. F., Schuetz, D. J., Balmer, M. F., Gushow, T. S., Park, C. N., and McKenna, M. J. 1980. A Two-Year Toxicity and Oncogenicity Study with Acrylonitrile Following Inhalation Exposure of Rats. Dow Chemical Co. Toxicology Research Laboratory, Midland, MI (unpublished).

Sakurai, H., Onodera, M., Utsunomiya, T., Minakuchi, H., Iwai, H., and Matsumuar, H. 1978. Health effects of acrylonitrile in acrylic fibre factories. *Br. J. Ind. Med.* 35:219–225.

U.S. EPA. 1991. Integrated Risk Information System (IRIS). Reference Concentration (RfC) for Inhalation Exposure for Acrylonitrile. Online. (08/15/91). Office of Health and Environmental Assessment, Environmental Criteria and Assessment Office, Cincinnati, OH.

ISOCYANATES AND NASAL TOXICITY: RISK ASSESSMENT ISSUES

M. M. Greenberg

Environmental Criteria and Assessment Office, U.S. Environmental Protection Agency, Research Triangle Park, North Carolina, USA

Hexamethylene diisocyanate (HDI) and diphenylmethane diisocyanate (MDI) are reactive and irritant chemicals of commercial importance that have nasal end points in chronic inhalation exposures of rodents. These end points are being used by the U.S. Environmental Protection Agency (EPA) to develop inhalation reference concentrations (RfCs) for chronic exposure situations. In humans, both compounds are known to induce asthmatic symptoms and are suspected as causative agents in chronic pulmonary function decline. Both substances are identified as hazardous air pollutants in the Clean Air Act Amendments of 1990. Characteristics of these chemicals underscore risk assessment issues that complicate the extrapolation of results from laboratory exposures to human beings. These issues are compound formulation, exposure conditions, and biological and chemical reactivity.

Exposure of humans typically involves exposure to aerosols of polymeric forms of HDI and to aerosols of polymeric MDI containing varying proportions of monomer. In contrast, chronic inhalation exposures of rodents involved pure monomer (HDI), and inhalation studies with MDI have involved aerosols with a particle size less than those expected during exposures of humans. Little is known of the nasal effects of isocyanate polymeric forms and their chemical reactivity. Extrapolation of results based solely on monomer or atypical aerosols may under- or overestimate the exposure level at which adverse nasal lesions occur. Nasal effects in laboratory rats exposed to these isocyanates exhibited an anterio-posterior gradient in which lesion severity remained minimal to mild across exposure concentrations. The nature and distribution of particular lesions and their lack of progression to adverse manifestations suggest that they be considered adaptive, not adverse.

COMPOUND FORMULATION AND EXPOSURE CONDITIONS

The U.S. EPA RfD/RfC Verification Work Group has derived an RfC for HDI based upon nasal toxicity (degeneration of olfactory cells) exhibited in a chronic inhalation study (Miles, Inc., 1989) in which rats were exposed to vapors from 99.8% pure monomer. Use of data from exposures to HDI monomer to estimate reference levels in human beings may be inappropriate. Human exposure typically involves exposure to aerosols of HDI polymers, not monomer. HDI-biuret and HDI-isocyanurate are polymeric forms (see Figure 1) typically used in commercial formulations in which the monomeric content is less than 1%. The commercial mixtures have very low volatility, and effects on the nasal tract are unknown. Use of these polymeric forms is dictated by commercial usage: spray painting (e.g., in automobile manufacture) and brush or spray applications of durable coatings. During spray painting, these compounds largely exist in droplet rather than vapor form (Janko et al., 1992). In the Miles, Inc. (1989) study, an exposure-related increase in the incidence of hyperplasia/metaplasia and hyperkeratosis was observed in anterior nasal regions. These lesions are judged adaptive because of a lack of progression to more adverse pathology with increasing exposure concentration. Damage to the olfactory region occurred only at the highest of the three concentrations investigated.

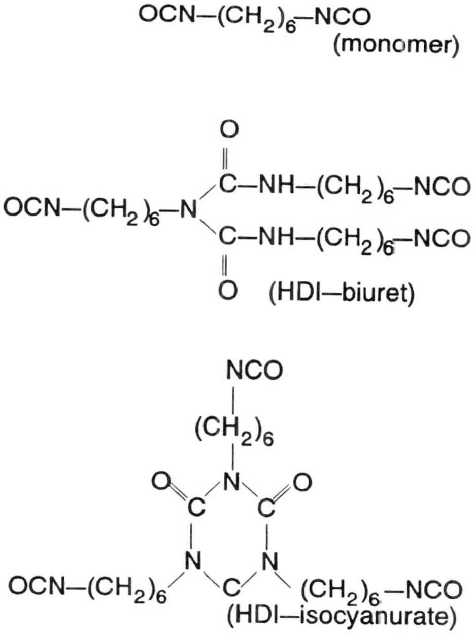

FIGURE 1. Structures of HDI monomers and prepolymers.

In human exposure scenarios involving aerosols of polymeric HDI, one might expect nasal effects to be even more confined to anterior regions as a consequence of aerosol deposition patterns. In addition, aerosol formation favors formation of micelles (Brochhagen & Grieveson, 1984), which are expected to react with water at a rate less than that of vapor (Gilbert, 1988). Thus, a no-observed-adverse-effect level (NOAEL) for nasal effects in humans exposed to polymeric HDI aerosols may be significantly higher than a NOAEL identified from the Miles, Inc. (1989) study results for rats. Chemical reactivity considerations, which also can play an important role in the occurrence and distribution of nasal lesions, are described in the next section.

MDI is generally used in foaming applications in prepolymer form in which the percentage of monomer may vary from 10 to 65% (Mobay Chemical Corporation, 1983). Polymeric MDI has a very low volatility and condenses readily to an aerosol in inhalation chambers with urethane particles providing nuclei for condensation (Dharmarajan, 1979). Monomeric MDI, a liquid at 38°C, solidifies upon contact with water (Gilbert, 1988); polymeric MDI upon reaction with water forms a hard crust composed of polyureas. In a chronic inhalation study (Reuzel et al., 1994), rats were exposed to an aerosol of polymeric MDI with a monomeric content of 47%. However, the aerosol generation system in this study generated aerosols in size ranges considerably different from those generated in commercial airless foaming operations. The mass median aerodynamic equivalent diameter (MMAD) in the Reuzel et al. (1994) study was 0.7 μm with a geometric standard deviation (GSD) of 2.46. The MMAD in spraying operations using commercial equipment was found to be 11 μm with a GSD of 3.6 (Dharmarajan, 1979). Since aerosol size in human exposure scenarios is likely to be significantly larger than those generated in animal exposures, there would likely be a shift from the posterior toward an anterior distribution of nasal lesions in humans and a shift toward adaptive lesions. Principal lesions (basal-cell hyperplasia) noted in the Reuzel et al. (1994) study were located in the olfactory region and were used to derive a human equivalent concentration for the lowest-observed-adverse-effect (LOAEL) level. There was no indication of an increased incidence of hyperplasia/metaplasia in anterior regions, as had been observed in HDI exposures. In human exposure scenarios involving MDI aerosols with larger MMAD, damage in olfactory regions could be expected to be less than that exhibited in the Reuzel et al. (1994) study, as a result of increased deposition in anterior regions.

CHEMICAL AND BIOLOGICAL REACTIVITY

Isocyanate groups are highly reactive with a wide variety of biological macromolecules (Brown, 1987). Some possible in vivo biological reactions of isocyanates are shown in Figure 2. Studies have shown that these reactions compete successfully with aqueous hydrolysis of the isocyanate functional groups (Brown et al., 1987). Limited information suggests that hydrolysis of monomeric HDI is slow (Berode et al., 1991). All diisocyanates, particularly polymeric forms, have very limited solubility in water (Kennedy & Brown, 1992). Thus, differences in the distribution of water content in various areas of the nasal tract are likely to be reflected in the extent to which biological macromolecules react with diisocyanates. The controlled use of water is the principal means by which polyurethanes are produced commercially. For example, it has been demonstrated that if neat toluene

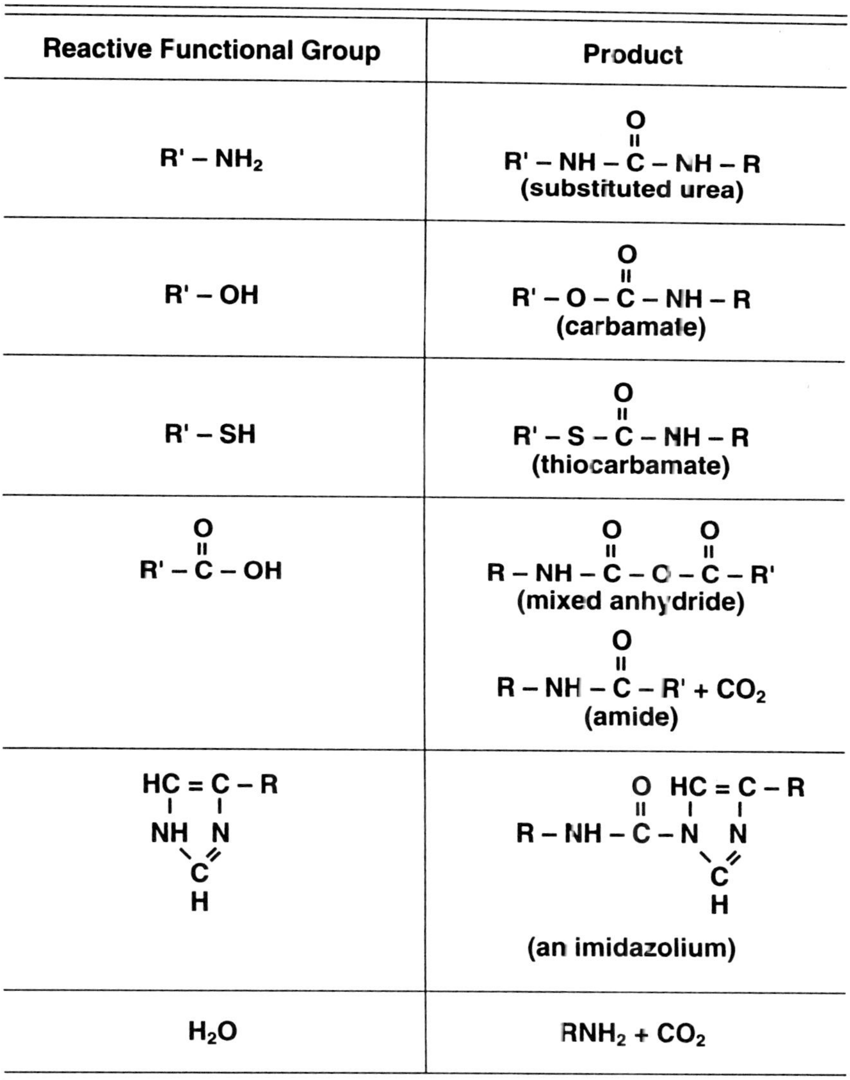

FIGURE 2. Possible in vivo reactions of isocyanates.

diisocyanate (TDI) is directly added to an aqueous solution, a polymer can rapidly form around unreacted TDI, limiting further availability for hydrolysis (Brochhagen & Grieveson, 1984). Reaction pH also can dictate reaction kinetics. Diamine formation is favored when pH is acidic or alkaline, whereas at neutral pH polymer formation is favored (Saunders & Frisch, 1962). In anterior regions of the nasal cavity where mucociliary clearance is minimal (Morgan et al., 1984), one might expect increased reaction with nasal tissue based on access of the isocyanates, the increased number of reactive groups, and deposition characteristics. However, deposition of polymeric isocyanates may lead to crust formation, blocking further access of the compounds to reactive sites in nasal tissue.

In contrast to monomeric HDI, HDI-isocyanurate and HDI-biuret each contain three isocyanate functional groups. Polymeric forms of MDI also will contain isocyanate groups in greater number than found in monomeric MDI. However, access of reactive MDI to nasal tissues may be limited by the formation of polyurethane film. This possibility is suggested by the findings with aerosols generated from pure monomeric MDI. Dharmarajan (1979) found that MDI reacted with water vapor in air prior to deposition on filters in the sampling apparatus used; MDI could not be detected on the filters because of polyurethane formation. In human exposure scenarios involving polymeric/monomeric MDI, deposition could also occur in the form of carbamates formed in the course of foam manufacture or as disubstituted ureas formed in situ after reaction with water to form amines. Both reaction products could limit further access of isocyanate functional groups to nasal tissue. This reaction course may explain the lack of nasal irritation in the Reuzel et al. (1994) study, as had occurred with HDI.

Subchronic (90-day) inhalation studies with rats involving aerosols of HDI-polyisocyanurate (MMAD = 1.5 μm; GSD = 1.4) suggest that nasal toxicity of this polymer is less than that of HDI monomer (Pauluhn & Desmodur, 1987). No histopathological evidence of nasal lesions was observed at mean exposure levels of 0.5, 3.3, or 26.4 mg/m^3 when rats were exposed for 6 h/day, 5 days/wk. Test material contained less than 0.1% HDI monomer. Squamous metaplasia and goblet-cell hyperplasia were observed in subchronic (90-day) exposure of rats to HDI monomer levels up to 1 mg/m^3, 6 h/day, 5 days/wk (Miles, Inc., 1988).

NASAL DEPOSITION AHD HEALTH END POINTS

Although the direct effects of isocyanates in the nasal region have important risk assessment applications, the nasal tract may even be more important in the induction of sensitization to isocyanates leading to occupational asthma. The role of nasal deposition in asthma induction and/or reinitiation of isocyanate-induced asthma has not been well studied.

Some studies of the sensitization potential of isocyanates in laboratory animals have involved intranasal instillation of nonphysiological concentrations of TDI in guinea pigs, a model of human asthma. The usefulness of such models has most recently been reviewed by Campos and Church (1992). Sugawara et al. (1993) have demonstrated that intranasal instillation of TDI sensitizes animals such that when TDI is applied to skin, a typical wheal and flare reaction results. Abe et al. (1992) provided suggestive evidence that intranasal instillation of TDI to guinea pigs resulted in histamine release from nasal mucosa, resulting in rhinorrhea and sneezing, although no breathing difficulty was observed. A characteristic response of guinea pigs exposed to TDI by inhalation is increased serum levels of immunoglobulin G (Karol et al., 1978). In humans sensitized to isocyanates, elevation in isocyanate-specific immunoglobulin E is a common observation in immediate-onset asthmatic reactions. One would expect that in humans the nasal tract would represent an important site with respect to asthma induction, given the large surface area and rapid access of inhaled materials to receptor sites for antigen formation. Varley et al. (1991) have shown that more than 80% of a 1.4-μm MMAD radiolabeled aerosol inhaled by conscious free-breathing guinea pigs is deposited in the nose. Thus, even in the absence of nasal toxicity, deposition of isocyanates in the nasal tract could have important implications for pulmonary deficits.

SUMMARY

HDI and MDI are two examples of isocyanates that have a wide range of biological and chemical activity. Identification of relevant threshold concentrations for nasal effects in human beings is complicated by the lack of appropriate formulation/exposure conditions used in laboratory animal studies. Studies that employ only monomer or atypical particle sizes may result in concentration-response relationships and lesion distributions quite different from human exposure scenarios. Particular attention should be directed to improved protocol design in chronic inhalation studies involving compounds of this nature, if extrapolation to the human situation is to be meaningful. Extrapolation of results from animal species to humans is further complicated unless the full range of chemical and biological reactivity of these compounds is considered. Polymeric forms of HDI and MDI may be less reactive than the monomers and may inhibit access of the reactive isocyanate groups to nasal tissue because of polymer film formation. The role of the nasal tract in sensitization reactions caused by isocyanates is an area for future research. Deposition characteristics of isocyanates and surface area in the nasal tract suggest that this region may play an important role in induction of occupational asthma.

REFERENCES

Abe, Y., Takeda, M., Irifune, M., Ogino, S., Kalubi, B., Imamura, H., Fukui, H., Wada, H., and Matsunaga, T. 1992. Effects of capsaicin desensitization on nasal allergy-like symptoms and histamine release in the nose induced by toluene diisocyanate in guinea pigs. *Acta Otolaryngol.* 112:703–709.

Berode, M., Testa, B., and Savolainen, H. 1991. Bicarbonate-catalyzed hydrolysis of hexamethylene diisocyanate to 1,6-diaminohexane. *Toxicol. Lett.* 56(1–2):173–178.

Brochhagen, F. K., and Grieveson, B. M. 1984. Environmental aspects of isocyanates in water and soil. *Cell. Polymers* 3:11–17.

Brown, W. E. 1987. The chemistry and biochemistry of isocyanates: An overview. In *Current topics in pulmonary pharmacology and toxicology*, Vol. 1, ed. M. A. Hollinger, pp. 200–225. New York: Elsevier.

Brown, W. E., Green, A. H., Cedel, T. E., and Cairns, J. 1987. Biochemistry of protein-isocyanate interactions: A comparison of the effects of aryl vs. alkyl isocyanates. *Environ. Health Perspect.* 72:5–11.

Campos, M. G., and Church, M. K. 1992. How useful are guinea-pig models of asthma. *Clin. Exp. Allergy* 22:665–666.

Dharmarajan, V. 1979. Occupational exposures to methylene bisphenylisocyanate (MDI): Gaseous or aerosol. *J. Environ. Pathol. Toxicol.* 2:1–8.

Gilbert, D. S. 1988. Fate of TDI and MDI in air, soil, and water. *J. Cell. Plastics* 24:178–192.

Janko, M., McCarthy, K., Fajer, M., and van Raalte, J. 1992. Occupational exposure to 1,6-hexamethylene diisocyanate-based polyisocyanates in the state of Oregon, 1980–1990. *Am. Ind. Hyg. Assoc. J.* 53(5):331–338.

Karol, M. H., Ioset, H. H., Riley, E. J., and Alarie, Y. 1978. Hapten-specific respiratory hypersensitivity in guinea pigs. *Am. Ind. Hyg. Assoc. J.* 39:546–556.

Kennedy, A. L., and Brown, W. E. 1992. Isocyanates and lung disease: Experimental approaches to molecular mechanisms. *Occup. Med. State of the Art Rev.* 7:301–329.

Miles, Inc.. 1988. 90-Day Inhalation Toxicity Study with 1,6-Hexamethylene Diisocyanate in Rats, with attached appendices and cover letter dated 1/18/89. TSCATS/401508. EPA/OTS document 86-890000080.

Miles, Inc. 1989. Chronic Inhalation Toxicity and Oncogenicity Study with 1,6-hexamethylene Diisocyanate (HDI) in Rats (Final Report). TSCATS/405187. EPA/OTS document 86-900000055. Study 83-241-01.

Mobay Chemical Corporation. 1983. Health and Safety Information for MDI Diphenylmethane Diisocyanate Monomeric, Polymeric, Modified, November.

Morgan, K. T., Jiang, X. Z., Patterson, D. L., and Gross, E. A. 1984. The nasal mucociliary apparatus: Correlation of structure and function in the rat. *Am. Rev. Respir. Dis.* 130:275–281.

Pauluhn, J., and Desmodur, N. 1987. Study of the Subchronic Inhalation Toxicity to Rats in Accordance with OECD Guideline No. 413. Report Br 16070, Bayer AG, Federal Republic of Germany, September 30.

Reuzel, P. G. J., Arts, J. H. E., Lomax, L. G., Kuijpers, M. H. M., Kuper, C. F., Gembardt, C., Feron, V. J., and Löser, E. 1994. Chrome inhalation toxicity and carcinogenicity study of respirable polymeric methylene diphenyl diisocyanate (polymeric MDI) aerosol in rats. *Fund. Appl. Toxicol.* 22:195–210.

Saunders, J. H., and Frisch, K. C. 1962. *Polyurethanes: Chemistry and technology*. New York: Interscience.

Suguwara, Y., Okamoto, Y., Sawahata, T., and Taaka, K. 1993. Skin reactivity in guinea pigs sensitized with 2,4-toluene diisocyanate. *Int. Arch. Allergy Immunol.* 100:190–196.

Varley, J. G., Perring, S., Seminario, M. C., Fleming, J., Holgate, S. T., and Church, M. K. 1991. Aerosol deposition in the guinea pig. A comparison of two methods of exposure. *Am. Rev. Respir. Dis.* 143:A708.

INDEX